**Conductive Atomic Force Microscopy**

# Conductive Atomic Force Microscopy

## Applications in Nanomaterials

*Edited by Mario Lanza*

WILEY-VCH

**Editor**

***Prof. Mario Lanza***
Soochow University
Inst. of Funct. Nano & Soft Materials
199 Ren-ai Road
215123 Suzhou
China

*Cover*
Cover image was kindly provided by the editor



**Library of Congress Card No.:**
applied for

**British Library Cataloguing-in-Publication Data**
A catalogue record for this book is available from the British Library.

**Bibliographic information published by the Deutsche Nationalbibliothek**
The Deutsche Nationalbibliothek lists this publication in the Deutsche Nationalbibliografie; detailed bibliographic data are available on the Internet at <http://dnb.d-nb.de>.



**Print ISBN:** 978-3-527-34091-0
**ePDF ISBN:** 978-3-527-69978-0
**ePub ISBN:** 978-3-527-69979-7
**Mobi ISBN:** 978-3-527-69980-3
**oBook ISBN:** 978-3-527-69977-3

**Cover Design** Formgeber, Mannheim, Germany
**Typesetting** SPi Global, Chennai, India
**Printing and Binding** Markono Print Media Pte Ltd, Singapore

Printed on acid-free paper

# Contents

# Oxide Films and Conduction AFM

*Sean Joseph O'Shea*

*A*STAR, Institute of Materials Research and Engineering (IMRE), Materials Processing and Characterization Department, 2 Fusionopolis Way, Singapore 138634, Singapore*

Conduction through oxides has always been part of the scanning probe microscopy (SPM) story. Indeed, the idea of the scanning tunneling microscope (STM) originates in part from how to locally study electrical properties of metal oxides [1]. However, a well-known limitation with STM is that it cannot be used on thick insulating surfaces, for which the sister technique atomic force microscopy (AFM) proved more practical in both research and industry.

One of the first SPM experiments on oxides was carried out by Mark Welland, in which he observed telegraph noise in thin $SiO_2$-on-Si with STM [2]. In early 1992, Martin Murrell joined Welland's STM group at Cambridge wanting to continue research on silicon oxides, especially the breakdown properties of very large-scale integration (VLSI) oxides. Given that the gate oxides at that time were far too thick (~10 nm) for STM, it was natural that attention turned to AFM. I was developing the AFM research in the group and Timothy Wong had constructed an all-digital SPM control system, which allowed great versatility in exploring different instrumentation methods. Jack Barnes and Sandy McKinnon provided much needed electrical engineering. Combining all this expertise, we implemented our version of Conductive atomic force microscopy (CAFM) by coating AFM cantilevers varnished with a 100-nm-thick film of titanium to make a conducting tip, programming the digital controller to output defined voltage ramps and limit the current flow to prevent catastrophic oxide breakdown, and then measuring current–voltage characteristics and images on 12-nm-thick, VLSI quality $SiO_2$ provided by Heyns and Verhaverbeke of IMEC [3].

The CAFM experiment worked without much trouble, and one of our conclusions, and indeed motivations, was that regions of the oxide could have far larger breakdown voltage than shown by conventional methods on large capacitor structures, presumably because the AFM probes nanoscale areas of the oxide, which are defect free. This conclusion was of course totally wrong! Our later work showed little difference between conventional device tests and CAFM [4]. It turned out a major issue was the applied electric field caused growth of material or contaminants under the tip, leading to higher than expected breakdown voltages. This induced surface oxidation was an annoyance for electrical characterization efforts, but proved very useful as a novel direct write lithography method [5]. Actually, the research community pursuing nanoscale lithography was applying quite similar methodologies [6] and closely paralleled the activity in CAFM at the time.

The silicon tips used in our original work were also part of the problem. Gold wire tips and conducting diamond tips (developed by Philipp Niedermann from Neuchatel) were much more useful [4]. The problem of the tip, which remains a critical issue even today, arises from the very high mechanical and electrical stress experienced in CAFM at the tip–sample contact, leading to tip wear or local oxidation, and a protocol using simultaneous force curve and current measurement was proposed as a means to test the suitability of tips for CAFM [7]. Further studies, with Mark Lantz and Ken Johnson, highlighted the interplay between the measured conduction and the contact mechanics at nanometer length scales [8]. Subsequent approaches have extensively explored the use of intermittent or tapping mode CAFM to circumvent the problem of high forces acting on the tip.

In tandem with the aforementioned efforts in CAFM, other AFM-based approaches for nanoscale electrical characterization of semiconductor materials were being actively pursued, notably Kelvin Probe microscopy [9] and CAFM spreading resistance measurement [10], and the utility of CAFM and related AFM methods in semiconductor research remains strong. This is indicated by the many material systems studied over the last 20 years, ranging from nanoscale devices, single defects and dopants, dielectric and insulating thin films, quantum dots, nanowires, and 2D materials. The CAFM method has also found applications in "soft" matter materials, particularly in molecular electronics [11] and thin organic films [12]. An area that deserves more effort is application in biology, such as understanding charge transfer in photosynthesis or across lipid membranes. Here the challenges of working in aqueous environments while maintaining robust electrical measurement are daunting, but methods based on non-contact AFM, for example, Kelvin Probe, appear feasible.

I believe there will always be a need for characterization using conduction SPM methods because not only are real-world electrical devices and sensors becoming increasingly smaller, but at a fundamental level there is always a drive to further understand the basic building blocks of materials, that is, the functional and transport properties of materials at an atomic or molecular scale. This book brings together many of the SPM electrical characterization technologies and indicates future challenges and directions to explore, no doubt requiring further development of instrumentation and novel techniques. The continuing evolution of this research field is illustrated in a final example. We started our story developing CAFM because gate oxides were thick and STM could not be used. Some 20 years later, gate oxides are so thin that tunneling is possible across them and we now routinely use STM for gate dielectric studies. We have come full circle!

## References


1 Binnig, G. and Rohrer, H. (1987) Scanning tunneling microscopy – from birth to adolescence. *Rev. Mod. Phys.*, **59** (3), 615–625.

2 Welland, M.E. and Koch, R.H. (1986) Spatial location of electron trapping defects on silicon by scanning tunneling microscopy. *Appl. Phys. Lett.*, **48** (11), 724–726.

3 (a) Murrell, M.P., Welland, M.E., O'Shea, S.J., Wong, T.M.H., Barnes, J.R., McKinnon, A.W., Heyns, M., and Verhaverbeke, S. (1993) Spatially resolved electrical measurements of $SiO_2$ gate oxides using atomic force microscopy. *Appl. Phys. Lett.*, **62** (7), 786–788;
(b) A similar idea was undertaken by Sugawara, Y., Fukano, Y., Nakano, A., Ida, T., and Morita, S. (1992) Oxidation site of polycrystalline silicon surface studied using scanning force/tunneling microscope (AFM/STM) in air. *Jpn. J. Appl. Phys.*, **31** (6A, Part 2), L725–L727.

4 O'Shea, S.J., Atta, R.M., Welland, M.E., and Murrell, M.P. (1995) Conducting AFM study of $SiO_2$ breakdown. *J. Vac. Sci. Technol. B*, **13**, 1945–1953.

5 (a) Snow, E.S. and Campbell, P.M. (1994) Fabrication of silicon nanostructures with an atomic force microscope. *Appl. Phys. Lett.*, **64** (15), 1932–1934;
(b) Day, H.C. and Allee, D.R. (1993) Selective area oxidation of silicon with a scanning force microscope. *Appl. Phys. Lett.*, **62** (21), 2691–2693.

6 (a) Mujumdar, A., Oden, P.I., Carrejo, J.P., Nagahara, L.A., Graham, J.J., and Alexander, J. (1992) *Appl. Phys. Lett.*, **61** (19), 2293–2295;
(b) Hosaka, S., Koyanagi, H., and Kikukawa, A. (1993) *Jap. J. Appl. Phys.*, **32** (3B, Part 2), L464–L467.

7 O'Shea, S.J., Atta, R.M., and Welland, M.E. (1995) Characterisation of tips for conducting AFM. *Rev. Sci. Instrum.*, **66**, 2508–2512.

8 Lantz, M.A., O'Shea, S.J., and Welland, M.E. (1997) Simultaneous force and conduction measurements in atomic force microscopy. *Phys. Rev. B*, **56**, 15345–15352.

9 Williams, C.C., Slinkman, J., Hough, W.P., and Wickramasinghe, H.K. (1989) Lateral dopant profiling with 200 nm resolution by scanning capacitance microscopy. *Appl. Phys. Lett.*, **55** (16), 1662–1664.

10 De Wolf, P., Snauwaert, J., Hellemans, L., Clarysse, T., Vandervorst, W., D'Olieslaeger, M., and Quaeyhaegens, D. (1995) Lateral and vertical dopant profiling in semiconductors by atomic force microscopy using conducting tips. *J. Vac. Sci. Technol. A*, **13** (3), 1699–1704.

11 Kelley, T.W., Granstrom, E.L., and Frisbie, C.D. (1999) *Adv. Mater.*, **11** (3), 261–264.

12 Specht, M., Ohnesorge, F., and Heckl, W.M. (1991) Simultaneous measurement of tunneling current and forces as a function of position through a lipid film on a solid substrate. *Surf. Sci.*, **257** (1–3), L653–L658.

## List of Contributors

***Fei Hui***
Soochow University
Collaborative Innovation Center of Suzhou Nanoscience and Technology
Institute of Functional Nano & SoftMaterials (FUNSOM)
199 Ren-Ai Road
Suzhou 215123
China

***Marco Allione***
King Abdullah University of Science and Technology (KAUST)
Physical Science and Engineering Division (PSE)
SMILEs Lab
Thuwal 23955-6900
Saudi Arabia

***Lidia Aguilera***
Technology Transfer Company
Knowledge Innovation Market
08005 Barcelona
Spain

***Guenther Benstetter***
Deggendorf Institute of Technology
Faculty of Electrical Media and Computer Engineering
94469 Deggendorf
Germany

***Patrizio Candeloro***
University Magna Graecia
Departement of Experimental Clinics
Bionem Lab Campus "Salvatore Venuta" Viale Europa
88100 Germaneto-Catanzaro
Italy

***Umberto Celano***
KU Leuven
Department of Physics and Astronomy (IKS)
Celestijnenlaan 200D
3001 Leuven
Belgium

*and*

IMEC
Kapeldreef 75
3001 Heverlee (Leuven)
Belgium

***Enzo Di Fabrizio***
King Abdullah University of Science and Technology (KAUST)
Physical Science and Engineering Division (PSE)
SMILEs Lab
Thuwal 23955-6900
Saudi Arabia

*and*

University Magna Graecia
Departement of Experimental Clinics
Bionem Lab Campus "Salvatore
Venuta" Viale Europa
88100 Germaneto-Catanzaro
Italy

***Rimma Dekhter***
Nanonics Imaging Ltd
Jerusalem
Israel

***Patrick Fiorenza***
Consiglio Nazionale delle Ricerche
Istituto per la Microelettronica e
Microsistemi (CNR-IMM)
Strada VIII, 5
95121 Catania
Italy

***Gabriele Fisichella***
Consiglio Nazionale delle Ricerche
Istituto per la Microelettronica e
Microsistemi (CNR-IMM)
Strada VIII, 5
95121 Catania
Italy

***Werner Frammelsberger***
Deggendorf Institute of Technology
Faculty of Mechanical Engineering
and Mechatronics
Dieter-Görlitz-Platz 1
94469 Deggendorf
Germany

***Filippo Giannazzo***
Consiglio Nazionale delle Ricerche
Istituto per la Microelettronica e
Microsistemi (CNR-IMM)
Strada VIII, 5
95121 Catania
Italy

***Andrea Giugni***
King Abdullah University of Science
and Technology (KAUST)
Physical Science and Engineering
Division (PSE)
SMILEs Lab
Thuwal 23955-6900
Saudi Arabia

***Giuseppe Greco***
Consiglio Nazionale delle Ricerche
Istituto per la Microelettronica e
Microsistemi (CNR-IMM)
Strada VIII, 5
95121 Catania
Italy

***Joan Grifoll-Soriano***
Technology Transfer Company
Knowledge Innovation Market
08005 Barcelona
Spain

***Enric Grustan-Gutierrez***
Soochow University
Collaborative Innovation Center of
Suzhou Nanoscience and Technology
Institute of Functional Nano & Soft
Materials (FUNSOM)
199 Ren-Ai Road
Suzhou 215123
China

***Alexander Hofer***
Deggendorf Institute of Technology
Faculty of Electrical Media and
Computer Engineering
94469 Deggendorf
Germany

***Vanessa Iglesias***
International Iberian Nanotechnology
Laboratory
Av. Mestre José Veiga
4715-330 Braga
Portugal

*Andrey Ignatov*
Nanonics Imaging Ltd
Jerusalem
Israel

*Xu Jing*
Soochow University
Collaborative Innovation Center of
Suzhou Nanoscience and Technology
Institute of Functional Nano & Soft
Materials (FUNSOM)
Suzhou 215123
China

*Oliver Krause*
Nanoworld Services GmbH
Schottkystraβe 10
91058 Erlangen
Germany

*Sasha Krol*
Nanonics Imaging Ltd
Jerusalem
Israel

*Mario Lanza*
Soochow University
Institute of Functional Nano & Soft
Materials
Suzhou 215123
China

*David Lewis*
Nanonics Imaging Ltd.
Har Hotzvim Hi-Tech Park
Beit Bynet
Jerusalem 97775
Israel

*Donping Liu*
Dalian Nationalities University
School of Physics and Materials
Engineering
Dalian, Development Zone
Dalian 116600
China

*Matteo Lorenzoni*
Campus de la Universitat Autònoma
de Barcelona
Microelectronics Institute of
Barcelona (IMB-CNM, CSIC)
NEMS & Nanofabrication group
E-08193 Bellaterra
Spain

*Yi Lv*
Fudan University
Department of Physics
Shanghai 200433
China

*Nicolas F. Martinez*
Concept Scientifique Instruments
(CSI)
17 Avenue des Andes
Les Ulis 91940
France

*Sean Joseph O'Shea*
A*STAR, Institute of Materials
Research and Engineering (IMRE)
Materials Processing and
Characterization Department
2 Fusionopolis Way
Singapore 138634
Singapore

*Louis Pacheco*
Concept Scientifique Instruments
(CSI)
17 Avenue des Andes
Les Ulis 91940
France

*Chengbin Pan*
Soochow University
Collaborative Innovation Center of
Suzhou Nanoscience and Technology
Institute of Functional Nano & Soft
Materials (FUNSOM)
199 Ren-Ai Road
Suzhou 215123
China

*Mathias Rommel*
Fraunhofer Institute for Integrated Systems and Device Technology IISB
Schottkystrasse 10
91058 Erlangen
Germany

*Francesc Pérez-Murano*
Campus de la Universitat Autònoma de Barcelona
Microelectronics Institute of Barcelona (IMB-CNM, CSIC)
NEMS & Nanofabrication group
E-08193 Bellaterra
Spain

*Gerardo Perozziello*
University Magna Graecia
Departement of Experimental Clinics
Bionem Lab Campus "Salvatore Venuta" Viale Europa
88100 Germaneto-Catanzaro
Italy

*Kin Leong Pey*
Singapore University of Technology and Design (SUTD)
Engineering Product Development (EPD)
Singapore 487372
Singapore

*Fabrizio Roccaforte*
Consiglio Nazionale delle Ricerche
Istituto per la Microelettronica e Microsistemi (CNR-IMM)
Strada VIII, 5
95121 Catania
Italy

*Albena Paskaleva*
Institute of Solid State Physics
Bulgarian Academy of Sciences
72 Tzarigradsko Chaussee
1784 Sofia
Bulgaria

*Yuanyuan Shi*
Soochow University
Collaborative Innovation Center of Suzhou Nanoscience and Technology
Institute of Functional Nano & Soft Materials (FUNSOM)
199 Ren-Ai Road
Suzhou 215123
China

*Shubhakar Kalya*
Singapore University of Technology and Design (SUTD)
Engineering Product Development (EPD)
Singapore 487372
Singapore

*Bo Song*
Soochow University
Suzhou Key Laboratory of Macromolecular Design and Precision Synthesis
Jiangsu Key Laboratory of Advanced Functional Polymer Design and Application
College of Chemistry
Chemical Engineering and Materials Science
Suzhou 215123
China

*Alina Strinkovsky*
Nanonics Imaging Ltd
Jerusalem
Israel

***Bruno Torre***
King Abdullah University of Science and Technology (KAUST)
Physical Science and Engineering Division (PSE)
SMILEs Lab
Thuwal 23955-6900
Saudi Arabia

***Wilfried Vandervorst***
KU Leuven
Department of Physics and Astronomy (IKS)
Celestijnenlaan 200D
3001 Leuven
Belgium

*and*

IMEC
Kapeldreef 75
3001 Heverlee (Leuven)
Belgium

***Rong Wu***
Fudan University
Department of Physics
Shanghai 200433
China

***Yijun Xia***
Soochow University
Suzhou Key Laboratory of Macromolecular Design and Precision Synthesis
Jiangsu Key Laboratory of Advanced Functional Polymer Design and Application
College of Chemistry
Chemical Engineering and Materials Science
Suzhou 215123
China

***Fei Xue***
Fudan University
Department of Physics
Shanghai 200433
China

***Xinju Yang***
Fudan University
Department of Physics
Shanghai 200433
China

***Shengli Zhang***
Fudan University
Department of Physics
Shanghai 200433
China

***Yifei Zhang***
Fudan University
Department of Physics
Shanghai 200433
China

# 1

# History and Status of the CAFM

*Chengbin Pan, Yuanyuan Shi, Fei Hui, Enric Grustan-Gutierrez, and Mario Lanza*

*Soochow University, Collaborative Innovation Center of Suzhou Nanoscience and Technology, Institute of Functional Nano & Soft Materials (FUNSOM), 199 Ren-Ai Road, Suzhou 215123, China*

## 1.1 The Atomic Force Microscope

The atomic force microscope (AFM, Figure 1.1) measures the interaction force that appears between an ultrasharp tip and a sample when the distance separating them is in the nanometric range [1]. The tip, which at the apex has a radius down to few nanometers, is located at the end of a cantilever. Its length, width, and thickness are typically of hundreds, tens, and few micrometers (respectively) [2]. The interaction force between the tip and the sample ($F_c$) produces a deflection of the cantilever according to Hooke's law [3] (see Eq. (1.1)), where $k_c$ and $\delta_c$ are the spring constant and the deflection of the cantilever.

$$F_c = -k_c \cdot \delta_c \tag{1.1}$$

In most AFMs, this deflection is detected using an optical system based on a laser beam focused on the top surface of the cantilever driving the reflection to the center of a photodiode (see Figure 1.2) [4]. When the tip is far from the surface, the interaction between tip and sample is null and, therefore, no deflection of the cantilever is observed (the laser spot stays at the center of the photodiode). When the tip is close enough to the sample, the cantilever flexes due to the interaction force that appears between both of them, and the laser spot is deflected, changing its position on the photodiode. By processing the position of the laser spot on the photodiode, the force that has provoked such deflection can be quantified. Since the force depends on the distance that the tip has deflected (Eq. (1.1)), information about the topography of the analyzed sample can be obtained. Following this working principle, if the AFM tip is moved laterally along the surface of the sample (in the $X$ and $Y$ axes), topographic information about an entire area (many point locations) can be collected. These data are sent to the computer which, using an image processing software, can depict a three-dimensional (3D) topographic map [5]. Usually a standard AFM topographic map contains a matrix of 256 × 256 pixels (positions). The number of lines per image and pixels per line, as well as many other live scan parameters, such as tip lateral speed, can be easily modified via software.

*Conductive Atomic Force Microscopy: Applications in Nanomaterials*, First Edition.
Edited by Mario Lanza.

**Figure 1.1** Photograph of the dimension icon AFM from Bruker. This is the most representative image of an AFM, as this equipment (and previous models with very similar structure) are by far the most widespread (sold) AFM configuration. (Reproduced with permission from [1]. Copyright Bruker 2015.)

However, this methodology entails certain risks: if the tip scans at a constant height in the $Z$-axis the presence of a high hillock on the surface of the sample can result in a dramatic collision, leading to unwanted tip/sample damage. Similarly, if the tip encounters a deep enough valley on the sample, the distance between them may become too large, leading to negligible tip/sample interaction and therefore failure to monitor the topography of the sample. To avoid these problems, the AFM uses an electronic feedback (controller) that continuously corrects the tip-to-sample distance in the $Z$-axis after measuring the height of each pixel (location) within the image (map), ensuring constant cantilever deflection during the whole scan [3] (see Figure 1.2). The cantilever deflection (also called the *deflection setpoint*) can be set by the user via software, and it controls the interaction force between the tip and the sample. The movement of

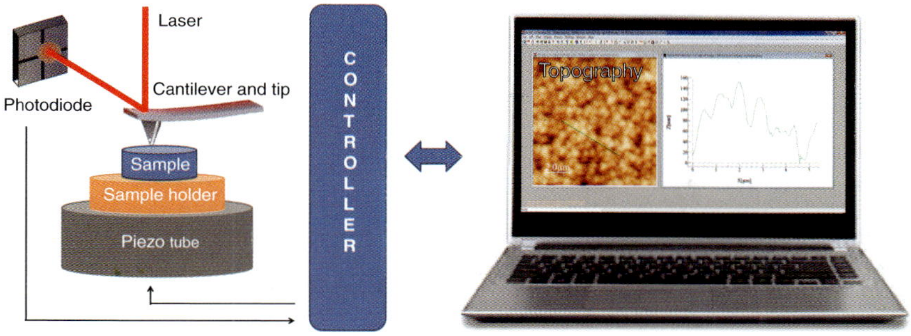


**Figure 1.2** Schematic displaying how the cantilever deflections in an AFM can be processed to obtain a topographic map. The deflection of the cantilever is detected with a laser, and the changes of the laser position in the photodiode are sent to the controller which corrects the position of the tip through the piezo tube. The data are used to build a topographic map using an image processing software compatible with AFMs. (Modified and reprinted with permission from [4], copyright by Jelena Živković 2013.)

the tip and/or sample in the $X$, $Y$, and $Z$ directions is normally applied through piezoelectric actuators. Most AFMs incorporate a piezotube that provides 3D movement to the tip and/or the sample (in Figure 1.2 it moves the sample). The AFM also requires a mechanical anti-vibration system that isolates it from external perturbations. This allow the AFM achieving a resolution down to 1 nm in the $X$, $Y$-directions and 0.1 nm in the $Z$-direction.

It is worth noting that some modern AFMs do not use an optical system (laser plus photodiode) to detect the changes on the tip deflection, instead they use a piezoelectric sensor attached to the cantilever. When the cantilever flexes, the resistance of the piezoelectric sensor changes quantifying the surface height. This system, called *tuning fork* [6], avoids mounting any hardware above the tip (e.g., photodiode) leading to an easy combination with other tools (e.g., Raman spectroscopy), as well as multiprobe configurations. This setup is analyzed in depth in Chapter 13.

Depending on the tip-sample distance during the measurements, different operation regimes can be described. For distances larger than 0.5 nm, the detected forces are mainly electrostatic, magnetic, and van der Waals, which result in an attractive regime [7, 8]. On the contrary, for distances smaller than 0.3 nm, the tip-sample interaction turns into a repulsive regime, in which it can be considered that the tip physically contacts the sample. The contact area between the tip and the sample ($A_c$) mainly depends on the contact force, the geometry of the tip, and the stiffness of both tip and sample, and it is widely accepted that it can range between 1 and 800 nm$^2$ [9, 10] (see also Chapter 3). Both operation regimes lead to the two classic operation modes of an AFM: the contact and the noncontact modes. The main difference between them is that in contact mode the vertical resolution is higher but, on the other hand, the lateral frictions with the surface of the sample are much larger, leading to undesired tip and/or sample wearing. Some alternative operation modes that combine the benefits from both of them (e.g., the tapping mode) [11] have been developed.

Finally, it is important to take into account that, when the AFM measurements are performed in air (without any environmental control system), a water layer gets deposited on the surface of both the tip and the sample owing to the ambient humidity. This water layer, which acts as a meniscus when the tip contacts the surface, introduces capillary forces between the tip and the sample, which must be also considered [12].

## 1.2 The Conductive Atomic Force Microscope

The conductive AFM (CAFM), also referred to in the literature as C-AFM, conductive probe AFM (CP-AFM), conductive scanning probe microscope (C-SPM), or conductive scanning force microscope (C-SFM), is basically an AFM that records the currents flowing at the tip/sample nanojunction simultaneously to the topography. The structure of a CAFM (see Figure 1.3) is very similar to that of the standard AFM, with only three main differences: (i) the probe tip must be conductive, (ii) a voltage source is needed to apply a potential difference between the tip and the sample holder, and (iii) a preamplifier is used to convert the (analogical) current signal into (digital) voltages that can be read by the computer. CAFM probes can be easily acquired from any manufacturer at competitive prices [2], the voltage source is located inside the AFM controller (no additional hardware is required), and the preamplifier can be purchased from the AFM manufacturer. In CAFM experiments, the sample is usually fixed on the sample holder using a conductive tape or paste, the most widely used being silver paints [13]. A Faraday cage is also convenient to isolate the sample from any external electrical interference. Using this setup, when a potential difference is imposed between the tip and the sample an electrical field is generated, which results in a net current flowing from the tip to the sample or vice versa. Therefore, the local electrical properties of the samples can be monitored at a very high nanometric resolution. The currents collected by the CAFM obey Eq. (1.2) [14], in which $I$ is the total current flowing through the tip/sample nanojunction, $J$ is the current density, and $A_{eff}$ is the effective emission area

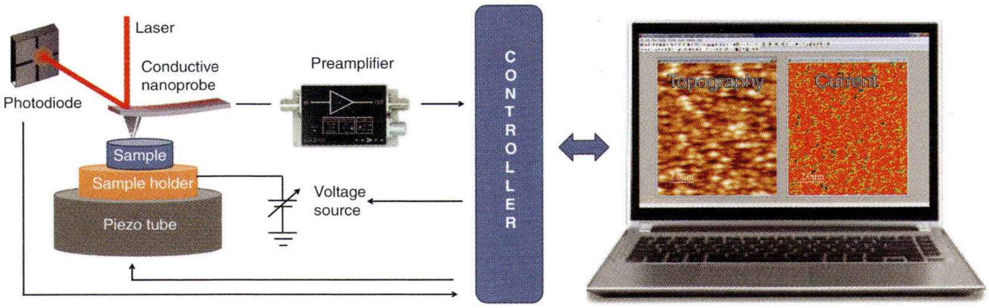


**Figure 1.3** Block diagram of a conventional conductive atomic force microscope. Compared to the AFM, the three new elements are the conductive tip, preamplifier, and sample bias.

through which electrons can flow (from now on we will refer to it just as effective area).

$$I = J \cdot A_{\mathrm{eff}} \tag{1.2}$$

The value of $J$ mainly depends on the conductivity of the tip/sample system and the voltage applied between them, and it is highly affected by intrinsic inhomogeneities in the samples, such as thickness fluctuations, local defects, and doping. The lateral resolution of the technique is defined by the term $A_{\mathrm{eff}}$, which can range from tenths of square nanometers up to thousands of square micrometers depending on many experimental factors, including the conductivity of the sample, the geometry of the tip, the tip/sample contact force, the stiffness of the tip and the sample, and even the relative humidity of the atmosphere in which the experiment is performed.

The most common mistake in CAFM research is to assume that the effective emission area ($A_{\mathrm{eff}}$) equals the physical contact area ($A_c$). Strictly, this assumption is erroneous because in many different tip/sample systems, the electrical field applied may propagate laterally (see Figure 1.4). For example, when the CAFM tip is placed on a metallic electrode, $A_{\mathrm{eff}}$ equals the entire area covered by the electrode, as its lateral electrical conductivity is very high [15, 16]. In order to provide a comprehensive definition, the effective area $A_{\mathrm{eff}}$ can be understood as the sum of all those infinitesimal spatial locations on the surface of the sample that are electrically connected to the CAFM tip (the potential difference is negligible). As such, $A_{\mathrm{eff}}$ is a virtual entity that summarizes all electrically relevant effects within the tip/sample contact system into a single value, over which the current density is assumed to be constant. The difference between contact area and effective emission area is explained in depth in Chapters 3 and 4. The small

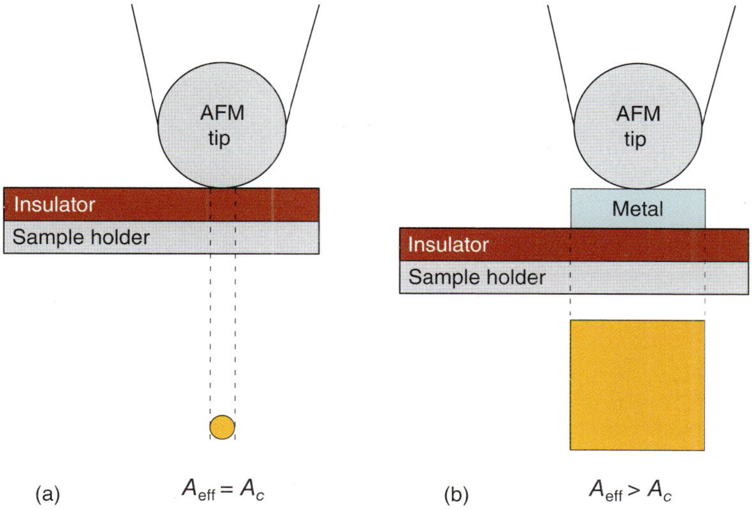


**Figure 1.4** Schematic of the effective emission area through which electrons can flow ($A_{\mathrm{eff}}$) in a CAFM when the tip is placed on (a) a flat insulating sample and (b) a flat metallic electrode deposited on an insulating sample.

dimensions of $A_{eff}$ make the current densities flowing through the tip/sample system extremely large. The smallest currents detected by a CAFM are defined by its electrical noise, which is (in the best cases) hundreds of femtoamperes. A 1 pA current flowing through a typical $A_{eff}$ of 100 $nm^2$ gives a current density of 1 $A/cm^2$. Such large current densities can dramatically accelerate the degradation of most CAFM probe tips, reducing the reliability of the measurements and increasing the cost of the research. This problem is further aggravated by the high lateral frictions present in the tip/sample system during the scans [17, 18].

The first types of conductive nanoprobes used in CAFM experiments, which are still widely used nowadays, are the standard silicon nanoprobes varnished with thin metallic films, including Pt, Au, Ru, Ti, and/or Cr (Figure 1.5). The thickness of the varnish should be thick enough to withstand the large current densities and frictions, and at the same time thin enough to not increase significantly the radius of the tip apex, maintaining its sharpness and ensuring a high lateral resolution during the measurements. As mentioned, the lifetime of conductive tips for CAFM experiments is much shorter than in any other AFM mode, mainly owing to metallic varnish melting and loss of tip mass during the scans. To solve this problem, new CAFM silicon tips coated with hard materials (e.g., phosphorous-doped diamond [19] and graphene [17, 18, 20–24]), as well as full conductive tips [25–27] have appeared (see Figure 1.5). Other factors related to the sample, such as stiffness, rugosity, stickiness, and conductivity play very important roles when deciding on the type of tip to be used in a CAFM analysis. Chapter 2 presents an in-depth description of the fabrication process and reliability of conductive nanoprobes for AFM. Advices on how to select the best CAFM tip for each experiment are provided in Chapter 2.

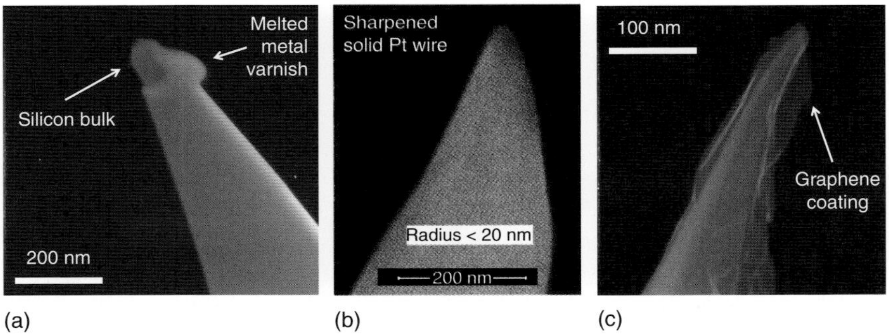


**Figure 1.5** (a) Scanning electron microscope images of (a) metal-varnished silicon nanoprobe, (b) a sharpened solid Pt wire compatible for CAFMs, and (c) a metal-varnished silicon nanoprobe coated with a sheet of single-layer graphene. The picture in (a) intentionally shows a tip with the metallic varnish worn off, so that the core bulk of silicon can be observed. (Panels (a) and (c) have been modified and reprinted with permission from [18], copyright from Royal Society of Chemistry 2016. Panel (b) has been reproduced with permission from [25], copyright American Institute of Physics 2004.)

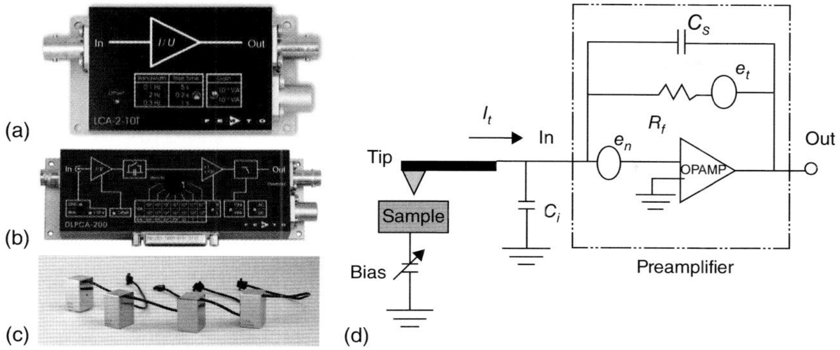


**Figure 1.6** (a, b) Photographs of two preamplifiers for CAFMs, the first one with fixed gain and the second with variable gain. (c) Photographs of different application modules for Bruker AFMs, including the CAFM module (which contains a preamplifier). (d) Simplified schematic of a conventional preamplifier used in CAFMs. The main parts are (i) an operational amplifier with high input impedance (OPAMP); (ii) a feedback resistor ($R_f$) and its parasite capacitor ($C_s$) and Johnson noise ($e_t$)-associated effects; (iii) a noise voltage source associated with the operational amplifier ($e_n$); and (iv) a capacitance associated with the input interconnections ($C_i$). (Panels (a) and (b) have been reprinted from [29], copyright FEMTO 2015. Panel (d) has been modified and reprinted with permission from [30], copyright American Institute of Physics 1990.)

The analogical current signals flowing through the tip/sample nanojunction are sent to the preamplifier Figure 1.6a,b, which transforms them into digital voltages that can be read by the data acquisition (DAQ) card of the computer (see Figure 1.6). Many manufacturers integrate the preamplifier in the so called "CAFM application module," which is a removable component that can be fixed to the AFM (usually very close to the tip to minimize electrical noise). Similarly, many other modules allow AFMs to perform other operations (see Figure 1.6c), such as scanning capacitance microscopy (SCM) or scanning spreading resistance microscopy (SSRM). In most CAFM experiments, the currents measured can range typically from few picoamperes to hundreds of microamperes, while the voltages readable by the DAQ card usually range between −3 and +3 V [31]. Therefore, the preamplifier needs to provide a very low noise and a high transimpedance (gain). Figure 1.6d shows the simplified schematic of a typical low-noise preamplifier for CAFM measurements [30], in which some elements can be distinguished: (i) an operational amplifier with high input impedance; (ii) a feedback resistor ($R_f$) and its parasite capacitor ($C_s$) and Johnson noise ($e_t$)-associated effects; (iii) a noise voltage source associated with the operational amplifier ($e_n$); and (iv) a capacitance associated with the input interconnections ($C_i$). A correct selection of the electrical components is essential to achieve good and reliable CAFM data acquisition. For example, the value of $R_f$ is not trivial: a very high value of $R_f$ improves the noise-signal ratio, while reduces the bandwidth of the preamplifier. Therefore, the value of $R_f$ should be chosen to provide enough bandwidth and a noise level below the current values that we want to measure. The parameter $e_n$ can be easily reduced by using a commercial low-noise operational amplifier. The capacitance associated with the connections ($C_i$) can be easily minimized by placing the preamplifier as near as possible to the conductive tip. The company FEMTO, one of the world's leading manufacturers of preamplifiers compatible with CAFMs, can provide devices with electrical noise as low as 3 fA and a gain up to $10^{13}$ V/A [29] (similar to those in Figures 1.6). Nevertheless, the main limitation of CAFM preamplifiers is their narrow current dynamic range, which usually allows collecting electrical signals only within three or four orders of magnitude (or even less). To solve this problem, preamplifiers with an adjustable gain can be used to focus on specific ranges [29] (similar to the one in Figure 1.6b). A more sophisticated solution for this problem is to combine the CAFM with a sourcemeter [17, 18], semiconductor parameter analyzer (SPA), or with a logarithmic preamplifier [28], which can capture the currents flowing through the tip/sample system at any range and with a high resolution. These methodologies are amply described in Chapters 10–12.

The currents flowing through the tip/sample system and converted by the preamplifier are (in almost every CAFM experiment) a consequence of the application of a voltage between the tip and the sample holder. Only a small portion of the works reported with CAFM did not require the application of a bias, that is, in the case of experiments in which the current is generated by other means, such as photoelectric [32] or piezoelectric [33] effects. Most CAFMs let the user select the value and polarity of the bias which should be applied on the tip while keeping the sample holder grounded, or vice versa. Therefore, the currents usually flow vertically through the sample [10], although lateral currents can be also measured by connecting the surface of the sample to the sample holder [34] (e.g., using silver paint or a wire bonder). Usually CAFM electronics allow applying voltages ranging from −10 to +10 V and, as mentioned, the currents that can be observed are always within three or four orders of magnitude, never smaller than 1 pA or larger than 10 μA. Apart from constant voltage stresses (CVS) applied during a scan, ramped voltage stresses (RVS) to collect current versus voltage ($I$–$V$) curves can also be performed by keeping the tip static at one single location (during the RVS). Despite this setup being more than enough for many experiments, some studies may require the use of enhanced electronic capabilities, such as the application of constant current stresses (CCS), the measurement of current versus time curves ($I$–$t$), the use of current limitations, or simply the application or measurement of larger voltages or currents. Many AFMs incorporate an input/output directly connected to the tip or the sample holder (or even both), allowing the use of an external sourcemeter or an SPA to apply/collect electrical signals. A detailed description of the combination of CAFM with a sourcemeter and SPA is presented in Chapter 10.

With this setup, many different kinds of experiments have been performed in recent years. The CAFM can be used to monitor the properties of materials, as well as to modify them with atomic resolution.

## 1.3 History and Status of the CAFM

While the first AFM was developed by G. Binning, C. F. Quate, and C. H. Gerber in 1886 at the IBM Research Laboratories in Zurich [35], the first modification of an AFM to perform current measurements was reported by M. P. Murrel at the Engineering Department of Cambridge University [36] in 1993. The setup was initially conceived to monitor local tunneling currents through 12 nm thick $SiO_2$ films. The authors evaporated 100 nm of titanium on a standard silicon cantilever from Nanoprobe, and monitored the current flowing from the sample to the tip using a picoammeter and a homemade digital-signal-processor-based AFM control system. The experiment consisted of measuring an $I$–$V$ curve at different locations on the sample (Figure 1.7a) distributed in a matrix form. By reporting the voltages at which tunneling current appeared at each location, the authors built up the first AFM image containing electrical information about the analyzed sample (each location represents a pixel of the image, Figure 1.7b). The data unequivocally show local variations of the conductivity in the $SiO_2$ film. By fitting the currents measured during the $I$–$V$ curves to the Fowler-Nordheim equation, the authors reported for the first time the value of $A_{eff}$, which was in that case $500 \pm 500$ nm$^2$, and they suggested a lateral resolution below 20 nm.

Over the following years, the CAFM became a very popular tool in the field of characterization of thin dielectrics. In 1995 and 1996, O'Shea *et al.* [37] and Ruskell *et al.* [38] further improved the lateral resolution of the CAFM technique, achieving values of 10 and 8 nm, respectively. This enhanced resolution allowed to observe the first topographic–current correlations, and the inhomogeneity

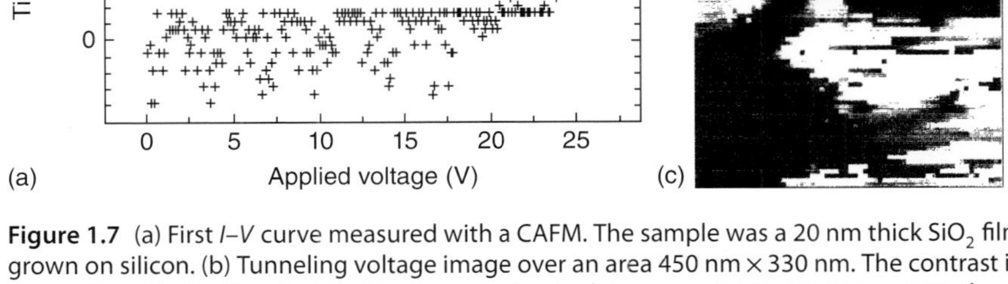


**Figure 1.7** (a) First $I$–$V$ curve measured with a CAFM. The sample was a 20 nm thick $SiO_2$ film grown on silicon. (b) Tunneling voltage image over an area 450 nm × 330 nm. The contrast is proportional to the threshold voltage required at each image point to generate a tunnel current of 0 pA. White to black corresponds to applied voltages of 15.0 and 40.5 V, respectively. (c) AFM topography of oxide surface recorded in the repulsive mode. Black to white correspond to a total vertical excursion of 90 nm. (Reprinted from [36]. Copyright American Institute of Physics.)

observed in the current maps was associated with the presence of local native defects in the oxide. Works by Olbrich [39–41] and Ebersberger [42] reported that, in $SiO_2$ films thinner than 5 nm, the tunneling current increases exponentially with thickness reductions. Consequently, thickness fluctuations of tenths of nanometer in the $SiO_2$ film could create electrically weak spots that reduce the reliability of the whole dielectric film, as dielectric breakdown (BD) is a stochastic process. The capability of the CAFM for determining the thickness of thin oxides was further demonstrated by Frammelsberger and co-workers [10, 43] who statistically analyzed more than 7200 $I$–$V$ curves, and reported $SiO_2$ thicknesses with a sensitivity of $\pm$0.3 nm. Other local phenomena such as charge trapping [44], trap-assisted tunneling [45–50], and stress-induced leakage current (SILC) [51] can be also easily monitored with CAFM. In general, the CAFM can monitor the effect of any process that introduces local changes in the structure of the dielectric, including thermal annealing [52–58], doping [59], and irradiation [60–62].

In particular, the CAFM is especially useful to determine which locations of the samples lead to premature BD, which can provide essential information about the reliability of the samples. In this direction, Iglesias *et al.* [63] demonstrated that in 5-nm-thick polycrystalline hafnium dioxide the grain boundaries are weaker from an electrical point of view, presenting larger currents in current maps (see Figure 1.8), and that the BD is potentially triggered with more ease in these sites. The images displayed in Figure 1.8 show unprecedented high resolution and topography-current correlation, as they were performed in ultrahigh vacuum (UHV) conditions (see Chapter 3). The CAFM also helped to confirm the percolation theory of the BD by experimentally proving that this is a very local phenomenon that occurs in small areas typically below 100 $nm^2$ [51]. Lateral propagations of the BD event can also be detected by CAFM [64–66]. By using an external SPA connected to the CAFM, the relationship between the current limitation and the size of the BD spot can also be assessed [67, 68], and the statistical analyses of these nanoscale data also fit the same distributions (Weibull) as those at the device level [69, 70]. The severity of the BD event

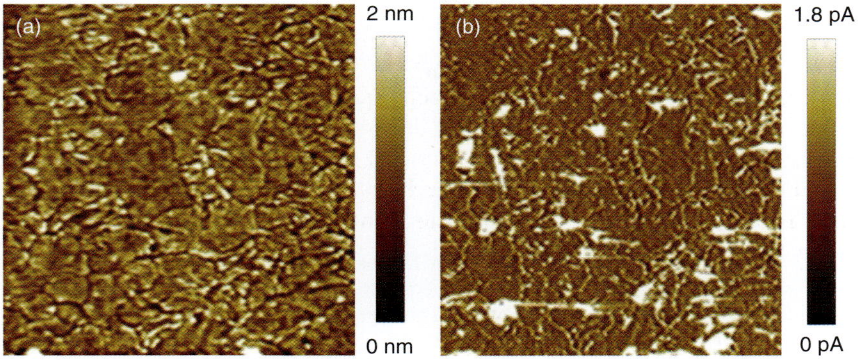


**Figure 1.8** (a) Topographical and (b) current images obtained at 6.5 V on the surface of a polycrystalline $HfO_2/SiO_2/Si$ stack (area 1 μm × 1 μm. The CAFM revealed that the crystals are more insulating than the grain boundaries. (Reprinted with permission of [63]. Copyright American Institute of Physics 2010.)

can also be studied from the BD-induced epitaxy [45, 71–73], which can be observed in subsequent topographic images collected with the CAFM after a voltage ramp. Similarly, the analysis of the BD recovery (resistive switching RS) can also be monitored by CAFM. In 2012, it was widely accepted that transition metal oxides such as $HfO_2$, $Al_2O_3$, and $TiO_2$ could show RS, enabling their use in resistive random access memories but the physics behind the RS phenomenon and the local features promoting this behavior were still unknown. By using a CAFM, Lanza *et al.* [74] demonstrated for the first time that, in atomic-layer-deposited $HfO_2$ films, RS only takes place at the grain boundaries of polycrystalline samples. All the capabilities of the CAFM for studying RS in dielectrics have been summarized in the review article of Ref. [75].

The use of CAFM for the characterization of thin dielectrics has been recently also extended to the field of 2D dielectrics. Lee *et al.* [76] analyzed for the first time the tunneling current through multilayer hexagonal boron nitride (*h*-BN) sheets using a CAFM. More specifically, he reported that the leakage current at low fields in *h*-BN thinner than three layers occurs by direct tunneling, while in thicker stacks it flows according to the Fowler-Nordheim tunneling. In fact, the CAFM has become very useful in these early stages of 2D materials research as they were firstly isolated by mechanical exfoliation, and the size of the nanosheets produced by this method is too small to place electrodes by photolithography (i.e., they can only be electrically characterized via CAFM). Ji *et al.* [77] analyzed the degradation process of multilayer *h*-BN and, by measuring sequences of CAFM $I$–$V$ curves she determined that *h*-BN is much more reliable than $HfO_2$ films with similar equivalent oxide thicknesses. Hattori *et al.* [78] analyzed the BD event in multilayer *h*-BN and, by combining $I$–$V$ curves with topographic/current maps, he determined that the BD takes place layer by layer. These properties of *h*-BN could be only obtained thanks to the capabilities of the CAFM [79, 80].

Apart from the study of thick dielectrics, the CAFM technique has expanded to many other fields of science, including physics, materials science, chemistry, and engineering. Table 1.1 classifies the number of CAFM-related papers (and their citations) in the main different fields of science. As can be seen, the CAFM has been successfully used to study nanoparticles [81], molecules [82], nanowires (NW) [33], carbon nanotubes [83], two dimensional materials [63, 84, 85], coatings [86], local oxidation [87], photoelectricity [88], and piezoelectricity [89]. In general, CAFM can be a very powerful technique to acquire valuable local electrical information of any kind of sample or device, and for this reason, CAFM research has spread rapidly. Since its invention in 1993 and as of June 14, 2016, the total number of CAFM-related research papers detected in the Web of Science[1] is 1618 [90], and Figure 1.9 shows the number of research papers published per year (and their citations). According to the number of citations, Table 1.2

1 Search criteria used: "Topic" = "conductive atomic force microscopy" or "conductive atomic force microscope" or "conducting atomic-force microscopy" or "conducting atomic-force microscope" or "conductive AFM" or "conductive-AFM" or "conductive probe AFM" or "conductiveprobe atomic force microscope" or "conductive probe atomic force microscopy" or "conductive scanning probe microscope" or "conductive scanning probe microscopy" or "conductive scanning force microscope" or "conductive scanning force microscopy" or "local conductive atomic force microscope" or "local conductive atomic force microscopy" or "local conductive AFM" or "LC-AFM."

**Table 1.1** Research area rankings in CAFM publications.

| Research area | Number of papers | Citations |
|---|---|---|
| Physics | 1 106 | 18 824 |
| Materials science | 702 | 12 080 |
| Science, technology, other topics | 511 | 12 832 |
| Chemistry | 439 | 11 252 |
| Engineering | 214 | 2 130 |
| Microscopy | 81 | 1 678 |
| Electrochemistry | 62 | 1 024 |
| Optics | 53 | 523 |
| Biochemistry, molecular biology | 51 | 1 389 |
| Physiology | 47 | 1 559 |

*Source:* Thomson Reuters ISI Web of Science; search dated June 14, 2016 using "Topic" = "conductive atomic force microscopy" or "conductive atomic force microscope" or "conducting atomic-force microscopy" or "conducting atomic-force microscope" or "conductive AFM" or "conductive-AFM" or "conductive probe AFM" or "conductive probe atomic force microscope" or "conductive probe atomic force microscopy" or "conductive scanning probe microscope" or "conductive scanning probe microscopy" or "conductive scanning force microscope" or "conductive scanning force microscopy" or "local conductive atomic force microscope" or "local conductive atomic force microscopy" or "local conductive AFM" or "LC-AFM."

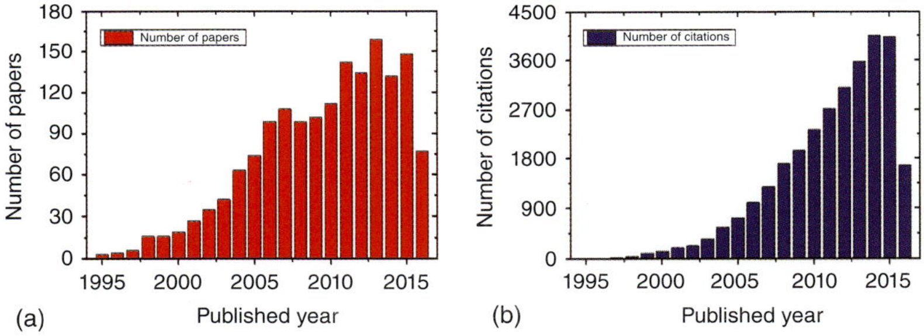


**Figure 1.9** Statistic analysis of the (a) publication number and (b) citation about CAFM in each year (Thomson Reuters ISI Web of Science; search dated September 20, 2015 using "Topic" = "conductive atomic force microscope" or "conductive atomic force microscopy" or "conducting atomic-force microscopy" or "conductive AFM"; 1325 records).

compiles the most influential research papers in CAFM research, and Table 1.3 shows the ranking of the most active countries in CAFM research, on the basis of the number of publications.

One of the main reasons for the versatility of the CAFM is the continuous development of commercial setups with enhanced capabilities. For example, the Veeco DI 3100 AFM (and its newer evolution, the Bruker's Dimension Icon®) provides a genuine large-size sample holder that remarkably reduces the setup preparation time if many samples need to be studied [91]. Oxford Instruments

**Table 1.2** Paper citation rankings in CAFM publications.

| Title | Author | Reference | Citations |
|---|---|---|---|
| Piezoelectric nanogenerators based on zinc oxide nanowire arrays | Z. L. Wang and J. H. Song | *Science*, **312**, 242–246, 2006 | 3077 |
| Switching the electrical resistance of individual dislocations in single-crystalline $SrTiO_3$ | K. Szot, W. Speier, G. Bihlmayer, and R. Waser | *Nat. Mater.*, **5**, 312–320, 2006 | 915 |
| Resistive switching mechanism of $TiO_2$ thin films grown by atomic-layer deposition | B. J. Choi, D. S. Jeong, S. K. Kim, C. Rohde, S. Choi, J. H. Oh, H. J. Kim, C. S. Hwang, K. Szot, R. Waser, B. Reichenberg, and S. Tiedke | *J. Appl. Phys.*, **98**, 033715, 2005 | 681 |
| Conduction at domain walls in oxide multiferroics | J. Seidel, L. W. Martin, Q. He, Q. Zhan, Y.-H. Chu, A. Rother, M. E. Hawkridge, P. Maksymovych, P. Yu, M. Gajek, N. Balke, S. V. Kalinin, S. Gemming, F. Wang, G. Catalan, J. F. Scott, N. A. Spaldin, J. Orenstein, and R. Ramesh | *Nat. Mater.*, **8**, 229–234, 2009 | 455 |
| A bond-fluctuation mechanism for stochastic switching in wired molecules | G. K. Ramachandran, T. J. Hopson,A. M. Rawlett, L. A. Nagahara, A. Primak, and S. M. Lindsay | *Science*, **300**, 1413–1416, 2003 | 341 |
| Nanoscale control of an interfacial metal-insulator transition at room temperature | C. Cen, S. Thiel, G. Hammerl, C. W. Schneider, K. E. Andersen, C. S. Hellberg, J. Mannhart, and J. Levy | *Nat. Mater.*, **7**, 298–302, 2008 | 275 |
| Probing the nature of defects in graphene by Raman spectroscopy | A. Eckmann, A. Felten, A. Mishchenko, L. Britnell, R. Krupke, K. S. Novoselov, and C. Casiraghi | *Nano Lett.*, **12**, 3925–3930, 2012 | 264 |
| Making electrical contacts to molecular monolayers | X. D. Cui, X. Zarate, J. Tomfohr, O. F. Sankey, A. Primak, A. L. Moore, T. A. Moore, D. Gust, G. Harris, and S. M. Lindsay | *Nanotechnology*, **13**, 5–14, 2002 | 256 |
| Electrochemical reduction of oriented fraphene oxide films: an in situ Raman spectroelectrochemical study | G. K. Ramesha and S. Sampath | *J. Phys. Chem. C*, **113**, 7985–7989, 2009 | 254 |

(*continued*)

**Table 1.2** (Continued)

| Title | Author | Reference | Citations |
|---|---|---|---|
| Piezoelectric and semiconducting coupled power generating process of a single ZnO belt/wire. A technology for harvesting electricity from the environment | J. H. Song, J. Zhou, and Z. L. Wang | *Nano Lett.*, **6**, 1656–1662, 2006 | 221 |

*Source:* Thomson Reuters ISI Web of Science; search dated 14 June 2016 using "Topic" = "conductive atomic force microscopy" or "conductive atomic force microscope" or "conducting atomic-force microscopy" or "conducting atomic-force microscope" or "conductive AFM" or "conductive-AFM" or "conductive probe AFM" or "conductive probe atomic force microscope" or "conductive probe atomic force microscopy" or "conductive scanning probe microscope" or "conductive scanning probe microscopy" or "conductive scanning force microscope" or "conductive scanning force microscopy" or "local conductive atomic force microscope" or "local conductive atomic force microscopy" or "local conductive AFM" or "LC-AFM."

**Table 1.3** Country rankings in CAFM publications.

| Country | Number of papers | Citations |
|---|---|---|
| USA | 359 | 13 549 |
| China | 227 | 5 617 |
| Japan | 203 | 2 496 |
| Germany | 167 | 4 331 |
| South Korea | 145 | 2 704 |
| France | 118 | 1 178 |
| Taiwan | 112 | 2 292 |
| Spain | 105 | 1 081 |
| Italy | 101 | 1 224 |
| England | 81 | 2 462 |

*Source:* Thomson Reuters ISI Web of Science; search dated 14 June 2016 using "Topic" = "conductive atomic force microscopy" or "conductive atomic force microscope" or "conducting atomic-force microscopy" or "conducting atomic-force microscope" or "conductive AFM" or "conductive-AFM" or "conductive probe AFM" or "conductive probe atomic force microscope" or "conductive probe atomic force microscopy" or "conductive scanning probe microscope" or "conductive scanning probe microscopy" or "conductive scanning force microscope" or "conductive scanning force microscopy" or "local conductive atomic force microscope" or "local conductive atomic force microscopy" or "local conductive AFM" or "LC-AFM."

(former Omicron) provides AFMs embedded in UHV systems for extremely high accuracy [67, 68] and spatial resolution [63, 92]. Uppal *et al.* [68] studied the pre- and post-degradation of $HfO_2$ dielectrics using an Omicron CAFM working in UHV, and observed a change in the conduction from Fowler-Nordheim to direct tunneling. When measuring electrical currents with the AFM tip, the *Resiscope* mode of Agilent AFMs provides unique extra-large voltage and current dynamic ranges thanks to its logarithmic current-to-voltage converter [93]. This provides a current window up to nine orders of magnitude (from picoamperes to milliamperes), while the electronics of most commercial AFMs saturate around some nanoamperes. Coll *et al.* [94] used this setup to monitor the hysteretic bipolar switching phenomenon in ultrathin $CeO_2$ layers grown on $La_{0.7}Sr_{0.3}MnO_3/SrTiO_3$ functional complex oxides. Newer manufacturers often provide modular structures compatible with other techniques. Stahl *et al.* [95] mounted an AFM head on the sample stage of a commercial SEM (Jeol JSM-820) to prove the functioning of piezoresistive cantilevers. Anderson [96] combined a Digital Instruments AFM with a Kaiser HoloProbe Raman instrument to enhance the Raman signal at the surface of a $C_{60}$ film solution cast on mica. Another interesting mode offered by many manufacturers is the possibility to collect current measuring in semicontact mode, which reduces notably the tip/sample frictions enhancing the lifetime of the tip and the reliability of the samples (see Chapter 3); this methodology also enables the use of the CAFM to study the electrical properties of soft materials such as polymers. The ultimate evolution of the AFM provided by Nanonics incorporates for the first time the possibility to perform multiprobe electrical and optical measurements [97] (up to four probes), thanks to the use of a piezoelectric sensor to detect the deflection of the cantilever [98], avoiding the use of fixed and nonmodular optical detectors (traditional laser and photodiode). This setup is discussed in detail in Chapter 13. Despite these advances, each one of these AFMs presents some important limitations. For example, the Dimension Icon measurements does not provide environmental chamber impeding to perform specific experiments, such as bipolar measurements in many kinds of samples (due to the apparition of local anodic oxidation); Omicron AFMs are complex, as they require long times to prepare the setup, and the UHV degrades the ultrasharp tips extremely fast owing to high frictions (the water layer on the surfaces of both tip and sample are removed); from our experience, Agilent AFMs are not so user-friendly and robust compared to the Veeco/Bruker AFMs, and Nanonics AFMs use self-made transparent and flexible cantilevers that are interesting for many applications, but their lateral resolution is usually smaller than that of traditional ultrasharp silicon cantilevers. Similarly, Park AFMs provide a very powerful software that allows a wide range of image processing, but the tip holder for electronic measurements requires soldering. More details about AFM comparatives can be found in review papers and specialized websites [99].

In the following chapters, we have invited selected top CAFM scientists to describe their work in different CAFM research fields that have had a special impact, not only case studies but also setup developments and their applications.

## 1.4 Editor's Choice: On the Use of CAFM to Study Nanogenerators Based on Nanowires

In this section, we present the use of CAFM for the investigation of piezoelectric NWs. The reason is that this work has had a special impact in nanosciences (see Table 1.2). The use of NWs for energy applications has garnered much attention during the last decade. For example, the fabrication of NWs has been a technique often used to enhance the surface area of a material, which can be beneficial in many energy conversion systems, such as photovoltaic systems [100] and photoelectrochemical water splitting solar cells [101]. In this kind of samples, the CAFM can be a powerful tool to analyze locally the currents generated by the NWs. For example, Pan *et al.* [89] used a standard CAFM to analyze local photo currents generated in arrays of vertically aligned ZnO NWs. Different samples were fabricated, and different performances were detected depending on the amount and size of the NWs clusters, which can be controlled with the insertion a polymer layer between the NWs. NW clustering reduces the light absorbed by the NWs. CAFM contributed to demonstrate that samples without nanoclusters produced larger photocurrents, observing more conductive spots with larger conductivity. Moreover, the currents collected strongly depended on the amount of light driven to the surface of the samples (i.e., percentage of power of the solar simulator). When studying photocurrents, it should be highlighted that the laser of the CAFM can be already used to excite the samples, making easy the observation of photocurrents in the absence of bias. Additional laser illumination could be also used to generate the currents. For example, Howell *et al.* [88] used a CAFM to map the efficiency of GaN/InGaN NW arrays illuminated from the bottom with an additional laser. Through complementing with Raman spectroscopy and finite difference time domain (FDTD) simulations, they noted the influence of NW structure to device performance, confirming that higher indium concentration within the photoactive regions and fewer defects on the NW are able to enhance the cell efficiency. Despite this setup has been also described as scanning photocurrent microscopy (SPCM), the working principle is very similar. It is worth noting that the current collection does not require the application of any bias, confirming the solar-to-electrical energy conversion. More details about how to modify a CAFM for photoelectric applications are given in Chapter 15.

Even more interesting is the characterization of the piezoelectric effect in specific types of NWs, as the tip of the CAFM can at the same time bend the (vertically aligned) NWs while the current is collected. This is a unique capability of the CAFM, as other techniques such as STM do not work in contact mode and, therefore, cannot apply mechanical stresses to the NWs. The first relevant work in this direction was reported in 2006 by Wang and Song [33], who used the CAFM to scan a single and a forest of semiconducting ZnO piezoelectric NWs. To date, a variety of methods have been developed to synthesize piezoelectric NW arrays, including chemical hydrothermal synthesis [102], microwave-assisted hydrothermal synthesis [103], thermal evaporation synthesis [104], vapor-confined face-to-face annealing [105], pulsed direct current magnetron sputtering [106], radiofrequency magnetron sputtering[107],

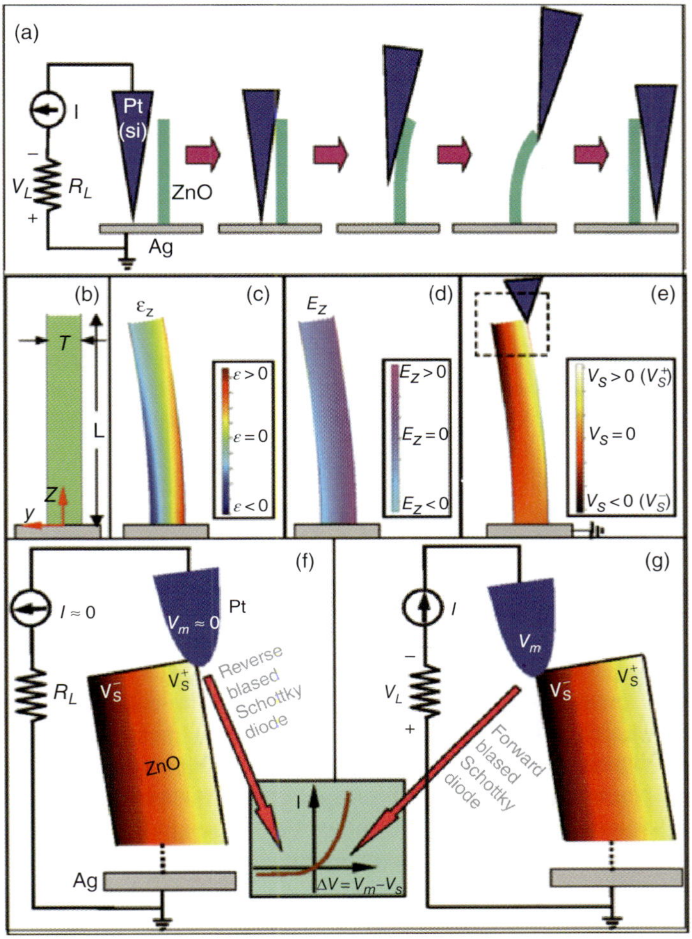


**Figure 1.10** (a) Experimental setup and the whole CAFM measurement procedures in contact mode. (b) Schematic of a freestanding nanowire. (c) Physical strain distribution along the $Z$-axis of the nanowire under transverse stress. (d) Corresponding electrical filed distribution. (e) Coincidentally potential distribution along the tested nanowire. (f, g) Schematic of forward and reversed Schottky barrier respectively.

electron-beam template lithography[108], and chemical vapor deposition[109]. During recent years, chemical hydrothermal synthesis has become a dominant solution owing to its speed, low cost, and operation temperature. In Wang's work [33], the NWs grown by the vapor–liquid–solid (VLS) process were scanned with the CAFM tip as displayed in Figure 1.10, realizing rapidly that, in the absence of bias, current peaks correlated systematically with the bending of the NW. The mechanical strain, and therefore the currents generated, could be controlled by altering the scan parameters such as tip/sample contact force, tip speed and integral/proportional gain. The authors complemented their experimental observations with simulations via finite element method, and reported that

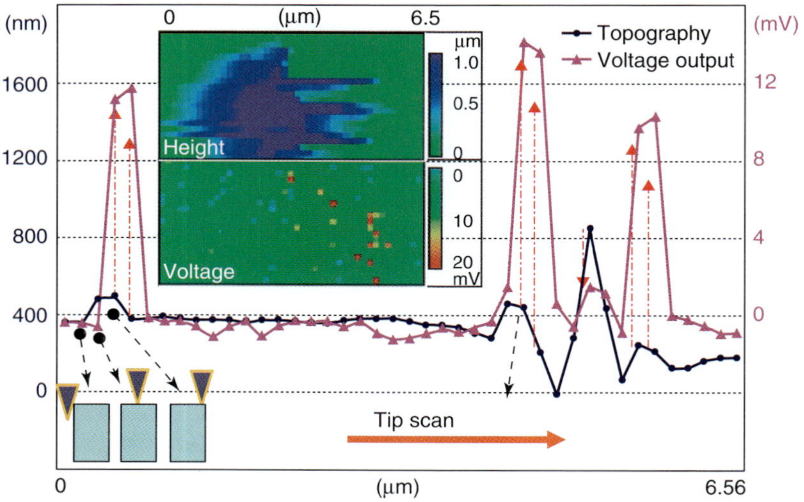


**Figure 1.11** The simultaneously collected profiles of the topograsphy and CAFM current map. The corresponding images of size of 6.5 µm × 3.2 µm are displayed in the inset. A schematic in the left-bottom corner depicts the accurate tip location based on the corresponding cross-section line of topography.

the current is only observable at the compressed surface of the NW, which is forward biased with respect to the (grounded) CAFM tip. From the point of view of energy transformation, during the first half of the bending process, the mechanical energy is first accumulated and stored in the piezoelectric NW. When the tip crosses the neutral line (right at the center of the NW), the stored energy starts to discharge and gradually converts into the electrical current. A detailed description of the experiment is shown in Figure 1.11.

Another behavior that further corroborates the piezoelectric origin of the currents detected is that they appear at different locations of the NW depending on their doping. Su *et al.* [110] observed that p-type GaN semiconducting NWs show the discharge (current peak) when the CAFM tip touches the elongated side of the NW, while for the aforementioned n-type ZnO NWs, that happens on the compressed side (see Figure 1.11). This observations are possible thanks to the unique capability of correlating the topography with the current signals in the CAFM setup. As the NWs usually present a very high aspect ratio, the profile of these structures are particularly easy to monitor. Lanza *et al.* [111] reported very good correlation between topography and current in highly populated arrays of vertically aligned ZnO NWs (Figure 1.12). Yang *et al.* [112] also observed such correlation in similar structures during a single line scan. The relationship between height/width and currents generated can be also studied via CAFM. A sequence of CAFM measurements in piezoelectric nanogenerators based on the GaN NW arrays was carried out by Hung [113]. Vertical n-type GaN NW arrays with different aspect ratios were grown on a GaN sample by inductively coupled plasma etching. Then, these NWs were sequentially scanned with

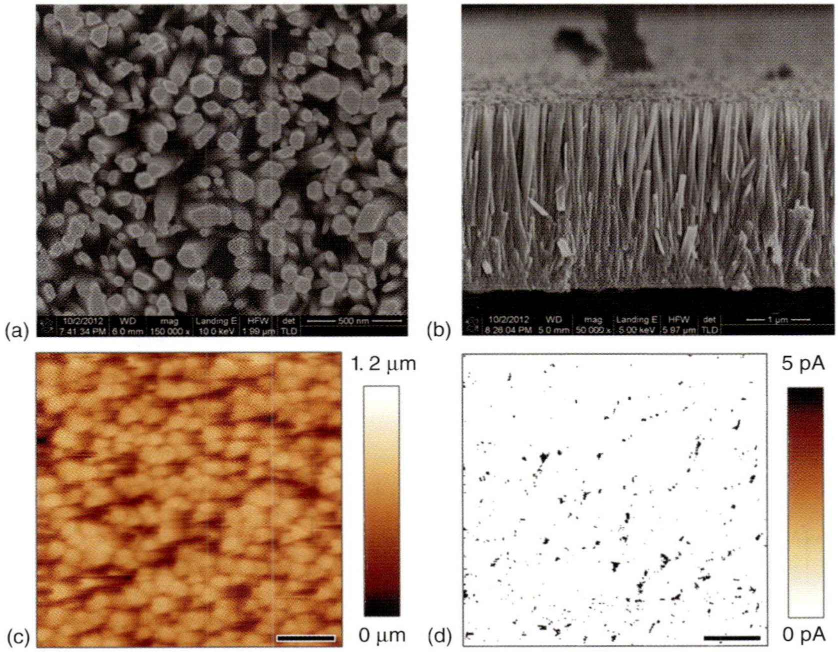


**Figure 1.12** Nanowire characterization. (a, b)Top and cross-sectional SEM images of the nanowire array. (c, d) Simultaneously collected topographic and current maps on the same sample using a Pt-Ir tip exerting a force of 0.1 nN and without bias. The scale bar for images (c, d) is 1 μm.

the CAFM using a conductive CrPt-coated nanoprobe. The measurements indicate that the piezo-current generated by the GaN NW starts to decrease as the aspect ratio increases, and that the output voltage could nearly approach zero when the aspect ratio is close to one.

It is very interesting that the current densities measured at the device level and at the nanoscale in nanogenerators made of piezoelectric NWs are dramatically different. When the NWs array is sandwiched between two electrodes and compressed, the current densities collected (typically using a sourcemeter or an SPA) are very modest, typically of the order of just a few micrometer per square centimeter [114]. On the contrary, the currents collected with the tip of the CAFM for similar structures are of few/tens of picoamperes. Considering the emission area ($A_{eff}$) in CAFM experiments (~100 $nm^2$) that implies a current density of 1 $A/cm^2$ (in the worst case, for a current of just 1 pA). Therefore, the potential for energy conversion of ZnO NWs is uniquely demonstrated thanks to the CAFM. The explanation for this current mismatch is related to the difficulty in driving the current out of the NWs array, as well as due to the formation of NW clusters. The effectiveness of the strategies used to solve this problem can also be studied by means of CAFM. For example, the poor interface between the flat, rigid, top electrode and the vertically aligned NWs can be improved by patterning pyramidal brushes at the bottom interface of the top electrode, which improves the contact with the NWs [115]. Another more sophisticated methodology is the insertion of a graphene sheet which, thanks to its large lateral conductivity and flatness can enhance the charge transport at the NWs array, and the yield of the whole device [111, 116]. On the other hand, the problem of NWs clustering can be solved by inserting a polymer between the NWs, which effectively separates them, avoiding short circuits [89].

## 1.5 Conclusions

The CAFM was invented in 1993 by M. P. Murrell and coauthors at the University of Cambridge, and it measures the local currents flowing between an ultrasharp tip and a sample. The CAFM is basically an AFM with three additional elements: a conductive probe tip, a voltage source, and a current/voltage preamplifier. The lateral resolution of the measurements is defined by the parameter $A_{eff}$, and it depends on the conductivity of the sample, the scanning parameters (e.g. contact force), the radius of the tip, and the environmental conditions in which the experiments are performed. The CAFM was initially used to measure the homogeneity and reliability of thin dielectrics, but later it expanded to many other fields of science. One of the main advantages compared to other nanoelectronic techniques is that the CAFM can simultaneously analyze the topography and conductivity of the sample scanned. This has allowed performing electromechanical measurements, such as inducing the movement of piezoelectric NWs and simultaneously collecting the current. New CAFM developments include multiprobe setups and their *in situ* combination with other techniques, which confirms a very promising panorama for CAFM based techniques.

## References

1 Bruker Dimension Icon Brochure, B068-RevB1, https://www.bruker.com/products/surface-analysis/atomic-force-microscopy/dimension-icon/learn-more.html (accessed 4 February 2016).

2 NanoWorld Nanoworld atomic force microscope probes catalogue, http://www.nanoworld.com/afm-probes-catalog (accessed 4 February 2016).

3 R.R.L. De Oliveira, D.A.C. Albuquerque, T.G.S. Cruz, F.M. Yamaji and F.L. Leite (2012) Measurement of the Nanoscale Roughness by Atomic Force Microscopy: Basic Principles and Applications 147–174. This is a chapter in the book: ATOMIC FORCE MICROSCOPY - IMAGING, MEASURING AND MANIPULATING SURFACES AT THE ATOMIC SCALE (Edited by Victor Bellitto). ISBN 978-953-51-0414-8. Published by InTech, in Janeza Trdine 9, 51000 Rijeka, Croatia (accessed 11 February 2017).

4 Zivkovic, J. (2013) AFM force spectroscopy of viral systems. PhD thesis. Radboud University.

5 Horcas, I., Fernández, R., Gómez-Rodríguez, J.M., Colchero, J., Gómez-Herrero, J., and Baro, A.M. (2007) A software for scanning probe microscopy and a tool for nanotechnology. *Rev. Sci. Instrum.*, **78**, 013705.

6 Tuning forks for ultrasensitive force spectroscopy measurements. Nanonics Imaging Ltd www.nanonics.co.il (accessed 13 February 2017).

7 Martin, Y., Williams, C.C., and Wickramasinghe, H.K. (1989) Atomic force microscope force mapping and profiling on an sub 100-a scale. *J. Appl. Phys.*, **61** (10), 4723–4729.

8 Sarid, D. and Elings, V. (1991) Review of scanning force microscopy. *J. Vac. Sci. Technol., B*, **9** (2), 431–437.

9 Israelachvili, J.N. (1992) *Intermolecular and Surface Forces*, Academic Press, London.

10 Frammelsberger, W., Benstetter, G., Kiely, J., and Stamp, R. (2007) C-AFM-based thickness determination of thin and ultra-thin SiO2 films by use of different conductive-coated probe tips. *Appl. Surf. Sci.*, **253** (7), 3615–3626.

11 Zhong, Q., Inniss, D., Kjoller, K., and Elings, V.B. (1993) Fractured polymer silica fiber surface studied by tapping mode Atomic-Force Microscopy. *Surf. Sci.*, **290**, 688–692.

12 Cappella, B. and Dietler, G. (1999) Force-distance curves by Atomic Force Microscope. *Surf. Sci. Rep.*, **34** (1-3), 1–104.

13 SPI supplies http://www.2spi.com/category/silver-paints-pastes (accessed 4 February 2016).

14 Lanza, M., Porti, M., Nafría, M., Aymerich, X., Whittaker, E., and Hamilton, B. (2010) Electrical resolution during conductive atomic force microscopy measurements under different environmental conditions and contact forces. *Rev. Sci. Instrum.*, **81**, 106110.

15 Rommel, M., Jambreck, J.D., Lemberger, M., Bauer, A.J., Frey, L., Murakami, K., Richter, C., and Weinzierl, P. (2013) Influence of parasitic capacitances on conductive AFM I-V measurements and approaches for its reduction. *J. Vac. Sci. Technol., B*, **31**, 01A108.

16 Yanev, V., Erlbacher, T., Rommel, M., Bauer, A.J., and Frey, L. (2009) Comparative study between conventional macroscopic IV techniques and advanced AFM based methods for electrical characterization of dielectrics at the nanoscale. *Microeletron. Eng.*, **86**, 1911–1914.

17 Lanza, M., Bayerl, A., Gao, T., Porti, M., Nafria, M., Jing, G., Zhang, Y., Liu, Z., and Duan, H. (2013) Graphene-coated atomic force microscope tips for reliable nanoscale electrical characterization. *Adv. Mater.*, **25**, 1440–1444.

18 Hui, F., Vajha, P., Shi, Y., Ji, Y., Duan, H., Padovani, A., Larcher, L., Li, X.R., Xu, J.J., and Lanza, M. (2016) Moving graphene devices from lab to market: advanced graphene-coated nanoprobes. *Nanoscale*, **8**, 8466–8473.

19 Diamond coated AFM tips, Nanoworld product catalogue, http://www.nanoworld.com/diamond-coated-afm-tips (accessed 13 February 2017).

20 Wen, Y.G. (2012) Multilayer graphene-coated atomic force microscopy tips for molecular junctions. *Adv. Mater.*, **24**, 3482.

21 Martin-Olmos, C., Rasool, H.M., Weiller, B.H., and Gimzewski, J.K. (2013) *ACS Nano*, **7**, 4164.

22 Gimzewski, J.K. (2013) Graphene MEMS: AFM probe performance improvement. *ACS Nano*, **7**, 4164.

23 Shim, W., Brown, K.A., Zhou, X.Z., Rasin, B., Liao, X., and Mirkin, C.A. (2012) Multifunctional cantilever-free scanning probe arrays coated with multilayer graphene. *Proc. Natl. Acad. Sci. U.S.A.*, **109**, 18312.
24 Lanza, M., Gao, T., Yin, Z.X., Zhang, Y.F., Liu, Z.F., Tong, Y.Z., Shen, Z.Y., and Duan, H.L. (2013) Nanogap based graphene coated AFM tips with high spatial resolution, conductivity and durability. *Nanoscale*, **5**, 10816.
25 Bussmann, E. and Williams, C.C. (2004) Sub-10 nm lateral spatial resolution in scanning capacitance microscopy achieved with solid platinum probes. *Rev. Sci. Instrum.*, **75**, 422.
26 Rocky Mountain Nanotechnology http://rmnano.com/ (accessed 4 February 2016).
27 Hantschel, T., Demeulemeester, C., Eyben, P., Schulz, V., Richard, O., Bender, H., and Vandervorst, W. (2009) Conductive diamond tips with sub-nanometer electrical resolution for characterization of nanoelectronics device structures. *Phys. Status Solidi A*, **206**, 2077.
28 Aguilera, L., Lanza, M., Porti, M., Grifoll, J., Nafría, M., and Aymerich, X. (2008) Improving the electrical performance of a conductive atomic force microscope with a logarithmic current-to-voltage converter. *Rev. Sci. Instrum.*, **79** (7), 073701.
29 FEMTO http://www.femto.de/en/ (accessed 4 February 2016).
30 Tiedje, T. and Brown, A. (1990) Performance limits for the scanning tunneling microscope. *J. Appl. Phys.*, **68** (2), 649–654.
31 Instriuments http://www.ni.com/data-acquisition/ (accessed 4 February 2016).
32 Ji, Y., Hui, F., Shi, Y., Iglesias, V., Lewis, D., Niu, J., Long, S., Liu, M., Hofer, A., Frammelsberger, W., Benstetter, G., Scheuermann, A., McIntyre, P.C., and Lanza, M. (2016) Characterization of the photocurrents generated by the laser of atomic force microscopes. *Rev. Sci. Instrum.*, **87**, 083703.
33 Wang, Z.L. and Song, J.H. (2006) Piezoelectric nanogenerators based on zinc oxide nanowire arrays. *Science*, **312** (5771), 242–246.
34 Muneer, A., Sang, A., Han, D., Hoang, T., Jongwan, J., and Yongho, S. (2011) Local conductance measurement of graphene layer using conductive atomic force microscopy. *J. Appl. Phys.*, **110**, 054307.
35 Binnig, G., Quate, C.F., and Gerber, C. (1986) Atomic-force microscope. *Phys. Rev. Lett.*, **56**, 930–933.
36 Murrell, M.P., Welland, M.E., O'Shea, S.J., Wong, T.M.H., Barnes, J.R., McKinnon, A.W., Heyns, M., and Verhaverbeke, S. (1993) Spatially resolved electrical measurements of SiO2 gate oxides using atomic force microscopy. *Appl. Phys. Lett.*, **62**, 786.
37 O'Shea, S. J., Atta, R. M., Murrell, M. P., Welland, M. E. (1995) Conducting atomic-force microscopy study of silicon dioxide breakdown. *J. Vac. Sci. Technol., B*, **13** (5), 1945-1952.
38 Ruskell, T.G., Workman, R.K., Chen, D., and Sarid, D. (1996) High resolution Fowler-Nordheim field emission maps of thin silicon oxide layers. *Appl. Phys. Lett.*, **68** (1), 93–95.
39 Olbrich, A., Ebersberger, B., Boit, C. (1998) Nanoscale electrical characterization of thin oxides with conducting Atomic Force Microscopy.

Proceedings of the IEEE International Reliability Physics Symposium, 163-168.
40 Olbrich, A., Ebersberger, B., and Boit, C. (1998) Conducting atomic force microscopy for nanoscale electrical characterization of thin SiO2. *Appl. Phys. Lett.*, **73** (21), 3114–3116.
41 Olbrich, A., Ebersberger, B., Boit, C., Vancea, J., and Hoffmann, H. (1999) A new AFM-based tool for testing dielectric quality and reliability on a nanometer scale. *Microelectron. Reliab.*, **39** (6-7), 941–946.
42 Ebersberger, B., Benzinger, G. (1996) Thickness mapping of thin dielectrics with emission microscopy and conductive atomic microscopy for assessment of dielectrics reliability. Proceedings of the IEEE International Reliability Physics Symposium, 126-130.
43 Frammelsberger, W., Benstetter, G., Kiely, J., and Stamp, R. (2006) Thickness determination of thin and ultra-thin SiO2 films by CAFM IV-spectroscopy. *Appl. Surf. Sci.*, **252** (6), 2375–2388.
44 Polspoel, W. and Vandervorst, W. (2007) Evaluation of trap creation and charging in thin SiO2 using both SCM and C-AFM. *Microelectron. Eng.*, **84** (3), 495–500.
45 Nasyrov, K.A., Shaimeev, S.S., and Gritsenko, V.A. (2009) Trap-assisted tunneling hole injection in SiO2: experiment and theory. *J. Exp. Theor. Phys.*, **109** (5), 786–793.
46 Fiorenza, P., Polspoel, W., and Vandervorst, W. (2006) Conductive atomic force microscopy studies of thin SiO2 layer degradation. *Appl. Phys. Lett.*, **88** (22), 222104.
47 Wu, Y.L. and Lin, S.T. (2006) Two-trap-assisted tunneling model for post-breakdown I-V characteristics in ultrathin silicon dioxide. *IEEE Trans. Device Mater. Reliab.*, **6** (1), 75–80.
48 Pakes, C.I., Ramelow, S., Prawer, S., and Jamieson, D.N. (2004) Nanoscale electrical characterization of trap-assisted quasibreakdown fluctuations in SiO2. *Appl. Phys. Lett.*, **84** (16), 3142–3144.
49 Degraeve, R., Kaczer, B., Schuler, F., Lorenzini, M., Wellenkens, D., Hendrickx, P., Van Houdt, J., Haspeslagh, L., Tempel, G., Groeseneken, G. (2001) Statistical model for SILC and pre-breakdown current jumps in ultra-thin oxide layers. Proceedings of the IEEE International Electron Devices Meeting, 121–124.
50 Porti, M., Nafria, M., Aymerich, X., Olbrich, A., and Ebersberger, B. (2001) Electrical characterization of stressed and broken down SiO2 films at a nanometer scale using a conductive atomic force microscope. *J. Appl. Phys.*, **91** (4), 2071–2079.
51 Tan, T.T., Liu, Z.T., Tian, H., and Liu, W.T. (2010) Low voltage stress-induced leakage current in $HfO_2$ dielectric films. *Mater. Sci. Eng., B*, **171**, 159–161.
52 Ang, D.S., Ong, Y.C., O'Shea, S.J., Pey, K.L., Tung, C.H., and Kawanago, T. (2008) Polarity dependent breakdown of the high-k/$SiO_X$ gate stack: a phenomenological explanation by scanning tunneling microscopy. *Appl. Phys. Lett.*, **92** (19), 192904.

53 Lu, X.B., Zhang, X., Huang, R., Liu, H.B., Chen, Z.H., Zhou, H.W., Wang, X.P., Nguyen, B.Y., and Wang, C.Z. (2004) Effect of post-annealing on the physical and electrical properties of LaAlO3 gate dielectrics. *Proc. Int. Conf. Solid-State Integr. Circuits Technol.*, **1** (3), 419–422.

54 Lanza, M., Porti, M., Nafría, M., Aymerich, X., Benstetter, G., Lodermeier, E., Ranzinger, H., Jaschke, G., Teichert, S., Wilde, L., and Michalowski, P. (2009) Crystallization and silicon diffusion nanoscale effects on the electrical properties Of Al2O3 based devices. *Microelectron. Eng.*, **86** (7-9), 1921–1924.

55 Lanza, M., Porti, M., Nafría, M., Aymerich, X., Benstetter, G., Lodermeier, E., Ranzinger, H., Jaschke, G., Teichert, S., Wilde, L., and Michalowski, P. (2011) Conductivity and charge trapping after electrical stress in amorphous and polycristalline Al2O3 based devices studied with AFM related techniques. *IEEE Trans. Nanotechnol.*, **10** (2), 344–351.

56 Bayerl, A., Lanza, M., Porti, M., Nafría, M., Aymerich, X., and Benstetter, G. (2011) Nanoscale and device level gate conduction variability of high-k dielectrics-based metal-oxide-semiconductor structures. *IEEE Trans. Device Mater. Reliab.*, **11** (3), 495–501.

57 Lanza, M., Iglesias, V., Porti, M., Nafría, M., and Aymerich, X. (2011) Polyrystallization effects on the variability of the electrical properties of high-k dielectrics at the nanoscale. *Nanoscale Res. Lett.*, **6**, 108.

58 Bayerl, A., Lanza, M., Aguilera, L., Porti, M., Nafría, M., Aymerich, X., and De Gendt, S. (2013) Nanoscale and device level electrical behavior of ALD annealed Hf-based gate stacks grown with different precursors. *Microelectron. Reliab.*, **53** (6), 867–871.

59 Muenstermann, R., Menke, T., Dittmann, R., Mi, S., Jia, C.L., Park, D., and Mayer, J. (2010) Correlation between growth kinetics and nanoscale resistive switching properties of SrTiO3 thin films. *J. Appl. Phys.*, **108**, 124504.

60 Wu, Y.L., Lin, S.T., Chang, T.M., and Liou, J. (2007) Nanoscale bias-annealing effect in postirradiated thin silicon dioxide films observed by conductive atomic force microscopy. *IEEE Trans. Device Mater. Reliab.*, **7** (2), 351–355.

61 Wu, Y.L., Lin, S.T., Chang, T.M., and Liou, J. (2007) Reliability study of ultrathin oxide films subject to irradiation-then-stress treatment using conductive atomic force microscopy. *Microelectron. Reliab.*, **47** (2-3), 419–421.

62 Porti, M., Nafria, M., Gerardin, S., Aymerich, X., Cester, A., Paccagnella, A., and Ghidini, G. (2009) Implanted and irradiated SiO2/Si structure electrical properties at the nanoscale. *J. Vac. Sci. Technol., B*, **27** (1), 421–425.

63 Iglesias, V., Porti, M., Nafría, M., Aymerich, X., Dudek, P., Schroeder, T., and Bersuker, G. (2010) Correlation between the nanoscale electrical and morphological properties of crystallized hafnium oxide-based metal oxide semiconductor structures. *Appl. Phys. Lett.*, **97**, 262906.

64 Blasco, X., Nafría, M., Aymerich, X., and Vandervorst, W. (2005) Comparison of SiO2 and HfO2/SiO2 gate stacks electrical behaviour at a nanometer scale with CAFM. *Electron. Lett*, **41** (12), 719–721.

65 Wu, Y.L. and Lin, S.T. (2008) Breakdown spots propagation in ultrathin SiO2 films under repetitive ramped voltage stress using conductive atomic force microscopy. *J. Phys. Chem. Solids*, **69** (2-3), 470–474.

66 Lanza, M., Porti, M., Nafría, M., Aymerich, X., Whittaker, E., and Hamilton, B. (2010) UHV CAFM characterization of high-k dielectrics: effect of the technique resolution on the pre-and post-breakdown electrical measurements. *Microelectron. Reliab.*, **50** (9–11), 1312–1315.

67 Uppal, H.J., Bernardini, S., Efthymiou, E., Volkos, S.N., Dimoulas, A., Markevich, V., Hamilton, B. and Peaker, A.R. (2008) Nanoscale electrical characterization of ultra thin high-*k* dielectric MOS stacks: A conducting AFM study. *Materials Science in Semiconductor Processing*, **11**, 250–253.

68 Uppal, H.J., Mitrovic, I.Z., Hall, S., Hamilton, B., Markevich, V., and Peaker, A.R. (2009) Breakdown and degradation of ultrathin Hf-based (HfO2(x)(SiO 2)(1-x). *J. Vac. Sci. Technol., B*, **27** (1), 443–447.

69 Sire, C., Blonkiwski, S., Gordon, M.J., and Baron, T. (2007) Statistics of electrical breakdown field in HfO2 and SiO2 films from millimeter to nanometer length scales. *Appl. Phys. Lett.*, **91** (24), 242905.

70 Fiorenza, P., Lo Nigro, R., and Raineri, V. (2007) Breakdown kinetics at nanometer scale of innovative MOS devices by conductive atomic force microscopy. *Microelectron. Eng.*, **84** (3), 441–445.

71 Tung, C.H., Pey, K.L., Lin, W.H., and Radhakrishnan, M.K. (2002) Polarity-dependent dielectric breakdown-induced epitaxy (DBIE) in Si MOSFETs. *IEEE Electron Device Lett.*, **23** (9), 526–528.

72 Polspoel, W., Favia, P., Mody, J., Bender, H., and Vandervorst, W. (2009) Physical degradation of gate dielectrics induced by local electrical stress using conductive atomic force microscopy. *J. Appl. Phys.*, **106** (2), 024101.

73 Porti, M., Nafria, M., Blum, M.C., Aymerich, X., and Sadewasser, S. (2003) Atomic force microscope topographical artifacts after the dielectric breakdown of ultrathin SiO2 films. *Surf. Sci.*, **532–535**, 727–731.

74 Lanza, M., Zhang, K., Porti, M., Nafría, M., Shen, Z.Y., Liu, L.F., Kang, J.F., Gilmer, D., and Bersuker, G. (2012) Grain boundaries as preferential sites for resistive switching in the HfO2 resistive random access memory structures. *Appl. Phys. Lett.*, **100**, 123508.

75 Lanza, M. (2014) A review on resistive switching in high-k dielectrics: a nanoscale point of view using conductive atomic force microscope. *Materials*, **7**, 2155–2182.

76 Lee, G.H., Yu, Y.J., Lee, C., Dean, C., Shepard, K.L., Kim, P., and Hone, J. (2011) Electron tunneling through atomically flat and ultrathin hexagonal boron nitride. *Appl. Phys. Lett.*, **99**, 243114.

77 Ji, Y., Pan, C., Zhang, M., Long, S., Lian, X., Miao, F., Hui, F., Shi, Y., Larcher, L., Wu, E., and Lanza, M. (2016) Boron nitride as two dimensional dielectric: reliability and dielectric breakdown. *Appl. Phys. Lett.*, **108**, 012905.

78 Hattori, Y., Taniguchi, T., Watanabe, K., and Nagashio, K. (2015) Layer-by-layer dielectric breakdown of hexagonal boron nitride. *ACS Nano*, **9** (1), 916–921.

79 Piquemal-Banci, M., Galceran, R., Caneva, S., Martin, M.B., Weatherup, R.S., Kidambi, P.R., Bouzehouane, K., Xavier, S., Anane, A., Petroff, F., Fert, A., Robertson, J., Hofmann, S., Dlubak, B., and Seneor, P. (2016) Magnetic tunnel junctions with monolayer hexagonal boron nitride tunnel barriers. *Appl. Phys. Lett.*, **108**, 102404.

80 Britnell, L., Gorbachev, R.V., Jalil, R., Belle, B.D., Schedin, F., Katsnelson, M.I., Eaves, L., Morozov, S.V., Mayorov, A.S., Peres, N.M.R., Neto, A.H.C., Leist, J., Geim, A.K., Ponomarenko, L.A., and Novoselov, K.S. (2012) Electron tunneling through ultrathin boron nitride crystalline barriers. *Nano Lett.*, **12**, 1707–1710.

81 Sze, J.Y., Tay, B.K., Pakes, C.I., Jamieson, D.N., and Prawer, S. (2005) Conducting Ni nanoparticles in an ion-modified polymer. *J. Appl. Phys.*, **98**, 066101.

82 Cui, X.D., Primak, A., Zarate, X., Tomfohr, J., Sankey, O.F., Moore, A.L., Moore, T.A., Gust, D., Harris, G., and Lindsay, S.M. (2001) Reproducible measurement of single-molecule conductivity. *Science*, **294** (5542), 571–574.

83 Zhou, C., Kong, J., and Dai, H. (2000) Electrical measurements of individual semiconducting single-walled carbon nanotubes of various diameters. *Appl. Phys. Lett.*, **76**, 1597.

84 Vecchio, C., Sonde, S., Bongiorno, C., Rambach, M., Yakimova, R., Raineri, V., and Giannazzo, F. (2011) Nanoscale structural characterization of epitaxial graphene grown on off-axis 4H-SiC (0001). *Nanoscale Res. Lett.*, **6**, 269.

85 Giannazzo, F., Sonde, S., Rimini, E., and Raineri, V. (2011) Lateral homogeneity of the electronic properties in pristine and ion-irradiated graphene probed by scanning capacitance spectroscopy. *Nanoscale Res. Lett.*, **6**, 109.

86 Tan, S., Tang, Z., Liang, X., and Kotov, N.A. (2004) Resonance tunneling diode structures on CdTe nanowires made by conductive AFM. *Nano Lett.*, **4** (9), 1637–1641.

87 Lanza, M., Wang, Y., Gao, T., Bayer, A., Porti, M., Nafria, M., Zhou, Y., Jing, G., Zhang, Y., Liu, Z., Yu, D., and Duan, H. (2013) Electrical and mechanical performance of graphene sheets exposed to oxidative environments. *Nano Res.*, **6** (7), 485–495.

88 Howell, S.L., Padalkar, S., Yoon, K., Li, Q., Koleske, D.D., Wierer, J.J., Wang, G.T., and Lauhon, L.J. (2013) Spatial mapping of efficiency of GaN/InGaN nanowire array solar cells using scanning photocurrent microscopy. *Nano Lett.*, **13**, 5123–5128.

89 Pan, C., Hu, J., Grustan-Gutierrez, E., Hoang, M.T., Duan, H., Yvonnet, J., Mitrushchenkov, A., Chambaud, G., and Lanza, M. (2016) Suppression of nanowire clustering in hybrid energy harvesters. *J. Mater. Chem. C*, **4**, 3646.

90 ISI Web of Knowledge, Thomson Reuters www.webofscience.com.

91 Lanza, M., Porti, M., Nafria, M., Benstetter, G., Frammelsberger, W., Ranzinger, H., Lodermeier, E., and Jaschke, G. (2007) Influence of the manufacturing process on the electrical properties of thin (<4 nm) Hafnium based high-k stacks observed with CAFM. *Microelectron. Reliab.*, **47** (9-11), 1424–1428.

92 Okada, Y., Miyagi, M., Akahane, K., Kawabe, M., Shigekawa, H. (2002) Self-organized InGaAs quantum dots grown on GaAs (3 1 1) B substrate studied by conductive atomic force microscope technique *J. Cryst. Growth* **245**, 212–8.

93 Wu, S.J., Hopkins, T. http://nano.tm.agilent.com/apps_material/AN-ResiScope_5990-8413.PQ.pdf (accessed 21 July 2011).

94 Coll, M., Palau, A., Gonzalez-Rosillo, J.C., Gazquez, J., Obradors, X., and Puig, T. (2014) Integration of atomic layer deposition CeO2 thin films with functional complex oxides and 3D patterns. *Thin Solid Films*, **553**, 7–12.

95 Stahl, U., Yuan, C.W., Delozanne, A.L., and Tortonese, M. (1994) Atomic force microscope using piezoresistive cantilevers and combined with a scanning electron microscope. *Appl. Phys., Lett.*, **65** (22), 2878–2880.

96 Anderson, M.S. (2000) Locally enhanced Raman spectroscopy with an atomic force microscope. *Appl. Phys. Lett.*, **76** (21), 3130–3132.

97 Nanonics, http://www.nanonics.co.il/products/multiview-4000-system.

98 Sader, J.E. and Jarvis, S.P. (2004) Accurate formulas for interaction force and energy in frequency modulation force spectroscopy. *Appl. Phys. Lett.*, **84** (10), 1801–1803.

99 Nano verk http://www.nanowerk.com/atomic_force_microscope_manufacturers.php.

100 Leschkies, K.S., Divakar, R., Basu, J., Enache-Pommer, E., Boercker, J.E., Carter, C.B., Kortshagen, U.R., Norris, D.J., and Aydil, E.S. (2007) Photosensitization of zno nanowires with cdse quantum dots for photovoltaic devices. *Nano Lett.*, **7** (6), 1793–1798.

101 Wang, G., Wang, H., Ling, Y., Tang, Y., Yang, X., Fitzmorris, R.C., Wang, C., Zhang, J.Z., and Li, Y. (2011) Hydrogen-treated TiO2 nanowire arrays for photoelectrochemical water splitting. *Nano Lett.*, **11** (7), 3026–3033.

102 Xu, S., Lao, C.S., Weintraub, B., and Wang, Z.L. (2008) Density-controlled growth of aligned zno nanowire arrays by seedless chemical approach on smooth surfaces. *J. Mater. Res.*, **23**, 2072–2077.

103 Ge, X., Hong, K., Zhang, J., Liu, L.Q., and Xu, M.X.A. (2015) Controllable microwave-assisted hydrothermal method to synthesize ZnO nanowire arrays as recyclable photocatalyst. *Mater. Lett.*, **139**, 119–121.

104 San, X.G., Wang, G.S., Liang, B., Song, Y.M., Gao, S.Y., Zhang, J.S., and Meng, F.L. (2015) Catalyst-free growth of one-dimensional ZnO nanostructures on SiO2 substrate and in situ investigation of their H2 sensing properties. *J. Alloys Compd.*, **622**, 73–78.

105 Nam, G., Park, Y., Ji, L., Kim, B., Lee, S.H., Kim, D.Y., Kim, S., Kim, S.O., and Leem, J.Y. (2015) Facile synthesis and enhanced ultraviolet emission of zno nanorods prepared by vapor-confined face-to-face annealing. *ACS Appl. Mater. Interfaces*, **7**, 873–879.

106 Kar, J.P., Kim, S., Shin, B., Park, K.I., Ahn, K.J., Lee, W., Cho, J.H., and Myoung, J.M. (2010) Influence of sputtering pressure on morphological, mechanical and electrical properties of al-doped zno films. *Solid-State Electron.*, **54**, 1447–1450.

**107** Yang, M., Kim, H.C., and Hong, S.H. (2014) Growth of ZnO nanorods on fluorine-doped tin oxide substrate without catalyst by radio-frequency magnetron sputtering. *Thin Solid Films*, **573**, 79–83.

**108** Nicaise, S.M., Cheng, J.J., Kiani, A., Gradečak, S., and Berggren, K.K. (2015) Control of zinc oxide nanowire array properties with electron-beam lithography templating for photovoltaic applications. *Nanotechnology*, **26**, 075303.

**109** Khranovskyy, V. and Yakimova, R. (2012) Morphology engineering of ZnO nanostructures. *Physica B*, **407**, 1533–1537.

**110** Su, W.S., Chen, Y.F., Hsiao, C.L., and Tu, L.W. (2007) Generation of electricity in GaN nanorods induced by piezoelectric effect. *Appl. Phys. Lett.*, **90**, 063110.

**111** Lanza, M., Reguant, M., Zou, G., Lv, P., Li, H., Chin, R., Liang, H., Yu, D., Zhang, Y., Liu, Z., and Duan, H. (2014) High performance piezoelectric nanogenerators using two-dimensional flexible top electrodes. *Adv. Mater. Interfaces*, **1** (5), 1300101.

**112** Yang, T., Hertenberger, S., Morkötter, S., Abstreiter, G., and Koblmüller, G. (2012) Size, composition, and doping effects on In(Ga)As nanowire/Si tunnel diodes probed by conductive atomic force microscopy. *Appl. Phys. Lett.*, **101**, 233102.

**113** Hung, S.C. (2011) Generation of piezoelectricity by deflecting nanorods vertically on gan template. *J. Electrochem. Soc.*, **158** (12), H1265–H1269.

**114** Choi, D., Choi, M.Y., Choi, W.M., Shin, H.J., Park, H., Seo, J.S., and Park, J. (2006) Fully rollable transparent nanogenerators based on graphene electrodes. *Adv. Mater.*, **22**, 2187–2192.

**115** Xu, S., Wei, Y.G., Liu, J., Yang, R.S., and Wang, Z.L. (2008) Integrated multilayer nanogenerator fabricated using paired nanotip-to-nanowire brushes. *Nano Lett.*, **8**, 4027–4032.

**116** Kumar, B., Lee, K.Y., Park, H.K., Chae, S.J., Lee, Y.H., and Kim, S.W. (2011) Controlled growth of semiconducting nanowire, nanowall, and hybrid nanostructures on graphene for piezoelectric nanogenerators. *ACS Nano*, **5**, 4197–4204.

# 2

# Fabrication and Reliability of Conductive AFM Probes

*Oliver Krause*

*Nanoworld Services GmbH, Schottkystraβe 10, 91058 Erlangen, Germany*

## 2.1 Introduction

Atomic force microscopy (AFM) [1] is a widely used technology to image the topography in high resolution on both conducting and insulating surfaces. One of the further developments of AFM is conductive or electrical measurement modes (C-AFM).

Evolving from the scanning tunneling microscope (STM) [2] where a conductive tungsten wire is scanned across a conductive surface detecting tunneling currents, AFM was focused initially on imaging the topography of nonconductive surfaces. One of the main features distinguishing AFM from STM is the usage of special probes in which a sharp tip is mounted at the end of a cantilever [3–5] instead of a sharpend end of a tungsten wire.

Depending on the definition of conductive AFM measurements, the first reported C-AFM experiments were carried out by Martin *et al.* 6, who recorded a capacitance–voltage spectra at each pixel of their scan of a silicon/resist-on-silicon surface using a tungsten STM tip in their AFM. The first experiments measuring a flux of charges through an AFM probe were made by Welland *et al.* [7, 8], who used metal-coated silicon AFM probes for their experiments.

Today the main methods in C-AFM are scanning capacitance microscopy (SCM) [6], tunneling AFM (TUNA) [9], and scanning spreading resistance microscopy (SSRM) [10, 11], which allow to measure a wide range of electrical properties on various materials with nanometer-scale resolution.

Of course, all electrical applications have the need for conductive AFM probes. Various kinds of probes for C-AFM have been studied up to now. Highly doped, freshly etched silicon probes, coated standard (silicon or silicon nitride) AFM probes, with metals such as Ag, Au, Co, or Pt [12, 13], alloys such as titanium nitride, tungsten- or boron-carbide [14], or diamond [15]. Besides, massive metal [16], or diamond [17] probes, or massive diamond tips on silicon cantilevers [18] have been described.

*Conductive Atomic Force Microscopy: Applications in Nanomaterials*, First Edition.
Edited by Mario Lanza.

## 2.2 Manufacturing of Conductive AFM Probes

A sharp tip mounted onto a cantilever attached to a holder structure [1] illustrates the principle of an AFM probe. Diamond slivers glued at the end of a gold foil cantilever were used as the first AFM probes, but those were handcrafted approaches and those AFM probes were neither reliable nor mass-producible.

With the development of AFM, novel probe-manufacturing techniques were developed: semiconductor and microelectromechanical system (MEMS)-manufacturing methods were used, which allowed to fabricate several hundred AFM probes simultaneously on one single silicon wafer. Additionally, those manufacturing methods were precise enough to ensure stable cantilever and tip dimensions. Two different basic methods of manufacturing AFM probes evolved: (i) by depositing and structuring thin films as a cantilever [3, 4] or (ii) by etching a silicon wafer until only a monolithically integrated holder chip with cantilever and tip remains [5].

Following these methods, there are three possibilities to create a conductive AFM tip. (i) The previously made AFM tip could be coated with a conductive film, (ii) a conductive material could be used as the thin-film cantilever material, and (iii) the tip material could be converted into a (more) conductive material. Of course, combinations of the three different approaches are possible.

### 2.2.1 Thin Film Cantilever

In the research group of Carl Quate, different methods of micromachining AFM cantilevers were performed. Albrecht *et al.* [3] deposited a thin film layer, and created a cantilever by photolithographic structuring and subsequent etching. To create the tip, two different approaches were followed. For the so called mold-and-release tips, a hole is etched in the supporting silicon wafer. Depending on the crystal planes, a pyramidal pit with a sharp apex is formed. The thin film – usually silicon nitride, also silicon oxide – is deposited onto the support wafer with the pyramidal etch pitch. The formation of a proper etch pit for the tip and the accurate deposition of the thin film at the very bottom of that pyramid are very critical process steps to obtain sharp, single-tip AFM probes. After defining the cantilever by photolithography and etching, the supporting silicon wafer is removed by etching and a freestanding silicon nitride cantilever with a pyramidal silicon nitride tip remains. Thin film technology allows the fabrication of cantilevers with very low force constants from about 0.01 up to 2 N/m. Often the low-force-constant cantilevers are not sufficient to establish a stable electric contact between the AFM tip and the sample. Figure 2.1 shows a mold-and-release silicon nitride thin-film probe.

### 2.2.2 Corner Tips

The second method to manufacture a thin film cantilever is to first etch a silicon membrane, which is then covered by a (silicon nitride) thin film at the backside in a second step. In the next step, the cantilever is defined by reactive ion etching out of the thin-film layer. Finally, an anisotropic wet-chemical etching of the membrane leaves a silicon corner tip on the (silicon nitride) thin film cantilever [4].

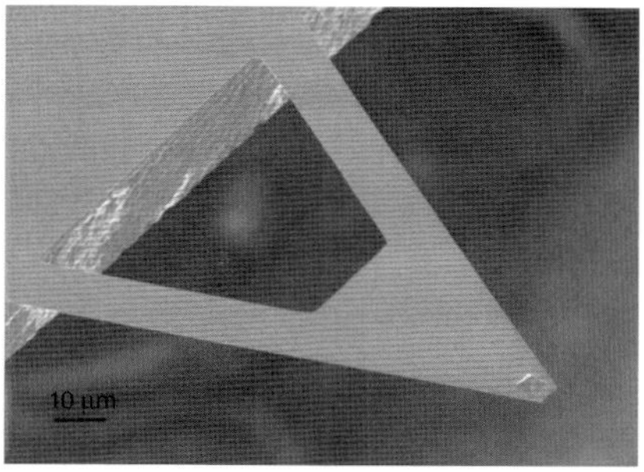

**Figure 2.1** SEM micrograph of a mold-and-release thin-film probe (NanoWorld® Pyrex Nitride Probe).

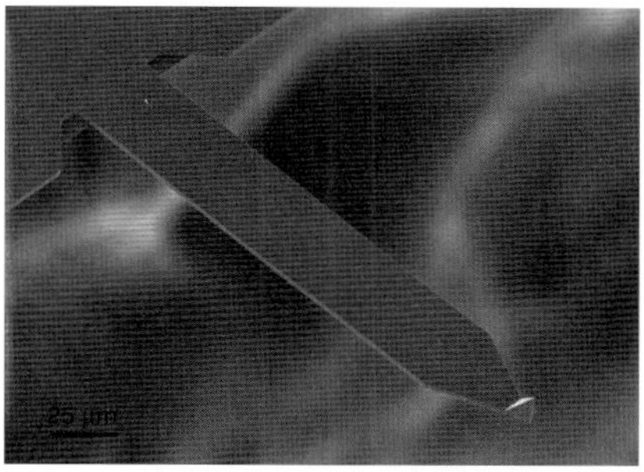

**Figure 2.2** SEM micrograph of a corner tip, fabricated by a silicon-on-insulator technology (MikroMasch® Opus™ AFM probe).

In a similar manner, corner tips could be produced on a silicon cantilever when SOI (silicon-on-insulator) wafers are used as the starting material [19] (Figure 2.2).

### 2.2.3 Etched Silicon Probes

Wolter *et al.* [5] from IBM invented a method to fabricate an AFM probe by etching crystalline silicon to form a monolithic, single crystalline holder chip, cantilever, and tip system: a silicon wafer is anisotropically etched from either the front or the back using aqueous potassium hydroxide solution. In several consecutive mask-and-etch cycles, the holder chip, cantilever, and tip are released. Together with the silicon corner tips on SOI, this probe manufacturing method is most reliable and allows the fabrication of AFM probes with cantilever in a range of stiffness from 0.1 to 1000 N/m and tip radii down to 2 nm. Until now, the etched silicon and silicon corner probes are the standard for AFM probes. Figure 2.3 shows a scanning electron microscopy (SEM) micrograph of a monolithic etched silicon AFM probe.

The silicon material which is used to manufacture these AFM probes is highly doped and therefore has semiconducting properties. Several research groups reported the use of bare silicon as conductive AFM probes, where the insulating native silicon oxide (growing while storing in normal atmosphere) has to be removed either by etching or an electric pulse in the C-AFM setup. Nevertheless, the use of bare silicon tips is reported as unpredictable, because of eventual tip blunting after pulsing and the ongoing oxidation of the silicon tip material [13, 14].

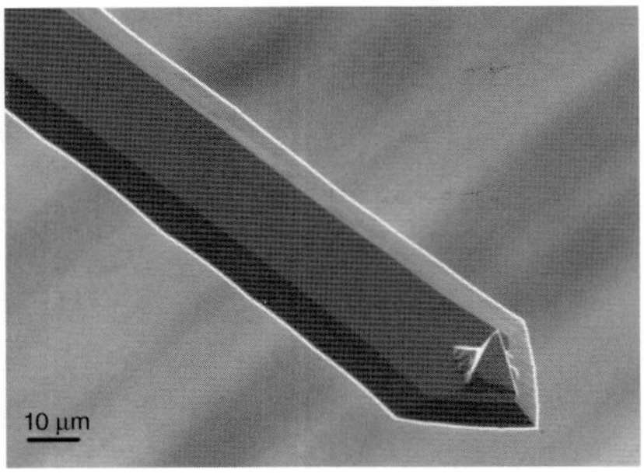


**Figure 2.3** SEM micrograph of an etched silicon AFM probe (NANOSENSORS™ PointProbe® Plus).

### 2.2.4 Coating of Probes

The most self-evident solution to create a conductive probe is to coat it with a conductive layer. Murrell *et al.* [8] coated a standard silicon AFM probe with 100 nm of titanium. Using the probe in contact mode, they applied a defined voltage to the sample and recorded the resulting current, receiving a tunneling current map of a silicon oxide surface ("local $I$–$V$ measurements", today known as *tunneling AFM* or *TUNA*) [9]. In a subsequent work O'Shea *et al.* [13] investigated different coatings (Ti, Pt, Au, diamond) and compared them with freshly etched, highly doped silicon tips, and Au wire tips.

As applying a coating to a readymade probe is basically a very easy method, this has been done by a large number of researchers. Etched silicon AFM probes have been coated with various metals (Au, Ag, Ti, Ni), alloys (TiN, CoCr), or carbides (WC, $B_{4C}$). These materials are mostly deposited by a sputtering or evaporation process. Here, the thickness of the coating and thereby the increase of the tip radius is scalable. Common material thicknesses are in the range of a few tens of nanometers. The resulting tip radii are in the same range (usually 30–100 nm). In Figure 2.4, the granular structure of a sputtered platinum coating[1] is visible. Metal-coated probes do show very good conductivity but wear out very fast [13, 16, 20]. Additionally, metal-coated probes can be affected by melting of the conductive material at the apex of the tip [16].

Using the same concept, diamond layers have been grown via a chemical vapor deposition process (CVD) [17]. When a tip is coated, the radius of the tip increases. For metal and alloy coatings, the thickness of the coating and thereby the increase of the tip radius is scalable. This is different when growing a conductive diamond layer: diamond is growing from a seed layer of nanocrystallites [15]. To obtain a closed and therefore conductive layer, the

1 Originally platinum-coated conductive probes were coated with pure platinum. Later, for the reason of greater hardness, a small amount of iridium (usually around 5%) was added. So, actually these probes should have be called *platinum–iridium* or $Pt_{0.95}Ir_{0.05}$-*coated* probes. For the sake of convenience, these probes are usually referred to as *platinum-coated* probes only. In this chapter, all probes referred to as *platinum coated*, were coated with a $Pt_{0.95}Ir_{0.05}$ alloy.

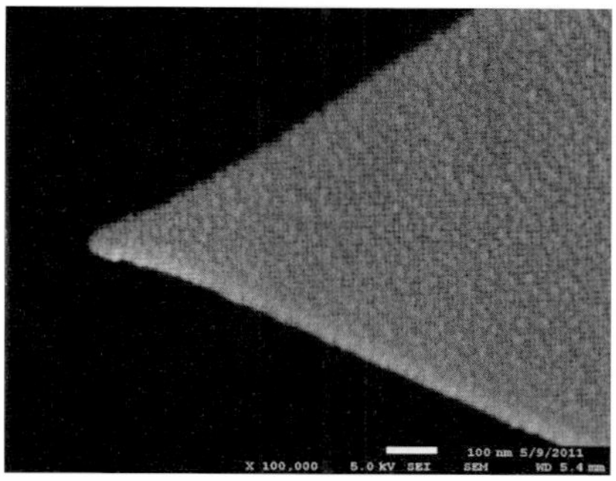


**Figure 2.4** High-resolution SEM image of a platinum-coated tip (NANOSENSORS™ Electrostatic Force Microscopy probe).

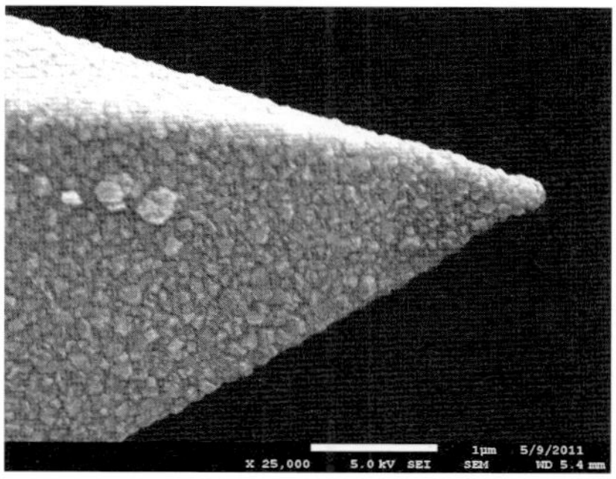


**Figure 2.5** SEM micrograph of a CVD diamond-coated tip (NANOSENSORS™ Conductive Diamond Tips).

nanocrystallites have to attain a certain grain size. A minimal thickness of about 100 nm is needed to form a closed (i.e., electrically conductive) diamond layer. This leads to tip radii above 100 nm [16, 17] (Figure 2.5). Mere diamond is not conductive, so during the deposition process, doping elements have to be added. On the one hand, the conductivity of diamond is limited by the doping process: it is relatively poor compared to the conductivity of metal-coated probes [16] and can vary over a decade [21]. On the other hand, introducing impurities into the diamond reduces the amount of $sp^3$-bindings and this will decrease the initial hardness of the diamond [22], thus increasing the wear-out compared to undoped diamond layers [23].

In principle, all types of AFM probes could be coated, but is has to be taken into account that an additional layer added to a freestanding cantilever could add significant stress and can cause a bending of the cantilever [13]. The thinner the cantilever the stronger it will be affected by the stress and bend. If the cantilever is bent too strongly, the adjustment of the reflected laser signal into the AFM's photodetector will not be possible any more; this would be the case for the cantilever shown in Figure 2.6. As thin film cantilevers are limited in thickness and force constants, most coated probes are based on standard etched silicon tips.

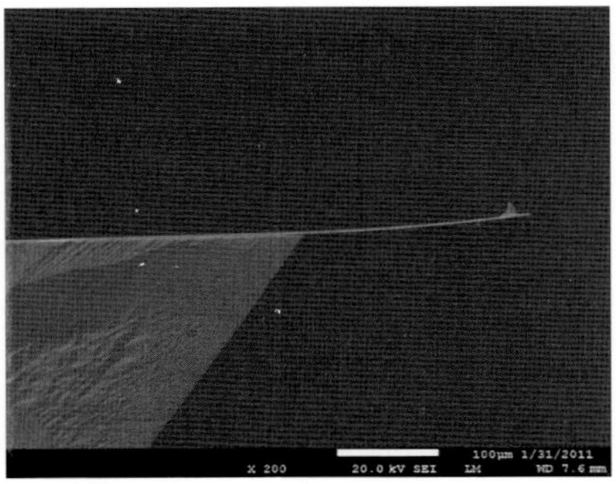


**Figure 2.6** AFM probe with extreme cantilever bending.

Today's standard silicon probes for C-AFM applications are either platinum or diamond coated.

### 2.2.5 Conductive Thin Film Probes

An alternative fabrication scheme to overcome the "thicker layer – larger radius" problem is the variation of the standard thin-film manufacturing approach [3]: silicon nitride or silicon oxide films were chosen to form cantilevers because of their existence in, or compatibility with, semiconductor and MEMS manufacturing technologies: there exist easy and reliable processes for depositing and structuring of the material and to resist the silicon-removal etching step. For the fabrication of standard thin-film probes, in a first step, an inverted pyramid is etched into the silicon. But instead of the standard silicon nitride thin film, a conductive material is deposited. As before, the deposited layer is structured to create a cantilever and the silicon support wafer is removed by wet etching to release the tip and the cantilever. With this method, solid-diamond [15] (Figure 2.7) and solid-metal tips [24] (Figure 2.8) have been created on diamond (Figure 2.7), metal [25], or silicon cantilevers [26] (Figure 2.9).

A common problem for all mold-and-release tips is to fill up the very bottom of the etched tip pyramid properly during the thin-film deposition process. This often results in blunt tips with large tip radii [16, 17]. Additionally, as

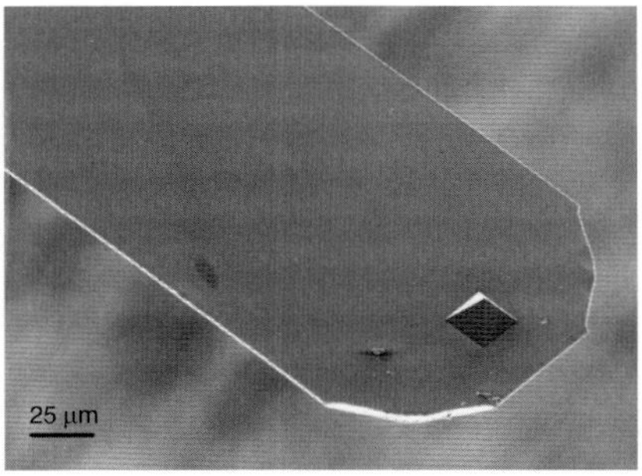


**Figure 2.7** SEM micrograph of a full-diamond AFM probe.

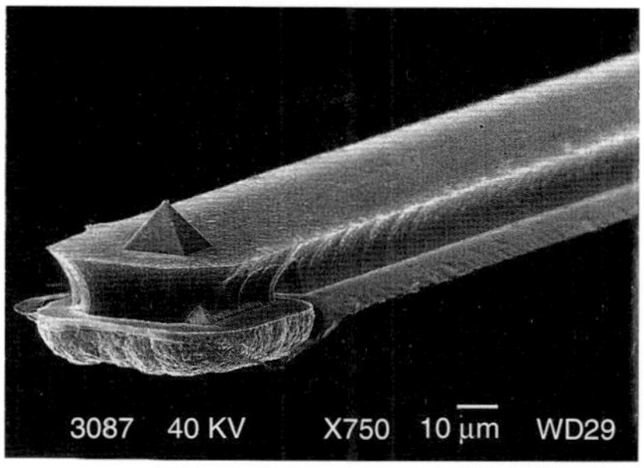


Figure 2.8 Metal tip on a silicon cantilever [24].

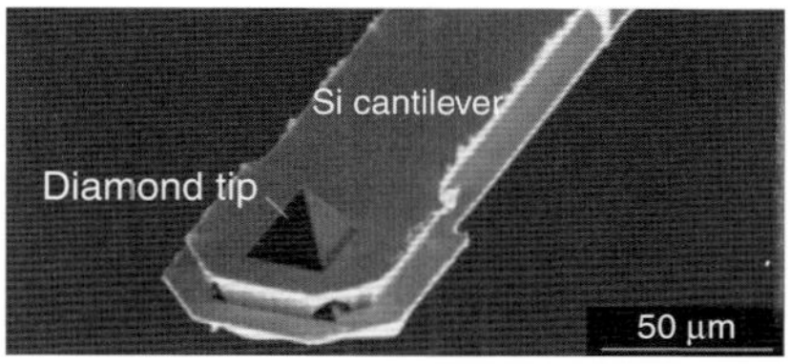


Figure 2.9 Diamond tip on a silicon cantilever [26].

for metal-coated probes, for massive metal probes with large tip radii, a fast wear-out behavior has been reported [16].

### 2.2.6 Material Conversion

Another approach followed was to convert silicon into another, more conductive, material. Various conductive thin film cantilever concepts have not been successful and silicon tips are still the most reliable AFM probes. Therefore, keeping the properties of a silicon tip (significantly smaller tip radius, a much higher aspect ratio, excellent mechanical properties, and a reliable manufacturing process compared to SiN thin-film tips) and making that kind of tip conductive would be big step toward a small radius, low wear-out C-AFM probe.

Doezema described a process commonly used in integrated circuit manufacturing: a metal is deposited on silicon followed by thermal annealing, which results in the formation of a so-called silicide [27]. In general, silicides show much better conductivity than silicon and higher hardness compared to the pure metal [28]. Akiyama *et al.* used PtSi in a demanding thin film cantilever manufacturing process [29]. Bhaskaran and colleagues picked up the idea of Doezema using a very simple approach by just depositing Pt around the tip pyramid of a standard silicon probe and then thermally annealing it, so a local PtSi layer was formed [30]. On the basis of this work, a whole monolithic silicon tip, cantilever, and chip system was coated and thermally treated in a way such that a completely PtSi-capsulated probe was made [31]. Figure 2.10 shows the tip apex of one such platinum silicide AFM probe.

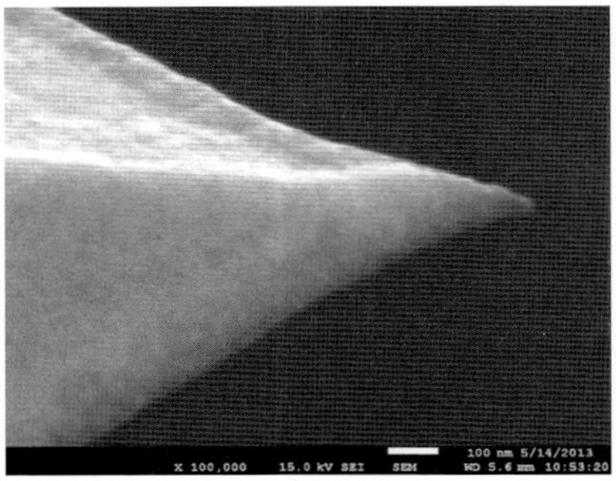


**Figure 2.10** High-resolution SEM image of a NANOSENSORS™ platinum silicide probe.

## 2.3 How to Choose Your C-AFM Tip

What kind of C-AFM tips should be used? Every AFM tip manufacturer provides platinum or cobalt–chromium-coated silicon probes. Originally, this kind of probes were invented for use in noncontact AFM operations such as electrostatic- or magnetic-force-microscopy. Owing to their high wear-rates in contact mode, they are not recommended for C-AFM applications by the manufacturers but they are used frequently here. Most companies also offer probes for conductive applications in contact mode: these are mostly diamond-coated silicon probes. However, platinum silicide AFM tips, thin-film diamond probes, and alternative coatings can also be found.

### 2.3.1 Cantilever Choice

How do you select your C-AFM probe? To answer this question, you have to find out first what forces between tip and sample your measurement mode and sample require. The range of forces starts around 50 nN in SCM and rises up to 100 μN for SSRM measurements.

The force a cantilever can exert on a sample surface is limited by the range of the AFM photodetector. A soft cantilever could be bended relatively easily, as only a small force would be needed for a bending that will lead the reflected laser light to the edge of the photodetector. If higher forces are required, a cantilever with a higher force constant has to be choosen to be chosen. In contrast, when using cantilevers with high force constants, small interaction forces can hardly be adjusted precisely. The interaction force between the tip and the sample has to be selected carefully and adapted to each individual combination of measurement mode and sample characteristic. If the interaction force is too low, no stable electric contact between the tip and the sample can be established. If the applied force is too high, the sample and the tip could be damaged.

### 2.3.2 Tip Material Choice

In a second step the coating material has to be selected. The user has to weigh resolution versus durability or wear. Another factor is whether a certain value of conductivity is needed.

**Table 2.1** Selection help: what type of AFM material should be used for which type of cantilever and for what kind of application.

| | Platinum | Platinum silicide | Diamond |
|---|---|---|---|
| Soft cantilever 0.2–2 N/m | SCM, CAFM, TUNA | SCM, CAFM, TUNA | CAFM, TUNA |
| Medium stiff cantilever 2–10 N/m | SCM, CAFM, TUNA | SCM, CAFM, TUNA | CAFM, TUNA |
| Stiff cantilever 10–40 N/m | Material too soft to withstand high forces | CAFM, TUNA | CAFM, TUNA, SSRM |

The general opinion is that, if high resolution or good conductivity is needed, a metal-coated tip has to be chosen. If durability is required, a low wear-out behavior or high contact forces are required, diamond-coated probes are the choice. The relatively new platinum silicide AFM tips are a good alternative as high-resolution, good-conducting, and low-wear tips. The conductivity of PtSi is in the range of metallic platinum and decades larger than those of doped diamond. Finally, the PtSi tip wear is significantly lower when compared to metal-coated probes, although the stability of diamond-coated probes could not be reached.

Table 2.1 provides a brief overview of what tip material has to be combined with which cantilever type to get a proper combination for a specific C-AFM method.

### 2.3.3 Resolution of C-AFM Tips

The resolution of an AFM tip is defined by its radius. As described earlier, for metallic coatings the radius is scalable (a 30 nm radius is common), whereas a diamond layer needs a certain thickness to form a closed layer. Here the macroscopic radius is around 100–200 nm but sometimes, when a diamond crystallite is prominent at the tip apex, a much better resolution can be achieved. Figure 2.11 shows the tip apices of diamond-coated silicon, platinum-coated silicon, and platinum silicide AFM probes. All three SEM images are captured with the same

(a) (b) (c)

**Figure 2.11** Direct comparison of tip apices of different C-AFM probes. (a) Diamond-coated silicon, (b) platinum-coated silicon, and (c) platinum silicide. All images are captured with identical magnifications. The scale bar size is 100 nm.

magnification. It can be seen clearly that the radius of the diamond-coated probe is much larger than those of the two other probes. Besides the greater hardness of the diamond itself, owing to its more massive apex, the diamond-coated tip can withstand much higher loads without breaking than the platinum or platinum silicide probe.

## 2.4 Tip Wear and Sample Damage: Applicable Forces and Currents in C-AFM

AFM probes and, of course, their conductive coatings are affected by different terms of failure. When the tip slides in contact mode across the surface, material – surface coating or tip-base material – will be removed, and the tip becomes blunt. This is referred to as *mechanical wear*. The user will notice this effect when there is loss of resolution. Additionally, the probe conductivity may degrade or the probe may lose its conductive properties completely. This is known as *electrical wear*. A probe might also lose its conductive properties if the current flow through the tip apex becomes very high. The apex might heat up and the coating material will melt or vaporize. Finally, when the tip–sample interaction forces are too high, the tip apex may break or the sample might be damaged as well.

In this chapter, three common conductive AFM probes – platinum ($Pt_{0.95}Ir_{0.05}$)-coated probes, platinum silicide probes, and doped diamond AFM tips – are compared in terms of electrical and mechanical wear and lifetime.

### 2.4.1 Tip Wear: Mechanical Wear – Varying Forces

When scanning in contact mode, a certain force is needed to ensure a stable electrical contact between the AFM tip and the sample surface. The degree of the force depends on the measurement mode, the measurement conditions, as well as on the needs for each specific sample. To see how a C-AFM tip apex will be affected when scanning with different interaction forces, SEM images of individual platinum-coated AFM probes were made. Then, the tips were engaged with defined forces of 200, 500, and 1000 nN, respectively. A contact-mode AFM scan with a total tip travel distance of 5 mm was performed. After the AFM scan, a second SEM image of each individual tip was made. Figure 2.12a–c shows the platinum coated silicon probes before and after the scan with varying forces. The SEM images are superimposed and aligned in such a way that the change undergone by the tip due to the AFM scan becomes clearly visible.[2] For the 200 nN scan in Figure 2.12a, very little material removal is visible. Here the reduction in tip height is less than 10 nm. In the 500 nN scan, in Figure 12.12b the height of the tip apex is reduced by about 40 nm by material removal, whereas after the

2 Within this overlay technique, SEM images of the same AFM tip can be precisely aligned at the very end of the tip's apex owing to the uniqueness of the coating crystallites. Further away from the tip apex, the overlay may vary because the tip cannot be aligned exactly with regard to its pyramid central axis. Here, the tip might be rotated very slightly and therefore appear wider.

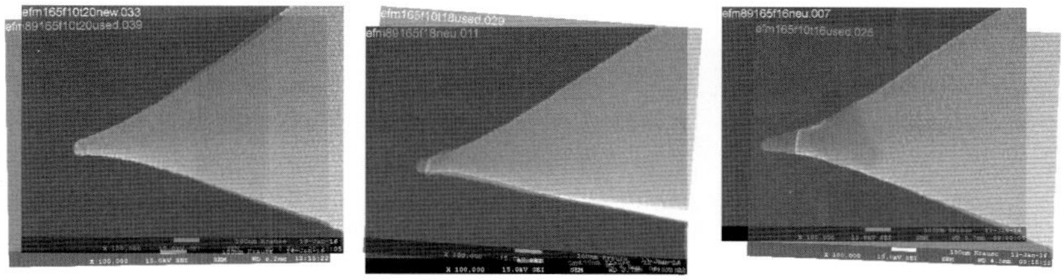

(a) (b) (c)

**Figure 2.12** Comparison of tip wear. Contact-mode scanning with a platinum tip with (a) 200 nN, (b) 500 nN, and (c) 1 μN interaction force. Overlay of SEM images before and after the scan.

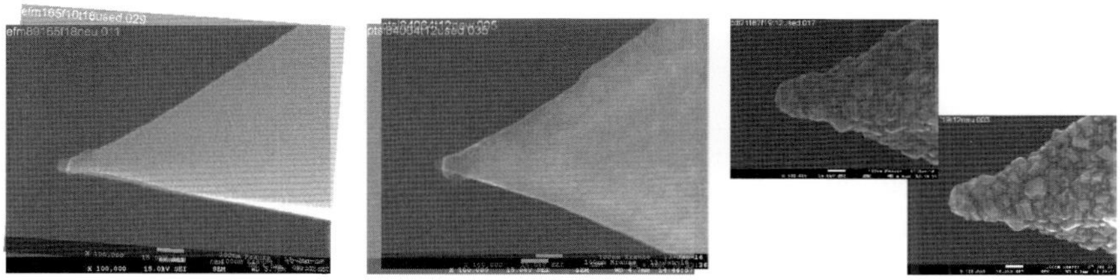

(a) (b) (c)

**Figure 2.13** Comparison of tip wear. Contact-mode scanning with a (a) platinum tip, (b) a platinum silicide tip, and (c) a conductive diamond tip with an interaction force of 500 nN. Overlay/comparison of SEM images before and after the scan.

1 μN force scan, Figure 2.12c, the tip is reduced by about 130 nm in height. Basically, this is easy to predict: the higher the applied force is, the larger the material removal will be.

### 2.4.2 Tip Wear: Mechanical Wear – Different Materials

The same experiment is performed with AFM probes coated differently but applying identical forces. The for a scan with 500 nN force are shown in Figure 2.13a–c. For the platinum-coated probe, the tip height is reduced by about 40 nm as already mentioned in the previous section. The removed height of the tip for the platinum silicide probe is significantly smaller, at about 25 nm. The reason for reduced material removal when comparing platinum with platinum silicide is the greater hardness of platinum silicide [28]. The conductive diamond-coated tip has by far the greatest hardness of all three compared materials. After the 5 mm scan, no wear or tip material removal is visible.

### 2.4.3 Tip Wear: Electrical Wear

For AFM probes with three different common coatings, platinum ($Pt_{0.95}Ir_{0.05}$), platinum silicide, and doped diamond, the loss of conductivity was investigated [31]: The tips were engaged on a hard conductive surface with an interaction force of 1 μN. A sample bias of −0.25 V was applied and the tips were scanned for 20 images (13 mm tip travel distance) over this surface. The current

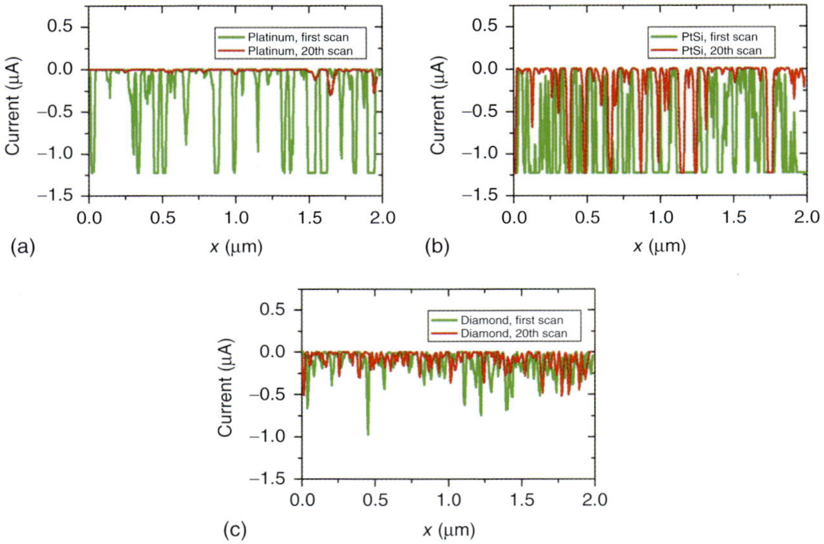


**Figure 2.14** Cross sections of current images of (a) a Pt probe, (b) a PtSi probe, and (c) a conductive diamond probe. Sample bias −0.25 V, force 1 μN, tip travel distance 13 mm. For each tip is shown a cross section out of the first and the last scan.

was measured. In Figure 2.14a–c, for each probe, a cross section of the first and the last image is shown. For the platinum and platinum silicide probes, in the first image, a maximum possible current of 1.2 μA is obtained. The diamond tips, though their conductivity is much poorer than that of the platinum and platinum silicide tips, do not reach the same current. After 20 images, the current flow through the platinum tip is nearly close to zero, whereas the current through the platinum silicide tip has decreased but is still significantly larger than that of the platinum tip. The smallest loss in conductivity could be observed for the diamond tip. For the diamond tip used for the scans in Figure 2.14c, after this experiment, no change in tip shape could be observed in SEM; the minor loss of conductivity could be related to the collection of dirt.

### 2.4.4 Tip Damage by Excess Voltage/High Currents

When voltage is applied to the tip and the sample, a current flow will occur when the tip is in contact with the sample surface. Owing to the small contact area (in the range of $10^{-3}$ to $10^{-2}$ $\mu m^2$), relatively small currents in the nano- and microampere regime will result in extremely high current densities. These high current densities could lead to extremely high temperatures in the tip–sample contact area, which may melt the conductive coating and lead to fusing of the tip apex.

Usually, in C-AFM modules, the voltage and current are limited (about ±10 V, 1 μA). However, even these values may result in tip damage or even in the destruction of the tip apex. Figures 2.15–2.18 show different images of different types of conductive tips to which voltage ramps were applied. In Figure 2.15, the apex of

Figure 2.15 Platinum-coated tip apex after a series of voltage rampings with a maximum current of about 0.5 μA.

Figure 2.16 Platinum silicide-coated tip apex after a series of voltage rampings with a maximum current of about 0.5 μA.

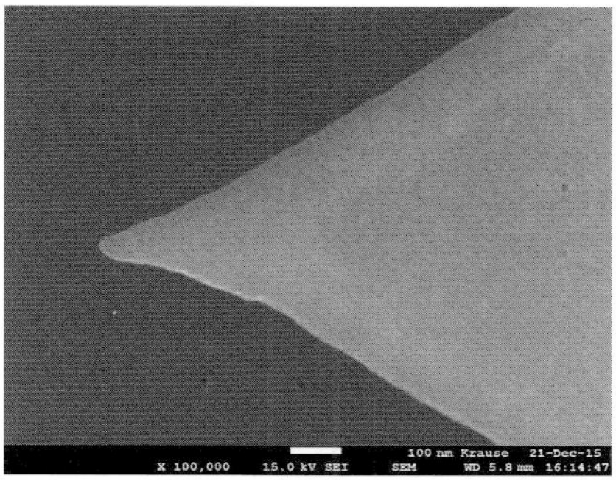


Figure 2.17 Platinum-coated tip apex after a series of voltage rampings with a maximum current of about 1 mA.

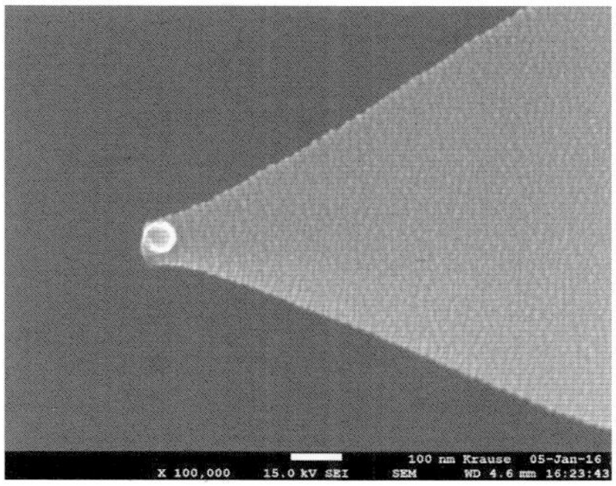


Figure 2.18 Platinum-coated tip apex after a series of voltage rampings with a maximum current of about 10 mA.

a platinum-coated tip after a series of voltage ramps with a maximum current of about 0.5 μA is shown. It can be seen that the very end of the tip apex has changed or is degenerated. Figure 2.16 shows the apex of a platinum silicide tip after an identical treatment. Here, no changes in tip morphology or texture are visible. The platinum silicide coating is able to withstand higher current impacts. In general, currents within the range normally permitted by the AFM software could change or damage the coating at the tip apex.

Higher currents as normally allowed by the AFM software do have the capability to destroy conductive AFM probes. In Figure 2.17, a voltage ramp resulting in currents of about 1 mA has been applied to a platinum-coated tip. The SEM image shows massive damage at the tip apex. The platinum coating melted owing to extreme current densities and solidified as a sphere after cooling down. In contrast, diamond-coated tips are able to withstand much higher current densities. In Figure 2.18, a diamond-coated tip was stressed with currents of about 10 mA. Even at these extremely high currents, no degradation or damage of the diamond coating is visible. This could be explained by an approximately 10 times bigger tip–surface contact area and a more than a decade higher thermal conductivity compared to platinum.

### 2.4.5 Damaging the Sample Surface

When an AFM probe is "scanned" in contact mode over a sample, the tip interacts with the surface. The amount of interaction is defined by the force applied to the cantilever. Depending on the materials of the tip and the sample and on the applied force, the sample (as well as the tip) might be damaged or scratched significantly. Especially when using SSRM on silicon surfaces, high forces are needed to penetrate the thin insulating native silicon oxide present. Figure 2.19a shows an AFM tapping-mode image of a silicon surface. In that area, a contact-mode scan with a scan size of 5 μm and an interaction force of 10 μN has been made with a diamond-coated tip before. It can be seen that the former silicon surface has changed massively in the area of high-force contact-mode scanning. Figure 2.19b shows a printed cross section of the scan. Especially at the reversal points of the

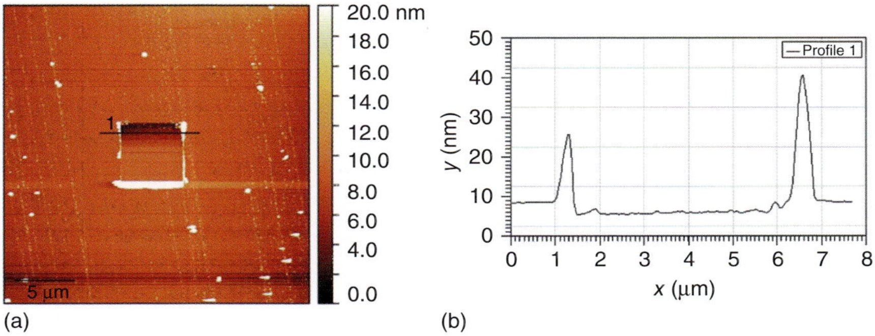


**Figure 2.19** (a) AFM tapping-mode image of a silicon surface which was previously scanned in contact mode with a diamond-coated probe with a 10 μN interaction force. (b) Cross section of scanned line.

tip, a massive pile-up of material occurred. Additionally, the tip pushed the material toward the end of the scan (lower edge of the image).

When performing C-AFM measurements either on soft samples or with high forces, the user should be aware that not only the C-AFM tip but also the sample surface could change significantly.

## 2.5 Conclusions

Unfortunately, there is no conductive AFM probe that will fit the needs for all applications and customer requests. We have to weigh between resolution, lifetime, and conductivity. Apart from very special applications, conductively coated silicon probes are the standard for C-AFM today. High spatial resolution is achieved with thin metallic coatings, which tends to have high wear rates and small lifetimes. Substituting metal-coated tips by silicided ones could set things right here. On the other hand, obtaining the highest possible lifetime needs diamond coatings. But diamond coatings will lead to large tip radii, and thus, poor spatial resolution.

Applying the appropriate force for each sample and application is essential for reliable electrical imaging. Too low a force may result in insufficient tip–sample interactions and will lead to bad and unstable conductance, whereas too high a force may damage the tip as well as the sample. The right choice of cantilever properties is essential here.

Finally, while in standard AFM application, the user "only" has to take care about tip convolution artifacts related to the topography, in C-AFM, "electrical artifacts" have also to be considered. The conductive coating of the tip may degenerate, dirt can be collected, the sample surface may be modified by the tip, or too high currents may cause changes in the surface or the tip. All these effects will influence and can falsify the electrical image.

## References


1 Binnig, G., Quate, C.F., and Gerber, C. (1986) *Phys. Rev. Lett.*, **56**, 930.

2 Binnig, G., Rohrer, H., and Gerber, C. (1982) *Phys. Rev. Lett.*, **49**, 57.

3 Albrecht, T.R., Akamine, S., Carver, T.E., and Quate, C.F. (1990) *J. Vac. Sci. Technol., A*, **8**, 3386.

4 Akamine, S., Barrett, R.C., and Quate, C.F. (1990) *Appl. Phys. Lett.*, **57**, 316.

5 Wolter, O., Bayer, T., and Greschner, J. (1991) *J. Vac. Sci. Technol., B*, **9**, 1353.

6 Martin, Y., Abraham, D.W., and Wickramasinghe, H.K. (1988) *Appl. Phys. Lett.*, **52**, 1103.

7 Welland, M.E. and Murrell, M.P. (1993) *Scanning*, **15**, 251.

8 Murrell, M.P., Welland, M.E., O'Shea, S.J., Wong, T.M.H., Barnes, J.R., McKinnon, A.W., Heyns, M., and Verhaverbeke, S. (1993) *Appl. Phys. Lett.*, **62**, 786.

9 O'Shea, S.J., Atta, R.M., Murrell, M.P., and Welland, M.E. (1995) *J. Vac. Sci. Technol., B*, **13**, 1945.

10 De Wolf, P. (1998) *J. Vac. Sci. Technol., B*, **16**, 355.
11 De Wolf, P., Geva, M., Hantschel, T., Vandervorst, W., and Bylsma, R.B. (1998) *Appl. Phys. Lett.*, **73**, 2155.
12 Thomson, R.E. and Moreland, J. (1995) *J. Vac. Sci. Technol., B*, **13**, 1123.
13 OShea, S.J., Atta, R.M., and Welland, M.E. (1995) *Rev. Sci. Instrum.*, **66**, 2508.
14 Ronning, C., Wondratschek, O., Büttner, M., Hofsäss, H., Zimmermann, J., Leiderer, P., and Boneberg, J. (2001) *Appl. Phys. Lett.*, **79**, 3053.
15 Niedermann, P., Hanni, W., Blanc, N., Christoph, R., and Burger, J. (1996) *J. Vac. Sci. Technol., A*, **14**, 1233.
16 Trenkler, T., Hantschel, T., Stephenson, R., De Wolf, P., Vandervorst, W., Hellemans, L., Malavé, A., Büchel, D., Oesterschulze, E., Kulisch, W., Niedermann, P., Sulzbach, T., and Ohlsson, O. (2000) *J. Vac. Sci. Technol., B*, **18**, 418.
17 Niedermann, P., Hänni, W., Morel, D., Perret, A., Skinner, N., Indermühle, P.-F., Buffat, P.A., and De Rooij, N.F. (1998) *Appl. Phys. A*, **66**, S31.
18 Hantschel, T., Niedermann, P., Trenkler, T., and Vandervorst, W. (2000) *Appl. Phys. Lett.*, **76**, 1603.
19 Toda, A. (1995) US5386110 (A), Jan. 31, 1995.
20 Lantz, M.A., O'Shea, S.J., and Welland, M.E. (1998) *Rev. Sci. Instrum.*, **69**, 1757.
21 Trenkler, T., Hantschel, T., Vandervorst, W., Hellemans, L., Kulisch, W., Oesterschulze, E., Niedermann, P., and Sulzbach, T. (1999). *Characterization of Conductive Probes for Atomic Force Microscopy.* in Proceedings of SPIE 3680, Design, Test, and Microfabrication of MEMS and MOEMS (eds B. Courtois, Selden B. Crary, W. Ehrfeld, H. Fujita, J.M. Karam, and K.W. Markus), p. 1168.
22 Liang, Q., Stanishevsky, A., and Vohra, Y.K. (2008) *Thin Solid Films*, **517**, 800.
23 Agrawal, R., Moldovan, N., and Espinosa, H.D. (2009) *J. Appl. Phys.*, **106**, 064311.
24 Hantschel, T., De Wolf, P., Trenkler, T., Stephenson, R., and Vandervorst, W. (1998) *Materials and Device Characterization in Micromachining*, (eds) C. R. Friedrich and Y. Vladimirsky, vol. 3512, Santa Clara, CA, p. 92.
25 Hantschel, T., Demeulemeester, C., Eyben, P., Schulz, V., Richard, O., Bender, H., and Vandervorst, W. (2009) *Phys. Status Solidi A*, **206**, 2077.
26 Hantschel, T. (1999) Proceedings of SPIE, Paris, France, 994, 1999.
27 Doezema, L.A., Kaszuba, P.V., Moszkowicz, L., Never, J.M., and Slinkman, J.A. (2001) US6198300 (B1).
28 Murarka, S.P. (1980) *J. Vac. Sci. Technol.*, **17**, 775.
29 Akiyama, T., Gullo, M.R., Rooij, N.F., Tonin, A., Hidber, H.R., Frederix, P.L.T.M., Engel, A., and Staufer, U. (2004) *Jpn. J. Appl. Phys.*, **43**, 3865.
30 Bhaskaran, H., Sebastian, A., and Despont, M. (2009) *IEEE Trans. Nanotechnol.*, **8**, 128.
31 Krause, O., Sebastian, A., and Despont, M. (2012) *Nanotechnology*, **1**, 125.

# 3

# Fundamentals of CAFM Operation Modes

*Guenther Benstetter[1], Alexander Hofer[1], Donping Liu[2], Werner Frammelsberger[3], and Mario Lanza[4]*

[1] *Deggendorf Institute of Technology, Faculty of Electrical, Media and Computer Engineering, Dieter-Görlitz-Platz 1, 94469 Deggendorf, Germany*
[2] *Dalian Nationalities University, School of Physics and Materials Engineering, No. 18, Liaohe West Road, Dalian, Development Zone, Dalian 116600, China*
[3] *Deggendorf Institute of Technology, Faculty of Mechanical Engineering and Mechatronics, Dieter-Görlitz-Platz 1, 94469 Deggendorf, Germany*
[4] *Soochow University, Institute of Functional Nano & Soft Materials, 199 Ren-Ai Road, Bld. 909, Office 3215, Suzhou 215123, China*

## 3.1 Introduction

The scope of this chapter is to discuss the fundamentals of conductive atomic force microscopy (CAFM) operation modes. The focus is on applied studies using commercially available scanning probe microscopes (SPMs) and tips operated primarily in ambient conditions. The chapter starts with an introduction to fundamental tip-sample interactions in contact mode operation, followed by a discussion on alternative operation modes and evaluation techniques.

Thanks to the easy sample preparation, the nondestructive measurement procedure, and the relatively simple experimental setup (which even enables operation at standard ambient conditions), CAFM has become a powerful method to study materials and surface properties in a wide field. If a bias voltage ($V_{tip}$) is applied between the sample and the tip (Figure 3.1a,d, with topography sections as shown in Figure 3.1b,e), the current flowing from the tip to the sample (or vice versa) can be recorded simultaneously with the sample topography. In the earliest applications [1–6], CAFM was utilized to study thin $SiO_2$ films on a Si substrate, for building a (MOS) or metal isolator semiconductor (MIS) structure with the conductive tip, the dielectric, and the substrate (Figure 3.1d). For a certain $V_{tip}$, the measured current is increased (Figure 3.1f ($x_1, x_2$)) for electrically weakened sites and vice versa (Figure 3.1f ($x_3$)). By scanning the tip along the sample surface, current maps can be acquired. Alternatively, the tip can be positioned at predefined points of interests and the local $I$–$V$ characteristics can be obtained by ramping the tip voltage and recording the corresponding current values ($I$–$V$ spectroscopy). An identical experimental setup can also be used to gain further insight into the inner micro- or nanostructure of semiconducting or conducting thin films or devices. As schematically depicted in Figure 3.1c,f, current variations

*Conductive Atomic Force Microscopy: Applications in Nanomaterials*, First Edition.
Edited by Mario Lanza.

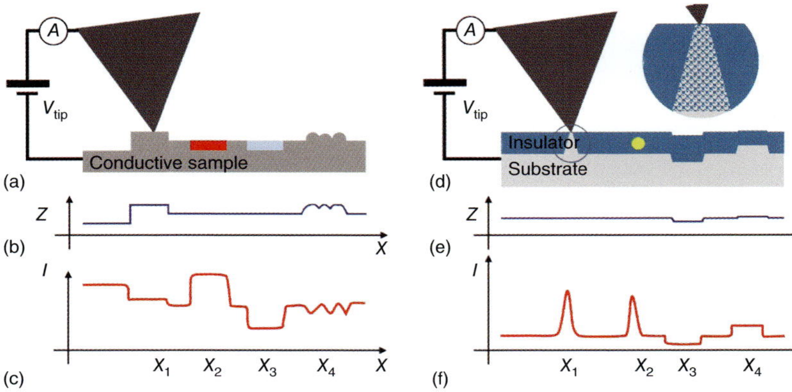


**Figure 3.1** Measurement setup for conductive samples (a) with topography section (b). Features on position $x_2$ or $x_3$ result in a higher or lower current and position $x_4$ showing a varying current due to particles (c). Measurement setup for measuring thin insulating layers (d) with topography section (e). Filament formation resulting in a high current at $x_1$, conductive inclusion at $x_2$, lower current due to higher insulator film thickness at $x_3$, and higher current due to lower insulator film thickness at $x_4$ (f).

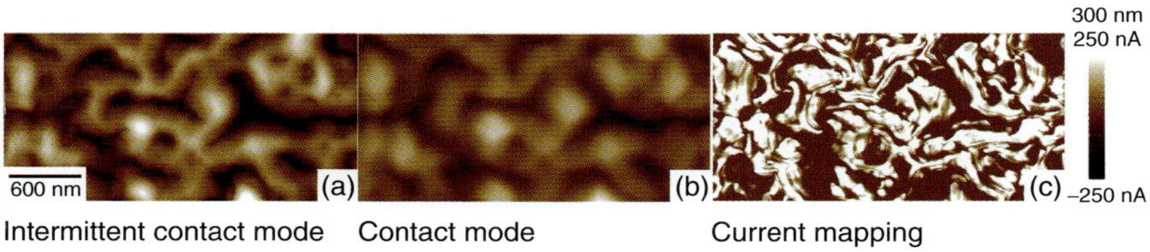


**Figure 3.2** Topography images of identical areas of $3 \times 1.5\ \mu m^2$ obtained in intermittent contact mode with high aspect ratio nonconductive tip (a) and in contact mode with low aspect ratio conductive tip (b) showing differences in spatial resolution. The comparison of the topography images (a or b) with the current mapping (c) reveals nanoscale features of the electrical conductivity that are not visible in the topography images.

can monitor local fluctuations in electrical conductivity or electron emission characteristics even if they are not observable in the topographical image.

In some cases, the current image differs from the surface topography image because the effective electron emission area is smaller than the mechanical contact area (Section 3.2). As an example, Figure 3.2 shows the topographic and current images of a thin film made of laser-sintered Ge nanoparticles [7, 8] (Figure 3.2b,c, respectively). As can be observed, the lateral resolution of surface structures in the current image is larger compared to the simultaneously recorded topographical image. Even an additional topographical image obtained with an ultrasharp nonconductive tip operated in intermittent contact AFM mode (IC-AFM) at the same location (Figure 3.2a) did not achieve a comparable resolution.

In CAFM experiments, the apex of the conductive probe tip is the only part of the microscope that physically contacts the sample. Therefore, the interaction between the probe tip and the sample is of utmost importance, as it can affect the resolution, quality, and reliability of the experimental results. Unfortunately, the tip-sample system is not a stable entity during measurements. During the experiments, large friction and electrical fields may produce wear of both the tip and the sample (predominately in contact mode), leading to morphological changes that could influence the experimental results. Moreover, the tip-sample junction can also be exposed to contamination. The following section provides an overview of fundamental conduction mechanisms and discusses important aspects of the tip-sample system related to the electrical conduction.

## 3.2 Tip-Sample Interaction: Contact Area, Effective Emission Area, and Conduction Mechanisms

The most accepted method to estimate the physical contact area ($A_c$) between the tip of a CAFM and the sample uses the Hertz contact theory (Figure 3.3) [3, 9–13]. $A_c$ is characterized by its contact radius ($r_c$), which depends on the contact force ($F_c$), the tip radius ($R_{\text{tip}}$), the elastic moduli ($E_{1/2}$), and the Poisson ratio ($\nu_{1/2}$) of the tip and the sample. Micro roughness of the tip and sample surfaces are neglected [13].

$$A_c = \pi r_c^2 = \pi \left( F_c R_{\text{tip}} \frac{3}{4} \left( \frac{1-\nu_1^2}{E_1} + \frac{1-\nu_2^2}{E_2} \right) \right)^{\frac{2}{3}} \tag{3.1}$$

The parameters needed for Eq. (3.1) can be usually found in the literature [14]. As the tip/sample structure is frequently a stack of different materials, which together are only a few tens of nanometers thick, it is uncertain which materials have the dominant influence on the overall elastic parameters [9]. Therefore, the following two assumptions are usually made [14]: (i) the mechanical properties of coated tips are determined by the coating material, and not by the bulk material of the tip; (ii) the mechanical properties of the sample are described exclusively either by the surface layer or by the substrate. In Figure 3.4, $A_c$ is

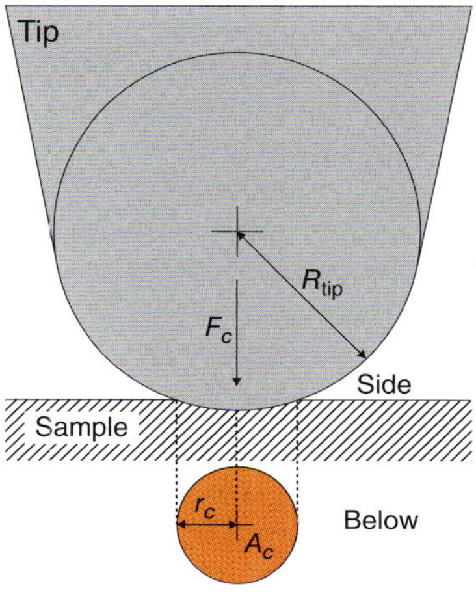


**Figure 3.3** Schematic of a hertzian contact between an AFM probe tip of radius $R_{\text{tip}}$ and a semi-infinite sample, following Bhushan [9] and Olbrich [10].

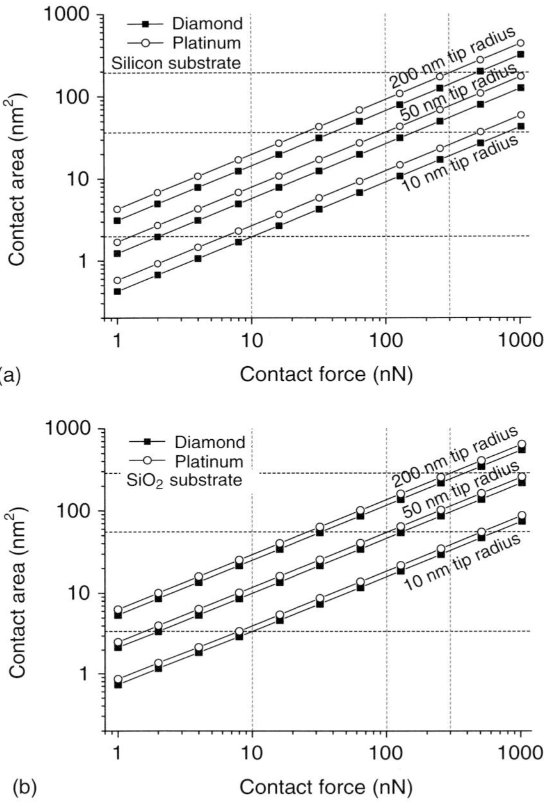


**Figure 3.4** Calculated extreme values of the contact area versus the contact force for $SiO_2$ substrate (a) and Si substrate (b), respectively. Tip material and tip radius are indicated. The dashed grid is for orientation. Material parameters, typical values: platinum, $E = 170$ GPa, $\nu = 0.36$; diamond, $E = 1000$ GPa, $\nu = 0.2$; thermal $SiO_2$, $E = 68$ GPa, $\nu = 0.2$; silicon, $E = 170$ GPa, $\nu = 0.2$.

plotted as a function of $F_c$ for a system consisting of a metal-coated silicon tip, assuming that the sample properties are either dominated by the Si substrate (a) or by the $SiO_2$ surface layer (b). The tip radii used are between $R = 10$ nm, which is below the value for PrIr-coated tips, and $R = 200$ nm, which is the largest value for phosphorous-doped diamond-coated tips. The comparatively small gaps between the graphs for platinum and diamond for a given tip radius indicate that different elasticity properties of the tip material are not dominant with respect to $A_c$. The tip radius and the applied force play a much larger role. Generally, $A_c$ is lower for Si substrate (Figure 3.4) and the gaps between platinum and diamond are increased, as the elasticity modulus for silicon is higher. A small contact area $A_c$ is required to achieve a good topographic, lateral resolution. It is important to note that $F_c$ comprises the force due to the cantilever deflection and additional forces. The additional forces are largely influenced by contamination, liquid layers, micro roughness of the tip and the sample, and the applied voltage [11, 13–16]. Therefore, calculating accurately the real value of $F_c$ in each experiment is not easy. For example, the voltage-induced increase of $F_c$ was measured to be in the order of 10–30 nN for an increase in tip voltage from 0 to 8 V (in air atmosphere) [14]. For MOS structures measured in room air atmospheres, values of $F_c$ were observed typically between

30 and 300 nN for different metal-coated tips [14]. Nevertheless, it has been demonstrated that the lateral resolution for topography measurements, and thus $A_c$, does not depend significantly on the measurement environment [17, 18]. On the contrary, as mentioned in Chapter 1, the currents measured by CAFM depend on the effective emission area ($A_{eff}$) through which charge carriers can flow from the tip to the sample (or vice versa). Different authors [1, 3, 10, 14, 16–24] have addressed the correlation between $A_c$ and $A_{eff}$. The main difference between them is that $A_c$ does not depend on the conductivity of the sample, as displayed in Eq. (3.1), while $A_{eff}$ is highly affected by it. Depending on the materials that form the tip and the surface of the sample, different fundamental $A_{eff}$ and conduction mechanisms may occur [25–29].

### 3.2.1 CAFM Tip on Metallic Surfaces

The simplest situation possible is when the conductive tip is placed on the surface of a metallic sample. Theoretically, under those circumstances, the CAFM tip becomes electrically connected to all points of the sample, as the voltage drop in a perfect metal is zero in all directions (laterally and vertically) [30]. The reason is that pure metals cannot withstand electrical fields. Therefore, $A_{eff}$ should equal the total area of the metallic sample. In many microelectronic experiments, the tip is not placed on a fully metallic sample, but it is located on a metallic electrode or pad, which is at the same time located on a semiconducting or insulating substrate. In this case, the metallic tip acts as a nanosized collector for current flowing through the whole area between the electrode and the substrate. This behavior is very similar to what happens when the tip of a probestation is placed on a metallic pad, with the only difference that, in CAFM, the radius of the tip is much smaller.

The electronic transport between the tip and the metallic sample or electrode should be (in theory) governed by the theory of electronic transport in metals [31]. When an electrical field is applied across a pure metal, the charge transport along it can happen in ballistic or diffusive regime, depending on the length scale. In ballistic regime, the electrons flow free of scattering and the total current only depends on the geometry of the conductor. Ballistic conduction has been widely used in micro- and nanoelectronics to describe the conduction along nanowires, carbon nanotubes, two-dimensional materials, and electrically broken dielectrics, and it has been well defined by the quantum point contact model [32]. At larger length scales, electrons end up colliding with the atoms in the material, leading to scattering and a reduced transmission coefficient, which leads to the diffusive regime. Nevertheless, the conduction through metals is highly affected by distorted interfaces (including the buckled plane and vacancy-rich interfaces), metallic and nonmetallic impurities, metallic interlayer interfaces, metallic multilayer interfaces, and interface alloys [30]. Furthermore, at the junction between the tip and the surface of the sample, there are additional factors that further increase the difficulty in charge transport. These are, for instance, (i) formation of a metallic heterojunction due to different work functions of the tip and the sample and (ii) presence of water molecules and impurities strictly between the tip and the sample, which introduces a layer with uncertain thickness and chemical composition between them. These phenomena remarkably increase the effective resistance at the junction. The company Budget sensors characterized the contact resistance of their conductive AFM tips (model: ElectriTrap-300-G) using a platinum thin film surface. The tips were made of silicon and included a metallic coating consisting on 5 nm chromium and 25 nm platinum. Their tests revealed a resistance of 300 Ω in best cases [33]. Spectroscopic $I$–$V$ curves collected with these tips on conductive samples typically show linear behavior [34].

The geometry of the metallic sample or electrode amply affects the collection of current. In CAFM science, the measurements of metallic samples reveal peaks of current at voids with reduced thickness (Figure 3.1a), owing to the lower distance the charge carriers need to overcome. Moreover, if the metallic sample or electrode is too thick and/or if the lateral size is too large, potential differences and capacitance effects within the metal may appear, limiting the total current collected. Sonde and co-workers [35–37] observed that when the tip is placed on a graphene flake, the total capacitance measured is much larger than that recorded on bare $SiO_2$, owing to the spreading of the area through which electrons can flow. It has been calculated that this area is proportional to the electron mean free path on the graphene sheet. In any case, the area spreading was smaller than the total area of the graphene sheet, indicating that the propagation effect was limited to a specific area. Detailed information about the use of a CAFM tip as a collector on metallic pads with different sizes is present in Section 4.2. It is worth noting that, under these circumstances, CAFM is not able to perform strictly local measurements with nanometric lateral resolution, and the currents collected apply to larger areas, as $A_{eff}$ spreads.

### 3.2.2 CAFM Tip on Semiconducting Surfaces

When the metallic CAFM tip is placed on a semiconductor sample, a depletion barrier may form in the semiconductor if the work function of the metal is larger than the electron affinity $\chi$ (for n-type semiconductor) or if it is smaller than electron affinity plus semiconductor gap (for p-type semiconductor, Figure 3.5). This depletion barrier behaves similarly to a diode generating an effective potential barrier between the tip and the bottom of the sample. If the voltage applied during the CAFM operation has the same polarity as the depletion barrier (forward bias), an ohmic contact is formed, and the current increases exponentially with the voltage (Eq. (3.2)). On the contrary, when a reverse voltage is applied, the current is dominated by Schottky emission [27] (Eq. (3.3)). Basically, Schottky emission follows the formula of thermionic emission [26] (Eq. (3.6)), which implies that the current is governed by carriers that are excited over the metal-insulator potential barrier. While for thermionic emission, a dielectric layer that is stressed by $E_{ox}$ (Eq. (3.6)) is present actually, in the case of Schottky emission, the depletion layer in reverse direction performs as an insulator, which is stressed by $E_m$ (Eq. (3.3)). Therefore, in the tip/semiconductor junction, in forward direction, the current increases with the voltage exponentially, while in reverse direction, the current remains comparatively small (Table 3.1).

In a metal/semiconductor junction made by the CAFM-tip/semiconducting-sample, the value of $A_{eff}$ is not as clear as in CAFM-tip/metallic-sample (in which

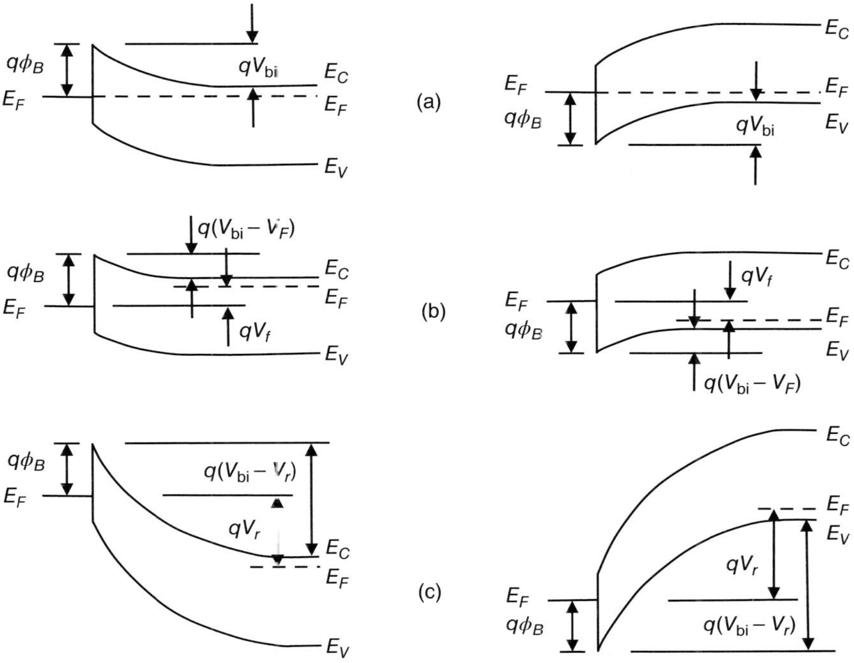


**Figure 3.5** Schematic energy-band diagrams of metal-semiconductor structures, n-type (left), p-type (right): (a) thermal equilibrium. (b) Forward bias. (c) Reverse bias. (Adopted from Ref. [26].)

**Table 3.1** Basic conduction processes in tip/semiconductor junctions.

| Process | Expression | Equations | References |
|---|---|---|---|
| Schottky diode (forward) | $I = I_0 \exp\left(\left(\frac{qV_f}{\eta kT}\right) - 1\right)$ | (3.2) | [26] (Sze, Ng) |
| Schottky emission (reverse) | $I_0 = A_{\text{eff}} A^{**} T^2 \exp\left(-\frac{q}{kT}\left(\phi_B - \sqrt{qE_m/4\pi\epsilon_s}\right)\right)$<br>$E_m = \sqrt{\frac{2qN}{\epsilon_s}\left(V_r + V_{\text{bi}} - \frac{kT}{q}\right)}$ | (3.3) | [26] (Sze, Ng) |

$I$ = Electrical current, $A_{\text{eff}}$ = effective emission area, $q$ = Electron charge, $\phi_B$ = Barrier height, $T$ = absolute temperature, $A^{**}$ = effective Richardson constant, $k$ = Boltzmann constant, $I_0$ = saturation current (Schottky emission), $V_f$ = forward bias, $\eta$ = ideality factor, near unity for $T \geq 300\,\text{K}, N \leq 10^{17}\,1\,\text{cm}^{-3}$ [38]. $E_m$ = electrical field maximum, $V_r$ = reverse bias, $V_{bi}$ = built in voltage, $\epsilon_s$ = permittivity of the semiconductor.

$A_{\text{eff}}$ spreads) or in CAFM-tip/insulating-samples (in which $A_{\text{eff}}$ approaches $A_c$, as will be explained in the next section). Although many works have been devoted to analyze Schottky junctions at the CAFM-tip/semiconducting-sample interface, most of them are focused on the charge transport, but not on the study of the parameter $A_{\text{eff}}$. Moreover, for many semiconducting samples analyzed with

CAFM (such as silicon), it is expected that its surface incorporates a native oxide (which is usually insulating), avoiding a lateral propagation and confining the current to the area strictly below the tip. Interestingly, most of the works published in this direction analyzed the currents driven from spectroscopic $I$–$V$ curves, in which local inhomogeneities are measured by placing the tip at different locations of the surface, but not collecting current maps. In any case, from the existing literature it can be concluded that the parameter $A_{eff}$ is not so critical when measuring semiconducting samples because, as mentioned above, material properties and voltages applied can strongly modulate the resistivity at the junction.

### 3.2.3 CAFM Tip on Insulating Surfaces

In the CAFM, the junction between the conductive tip and an insulating sample is the most commonly studied, as it is the one that provides the largest lateral resolution. In these systems, the sample surface is usually extremely flat and the contact area can in some cases be as small as 1 nm$^2$. Therefore, an accurate evaluation of the tip/sample interaction is critical for a correct interpretation of the data measured, and many external factors may have a dramatic effect on the value of $A_{eff}$. Among them, the one that has shown to have a greater influence is the relative humidity of the environment, which produces a water meniscus at the tip/insulator nanojunction (Figure 3.6). This phenomenon, which does not play a significant role in tip/metal and tip/semiconductor junctions, can increase $A_{eff}$ by a factor up to 800. For this reason, in the following, we analyze the conduction with and without considering the effect of humidity. At room conditions, humidity is always an issue, whereas measuring in ultrahigh vacuum (UHV) or inert gas (like dry nitrogen) may control or even prevent humidity.

For a nonperturbed tip/insulator system the value of $A_{eff}$ equals $A_c$, as the only pathway for current flow is that in which the tip strictly contacts the material

**Figure 3.6** Water capillary bridge at the interface between a pyramid-shaped AFM probe tip, as observed in an environmental scanning electron microscope (ESEM). (Reprinted from Ref. [39].)

under test. Typical values of $A_{eff}$ achieved in UHV are $\leq 10\,nm^2$ [17, 18, 24], similar to the values of $A_c$ calculated with the hertz contact theory. Lanza *et al.* [18] (2 nm $HfO_2$/1 nm $SiO_2$/n-Si) observed good correlation between $A_c$ and $A_{eff}$ for comparatively low load forces (1–5 nN). Celano *et al.* [24] (1.5 nm $SiO_2$/n-Si) report that the hertz contact theory based on the nominal tip radius of 30 nm overestimates $A_{eff}$ ($\approx 4\times$ at 80 nN; $\approx 2.4\times$ at 260 nN; $\approx 1.2\times$ at 490 nN). They suspect asperities of the tip and the sample, which create points of higher pressure and thus, higher conductivity, at these spots to be the reason. Values of $A_c$ for the Hertz contact theory were observed closer to $A_{eff}$ than values for the Derjaguin-Muller-Toporov (DMT) model. Celano *et al.* [24] estimated $A_{eff}$ assuming a given oxide thickness measured by ellipsometry and X-ray photoemission spectroscopy. A difference of factor 4 (2) for $A_{eff}$ represents a difference of $\approx 0.15\,nm$ ($\approx 0.08\,nm$) [14] in the oxide thickness and vice versa, which is in the range of natural oxide thickness variations. These experimental results indicate that for very thin surface layers, here 1.5 nm $SiO_2$, it is feasible to use the mechanical properties of the substrate rather than of the film to calculate $A_c$. The $A_c$ readings for silicon in Figure 3.4 are smaller than those for $SiO_2$ and thus closer to the experiment. In UHV, the current measured is linked to the force applied to the tip [17, 18, 24]. With increasing force, the current increases as $A_c$ and hence $A_{eff}$ increases (and vice versa). In other words, in UHV, $A_{eff}$ and $I$ are linked to $A_c$ and hence to the tip radius and load force.

In tip/insulator nanojunctions exposed to ambient environment, the grade of correlation between $A_c$ and $A_{eff}$ may depend strongly on the measurement environment. Typical values of $A_{eff}$ were reported to be between 20 and 300 $nm^2$ [1, 3, 10, 14, 17–22]. Lanza *et al.* [17, 18] observed that $A_{eff}$ is increased in normal air (44% relative humidity) and nitrogen with controlled humidity (0.5% relative humidity) when compared to UHV. In addition, they report that, in contrast to UHV, in humid environment, the force applied to the tip has no significant influence on the current measured in spectroscopic $I$–$V$ curves, which indicates that $A_{eff}$ is not linked to $A_c$. Simulations of Kremmer *et al.* [23] (5 nm $SiO_2$/n-Si) show that the electrical field broadens at the $SiO_2$/Si interface due to a water film (3 nm thick) at the sample surface and $A_{eff}$ increases to approximately 300–700 $nm^2$. Frammelsberger *et al.* [14] presented typical values for $A_{eff}$ for different tip types under ambient conditions between 30 and 200 $nm^2$ (using a sample that consisted of 1.7–8.3 nm $SiO_2$/p-Si). $A_{eff}$ was observed to be not ultimately linked to the respective nominal tip radius and thus to $A_c$. In other words, under air environment, $A_{eff}$ is increased and widely insensitive to $A_c$, even for comparatively low relative humidity. Therefore, while the lateral resolution for topography measurements (and thus $A_c$) does not depend significantly on the measurement environment [17, 18], the lateral electrical resolution is decreased and the current is widely insensitive to tip radius and load force.

Tip-sample structures without purposely grown dielectric layers and involving comparatively rough samples may show a significantly different behavior (Figure 3.2). The tip and the sample form Schottky or ohmic contacts, for which a much lower electrical field is necessary to drive a measurable current. In addition, the mechanical contact may not occur exclusively on a circular region at the tip apex but may be dominated by tip and sample asperities. Therefore,

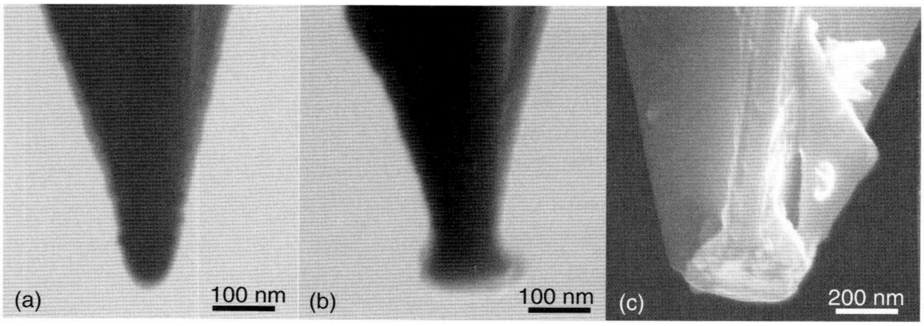


**Figure 3.7** Scanning electron micrographs of CoCr-coated tips in the initial state (a) and after approximately 10 sequences of 25 *I–V* measurements each (b). The tip radius increased owing to wear from approximately 50 to 300 nm. Severe wear may lead to fatal damage (c).

the influence of a water layer in air may be negligible and the effective electrical contact area depends on additional parameters such as micro/nanostructure and corresponding variations in local conductivity, local work function, surface roughness, and pressure distribution. In addition, variations of the tip force may yield in current confinement because of phase transitions [38, 40–45] or local penetration of thin oxide films. As a result, $A_{eff}$ may even decrease, while the theoretic $A_c$ is constant or increases. Even for almost ideal samples, nominal tip radius and spring constant may not apply for the tip used and the correct mechanical parameters of the tip-sample system may not be assessed in full complexity. Asperities of the tip and the sample, wear effects during measurement, and the influence of the nanometer-thin tip and sample coatings or nanocrystallites at the tip-sample interface may play a role. Wear is predominately a problem in UHV, but is also an issue in normal air (Figure 3.7), albeit the lubrication effect of a water layer. Wear increases the tip radius and may finally lead to a partial or complete loss of the conductive layer at the tip apex. In addition, in normal air, electrical stress-induced oxidation effects may also play a role [16, 23, 46]. They may change the properties of the tip-sample interface and thus, may influence $I$. Furthermore, contamination of the tip and the sample could also change the conductive properties of the tip-sample system. Sources of contamination are, for example, the environment, when working in normal air, debris at the sample surface or the tip, as well as residuals of tip wear and possible sample damage. In essence, $A_c$ may be estimated by the hertz contact theory with sensible results for standard CAFM measurements. As several parameters of the tip-sample system are not known exactly and are subject to dynamic changes during measurements, the benefit of more complex contact models may be doubted. This is even more the case, as researchers frequently do not explicitly determine or monitor the actual values of, for example, tip radius, spring constant, or the influence of additional forces and surface roughness.

Apart from the size of $A_{eff}$, the total current flowing through the tip/insulator system also strongly depends on the conduction mechanism, which is mainly dominated by the geometry and composition of the insulating film, as well as by the electrical field applied. Usually the tip/insulating system consists of a metal-coated silicon tip in contact with a $SiO_2$/Si stack, in which the $SiO_2$ has

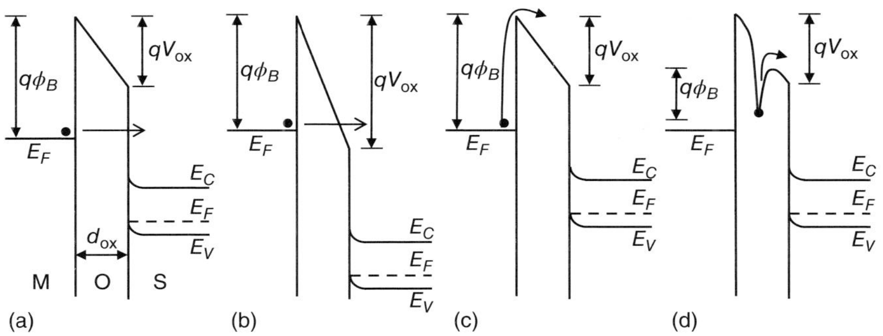


**Figure 3.8** Different fundamental conduction systems. MOS (MIS)-structures: (a) direct tunneling, (b) Fowler-Nordheim tunneling, (c) thermionic emission, and (d) Frenkel-Poole emission.

a thickness of a few nanometers [1–3, 14, 16–18, 23, 46]. Many other types of insulators such as high-*k* dielectrics [17, 40–42] and even twodimensional layered dielectrics [47, 48] have been also widely investigated by CAFM.

For highly insulating and almost ideal dielectric films, such as $SiO_2$, the tunneling current can be described by Fowler-Nordheim tunneling (FNT) (Eq. (3.4)) or direct tunneling (DT) (Eq. (3.5)). For FNT and DT the carriers tunnel quantum-mechanically through a trapezoid or a rectangular barrier, respectively (Figure 3.8). The models provided in Eqs (3.4) and (3.5) have shown their practical applicability for both CAFM [25] and device-level measurements [28, 29]. When carriers are excited over a metal-insulator or insulator-semiconductor barrier, thermionic emission takes place (Eq. (3.6)). For Schottky emission (Eq. (3.6)), the semiconductor parameters of Eq. (3.3) are replaced by the respective parameters of the oxide. Frenkel-Poole (FP) emission (Eq. (3.7)) is due to emission of trapped charge carriers (Figure 3.8) by thermal excitation [26]. At low voltages and high temperature, insulators may show ohmic-like behavior (Eq. (3.8)), also termed *hopping conduction* [26, 27].

Charge carriers in the conduction and the valence band, respectively, of metals, semiconductors, or insulators cause ohmic conduction (Eq. (3.8)). Here, $E_{ox}$ figuratively also applies for the electrical field in metals and semiconductors. Ohmic conduction and FP emission are related, but they differ in the calculation of the carrier density. Generally, the emission regime observed depends strongly on a selection of parameters and may constitute a superposition of the basic conduction processes (Table 3.2). Critical parameters include temperature, substrate doping, voltage applied as well as dielectric thickness, structure, content, and manufacturing specifications [26, 27, 50–53]. Even local variations were observed by CAFM. FP emission was reported to be the dominant conduction regime for $Ta_2O_5$ in the core of the crystalline grains, while near edges, trap-assisted tunneling, a combination of FP and FNT/DT, was observed [52].

It is important to keep in mind that the voltage over the insulator may not equal the applied voltage. Work function differences between the electrodes and band bending of semiconductors play a role and these are discussed further in Section 3.3. Electronic devices or macroscopic test structures are comparatively

**Table 3.2** Basic conduction processes in tip/semiconductor nanojunctions.

| Process | Expression | Equations | References |
|---|---|---|---|
| Fowler-Nordheim tunneling (FNT) | $I = A_{\text{eff}} \frac{q^3}{16\pi^2\hbar} \frac{m_e}{m_{\text{ox}}} \frac{E_{\text{ox}}^2}{\phi_B} \exp\left(-\frac{4}{3} \frac{\sqrt{2m_{\text{ox}}}}{q\hbar} \frac{\phi_B^{\frac{3}{2}}}{E_{\text{ox}}}\right)$ | (3.4) | [25, 49] |
| Direct tunneling (DT) | $I = I_{\text{FNT}} \exp\left(1 - \frac{qV_{\text{ox}}}{\phi_B}\right)^{3/2}$ | (3.5) | [25, 26] |
| Thermionic emission | $I = A_{\text{eff}} A^{**} T^2 \exp\left(-\frac{q}{kT}\left(\phi_B - \sqrt{qE_{\text{ox}}/4\pi\epsilon_{\text{ox}}}\right)\right)$ | (3.6) | [26, 27] |
| Frenkel-Poole (FP) emission | $I = A_{\text{eff}} q\mu N_{\text{cv}} \exp\left(-\frac{q}{kT}\left(\phi_B - \sqrt{qE_{\text{ox}}/\pi\epsilon_{\text{ox}}}\right)\right) E_{\text{ox}}$ | (3.7) | [26, 27] |
| Ohmic | $I = A_{\text{eff}} q\mu N_{\text{cv}} \exp\left(-\frac{\Delta E_{\text{ae}}}{kT}\right) E_{\text{ox}} = Aq\mu nE_{\text{ox}}$ | (3.8) | [26, 27] |

$I$ = Electrical current, $A_{\text{eff}}$ = effective emission area, $m_{\text{ox}}$ = effective charge carrier mass in the oxide, $m_e$ = free electron mass, $\hbar$ = reduced Planck constant, $q$ = electron charge, $\phi_B$ = barrier height, $E_{\text{ox}} = V_{\text{ox}}/d_{\text{ox}}$ = electrical field across the oxide, $V_{\text{ox}}$ = effective oxide voltage, $d_{\text{ox}}$ = effective oxide thickness, $\Delta E_{\text{ae}}$ = carrier activation energy difference, $T$ = absolute temperature, $A^{**}$ = effective Richardson constant, $\epsilon_{\text{ox}}$ = permittivity of the oxide, $k$ = Boltzmann constant, $N_{\text{cv}}$ = density of states, $n$ = charge carrier density, $\mu$ = carrier mobility, $I_0$ = saturation current (Schottky emission). $V_f$ = forward bias, $\eta$ = ideality factor, near unity for $T \geq 300\,\text{K}$, $N \leq 10^{17}\,1\,\text{cm}^{-3}$ [26]. $E_m$ = Electrical field maximum, $V_r$ = reverse bias, $V_{\text{bi}}$ = built in voltage, $\epsilon_s$ = permittivity of the semiconductor, $N$ = doping density.

well defined. This is different for the microscopic CAFM tip-sample system, as mentioned earlier. Therefore, critical conduction parameters such as emission area, oxide thickness, and the effective oxide voltage ($V_{\text{ox}}$) are regarded as electrically effective values for CAFM application and are indicated accordingly in Table 3.2 [16, 25]. In good agreement with CAFM measurements, the tunneling current of microscopic MOS structures may be calculated using a single tunneling current model for FNT and DT [25]. The tunneling current model comprises Eqs (3.4) and (3.5) and associated parameter sets [14, 16]. If the correction factor $(1 - (1 - qV_{\text{ox}}/\phi_B)^{3/2})$ (Eq. (3.5)) is maintained at unity for $V_{\text{ox}} > \phi_B/q$, the current may be calculated in a continuous manner from the DT to the FNT regime with a single parameter set [14, 16, 25]. The tunneling parameters, $m_{\text{ox}}$ and $\phi_B$, are not further discussed here because they may be considered as constant as they solely depend on the properties of the dielectric layer. The so called tip parameters comprise $A_{\text{eff}}$ and a value termed *offset voltage* ($V_{\text{off}}$) (Section 3.3), which takes into account that the tip voltage applied ($V_{\text{tip}}$) may not equal $V_{\text{ox}}$. Both values depend on tip and substrate materials and are critical not solely for DT and FNT but figuratively for the other emission regimes presented in Table 3.2.

## 3.3 Work Function Difference and Offset Voltage

In CAFM, the current measured is closely linked to $V_{\text{tip}}$. However, if the tip and the sample have different work functions and band bending plays a role, which is frequently the case for MOS (MIS) and metal/semiconductor structures, the voltage that actually defines $I$ may be different from $V_{tip}$. Therefore, a meaningful interpretation of CAFM measurements requires a proper relation between $V_{ox}$ and $V_{tip}$. For MOS (MIS) structures, $I$ depends strongly on $V_{ox}$ (Eqs (3.4) and (3.5)). Frammelsberger *et al.* [14, 25] introduced a constant offset voltage ($V_{off}$) for CAFM application (Eq. (3.9)). Equations (3.10) and (3.11) apply for p-type, $V^p_{off}$, and n-type, $V^n_{off}$, semiconductors, respectively.

$$V_{ox} = V_{tip} - V_{off} = V_{tip} - (V_{FB} + V_S) \tag{3.9}$$

$$V^p_{off} = \phi_m - \left(\chi + \frac{E_g}{2q} + \psi^p_S\right) + 2\psi^p_S = \phi_m - \chi - \frac{E_g}{2q} + \psi^p_S \tag{3.10}$$

$$V^n_{off} = \phi_m - \left(\chi + \frac{E_g}{2q} + \psi^n_S\right) + \left(\frac{E_g}{2q} + \psi^n_S\right) = \phi_m - \chi \tag{3.11}$$

The offset voltage comprises the flat-band voltage ($V_{FB}$) for ideal conditions and the voltage drop in the Si substrate ($V_S$). The metal work function is $\phi_m$, $\psi_S$ is the respective Fermi-level shift due to the dopant concentration, and $\chi$ is the electron affinity of the semiconductor. The energy level $E_c$ is the conduction band, $E_v$ is the valence band, and $E_g$ is the corresponding band gap of the semiconductor.

From Eqs (3.10) and (3.11), Figure 3.9 shows the theoretical offset voltage as a function of the substrate dopant density, p-type Si (a) and n-type Si (b), for

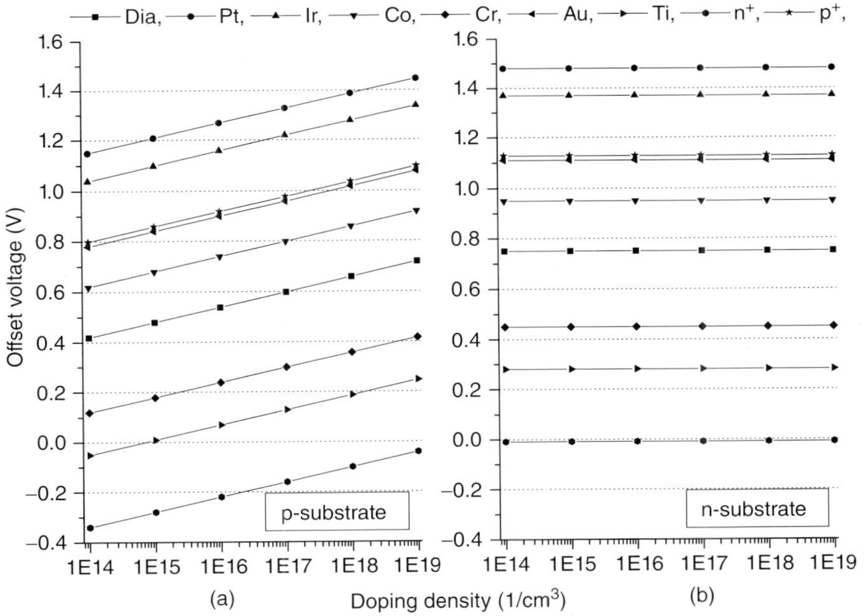


**Figure 3.9** Offset voltage as a function of substrate dopant density for different tip materials for p-type Si (a) (Eq. (3.10)) and n-type Si (b) (Eq. (3.11)) for substrate emission. In addition, the values for highly doped ($5 \times 10^{19}$ cm$^{-3}$) poly-Si gates, n$^+$ and p$^+$, are provided. Material parameters [10, 26, 54, 55], typical values: $\chi = 4.05$ eV, $E_g = 1.12$ eV, $\phi_m^{Dia} = 4.8$ eV, $\phi_m^{Pt}\{110\} = 5.53 \pm 0.13$ eV, $\phi_m^{Ir}\{110\} = 5.42 \pm 0.32$ eV, $\phi_m^{Co} = 5.0$ eV, $\phi_m^{Cr} = 4.5$ eV, $\phi_m^{Au}\{110\} = 5.16 \pm 0.22$ eV, $\phi_m^{Ti} = 4.3$ eV, $\phi_m^{n^+} = 4.0$ eV, $\phi_m^{p^+} = 5.2$ eV.

different tip materials. Interestingly, work function values of pure metals in the literature depend significantly on measurement method and crystal orientation [54]. Hence, it is not clear which value actually applies for CAFM application. For example, Derry *et al.* [54] report values between 5.4 and 6.1 eV for platinum, 5.42 and 6.15 eV for iridium, and 5.12 and 5.6 eV for gold. The numbers for platinum, iridium, and gold provided in Figure 3.9 represent the minimum of the recommended values for different crystal orientations reported by Derry *et al.* [54].

Frammelsberger *et al.* [14] (p-Si, $2 \pm 1 \times 10^{15}$ cm$^{-3}$) presented $V_{off}$ values for different metal-coated tips. The values reported, for example, for Co-Cr-coated tips (0.3 V), CoCr-coated tips (0.6 V), and PtIr /Pt-Ir-coated tips (1.3 V), are in the order of the theoretical values, but differences are visible (Figure 3.9). As $A_{eff}$ and $V_{off}$ show some interdependence and both values are influenced by the water layer in air, the values may be regarded as electrically effective rather than physically correct. In addition, other different effects are able to influence $V_{off}$, which are discussed below.

Kelvin probe force microscopy (KPFM) [56] and related techniques can be used to measure the contact potential difference (CPD) between the tip and the sample (Chapter 14). In UHV and in the absence of oxide layers, surface states, contamination, or other obstacles, the CPD value basically represents the work function difference. It is possible to determine the tip work function experimentally, facilitating a reference sample with known work function. However, for semiconductor samples, band bending needs to be taken into account. Melitz *et al.* [56] report that band bending is sensitive to illumination and that air exposure of semiconductor samples may change the CPD value, even though the measurements are performed in UHV. Native oxide and adsorbate induce or passivate surface states and are among the reasons that affect the surface potential. For measurements in air, the situation becomes even more critical. The Tip and the sample are continually exposed to air and a humidity-induced water layer shields the sample surface. Therefore, the CPD reading achieved in air does not represent the true work function difference of the tip and the sample. Nonetheless, KPFM in air is frequently employed because of its convenient handling. Even though the absolute CPD values may be different when compared to UHV, the relative information provides valuable insight into electronic structures of junctions and heterostructures including band alignment, failure mechanisms, and photon-induced surface electronic changes [56]. Moreover, Berthold *et al.* [57] employed PeakForce KPFM successfully in air to discriminate different types of copper oxides by the characteristic CPD values.

Celano *et al.* [24] determined the tip work function of Pt/Ir-coated tips and diamond-coated tips by KPFM in UHV, facilitating a freshly cleaved highly ordered pyrolytic graphite (HOPG) sample. For the Pt/Ir tip, they arrived at a work function of $4.75 \pm 0.05$ eV and for the diamond tip at $4.1 \pm 0.05$ eV. The tip voltage shift, representing $V_{off}$, was 0.7 eV for Pt/Ir and 0.05 eV for diamond, respectively ($\chi = 4.05$ eV, n-Si). When compared with the theoretical values in Figure 3.9, tip work function and thus $V_{off}$ (n-Si) were smaller by 0.68 eV for platinum and 0.7 eV for diamond. Using the theoretical $V_{off}$ values for Pt/Ir (i) and diamond tips (ii), respectively, would overestimate $d_{ox}$ by (i) 0.48 nm (FNT), 0.20 nm (DT), and by (ii) 0.49 nm (FNT), 0.21 nm (DT), respectively [14].

Therefore, if possible, it appears advisable to determine the tip work function experimentally by KPFM in UHV. However, possibility that the tip work function changes during CAFM measurement cannot be excluded. Even new tips may possess contamination or oxide layers from manufacturing or air exposure that influence the CPD value measured by KPFM. These layers can wear off easily during CAFM measurement leading to a different work function than determined experimentally. Furthermore, as far as stacked conductive coatings are involved, the outer layer may wear off during CAFM measurements and the inner layer may dominate the electrical behavior.

In air, electrical stress-induced chemical reactions may play an important role. Several authors [14, 16, 58, 59] reported significant shifts between successive $I$–$V$ curves or observed significant electrical stress-induced changes of the surface morphology after CAFM measurements [16, 21, 46, 60] for positive tip voltages. Figure 3.10 presents examples of stress-induced surface modifications. The image was recorded by tapping mode using a new uncoated silicon tip with a substantially stiffer tip cantilever than used for CAFM measurements, in order to increase the resolution and to widely exclude electrostatic artifacts [16]. Biased scan, local $I$–$V$ measurements ($3 \times 3$), and local breakdown spots elevated the sample surface at the respective stressed area. Occasionally, mobile debris was arranged around the scan area. Dielectric breakdown occurred only at the onset of the current flow, which may indicate that the CAFM tip may have been subject to anodic oxidation that reduced the current. The tip, sample, and water meniscus under electrical stress constitute a electrochemical nano cell [60]. Depending on the polarity, either the tip or the sample is exposed to

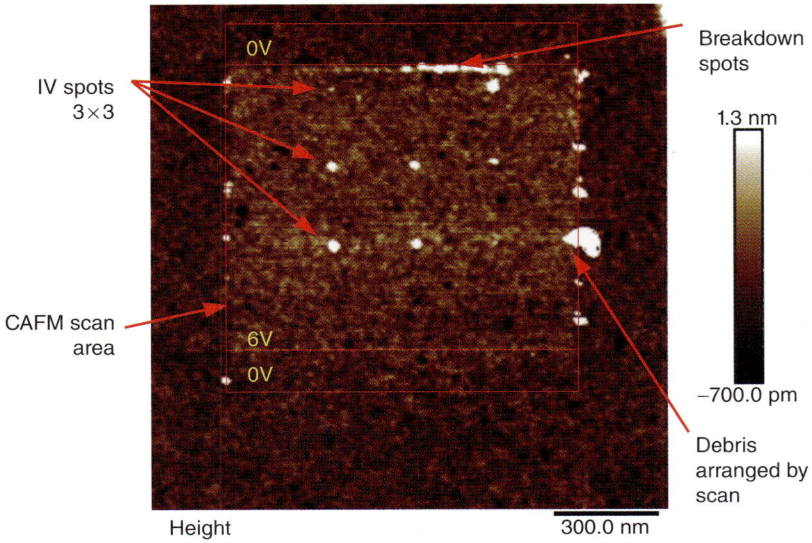


**Figure 3.10** Topography image (tapping mode) of a 1.3 nm $SiO_2$/n-Si sample subject to CAFM stress. The center area was stressed by $V_{tip} = +6$ V, 0,5 Hz scan rate. The surface protrusions in the center are a result of local $I$–$V$ measurements ($V_{tip} = 0 \ldots + 7$ V, 0.3 Hz ramp rate). CAFM: Cont-Pt (PtIr5) tip, $\approx 0.28$ N/m. Tapping mode: (uncoated Si) tip, $\approx 48$ N/m.

anodic oxidation. Local anodic oxidation [23, 60, 61] of sample surfaces in wet environment is well known, but anodic oxidation of conductive probe tips has rarely been investigated. Even noble metals [62] and diamond [63] have been reported to build up oxides under measurement conditions comparable with CAFM measurements. The tip oxide can influence the overall electrical oxide thickness and, also important, the voltage offset via a change of the surface potential. Electrical stress in combination with wear due to tip movement cause a dynamic growth and removal of tip oxide and thus affect the observed $I$–$V$ curve shifts. Polspoel *et al.* [46] demonstrated that protrusions after local $I$–$V$ measurements, biased scans, and dielectric breakdown events for negative sample bias (Figure 3.10) consist of $SiO_2$. They associated the size and shape of the protrusions with the local heat dissipation in the oxide. The higher the heat dissipation in the oxide, the more pronounced are the protrusions due to stronger melting and oxidation processes. The heat dissipation depends on oxide resistance and tip contact resistance and increases with increasing oxide resistance relative to the tip contact resistance and vice versa [46]. The magnitude of the tip contact resistance may change owing to dynamic tip oxidation and wear processes [46], which explains the observed inhomogeneity of CAFM-induced surface modifications.

In conclusion, the offset voltage, $V_{off}$, is an important parameter for CAFM measurements as the tip and sample work functions may be different and thus the voltage that determines the current flow, here the oxide voltage, $V_{ox}$, is different to the tip voltage, $V_{tip}$. Theoretic work function values for bulk materials depend on the crystal orientation and measurement method applied. They may be used in a first approach for CAFM application, but do not necessarily reflect the electrically effective values at the tip apex. In addition, air exposure alone of the tip or the sample influences the work function. Unfortunately, even though $V_{off}$ may be determined experimentally on the basis of the CPD measured by KPFM in UHV, the value may change during measurement. In UHV, this may occur predominately owing to wear effects. In air, electrochemical reactions may change dynamically the electrically effective tip conditions in combination with wear. Thus, in practice, $V_{off}$ remains difficult to define and may not be regarded as a stable entity in the interpretation of CAFM measurements.

## 3.4 Operation Modes

The previous chapter describes conduction mechanisms for different tip-sample material combinations in steady-state condition. Even though this fundamental application of CAFM is frequently used, it is not suitable for some new materials and applications. Therefore, new operation modes, continuously extending the field of CAFM applications, have been introduced. In general, different operation modes can be distinguished according to the contact formed between tip and sample. Besides the standard contact mode, characterized by a permanent physical tip-sample contact, the cantilever and the tip may oscillate close to the sample surface. The latter is called a *dynamic operation mode.* Several subtechniques have been developed and published under

**Table 3.3** Alternative names and subtechniques of CAFM: Conductive scanning probe microscopy (C-SPM) [6, 7], conductive scanning force microscopy (C-SFM) [8], tunneling AFM (TUNA) [11], TR TUNA [18–23], scanning conductive torsion mode microscopy (SCTMM) [8, 24–26] ).

| Nomenclature in this work | Alternative names or subtechniques | Typical current range | Typical applications | Oscillation frequency | Tip motion with respect to scan axis | Tip operation |
|---|---|---|---|---|---|---|
| CAFM | C-SFM or C-SPM | pA to μA | Semiconductors, metals, dielectrics | No | No | In permanent contact with surface |
| | TUNA | pA to nA | Thin dielectrics | | | |
| | C-AFM | Sub-nA to μA | Semiconductors, metals | | | |
| PF CAFM | PeakForce TUNA | pA to μA | Organic semiconductors, materials for energy storage | 1 kHz | Vertical oscillation | Dynamic |
| TR CAFM | TR TUNA<br>SCTMM | | Organic semiconductors, nanorods, nanopillars | 500 kHz<br>1.5 MHz | Torsional oscillation | |

different, sometimes synonymous, names. As indicated in Table 3.3, in this chapter, we use the nomenclature CAFM for referring to any conductive mode independent from the current range or the physical cause of the current. In addition, alternative names used in the literature to refer to the same mode are also listed.

### 3.4.1 Contact Mode

The most widely used operation mode for CAFM measurements is the contact mode, as it offers the simplest experimental setup compared to other operation modes. In the contact mode, the tip is in physical contact with the sample and the force exerted to the tip is controlled by the bending of the cantilever. Although the force can be adjusted and calculated using the cantilever spring constant, elaborate external calibration procedures are necessary to obtain precise values due to thermal drift and inaccuracies in laser beam adjustment. In addition, external forces may act on the tip depending on, for example, the environment and electrical stress conditions. In the contact mode, it is important to balance the conflicting demands of a stable tip-sample contact and preferably low contact forces. The lower the contact forces, the lower are the lateral forces between the tip and the sample that are predominantly responsible for wear and even damage of the tip and the surface. However, the CAFM operation requires a minimum force to establish a stable tip-sample current that is usually higher than the force needed for sole topographical imaging. The use of large contact forces during a scan may lead to undesired damage of both the tip and the sample, such as nano scratching of the surface of the sample, modification of the geometry and electrical properties of the tip, and damage or wearing of the metallic tip coating. In general, CAFM in contact mode is not suitable for delicate samples and/or sensitive tips owing to the comparatively high forces exerted to tip and sample.

Scanning spreading resistance microscopy (SSRM) is an additional conductive contact mode application with much higher contact forces (typically ≥μN) than for standard CAFM (typically nN regime). Therefore, doped diamond-coated or solid diamond tips are frequently employed. Owing to the high contact forces, the tip penetrates thin oxide layers or other insulators at the surface, and local phase transformation in nanometer-sized pockets right underneath the tip apex takes place for silicon and germanium surfaces [41–43]. Therefore, the current measured for SSRM is dominated by the local spreading resistance rather than the contact resistance or other serial resistances, as is the case for standard CAFM [40–42]. For this reason, SSRM is not considered as a CAFM-related technique (it is has not been included in Table 3.1), and it is not described in this book. For more information about SSRM, please refer to other works [38, 40–45].

### 3.4.2 PeakForce Mode

PeakForce CAFM (PF CAFM), also known as *PeakForce TUNA* is a relatively new technique. It is usually employed to investigate mechanical and electrical properties of very delicate samples such as nanotubes [64], nanowires [65], or organic semiconductors [66, 67]. PF CAFM is based on the so-called pulsed force mode and it was introduced by Rosa-Zeiser and co-workers [68]. In PF mode, the $z$-piezo excitation generates a sinusoidal tip movement at a frequency of 1 kHz that results in an intermittent tip-sample contact. During a scan, successive force-distance curves at each pixel of the image are recorded, as represented in Figure 3.11. The maximum force can be modified and should be adjusted so that a sufficient deformation of the sample is achieved. The force-distance curve is evaluated using the DMT or Sneddon model to extract valuable information about the sample, for example, local adhesion, deformation, or Young's modulus [13, 69]. The spring constant of the cantilever employed and the sample hardness need to match. Excessive cantilever stiffness may damage the sample by creating holes, whereas insufficient cantilever stiffness may result in insufficient contact force and unstable current reading. Compared to contact mode, tip wear is decreased significantly thanks to the shorter contact time, resulting in comparatively low lateral forces. Consequently, the repeatability of topography and current mappings is very high. The measurement setup for the PF CAFM is similar to the CAFM setup. However, compared to the contact mode CAFM, the evaluation of the measured current is more complex, because the tip-sample interaction and the resulting current values are time dependent [64]. As can be seen in Figure 3.11b,c, PF CAFM allows collecting two different current signals. First, the contact current represents the average current while the tip is in physical contact with the surface of the sample, beginning with the snap-on ($B$) and ending at the pull-off point ($D$). Second, the peak current represents the current at maximum force at the inflection point ($C$) of the force-distance curve. This peak current is not necessarily the highest current value occurring within one force-distance cycle.

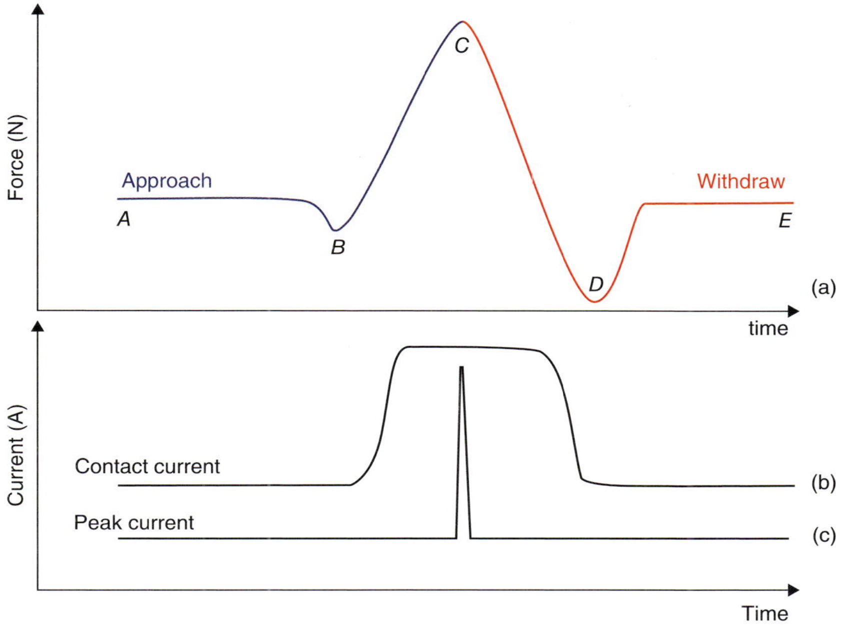


**Figure 3.11** Schematic of a time-dependent force curve for one complete PF cycle from *A* to *E* (a). Simplified current reading for the contact current (b) representing the current while the tip is in physical contact with the surface between point *B* and *D*. The peak current (c) is recorded at the maximum force at point *C*.

### 3.4.3 Torsional Resonance Mode

Another dynamic mode that reduces the tip-sample interaction forces compared to standard contact mode CAFM is torsional resonance conductive atomic force microscopy (TR CAFM), which has been also referred to by other names in the literature (Table 3.3). From an electrical point of view, similarly to CAFM and PF CAFM, in TR CAFM, a voltage is applied between the tip and the sample and the current flowing across the tip-sample junction is monitored (Figure 3.12a). The main difference between TR CAFM and other CAFM modes is the way in which the tip is moved near the surface of the sample. In TR CAFM, the base of the rectangular cantilever is excited laterally by two piezoelectric actuators, producing an oblique vibration at its torsional resonance frequency [70, 71]

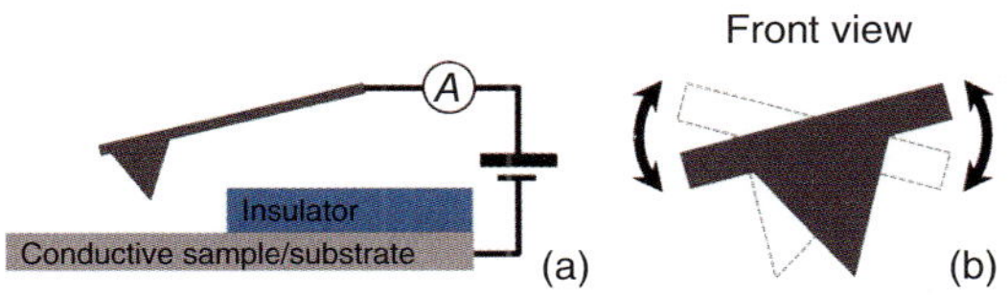


**Figure 3.12** Measurement setup for TR CAFM (a) and the front view of the cantilever and tip with indicated torsional movement (b). As shown schematically, the sample can be conductive or consist of a thin insulating surface layer.

(Figure 3.12b). The torsional resonance frequency of a cantilever can be up to 10-fold of its vertical resonant frequency [72], because the torsional bending stiffness of a rectangular cantilever is typically two orders of magnitude higher than the vertical one [71]. When a tip, oscillating in torsional mode, is placed very close to the surface of a sample (typically 0.3–2 nm [71, 73]), the tip-sample interaction reduces the amplitude of the torsional vibration. This signal change can be used as feedback for monitoring the tip-sample separation [70]. Therefore, the tip-sample interaction is more complex than in CAFM and PF CAFM, and it strongly depends on the tip-sample separation, sample surface properties, and resulting effective contact area and contact time. Moreover, in general, the TR CAFM setup is more elaborate compared to contact or standard intermittent operation modes because cross-talk from lateral frequencies needs to be avoided.

In studies about carbon nanowalls by Vetushka *et al.*, TR CAFM is described as a dynamic noncontact imaging mode enabling the detection of a tunneling current because of the small tip-sample separation in the range of 1–2 nm [73]. A continuous current is reported from Weber and Berger, when investigating the mechanical and electrical tip-sample contact on soft conducting polymer blends [70]. They assume it is very likely that the tip is in physical contact with the sample during imaging and that, in the case of organic semiconductors, the current is limited by the transport properties of the organic film rather than by the tip-sample contact resistance.

## 3.5 Case Studies

In Section 3.2, the tip-sample interaction is discussed for the contact mode operation, focusing on the characterization of tunneling currents through thin dielectrics. To demonstrate the wide range of CAFM applications three more case studies are introduced briefly.

### 3.5.1 Defects in SiC after Plasma Exposure in Fusion Reactors

A first case study describes recent application of conductive CAFM in the field of materials characterization after plasma-wall interaction experiments [74]. Silicon carbide (SiC), as one of the first walls or structural materials in fusion reactors, is subjected to strong heat flux and neutron radiation, as well as bombardment by energetic ions [75]. The SiC crystal exhibits high thermal conductivity, high-temperature stability, chemical inertness, and good resistance against irradiation [76, 77]. However, the ion and neutron irradiations induce defects in the material and cause degradation of material properties [78]. Therefore, the fundamental understanding of irradiation damage behavior in SiC is important for the development of fusion reactors [79, 80]. The ion implantation process results in creation of atomic-level defects whose accumulation can cause disordering, amorphization, and gas bubbles in implanted materials. Transmission electron microscopy (TEM) has been successfully used to observe gas bubbles along the cross-section of irradiated materials [81, 82]. However, the specimen preparation for TEM analysis is rather complicated, which typically includes polishing and ion milling processes. Specimens are usually changed or destroyed before TEM observation. A direct comparison between the sizes of defect points, and conductivity and surface topography of irradiated materials is necessary for understanding the damage formation, defect accumulation, and growth of gas bubbles. The CAFM technique is nondestructive, and it does not cause any damage to the irradiated materials.

PtIr-coated tips are used for the contact mode CAFM measurement performed at room temperature under ambient conditions. A constant voltage $V_{tip}$ is applied between the conductive tip and the irradiated sample and electron emission images through the irradiated samples are obtained simultaneously with topography at a typical scan rate of 0.30 Hz. Ion implantations into irradiated materials usually induce structural damages in the crystal and this can affect their electronic properties. When the conductive tip is scanned across the surface as schematically depicted in Figure 3.1a, the difference in the electron emission intensity across the surface of the irradiated sample can be observed from the current image. This method can be used to efficiently compare the surface distribution of microstructures with one of defects or bubbles in the destroyed layer. Figure 3.13 compares the surface topography (a) and current (b)

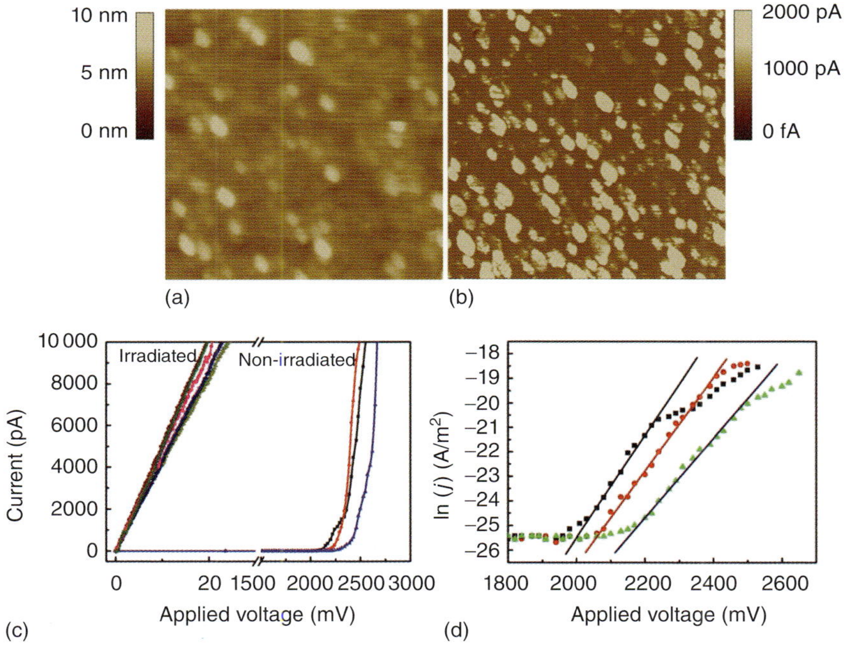


**Figure 3.13** (a) Topography and (b) simultaneously measured current image obtained at the 6H-SiC sample irradiated at the $He^+$ ion fluence of $8 \times 10^{20}$ ions/m$^2$. The 3.0 mV tip voltage is retained for the CAFM measurement. (c) Local *I–V* characteristics of nonirradiated and He-ion irradiated ($8 \times 10^{20}$ ions/m$^2$) 6H-SiC samples. (d) The ln($j$) values for nonirradiated 6H-SiC sample as a function of applied voltage. The straight lines represent the Schottky emission fits to the experimental data. The intersection with the *y*-axis gives the value for the saturation current density $J_0 = A^{**}T^2 \exp\left(-\frac{q\Phi_{SB}}{k_BT}\right)$. These fits lead to the Schottky barrier height ($\Phi_{SB}$) of $1.30 \pm 0.05$ eV.

images obtained at the same location of the 6H-SiC sample. The 6H-SiC sample was irradiated at a 100 keV $He^+$ ion fluence of $8.0 \times 10^{20}$ ions/m$^2$. The tip voltage (3.0 mV) is kept positive, which avoids the anodic oxidation of 6H-SiC samples as the electron injection occurs from the 6H-SiC sample. The characteristic shape of electron emission is obviously consistent with the surface topography image. This shows that the surface swelling of irradiated 6H-SiC samples is well related to the growth of the nanometer-sized defects in the destroyed layer. The electron emission image is more distinct than the surface topography image, which indicates that the electron emission is very sensitive to the inner microstructure in the destroyed layer. Plenty of elliptic defects can be observed from the current image, as shown in Figure 3.13b, and their semiprincipal axes of length ($a$ and $b$) are typically in the range of 15–38 nm. Clearly, this CAFM method for detecting He bubbles in SiC samples is rather simple and nondestructive.

In order to evaluate the effect of He ion implantation on the electron emission characteristics, local $I$–$V$ curves are obtained on nonirradiated and irradiated SiC samples, as shown in Figure 3.13c. The electron field emission of the irradiated SiC sample is two orders of magnitude lower than the one of nonirradiated SiC samples. The $I$–$V$ characteristics for the nonirradiated SiC samples may be analyzed in terms of Schottky barrier formation between the PtIr tip and the n-type SiC sample. The values of the Schottky barrier heights (SBHs) can be determined using the diode Eq. (3.12) [26]. $J_0$ is the saturation current density and can be calculated by $J = \frac{I}{A_{\text{eff}}}$. The $I$–$V$ characteristics measured from the nonirradiated SiC samples have been linearized in ln($J$) versus $V$ coordinates (Figure 3.13d), where $A_{\text{eff}}$ of the PtIr tip is supposed to be about 50 nm$^2$ [1, 83]. The intersection with the $y$-axis gives the value for the saturation current density, which is defined (Table 3.2) as

$$J_0 = A^{**}T^2 \exp\left(-\frac{q\Phi_{\text{SB}}}{k_B T}\right) \tag{3.12}$$

where $A^{**} = 146$ Å/cm$^2$/K$^2$ [84] is the Richardson constant for SiC, and $\Phi_{\text{SB}}$ is the SBH. These fittings lead to $\Phi_{\text{SB}}$ of $1.30 \pm 0.05$ eV, which are quite consistent with the barrier height of $1.39 \pm 0.21$ eV for Au/n-type SiC diodes of radius 5 μm [85]. The emission current for the He ion-irradiated SiC sample linearly increases with the applied voltage. The degradation of the Schottky characteristics at the relatively high temperature could be ascribed to the aggregation of carbon clusters, as observed by TEM analysis [86].

Ion implantation can break the C—C or C—Si bonds to form stable sp$^2$ hybridized carbon and electrically active carbon clusters [87]. The sp$^2$ hybridized carbon clusters can contribute to the formation of filament-like emission channels with a relatively high local field-enhancing factor and be the prominent electron emitter [83, 88]. The formation of nanometer-sized He bubbles can be well related to the agglomeration of electrically active carbon clusters leading to an increase in the leakage current. Diffusion of implanted He atoms along the surface results in an increase in the size of flattened He bubbles, and eventually a bubble layer with larger planar clusters of bubbles is formed.

The application of CAFM has been successfully extended to study the degradation of alternative plasma-facing materials such as polycrystalline tungsten or molybdenum. It has been clearly demonstrated that even low-energy ions play an important role in the material degradation after plasma exposure [74, 89, 90]

### 3.5.2 Electrical Conductivity of Dislocations in GaN

In this case study, the results for CAFM in contact mode, torsional resonance (TR) mode and PeakForce (PF) mode are compared. The different operation modes were applied to measure the microstructure and local electrical conductivity on a c-plane GaN epilayer (sample A) [91]. Pits, which are caused by threading dislocations originating from the free-standing GaN substrate, are apparent on top of the sample surface. Increased current is measured at the pit locations on the surface when compared to off-pit locations. Several effects such as carrier scattering or nonradiative recombination are responsible for this occurrence. The average diameter of the surface pits is approximately 200 nm. The PtIr-coated tips used enabled a good topography image showing the hexagonal shape of the surface pits. In order to compare the different operation modes, the scans were performed on the same position and with identical tips. Figure 3.14a depicts the topography image acquired in the PF mode. It shows a higher topographical resolution when compared to the contact and TR modes (not shown in Figure 3.14). The lattice planes are clearly visible in the PF mode but appear blurred in the TR mode and may not be resolved in the contact mode. In the contact mode, the surface pits appear enlarged compared to the PF images because of higher tip force and increased physical contact area. However, the current mapping shows the lowest noise level in the contact mode. Typical artifacts for contact mode CAFM are current tails (Figure 3.14b) observed in the scan direction and resulting from high current spots. For the contact and PF mode (Figure 13.1d), current spots can clearly be assigned to surface pit locations, whereas they are not visible for the TR mode (Figure 13.14c). The PF CAFM current mapping contains a median noise level, which can be reduced significantly by fast Fourier transformation filtering. Local $I$–$V$ characteristics

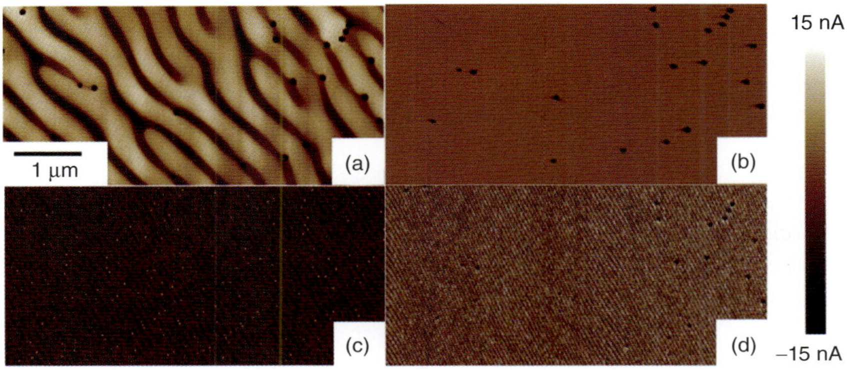


**Figure 3.14** Topography using the PF mode for an area of $1 \times 0.5\ \mu m^2$ (a) and current mappings of sample A for different operation modes: contact mode (b), TR mode (c), and PF mode (d). In this case study, the contact mode and PF CAFM deliver the best current mappings, whereas in the TR mode, currents on the surface pits locations could not be resolved.

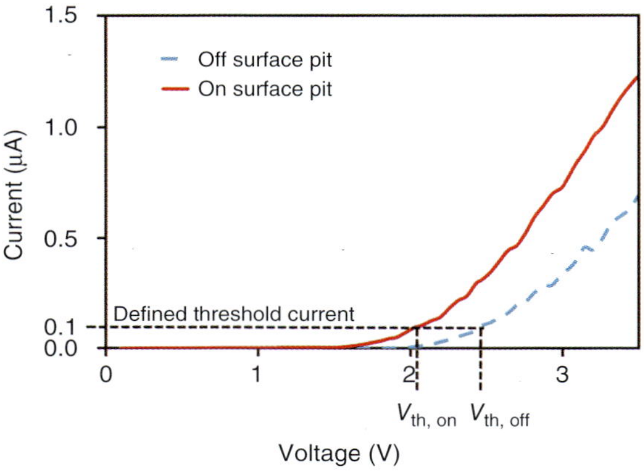


**Figure 3.15** Local *I–V* characteristics on sample A in forward bias taken on and off surface pits. The dotted line represents an arbitrary threshold current of 100 nA leading to a turn-on voltage shift of about 400 mV between $V_{th,on}$ and $V_{th,off}$.

have been measured, showing an onset voltage difference of 400 mV at surface pit positions when compared to off-pit positions (Figure 3.15) [91].

An additional GaN test structure (sample B) containing surface pits with dimension in the range of 30 nm was used. Here, it is important to note that the surface pit size is in the same range as the tip radius of the conductive coated tip used for contact mode CAFM. Therefore, in the contact mode, several surface pits did not appear in the topography image, even though high current spots indicated such pits at several sites. In order to overcome this drawback, high-resolution tapping mode topography imaging, facilitating comparatively sharp uncoated tips and contact mode CAFM results were combined (Figure 3.16) for the same site. By this means, it was possible to clearly relate the high current spots to corresponding surface pits. Despite the comparatively large tip radius of the conductive tip, the electrical resolution is obviously sufficient for the detection of high conductive dislocations. Furthermore, it may be observed (Figure 3.16) that several surface pits, which did not even appear in contact mode CAFM, do not show an increased current. This observation indicates that the surface pits may be caused by different types of dislocations which show different conduction mechanisms [92].

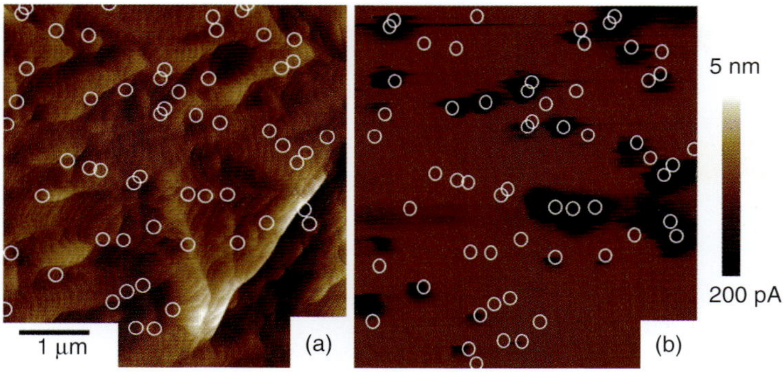


**Figure 3.16** Superposition of a 5 × 5 μm² IC-AFM topography (a) on sample B and indicated surface pits with a CAFM current mapping (b) at 22 V forward bias.

### 3.5.3 Microstructure and Local Electrical Conductivity of Laser-Sintered Nanoparticles

Micro- and nanostructured materials based on group-IV semiconductors (silicon and germanium) have gained growing importance for thermoelectric applications. One possible approach for the fabrication of small-scale thermoelectric generators is the laser sintering of highly crystalline germanium nanoparticles [7]. To further increase the efficiency of such devices, the nanoparticle grain size and the local conductivity behavior need to be analyzed besides other physical properties such as thermal conductivity or the Seebeck coefficient [8, 93]. Contact mode CAFM was used to study the local conductivity behavior. Figure 3.17a shows the corresponding topography image. The topography appears blurred because of the wide tip angle of the doped, full-diamond tip [94], whereas the current map simultaneously recorded (Figure 3.17b) shows a more detailed image. Domains with the same local conductivity cannot be correlated to topography features. For rough surfaces, this circumstance is very important as topography features frequently change $A_{eff}$, affecting the local conductivity. As discussed earlier, the electrical resolution can be much higher than the topographical resolution. Even grains with a size of 10 nm may be imaged by the current. Local $I$–$V$ measurements exhibit in most cases an almost ohmic characteristic, indicated by yellow and blue arrows/characteristics in Figure 3.17b,c. On comparatively low-conductive grain diode or Schottky behavior occurs (Figure 3.17b,c, red arrows/characteristics).

To monitor the effect of different load forces between the tip and the sample, the force was increased repeatedly from scan to scan and the current was recorded. Figure 3.18a shows the topography, while Figure 3.18b–d presents the current maps with increasing force level. At comparatively low forces of 600 nN (Figure 3.18b), small grains dominate the current map. At a load force of 2.2 μN, which is more typical for SSRM applications, the grains appear to merge

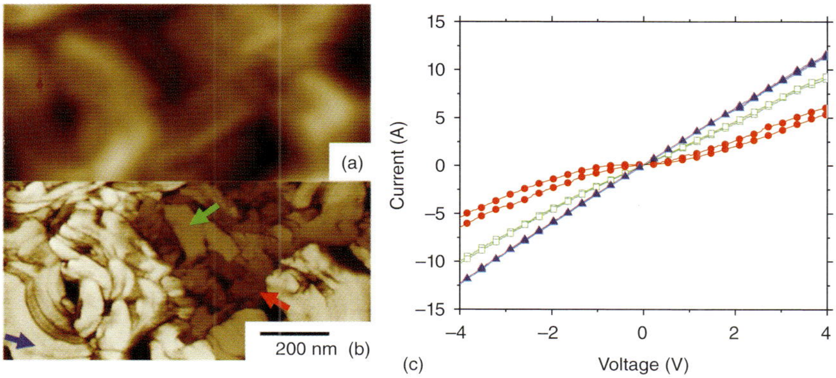


**Figure 3.17** Topography image of area $1 \times 0.5\ \mu m^2$. (a) Blurred picture of the surface (contact mode). current mapping (b) containing information that cannot be assigned to the topographical features. The $I$–$V$ characteristics exhibit different conduction regimes (c). Most of the characteristics show an ohmic behavior (blue and yellow arrows/characteristics), whereas on low conductive areas, a nonlinear characteristic occurs (red arrow/characteristics).

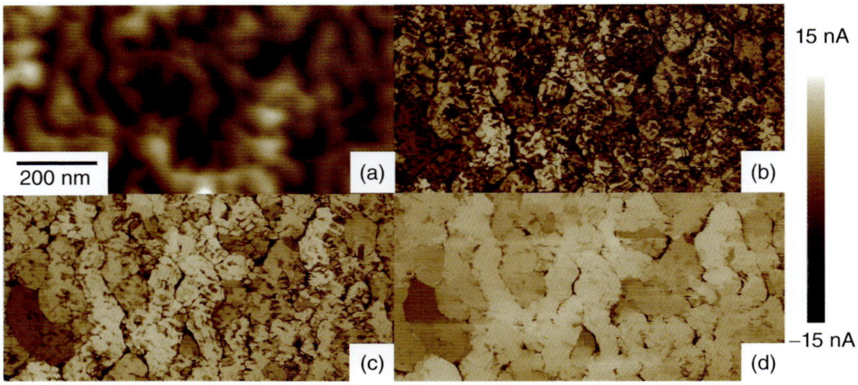


**Figure 3.18** Topography of an area of $1 \times 0.5\ \mu m^2$ (a). Current mappings for different contact forces reaching from 600 nN (b), 2.2 μN (c), and 5.5 μN (d). The tip voltage was 20 mV. Low forces lead to a depiction of the immediate surface structures with a very high lateral resolution. At forces above 1 μN (c and d) a germanium phase transition beneath the tip leads to a significant resistance drop and the meander structure underneath the immediate surface structures dominates the current map.

and larger areas of equal conductivity may be seen (Figure 3.18c), while smaller grains disappear. This is in agreement with the work by Schulze *et al.* [95], as they observed a significant drop of resistance at load forces around 1 μN, which is a result of a germanium phase transition beneath the tip. Increasing the force to a value of 5.5 μN gives rise to a tip penetration of immediate surface structures and the current mapping exhibits the meander structure underneath, which was formed during the laser sintering process (Figure 3.18c).

## 3.6 Conclusion and Future Perspectives

The intention of this chapter was to give a brief overview of the most important CAFM operation modes, focusing on a hands-on approach. Beginning with standard contact mode CAFM measurements and extending the applications by use of dynamic modes such as PF CAFM and TR CAFM, samples with a wide variety of surface properties, ranging from soft and fragile to ultrahard can be analyzed. The ability to simultaneously map topography and current in different operation modes and the additional option to acquire local $I$–$V$ characteristics make the CAFM a unique tool for current and future characterization challenges in nanoscience and nanotechnology.

## Acknowledgment

This work has been partially supported by the Bavarian State Ministry of the Environment and Consumer Protection via the project "UMWELTnanoTECH" and the Federal Ministry of Education and Research (BMBF), Germany.

## References


1 Olbrich, A., Ebersberger, B., and Boit, C. (1998) Conducting atomic force microscopy for nanoscale electrical characterization of thin SiO[sub 2]. *Appl. Phys. Lett.*, **73** (21), 3114.

2 Olbrich, A., Ebersberger, B., and Boit, C. (1998) Nanoscale electrical characterization of thin oxides with conducting atomic force microscopy. IEEE International Reliability Physics Symposium Proceedings 36th Annual, Reno, NV, USA, pp. 163–168.

3 Ruskell, T.G., Workman, R.K., Chen, D., Sarid, D., Dahl, S., and Gilbert, S. (1996) High resolution Fowler-Nordheim field emission maps of thin silicon oxide layers. *Appl. Phys. Lett.*, **68** (1), 93.

4 Murrell, M.P., Welland, M.E., O'Shea, S.J., Wong, T.M.H., Barnes, J.R., McKinnon, A.W., Heyns, M., and Verhaverbeke, S. (1993) Spatially resolved electrical measurements of SiO2 gate oxides using atomic force microscopy. *Appl. Phys. Lett.*, **62** (7), 786.

5 O'Shea, S.J., Atta, R.M., and Welland, M.E. (1995) Characterization of tips for conducting atomic force microscopy. *Rev. Sci. Instrum.*, **66** (3), 2508.

6 O'Shea, S.J. (1995) Conducting atomic force microscopy study of silicon dioxide breakdown. *J. Vac. Sci. Technol., B*, **13** (5), 1945.

7 Petermann, N., Stein, N., Schierning, G., Theissmann, R., Stoib, B., Brandt, M.S., Hecht, C., Schulz, C., and Wiggers, H. (2011) Plasma synthesis of nanostructures for improved thermoelectric properties. *J. Phys. D: Appl. Phys.*, **44** (17), 174034.

8 Stoib, B., Langmann, T., Petermann, N., Matich, S., Sachsenhauser, M., Wiggers, H., Stutzmann, M., and Brandt, M.S. (2013) Morphology, thermoelectric properties and wet-chemical doping of laser-sintered germanium nanoparticles. *Phys. Status Solidi A*, **210** (1), 153–160.

9 Bhushan, B. (2004) *Springer Handbook of Nanotechnology*, Springer, Berlin, NY.

10 Olbrich, A. (1999) Charakterisierung dünner Dielektrika mittels modifizierter Rasterkraftmikroskopie. Dissertation. Universität Regensburg.

11 Cappella, B. and Dietler, G. (1999) Force-distance curves by atomic force microscopy. *Surf. Sci. Rep.*, **34** (1-3), 1–104.

12 Born, A. (2000) Nanotechnologische Anwendungen der Rasterkapazitätsmikroskopie und verwandter Rastersondenmethoden. Dissertation. Universität Hamburg.

13 Butt, H.-J., Cappella, B., and Kappl, M. (2005) Force measurements with the atomic force microscope: technique, interpretation and applications. *Surf. Sci. Rep.*, **59** (1-6), 1–152.

14 Frammelsberger, W., Benstetter, G., Kiely, J., and Stamp, R. (2007) C-AFM-based thickness determination of thin and ultra-thin SiO2 films by use of different conductive-coated probe tips. *Appl. Surf. Sci.*, **253** (7), 3615–3626.

15 Yoon, E.-S., Yang, S.H., Han, H.-G., and Kong, H. (2003) An experimental study on the adhesion at a nano-contact. *Wear*, **254** (10), 974–980.

16 Frammelsberger, W., Benstetter, G., Kiely, J., and Stamp, R. (2006) Thickness determination of thin and ultra-thin SiO2 films by C-AFM IV-spectroscopy. *Appl. Surf. Sci.*, **252** (6), 2375–2388.

17 Lanza, M., Porti, M., Nafría, M., Aymerich, X., Whittaker, E., and Hamilton, B. (2010) UHV CAFM characterization of high-k dielectrics: effect of the technique resolution on the pre- and post-breakdown electrical measurements. *Microelectron. Reliab.*, **50** (9-11), 1312–1315.

18 Lanza, M., Porti, M., Nafría, M., Aymerich, X., Whittaker, E., and Hamilton, B. (2010) Note: Electrical resolution during conductive atomic force microscopy measurements under different environmental conditions and contact forces. *Rev. Sci. Instrum.*, **81** (10), 106110.

19 Porti, M. (2004) Impact of conductivity and size of the breakdown spot on the progressive-breakdown of thin SiO2 gate oxides. *Microelectron. Eng.*, **72** (1-4), 29–33.

20 Porti, M., Nafria, M., and Aymerich, X. (2003) Current limited stresses of SiO/sub 2/ gate oxides with conductive atomic force microscope. *IEEE Trans. Electron Devices*, **50** (4), 933–940.

21 Porti, M., Nafria, M., Aymerich, X., Olbrich, A., and Ebersberger, B. (2001) Nanoscale observations of the electrical conduction of ultrathin SiO/sub 2/ films with conducting atomic force microscopy. IEEE International Reliability Physics Symposium Proceedings. 39th Annual, Orlando, FL, USA, 30 April-3 May 2001, pp. 156-162.

22 Porti, M., Nafría, M., Aymerich, X., Olbrich, A., and Ebersberger, B. (2002) Electrical characterization of stressed and broken down SiO[sub 2] films at a nanometer scale using a conductive atomic force microscope. *J. Appl. Phys.*, **91** (4), 2071.

23 Kremmer, S., Peissl, S., Teichert, C., Kuchar, F., and Hofer, H. (2003) Modification and characterization of thin silicon gate oxides using conducting atomic force microscopy. *Mater. Sci. Eng., B*, **102** (1-3), 88–93.

24 Celano, U., Hantschel, T., Giammaria, G., Chintala, R.C., Conard, T., Bender, H., and Vandervorst, W. (2015) Evaluation of the electrical contact area in contact-mode scanning probe microscopy. *J. Appl. Phys.*, **117** (21), 214305.

25 Frammelsberger, W., Benstetter, G., Stamp, R., Kiely, J., and Schweinboeck, T. (2005) Simplified tunnelling current calculation for MOS structures with ultra-thin oxides for conductive atomic force microscopy investigations. *Mater. Sci. Eng., B*, **116** (2), 168–174.

26 Sze, S.M. and Ng, K.K. (2007) *Physics of Semiconductor Devices*, 3rd edn, Wiley-Interscience, Hoboken, N.J..

27 Chiu, F.-C. (2014) A review on conduction mechanisms in dielectric films. *Adv. Mater. Sci. Eng.*, **2014** (7), 1–18.

28 Schuegraf, K.F. (1994) Chenming Hu, Hole injection SiO2 breakdown model for very low voltage lifetime extrapolation. *IEEE Trans. Electron Devices*, **41** (5), 761–767.

29 Schuegraf, K.F. and Hu, C. (1994) Reliability of thin SiO2. *Semicond. Sci. Technol.*, **9** (5), 989–1004.

30 Elabd, M.M.F. (2010) *Transport Properties of Metal-Metal and Metal-Insulator Heterostructures*. Lehrstuhl für Theoretische Physik II Universität Augsburg, Oral examination: 9 June.

31 Nygard, J. (2005) *Tutorial on Electronic Transport*, Gotheburg, Sweden.

32 Miranda, E.A., Walczyk, C., Wenger, C., and Schroeder, T. (2010) Model for the resistive switching effect in HfO2 MIM structures based on the transmission properties of narrow constrictions. *IEEE Electron Device Lett.*, **31** (6), 609–611.

33 Budget sensors, AFM probe catalogue, model: *Electri*Tap300-G, https://www.budgetsensors.com/tapping_mode_afm_probes_electri.html (accessed 21 April 2017).

34 Hui, F., Vajha, P., Shi, Y., Ji, Y., Duan, H., Padovani, A., Larcher, L., Li, X.R., Xu, J.J., and Lanza, M. (2016) Moving graphene devices from lab to market: advanced graphene-coated nanoprobes. *Nanoscale*, **8** (16), 8466–8473.

35 Vecchio, C., Sonde, S., Bongiorno, C., Rambach, M., Yakimova, R., Raineri, V., and Giannazzo, F. (2011) Nanoscale structural characterization of epitaxial graphene grown on off-axis 4H-SiC (0001). *Nanoscale Res. Lett.*, **6** (1), 269.

36 Sonde, S., Giannazzo, F., Vecchio, C., Yakimova, R., Rimini, E., and Raineri, V. (2010) Role of graphene/substrate interface on the local transport properties of the two-dimensional electron gas. *Appl. Phys. Lett.*, **97** (13), 132101.

37 Giannazzo, F., Sonde, S., Rimini, E., and Raineri, V. (2011) Lateral homogeneity of the electronic properties in pristine and ion-irradiated graphene probed by scanning capacitance spectroscopy. *Nanoscale Res. Lett.*, **6** (1), 109.

38 Doering, S., Rudolf, R., Pinkert, M., Roetz, H., Wagner, C., Eckl, S., Strasser, M., Wachowiak, A., and Mikolajick, T. (2014) Scanning spreading resistance microscopy for failure analysis of nLDMOS devices with decreased breakdown voltage. *Microelectron. Reliab.*, **54** (9-10), 2128–2132.

39 van Honschoten, J.W., Brunets, N., and Tas, N.R. (2010) Capillarity at the nanoscale. *Chem. Soc. Rev.*, **39** (3), 1096–1114.

40 de Wolf, P., Geva, M., Hantschel, T., Vandervorst, W., and Bylsma, R.B. (1998) Two-dimensional carrier profiling of InP structures using scanning spreading resistance microscopy: SSRM an InP. *Appl. Phys. Lett.*, **73** (15), 2155.

41 Giannazzo, F., Fiorenza, P., and Raineri, V. (2008) in *Applied Scanning Probe Methods X* (eds B. Bhushan, M. Tomitori, and H. Fuchs), Springer, Berlin Heidelberg, pp. 63–103.

42 Eyben, P., Vandervorst, W., Alvarez, D., Xu, M., and Fouchier, M. (2007) in *Scanning Probe Microscopy* (eds S.V. Kalinin, A. Gruverman, A. Foster, and W. Hofer), Springer, New York, pp. 31–87.

43 Eyben, P., Clemente, F., Vanstreels, K., Pourtois, G., Clarysse, T., Duriau, E., Hantschel, T., Sankaran, K., Mody, J., Vandervorst, W., Mylvaganam, K., and Zhang, L. (2010) Analysis and modeling of the high vacuum scanning spreading resistance microscopy nanocontact on silicon. *J. Vac. Sci. Technol., B*, **28** (2), 401.

44 Clarysse, T., Eyben, P., Janssens, T., Hoflijk, I., Vanhaeren, D., Satta, A., Meuris, M., Vandervorst, W., Bogdanowicz, J., and Raskin, G. (2006) Active

dopant characterization methodology for germanium. *J. Vac. Sci. Technol., B*, **24** (1), 381.

45 Doering, S., Wachowiak, A., Rochel, M., Nowak, C., Hoffmann, M., Winkler, U., Richter, M., Roetz, H., Eckl, S., and Mikolajick, T. (2015) Polycrystalline silicon gate originated CMOS device failure investigated by Scanning Spreading Resistance Microscopy. *Microelectron. Eng.*, **142**, 40–46.

46 Polspoel, W., Favia, P., Mody, J., Bender, H., and Vandervorst, W. (2009) Physical degradation of gate dielectrics induced by local electrical stress using conductive atomic force microscopy. *J. Appl. Phys.*, **106** (2), 24101.

47 Ji, Y., Pan, C., Zhang, M., Long, S., Lian, X., Miao, F., Hui, F., Shi, Y., Larcher, L., Wu, E., and Lanza, M. (2016) Boron nitride as two dimensional dielectric: Reliability and dielectric breakdown. *Appl. Phys. Lett.*, **108** (1), 12905.

48 Piquemal-Banci, M., Galceran, R., Caneva, S., Martin, M.-B., Weatherup, R.S., Kidambi, P.R., Bouzehouane, K., Xavier, S., Anane, A., Petroff, F., Fert, A., Robertson, J., Hofmann, S., Dlubak, B., and Seneor, P. (2016) Magnetic tunnel junctions with monolayer hexagonal boron nitride tunnel barriers. *Appl. Phys. Lett.*, **108** (10), 102404.

49 Weinberg, Z.A. (1982) On tunneling in metal-oxide-silicon structures, *J. Appl. Phys.*, **53**, 5052.

50 Cheong, K.Y., Moon, J.H., Kim, H.J., Bahng, W., and Kim, N.-K. (2008) Current conduction mechanisms in atomic-layer-deposited HfO[sub 2]/nitrided SiO[sub 2] stacked gate on 4H silicon carbide. *J. Appl. Phys.*, **103** (8), 84113.

51 Paskaleva, A., Bauer, A.J., Lemberger, M., and Zürcher, S. (2004) Different current conduction mechanisms through thin high-k Hf[sub x]Ti[sub y] Si[sub z]O films due to the varying Hf to Ti ratio. *J. Appl. Phys.*, **95** (10), 5583.

52 Celano, U., Chintala, R.C., Adelmann, C., Richard, O., and Vandervorst, W. (2013) The unexpected effects of crystallization on Ta2O5 as studied by HRTEM and C-AFM. *Microelectron. Eng.*, **109**, 318–321.

53 Martin, D., Grube, M., Reinig, P., Oberbeck, L., Heitmann, J., Weber, W.M., Mikolajick, T., and Riechert, H. (2011) Influence of composition and bottom electrode properties on the local conductivity of TiN/HfTiO[sub 2] and TiN/Ru/HfTiO[sub 2] stacks. *Appl. Phys. Lett.*, **98** (1), 12901.

54 Derry, G.N., Kern, M.E., and Worth, E.H. (2015) Recommended values of clean metal surface work functions. *J. Vac. Sci. Technol., A*, **33** (6), 60801.

55 Hori, T. (1997) *Gate Dielectrics and MOS ULSIs: Principles, Technologies, and Applications*, Springer, Berlin, NY.

56 Melitz, W., Shen, J., Kummel, A.C., and Lee, S. (2011) Kelvin probe force microscopy and its application. *Surf. Sci. Rep.*, **66** (1), 1–27.

57 Berthold, T., Benstetter, G., Frammelsberger, W., Rodríguez, R., and Nafría, M. (2015) Nanoscale characterization of CH3-terminated Self-Assembled Monolayer on copper by advanced scanning probe microscopy techniques. *Appl. Surf. Sci.*, **356**, 921–926.

58 Peterson, C.A., Workman, R.K., Sarid, D., Vermeire, B., Parks, H.G., Adderton, D., and Maivald, P. (1999) Effects of moisture on Fowler–Nordheim characterization of thin silicon-oxide films. *J. Vac. Sci. Technol., A*, **17** (5), 2753.

59 Kremmer, S., Peissl, S., Teichert, C., Kuchar, F., and Hofer, H. (2002), Conducting atomic-force microscopy investigations on thin silicon gate oxides: influence of Tip shape and humidity. Proc. 28th Int. Symposium for Testing and Failure Analysis 473–482.

60 Bloeß, H., Staikov, G., and Schultze, J.W. (2001) AFM induced formation of SiO2 structures in the electrochemical nanocell. *Electrochim. Acta*, **47** (1-2), 335–344.

61 Heun, S., Kremmer, S., Ercolani, D., Wurmbauer, H., and Teichert, C. (2005) LEEM and XPEEM studies of C-AFM induced surface modifications of thermally grown SiO2. *J. Electron. Spectrosc. Relat. Phenom.*, **144-147**, 1163–1166.

62 Conway, B.E. (1995) Electrochemical oxide film formation at noble metals as a surface-chemical process. *Prog. Surf. Sci.*, **49** (4), 331–452.

63 Loh, K., Xie, X., Lim, Y., Teo, E., Zheng, J., and Ando, T. (2002) Surface oxygenation studies on -oriented diamond using an atom beam source and local anodic oxidation. *Surf. Sci.*, **505**, 93–114.

64 Desbief, S., Hergué, N., Douhéret, O., Surin, M., Dubois, P., Geerts, Y., Lazzaroni, R., and Leclère, P. (2012) Nanoscale investigation of the electrical properties in semiconductor polymer-carbon nanotube hybrid materials. *Nanoscale*, **4** (8), 2705–2712.

65 Geydt, P., Alekseev, P.A., Dunaevskiy, M., Lähderanta, E., Haggrén, T., Kakko, J.-P., and Lipsanen, H. (2015) Observation of linear *I–V* curves on vertical GaAs nanowires with atomic force microscope. *J. Phys. Conf. Ser.*, **661**, 12031.

66 Drijkoningen, J., Kesters, J., Vangerven, T., Bourgeois, E., Lutsen, L., Vanderzande, D., Maes, W., D'Haen, J., and Manca, J. (2014) Investigating the role of efficiency enhancing interlayers for bulk heterojunction solar cells by scanning probe microscopy. *Org. Electron.*, **15** (6), 1282–1289.

67 Gutierrez, J., Mondragon, I., and Tercjak, A. (2014) Quantitative nanoelectrical and nanomechanical properties of nanostructured hybrid composites by PeakForce tunneling atomic force microscopy. *J. Phys. Chem. C*, **118** (2), 1206–1212.

68 Rosa-Zeiser, A., Weilandt, E., Hild, S., and Marti, O. (1997) The simultaneous measurement of elastic, electrostatic and adhesive properties by scanning force microscopy: pulsed-force mode operation. *Meas. Sci. Technol.*, **8** (11), 1333–1338.

69 Dokukin, M.E. and Sokolov, I. (2012) Quantitative mapping of the elastic modulus of soft materials with HarmoniX and PeakForce QNM AFM modes. *Langmuir*, **28** (46), 16060–16071.

70 Weber, S.A.L. and Berger, R. (2013) Electrical tip-sample contact in scanning conductive torsion mode. *Appl. Phys. Lett.*, **102** (16), 163105.

71 Song, Y. and Bhushan, B. (2008) Atomic force microscopy dynamic modes: modeling and applications. *J. Phys. Condens. Matter*, **20** (22), 225012.

72 Huang, L. and Su, C. (2004) A torsional resonance mode AFM for in-plane tip surface interactions. *Ultramicroscopy*, **100** (3-4), 277–285.

73 Vetushka, A., Itoh, T., Nakanishi, Y., Fejfar, A., Nonomura, S., Ledinský, M., and Kočka, J. (2012) Conductive atomic force microscopy on carbon nanowalls. *J. Non-Cryst. Solids*, **358** (17), 2545–2547.

74 Fan, H., Li, R., Yang, D., Wu, Y., Niu, J., Yang, Q., Zhao, J., and Liu, D. (2013) Observation of He bubbles in ion irradiated fusion materials by conductive atomic force microscopy. *J. Nucl. Mater.*, **441** (1-3), 54–58.

75 Wu, C.H., Alessandrini, C., Bonal, J.P., Davis, J.W., Haasz, A.A., Jacob, W., Kallenbach, A., Keinonen, J., Kornejew, P., Moormann, R., Philipps, V., Roth, J., Scaffidi-Argentina, F., and Würz, H. (2001) Progress of the European R&D on plasma–wall interactions, neutron effects and tritium removal in ITER plasma facing materials. *Fusion Eng. Des.*, **56-57**, 179–187.

76 Giancarli, L., Bonal, J., Caso, A., Le Marois, G., Morley, N., and Salavy, J. (1998) Design requirements for SiC/SiC composites structural material in fusion power reactor blankets. *Fusion Eng. Des.*, **41** (1-4), 165–171.

77 Bockstedte, M., Marini, A., Pankratov, O., and Rubio, A. (2010) Many-body effects in the excitation spectrum of a defect in SiC. *Phys. Rev. Lett.*, **105** (2), 26401.

78 Zhou, H.-B., Jin, S., Zhang, Y., Lu, G.-H., and Liu, F. (2012) Anisotropic strain enhanced hydrogen solubility in bcc metals: the independence on the sign of strain. *Phys. Rev. Lett.*, **109** (13), 135502.

79 Liu, Y., Wang, G., Wang, S., Yang, J., Chen, L., Qin, X., Song, B., Wang, B., and Chen, X. (2011) Defect-induced magnetism in neutron irradiated 6H-SiC single crystals. *Phys. Rev. Lett.*, **106** (8), 87205.

80 Nozawa, T., Hinoki, T., Hasegawa, A., Kohyama, A., Katoh, Y., Snead, L.L., Henager, C.H., and Hegeman, J. (2009) Recent advances and issues in development of silicon carbide composites for fusion applications. *J. Nucl. Mater.*, **386-388**, 622–627.

81 Zhang, L., Ohuchi, K., Adachi, K., Ishimaru, K., Takayanagi, M., and Nishiyama, A. (2007) High-resolution characterization of ultrashallow junctions by measuring in vacuum with scanning spreading resistance microscopy: SSRM iun Vacuum. *Appl. Phys. Lett.*, **90** (19), 192103.

82 Oliviero, E., Beaufort, M.F., Pailloux, F., and Barbot, J.F. (2004) Damage formation in high energy helium implanted 4H-SiC. *Nucl. Instrum. Methods Phys. Res., Sect. B*, **218**, 391–395.

83 Liu, D., Benstetter, G., and Frammelsberger, W. (2003) The effect of the surface layer of tetrahedral amorphous carbon films on their tribological and electron emission properties investigated by atomic force microscopy. *Appl. Phys. Lett.*, **82** (22), 3898.

84 Saxena, V., Nong Su, J., and Steckl, A.J. (1999) High-voltage Ni- and Pt-SiC Schottky diodes utilizing metal field plate termination. *IEEE Trans. Electron Devices*, **46** (3), 456–464.

85 Eriksson, J., Weng, M.H., Roccaforte, F., Giannazzo, F., Leone, S., and Raineri, V. (2009) Toward an ideal Schottky barrier on 3C-SiC. *Appl. Phys. Lett.*, **95** (8), 81907.

86 Zhang, Y. (2004) Effects of implantation temperature on damage accumulation in Al-implanted 4H–SiC. *J. Appl. Phys.*, **95** (8), 4012.

**87** Calcagno, L., Ruggiero, A., Roccaforte, F., and La Via, F. (2005) Effects of annealing temperature on the degree of inhomogeneity of nickel-silicide/SiC Schottky barrier. *J. Appl. Phys.*, **98** (2), 23713.

**88** Panda, K., Sundaravel, B., Panigrahi, B.K., Magudapathy, P., Nandagopala Krishna, D., Nair, K.G.M., Chen, H.-C., and Lin, I.-N. (2011) Structural and electronic properties of nitrogen ion implanted ultra nanocrystalline diamond surfaces. *J. Appl. Phys.*, **110** (4), 44304.

**89** Ni, W., Yang, Q., Fan, H., Liu, L., Berthold, T., Benstetter, G., and Liu, D. (2015) Ordered arrangement of irradiation-induced defects of polycrystalline tungsten irradiated with low-energy hydrogen ions. *J. Nucl. Mater.*, **464**, 216–220.

**90** Yang, Q., You, Y.-W., Liu, L., Fan, H., Ni, W., Liu, D., Liu, C.S., Benstetter, G., and Wang, Y. (2015) Nanostructured fuzz growth on tungsten under low-energy and high-flux He irradiation. *Sci. Rep.*, **5**, 10959.

**91** Hofer, A., Benstetter, G., Biberger, R., Leirer, C., and Brüderl, G. (2013) Analysis of crystal defects on GaN-based semiconductors with advanced scanning probe microscope techniques. *Thin Solid Films*, **544**, 139–143.

**92** Hsu, J.W.P., Manfra, M.J., Molnar, R.J., Heying, B., and Speck, J.S. (2002) Direct imaging of reverse-bias leakage through pure screw dislocations in GaN films grown by molecular beam epitaxy on GaN templates. *Appl. Phys. Lett.*, **81** (1), 79.

**93** Stoib, B., Langmann, T., Matich, S., Antesberger, T., Stein, N., Angst, S., Petermann, N., Schmechel, R., Schierning, G., Wolf, D.E., Wiggers, H., Stutzmann, M., and Brandt, M.S. (2012) Laser-sintered thin films of doped SiGe nanoparticles. *Appl. Phys. Lett.*, **100** (23), 231907.

**94** Hantschel, T., Demeulemeester, C., Eyben, P., Schulz, V., Richard, O., Bender, H., and Vandervorst, W. (2009) Conductive diamond tips with sub-nanometer electrical resolution for characterization of nanoelectronics device structures. *Phys. Status Solidi A*, **206** (9), 2077–2081.

**95** Schulze, A., Verhulst, A.S., Nazir, A., Hantschel, T., Eyben, P., and Vandervorst, W. (2013) A comprehensive model for the electrical nanocontact on germanium for scanning spreading resistance microscopy applications. *J. Appl. Phys.*, **113** (11), 114310.

# 4

# Investigation of High-*k* Dielectric Stacks by C-AFM: Advantages, Limitations, and Possible Applications

*Mathias Rommel*[1] *and Albena Paskaleva*[2]

[1] *Fraunhofer Institute for Integrated Systems and Device Technology IISB, Schottkystrasse 10, 91058 Erlangen, Germany*
[2] *Institute of Solid State Physics, Bulgarian Academy of Sciences, 72 Tzarigradsko Chaussee, 1784 Sofia, Bulgaria*

## 4.1 Introduction

Electrical properties of $SiO_2$ and high-*k* dielectric films are usually characterized by the use of metal–insulator–semiconductor (MIS) or metal–insulator–metal gate stacks using standard macroscopic current–voltage ($I$–$V$) and capacitance–voltage ($C$–$V$) techniques. However, macroscopic measurements provide electrical parameters averaged over the whole gate area, which hinders distinguishing local inhomogeneities as well as nanoscale phenomena. This is particularly important in the case of high-*k* dielectrics which have a much more complicated electronic and bonding structure compared to $SiO_2$. As a consequence, the defects and traps and all related phenomena as well as their specific structural, electrical, and dielectric behavior are also quite different. The large amount of traps in high-*k* materials defines the substantial difference from $SiO_2$ with regard to electrical conductivity and the dominating conduction mechanism(s) through the dielectric. Generally, high-*k* dielectrics are trap-rich materials and the conduction mechanisms are usually trap related. In addition, in contrast to thermally grown $SiO_2$, high-*k* materials may also show varying morphologies in each film depending on process conditions (i.e., amorphous and/or crystalline phases) and high surface roughness (which depends on film morphology, thickness, and processing parameters). A polycrystalline structure of high-*k* material introduces conductivity inhomogeneities at the nanoscale, which are reflected in the electrical behavior of the device. Therefore, local variations of physical and electrical characteristics at the nanoscale are common in high-*k* dielectrics and they require dedicated characterization methods with high lateral resolution. In addition, the continuous scaling pushes the lateral dimensions of a complementary metal–oxide–semiconductor transistor to a few tens of nanometers and below, which is another reason that necessitates a detailed investigation of the nanoscale current distribution and determination of the relevant physical mechanisms of current flow. Therefore, together with conventional characterization techniques, new advanced methods for both physical

*Conductive Atomic Force Microscopy: Applications in Nanomaterials*, First Edition.
Edited by Mario Lanza.

and electrical evaluation with extremely high lateral resolution are required for a better understanding and study of high-*k* materials. Among different local scanning probe microscopy (SPM) techniques for electrical characterization, conductive atomic force microscopy (C-AFM) and tunneling atomic force microscopy (TUNA) as well as scanning capacitance microscopy (SCM) have attracted significant interest and are the most commonly used methods that offer a wide variety of characterization possibilities for advanced thin dielectric layers and yield new insights into material and electrical properties [1–4] as they allow for a better understanding and a locally resolved studying of current conduction mechanisms on an area of only a few hundreds or even tens of square nanometers [5]. Both C-AFM and TUNA measure leakage currents and they are, in principle, the same technique. However, for some commercial equipment suppliers, current amplifiers with different current regimes are offered and the corresponding measurement mode is either called C-AFM or TUNA. The TUNA current amplifier is very sensitive to low currents (measurement range up to 100 pA), whereas the C-AFM module is developed for higher currents (up to about 500 nA).

C-AFM and TUNA have been demonstrated as indispensables techniques for advanced gate dielectric analysis to investigate local defects, breakdown (BD), and post-BD phenomena [6–9]. C-AFM/TUNA was successfully applied to study degradation phenomena and BD in thin $SiO_2$ films at the same dimensional scale at which they take place. The phenomenology observed with C-AFM provided experimental evidence that the dielectric BD in thin $SiO_2$ is an extremely local phenomenon occurring in an area of about $10^{-12}\,cm^2$ ($100\,nm^2$), which however propagates laterally to larger areas [8, 10, 11]. The possibility to measure simultaneously topography and conductivity allowed to associate the observed negative charge present in $SiO_2$ with the immediate structural damage caused in the oxide during degradation and BD [12]. Next, the technique has been applied for the characterization of a number of high-*k* dielectrics, addressing a number of additional issues related to their specific structural and electronic properties such as current uniformity and variability of electrical properties across the layer depending on morphology, thickness or post-deposition annealing [3, 13–20], and influence of the ultrathin interfacial $SiO_2$ layer [21]. Also, charge trapping in thin $SiO_2$ and high-*k* films has been successfully evaluated using C-AFM/TUNA [22–24]. Moreover, the leakage current conduction mechanisms for $SiO_2$, $ZrO_2$, and $HfO_2$ thin films could be determined [13, 22, 25]. Grain boundaries (GBs) in polycrystalline films have been identified as a major limitation in reliability as they are the primary location where a large number of defects and electrically active centers are concentrated, which on its turn facilitates the degradation and the BD in high-*k* stacks. Different degradation kinetics at GBs and in grains have been established [26–29].

In this chapter, the capability and the accuracy of TUNA and C-AFM by comparing measurement results with data from conventional macroscopic $I$–$V$ methods is demonstrated. It is shown that, in particular, C-AFM/TUNA characterization complements conventional $I$–$V$ measurements on MIS structures by accessing higher current densities. Several limitations (e.g., parasitic capacitances in the pF range resulting from the cantilever of the probe and the probe holder and displacement current) which influence sensitivity and may be a possible obstacle for correct analysis of experimental data are also addressed. The feasibility of these techniques and their ability to study phenomena at nanoscale is demonstrated by several applications. The great potential of C-AFM/TUNA technique to investigate changes in high-*k* film morphology is shown. An approach to evaluate the thickness of the thin interfacial $SiO_2$ layer in high-*k* stacks and its change with the processing conditions is presented. It is shown how the evolution of local C-AFM/TUNA *I*–*V* curves and their hysteresis can be used to obtain information about the degradation mechanisms and processes, which are forerunners of BD in the high-*k* gate stacks. The gate oxide integrity of different thin films (silicon dioxide, silicon nitride, and hafnium oxide) is also addressed by performing constant voltage stress (CVS) at the nanoscale using C-AFM. Time to BD extracted from conventional current–voltage measurements and from C-AFM are compared.

## 4.2 Comparison Between Macroscopic *I–V* Measurements and C-AFM

There are many studies which evidence the applicability of C-AFM for electrical characterization of thin insulating films and correlate the results at a nanoscale with the data from standard *I*–*V* measurements on macroscopic MIS structures [6, 7, 30–32]. For example, in Figure 4.1a, the difference in the macroscopic *I*–*V* characteristics of metal-insulator-metal (MIM) capacitors with differently deposited $TiO_2$ layers is clearly visible. Sample $TiO_2$(TTIP) shows smaller leakage current density than sample $TiO_2(TiCl_4)$ does. The C-AFM leakage current distribution maps (Figure 4.1b) reveal that the overall leakage currents

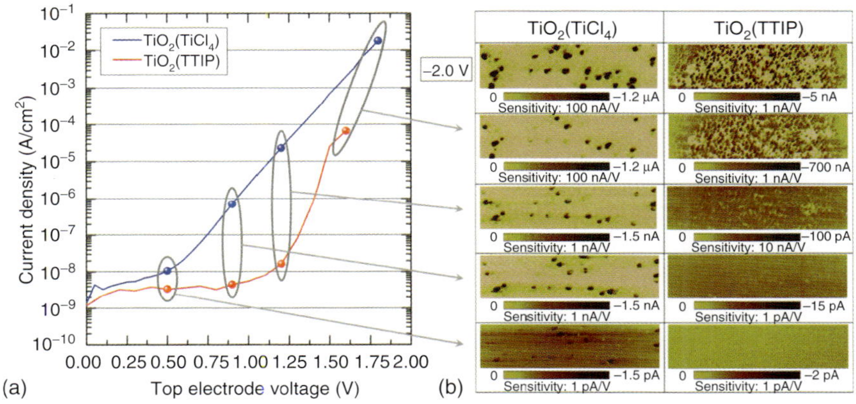


**Figure 4.1** Comparison of macroscopic and nanoscale current characteristics. (a) Macroscopic current density voltage curves for $TiO_2$ MIM structures using a $TiCl_4$ and a TTIP (Ti-tetra-isopropoxide) precursor, respectively. The symbols indicate the points of the curves where C-AFM maps with the same applied bias are shown in (b). (b) Leakage current distribution maps of samples $TiO_2(TiCl_4)$ and $TiO_2$(TTIP) with various applied bias voltages (in the top row, maps are shown for a voltage of 2.0 V). Note the different current scales. The scan area of all maps is $2 \times 0.5\ \mu m^2$ [30].

for sample $TiO_2(TiCl_4)$ are significantly larger than those for sample $TiO_2$(TTIP) with the leakage currents through the leaky grains of the former sample being larger by up to more than three orders of magnitude compared to the latter sample. Thus, the nanoscale leakage current characteristics show very good qualitative correlation with the macroscopic leakage current characteristics. In this section, the capability and the accuracy of C-AFM is demonstrated by performing a dedicated study that compares the C-AFM measurement results with data from conventional macroscopic $I$–$V$ methods using two approaches. First, $I$–$V$ measurements were performed on MIS capacitor structures, that is, the macroscopic manipulator probe (used in the conventional measurements) and the conductive AFM tip (used in the C-AFM experiments) were directly in contact with the gate electrode of an MIS capacitor. Thus, a one-to-one comparison of the results was possible. Secondly, standard C-AFM $I$–$V$-spectroscopy was performed directly on the insulator films (i.e., the AFM tip acted as the gate electrode) and the results were compared with data from macroscopic measurements via the corresponding current densities. These experiments were performed for both $SiO_2$ and high-$k$ films. In addition, it should be noted that for p-type silicon substrates, all C-AFM measurements with the tip directly on the insulator were carried out in inversion due to anodic oxidation occurring for C-AFM measurements performed in accumulation [33].

C-AFM $I$–$V$-spectroscopy was realized either using a Veeco Dimension 5000 SPM system with TUNA (higher sensitivity) or C-AFM (higher currents) modules, respectively, for both voltage directions [32] or by using a Bruker Icon system with a so-called extended TUNA module [31]. All electrical AFM measurements were performed with Pt/Ir-coated probes. The work function difference between Pt/Ir and the metallization of the conventional MIS structures is taken into account in the evaluation of the experiments. For such probes, the effective contact area between the tip and the sample surface can vary between 20 and 800 $nm^2$ [34]. It should also be noted that during the C-AFM experiments, the sample surface was always illuminated by the focused laser beam of the AFM laser head (system). Therefore, conventional $I$–$V$ measurements were also performed under illumination.

For $SiO_2$ samples, an excellent agreement between C-AFM $I$–$V$ s and conventional $I$–$V$ s was observed in accumulation (i.e., for positive substrate bias, see Figure 4.2). In inversion, a small shift of about 0.3 V between the conventional $I$–$V$ curves, on the one hand, and the C-AFM $I$–$V$ curves, on the other, can be observed. It should be noted that in inversion with C-AFM, much higher currents can be recorded than with conventional measurements where saturation at about 10 nA is observed. It is assumed that this effect occurs owing to the strongly localized laser illumination of the sample during C-AFM measurements. Therefore, a massive excess of electron–hole pairs is generated around the small metal contacts which locally increase the minority carrier concentration thus allowing for currents higher than 10 nA. In the conventional setup, however, the lack of minority carrier generation limits the tunneling current at high voltages, even though a microscope illumination consisting of a strong halogen lamp was used during those measurements. These results evidence the comparability of the $I$–$V$ characteristics obtained with the applied techniques on macroscopic MIS

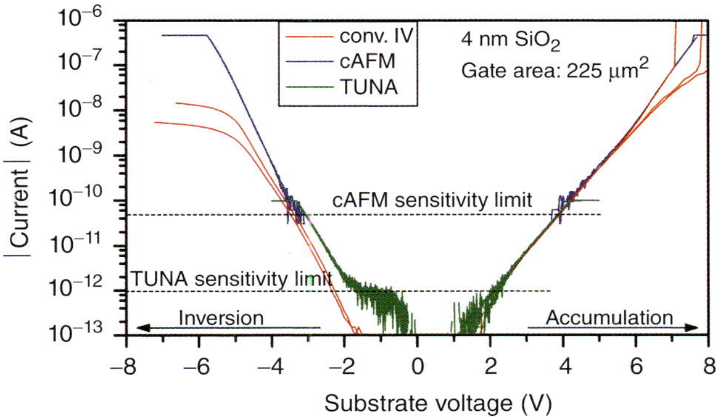


**Figure 4.2** Comparison of conventional *I–V* characteristics with *I–V* curves measured using C-AFM on samples with $SiO_2$-gate dielectric on p-type silicon; all measurements were performed on MIS capacitor structures. (Reprinted from [32], Copyright 2009, with permission from Elsevier.)

capacitor structures for the $SiO_2$ samples. Thus, AFM-based techniques could allow for probing of simple test structures with very small gate contacts.

Next, TUNA *I–V* curves measured with the tip directly in contact with $SiO_2$ surfaces (standard TUNA measurements) were compared with the results obtained on the MIS structures. Therefore, current densities versus voltage (*J–V*) characteristics were compared for all measurements (Figure 4.3a). Standard TUNA measurements directly on the dielectric layer reveal much higher current densities than measurements on MIS test structures and therefore complement conventional *I–V* techniques. The measured standard TUNA curves quantitatively agree quite well with *J–V* curves from measurements on test structures extrapolated to current densities in the range of 1–10 A/cm$^2$ (extrapolation

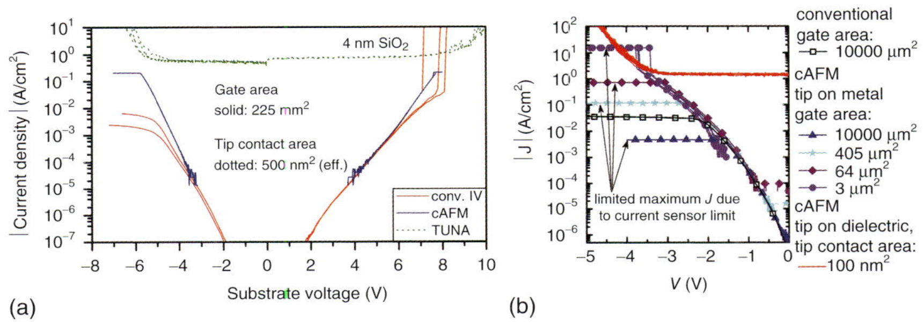


**Figure 4.3** (a) Comparison of *J–V* characteristics as obtained from standard C-AFM *I–V* measurements (tip in direct contact with the $SiO_2$ film) and characteristics from conventional and C-AFM measurements on MIS test structures from Figure 4.2. (Reprinted from [32], Copyright 2009, with permission from Elsevier). (b) Comparison of *J–V* characteristics for a 4 nm $SiO_2$ layer on p-type silicon with MIS capacitor test structures of different gate areas. Conventional: conventional probe station and semiconductor parameter analyzer; C-AFM tip on metal: C-AFM tip on metal gate and C-AFM current amplifier; C-AFM tip on dielectric: C-AFM tip on dielectric and C-AFM current amplifier. (Reproduced with permission from [31]. Copyright 2013, American Vacuum Society.)

not shown). This demonstrates the applicability and the accuracy of the TUNA techniques for measurements at the nanoscale.

The aforementioned experiments were repeated by extending the range of metal gate areas to rather small areas (i.e., down to 3 μm$^2$) to enable an even better correlation of C-AFM and conventional *I–V* measurements [31]. As can be seen from Figure 4.3b, the coincidence of C-AFM *J–V* and *J–V* curves from conventional MIS structures is directly visible for the range 1–100 A/cm$^2$ (here, assuming a contact area of approximately 100 nm$^2$ as a new probe was used which was also inspected in an SEM earlier).

Similar measurements for $HfO_2$ (high-*k*) layers on $SiO_2$ were also performed and evaluated. Again, from the *J–V* plots shown in Figure 4.4, a good agreement between conventional *J–V* results and C-AFM curves obtained on MIS structures is apparent. In addition, a theoretical calculation of direct and Fowler-Nordheim tunneling currents is also depicted in Figure 4.4. In inversion, tunneling of electrons from the substrate through the $SiO_2$ layer is the dominant charge transport mechanism and the calculation coincides well with measurement data. In accumulation, Fowler-Nordheim tunneling of electrons from the gate electrode is strongly suppressed by the low electric field in the high-*k* layer, and the source of the leakage current must be different. The calculated current densities in accumulation are due to Fowler-Nordheim tunneling of holes from the substrate [35]. Additionally, the high trap density in hafnium-based oxides gives rise to field-enhanced thermal emission of electrons (Poole-Frenkel conduction) [35].

Again, C-AFM measurement results with the tip directly in contact with $HfO_2$ surfaces were compared with the data obtained on the MIS structures (also shown in Figure 4.4). As observed for $SiO_2$ films, standard TUNA

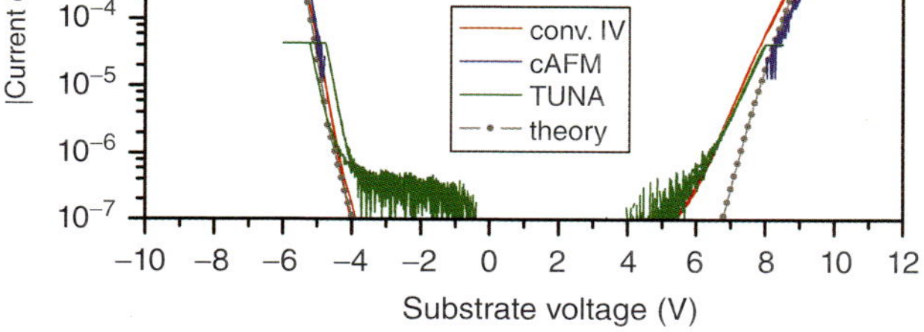


**Figure 4.4** Comparison of conventional *J–V* characteristics with *J–V* curves measured using C-AFM on samples with a $HfO_2/SiO_2$-gate dielectric; solid lines represent measurements performed on MIS capacitor structures, dotted lines represent measurements with tip in direct contact with the $HfO_2$ film. (Reprinted from [32], Copyright 2009, with permission from Elsevier.)

measurements address much higher current densities. Also, in inversion, the TUNA measurements quantitatively agree with the extrapolated *J*–*V* curves (extrapolation not shown) from measurements on test structures (although a deviation from the theoretical curve can be observed for both measurements at high current densities).

## 4.3 Influence of Displacement Currents on the Sensitivity of C-AFM Measurements

As already discussed, C-AFM allows for the localized evaluation of leakage currents (e.g., to compare between grains and GBs of high-*k* films), which is especially important to understand the nature of tunneling or leakage currents in crystalline or polycrystalline high-*k* layers and, thus the local current conduction mechanisms. However, the sensitivity of acquired *I*–*V* curves by C-AFM is usually limited by relatively high parasitic currents especially for small bias voltages (i.e., minimum current densities (*J*) of C-AFM characteristics which can surely be attributed to tunneling currents through the dielectric are very high, that is, in the range of 1 A/cm$^2$ and above, see Figure 4.3b [31] and Yanev *et al.* [32]). In order to address considerably smaller current densities which enable a simpler and more direct comparison between results obtained by microscopic or macroscopic test structures and C-AFM, the sensitivity of C-AFM has to be improved significantly. One important step toward this goal is to properly identify the most relevant effects or current contributions which limit the sensitivity of today's equipment [31]. In contrast to conventional *I*–*V* acquisition, local *I*–*V* measurements by C-AFM are performed applying a voltage sweep with a user-defined voltage sweep rate *sr* (which is d*V*/d*t*). Typically, a forward and a backward sweep are performed within one measurement sequence where, for both directions, different measurement parameters can be defined. Here, always the same (absolute value of) *sr* is used for both directions. Thus, only the sign of *sr* changes for forward and backward direction (i.e., d*V*/d*t* is negative in forward direction (e.g., from 0 to −10 V) and positive in the negative direction). Then, the overall current $I_{total}$ is composed (see Eq. (4.1)) of the tunneling current $I_{tunnel}$ through the dielectric (i.e., the current of interest), the displacement current $I_{disp}$ in case of any effective capacitance $C_{eff}$ to be charged according to Eq. (4.2), the offset current $I_{offset}$ due to the measurement setup, any noise as given by $I_{noise}$ and additional currents $I_{trans}$ due to transient effects which, for example, account for the response of the circuitry on abrupt changes in current (i.e., at the beginning of both, forward and backward direction). The effective overall capacitance $C_{eff}$ is simply expressed as the sum of the capacitance $C_{tip}$ of the tip/dielectric/semiconductor system and any other additional parasitic capacitances summarized as $C_{par}$ in Eq. (4.2).

$$I_{total} = I_{tunnel} + I_{disp} + I_{offset} + I_{noise} + I_{trans} \tag{4.1}$$

$$I_{disp} = C_{eff}\frac{dV}{dt} = (C_{tip} + C_{par})\frac{dV}{dt} \tag{4.2}$$

Here, the influence of the parasitic capacitance $C_{par}$ due to both the cantilever of the C-AFM probe as well as the probe holder are investigated.

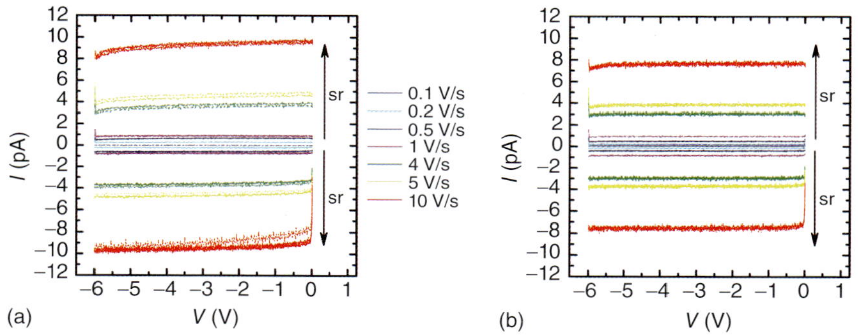


**Figure 4.5** Local C-AFM *I–V* curves at the edge of a thermally oxidized silicon sample ($SiO_2$ thickness is about 5 nm) using (a) standard probes and (b) shielded coplanar probes with different sweep rates *sr* (as indicated in the figure); negative (positive) currents were measured in forward (backward) direction with, for example, forward direction from 0 to −6 V. (Reproduced with permission from [31]. Copyright 2013, American Vacuum Society.)

If not stated otherwise, Pt/Ir Electrostatic force microscopy (EFM) probes (i.e., NANOSENSORS™ PPP-EFM100, in the following referred to as *standard probes*) on a standard shielded holder (i.e., Bruker DTRCH-AM holder) were used together with a so-called extended TUNA module together with a Bruker Icon tool. For SCM, the effect of $C_{par}$ has been intensively studied. However, for C-AFM this is not the case. From SCM studies, it is known that $C_{par}$ is typically in the pF range [36, 37], whereas the relevant tip-dielectric-semiconductor capacitance $C_{tip}$ is extremely small, that is, in the aF range [38]. Thus, when C-AFM *I–V* curves are obtained by applying a voltage sweep with a sweep rate $sr = dV/dt$, $C_{par}$ might contribute significantly to the sensitivity limit of C-AFM measurements (see Figure 4.5). This becomes obvious from the linear dependencies of the current levels from Figure 4.5 on *sr* (see Figure 4.6) which result in an overall effective capacitance of around 0.75–0.95 pF which is in good agreement with values of $C_{par}$ as known from SCM studies. It should be noted that the measurements shown in Figure 4.5 were performed such as to avoid significant leakage current through the dielectric. In addition, $C_{par}$ slightly increases when the measurement is performed with the probe support chip above the sample, which is the standard use case (see Figure 4.7 [31]). In that case, however, transient effects ($I_{trans}$) also influence the overall current for conventional probes.

Obviously, one option to decrease the displacement current is to use shielded probes which directly reduce $C_{par}$ (see Figures 4.5 and 4.7 [31]). However, such probes are rarely commercially available yet (besides, only for dedicated AFM systems) and rather expensive. Thus, they would lead to much higher costs for C-AFM measurements. Another option would be to apply very small sweep rates *sr* as this would directly decrease the displacement current $I_{disp}$. However, low sweep rates result in enhanced electrical stress and nonreproducible *I–V* characteristics when addressing the current regime of interest, that is, the current regime which is dominated by the leakage or tunneling current through the dielectric (see Figure 4.8).

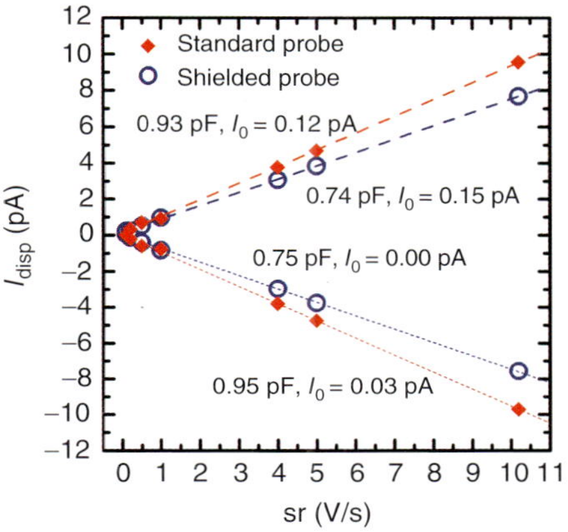


**Figure 4.6** $I_{disp}$ in dependence on *sr*. Symbols: evaluated data from measurements shown in Figure 4.5; lines: corresponding regression fits with the resulting effective capacitance and offset current $I_0$, respectively. (Reproduced with permission from [31]. Copyright 2013, American Vacuum Society.)

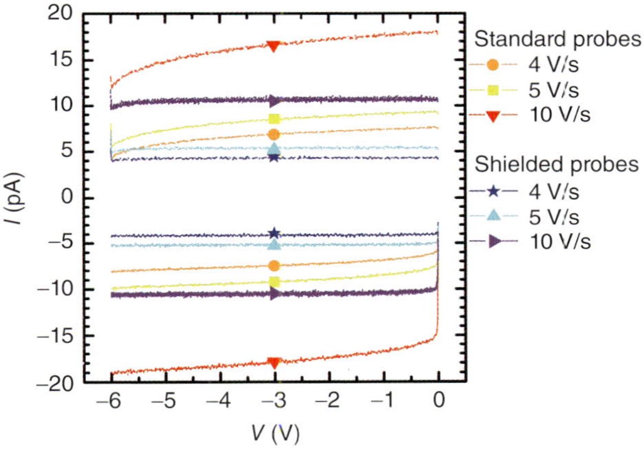


**Figure 4.7** Local C-AFM *I–V* curves at the center of the same sample as used for the measurements shown in Figure 4.5 using different sweep rates *sr* (as indicated in the figure) and both standard probes and shielded coplanar probes. Symbols are only used for indication purposes. (Reproduced with permission from [31]. Copyright 2013, American Vacuum Society.)

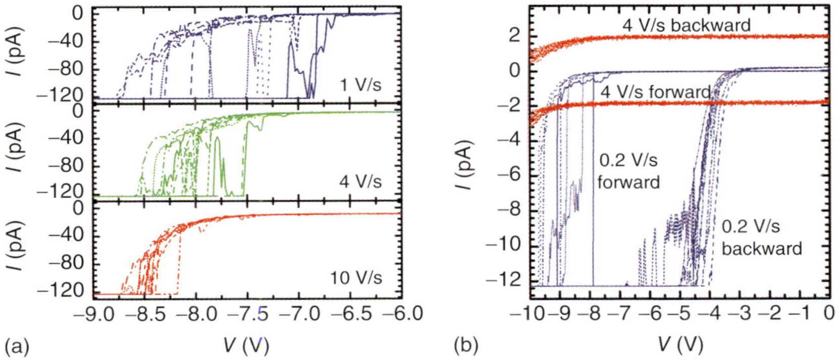


**Figure 4.8** (a) Forward *I–V* characteristics of the same thermally oxidized silicon sample as that in Figure 4.5 for different sweep rates as indicated in the figures (always at least eight representative characteristics are shown) using shielded probes. (b) *I–V* characteristics of a sample with a 10 nm $HfO_2$ ALD layer for 0.2 and 4 V/s using shielded probes. (Reproduced with permission from [31]. Copyright 2013, American Vacuum Society.)

Here, therefore, another approach is presented which aims at correcting measured data for the displacement current by a two-step measurement procedure and a simple data correction procedure [31]. If transient effects are negligible (i.e., $I_{trans} \approx 0$ or at least $I_{trans}$ significantly smaller than all other current contributions of Eq. (4.1), which for the experiments discussed in this work is the case for $sr \leq 1$ V/s using standard probes and $sr \leq 10$ V/s using shielded probes) and tunneling currents are significantly smaller than displacement and offset currents (see Eq. (4.1)), a simple data evaluation procedure can be proposed to correct for the displacement current. The procedure is based on the change of sign of the displacement current for forward and backward directions, respectively. Thus, simply adding forward and backward currents for each voltage in an appropriate voltage regime cancels out $I_{disp}$. If the resulting summed currents are constant in that voltage regime (neglecting any unavoidable noise), the data can be considered as $2 \cdot I_{offset}$. In that case, next to correcting for $I_{disp}$, even the offset current $I_{offset}$ of the overall setup can be corrected for. As $I_{offset}$ – from our experience – is not a constant, which can be determined once and then can be accounted for the given measurement system but rather changes with time and measurement range, such an approach would additionally improve reproducibility and sensitivity of C-AFM $I$–$V$ measurements. For such data, a voltage regime exists where both forward and backward directions exhibit constant currents (e.g., between −1 and −5 V for all $sr$ in Figure 4.5b), which is a requirement for the proposed procedure to be applicable. However, if the $I$–$V$ measurement in forward direction already leads to a significant tunneling current through the dielectric or even a local dielectric BD, the backward direction measurement often will be strongly distorted (e.g., [28]) and the computation procedure cannot be applied. Therefore, it is proposed to perform a forward and backward direction measurements with moderate voltage range (so that no significant tunneling current through the dielectric is observed, for example, up to −5 V for the experimental data shown in Figure 4.9a) first and subsequently perform the measurement at the same position with the voltage range of interest (e.g., up to

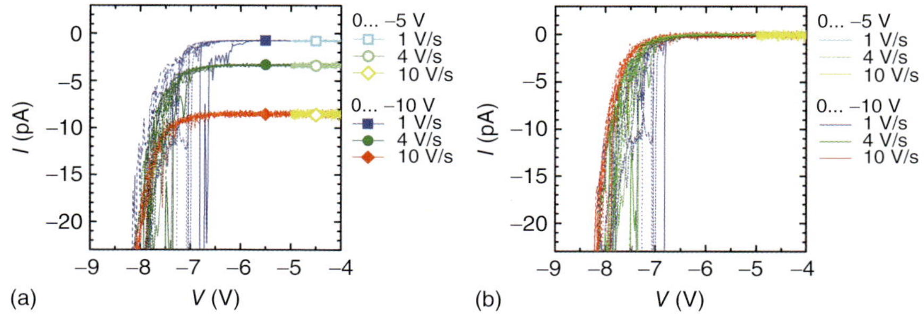


**Figure 4.9** (a) Forward $I$–$V$ characteristics of the same thermally oxidized silicon sample as that in Figure 4.5 for different sweep rates and voltage ranges as indicated using shielded probes (the data for voltages up to −10 V is the same as shown in Figure 4.8a). Symbols are only used for indication purposes. (b) Forward $I$–$V$ characteristics shown in (a) but corrected for both $I_{disp}$ and $I_{offset}$ using the proposed method described in the text. (Reproduced with permission from [31]. Copyright 2013, American Vacuum Society.)

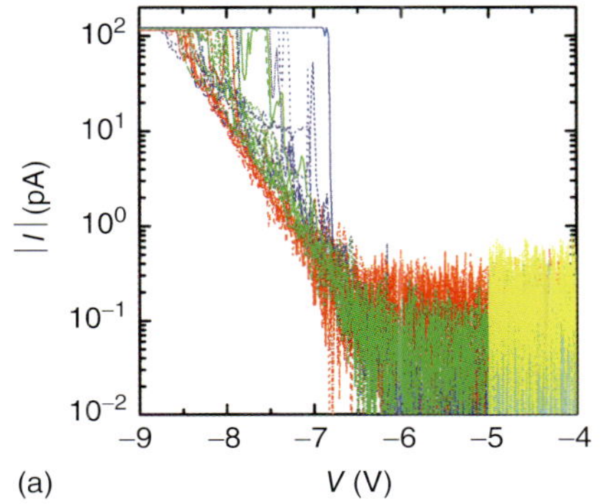


**Figure 4.10** (a) Forward *I–V* characteristics as in Figure 4.9b but presented in logarithmic current scale; (b) same data as in (a) but with reduced noise by applying a simple moving-average filter using 25 data points (Reproduced with permission from [31]. Copyright 2013, American Vacuum Society.)

−10 V for the experimental data shown in Figure 4.9a; note that for the sake of clarity, only forward directions are presented in Figures 4.9 and 4.10). From the first sweep at moderate voltages, $I_{disp}$ and potentially $I_{offset}$ can be determined so that the final $I$–$V$ curves can be properly corrected (see Figure 4.9b). Obviously, the suggested correction procedure significantly decreases the minimum leakage current density which can be detected. Following correction of the $I$–$V$ curves, however, the sensitivity is mainly dominated by noise (i.e., $I_{noise}$, see Figure 4.10a). From Fourier transformation analysis, it can be observed that the noise has peaks for 50 and 150 Hz, respectively, which should be avoidable by proper system design by the tool supplier. Here, the noise was decreased again by simply applying a moving average filter which results in a noise level below 100 fA (see Figure 4.10b).

Summarizing, even without improved hardware solutions C-AFM measurement sensitivity can significantly be improved by a two-step measurement procedure followed by an appropriate simple numerical data correction procedure. However, it should be mentioned that only for extremely thin dielectric layers, it might not be possible to find a voltage regime with a nearly constant current at all. However, in such cases, $I_{tunnel}$ dominates the overall current for the whole voltage regime and the correction is not required, anyway.

## 4.4 Applications of C-AFM

### 4.4.1 Morphology of Thin Dielectric Films

In this section, the great potential of the C-AFM/TUNA technique to investigate high-$k$ film morphology and its changes is demonstrated [30, 39]. The possibility to measure the leakage or tunneling current through the film with extremely high sensitivity simultaneously with the surface topography makes TUNA an indispensable tool to correlate morphology to electrical conduction and

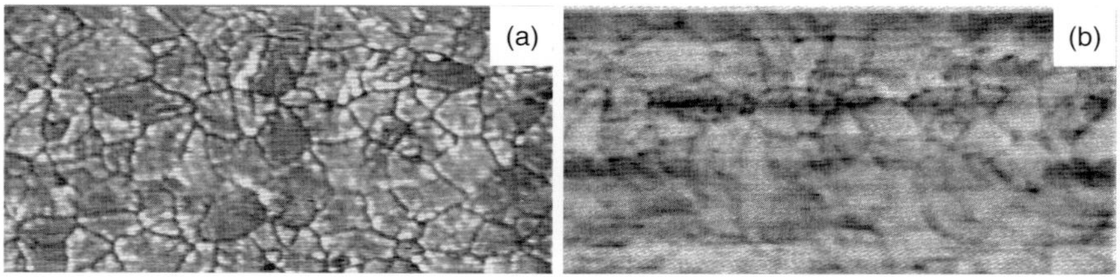


**Figure 4.11** (a) TUNA topography map (rms roughness of about 0.5 nm) and (b) the corresponding TUNA current map of nanocrystalline $HfSi_xO_y$ samples, scan size: $2 \times 1\ \mu m^2$, diamond-coated silicon tip, applied bias voltage: −8.2 V, topography scale: 0 nm (dark) −7 nm (bright), current scale: 0 pA (bright) to −5 pA (dark). (Reproduced with permission from [39]. Copyright American Institute of Physics 2008.)

to observe diverse phenomena and interesting relations between them. For example, a higher leakage current through GBs is widely suggested to be the main reason for the increased conductivity of crystalline films compared to their amorphous counterparts. C-AFM is the first technique to give experimental evidence for this phenomenon. In Figure 4.11, the topography and the corresponding current map of a $HfSi_xO_y$ layer annealed at 1000 °C are shown [39]. Crystalline grains can clearly be observed in the topography map. In the TUNA current map (obtained using diamond-coated Si probes), current structures (dark color) are visible which indicate larger current densities. From a detailed correlation of both maps, it can be concluded that most of the current structures in the TUNA current map correspond to GBs determined in the topography (i.e., higher leakage currents actually flow through the GBs). It should be noted that in this case $HfSi_xO_y$ grains and GBs, respectively, are clearly visible in both the topography and the TUNA current map.

By using TUNA, it is possible to differentiate between amorphous and nanocrystalline high-$k$ films even when the topography imaging fails to detect it. Several samples with $ZrO_2$ high-$k$ layers obtained by atomic layer deposition on native oxide and subjected to rapid thermal annealing (RTA) at various annealing conditions (450 °C for 30 s, 1 min, and 5 min, and 600 °C for 30 s and 1 min) were characterized by simultaneous TUNA current and topography mapping, using Pt/Ir-coated conductive probes (tip radius of about 25 nm). The TUNA topography maps of all $ZrO_2$ samples (as-deposited as well as RTA samples) show smooth surfaces (rms roughness values of about 0.2 nm) and no changes in the surface morphology after annealing. No significant structures, patterns, or crystalline grains could be observed for all topography maps. This was verified also by high-resolution AFM topography measurements in tapping mode using standard silicon probes (tip radius of about 10 nm). Figure 4.12 represents the topography of the sample annealed at 450 °C for 30 s [39]. The TUNA current maps, however, reveal significant differences (see Figure 4.13, where TUNA current maps for the as-deposited and two differently annealed $ZrO_2$ samples are shown). Relatively low voltages (−2.7 V) were applied during TUNA mapping

**Figure 4.12** Topography map of the $ZrO_2$ sample annealed at 450 °C for 30 s, AFM operated in tapping mode, scan size: $500 \times 500\ nm^2$; rms roughness: 0.2 nm. (Reproduced with permission from [39]. Copyright American Institute of Physics 2008.)

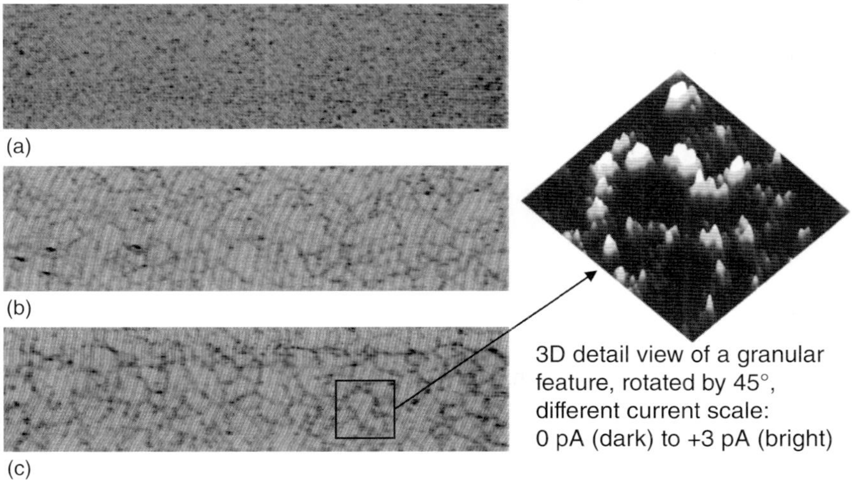


**Figure 4.13** TUNA current maps of $ZrO_2$ samples: (a) as-deposited, (b) annealed at 450 °C for 30 s, and (c) annealed at 450 °C for 1 min, scan size: $2 \times 0.5\ \mu m^2$, Pt/Ir coated silicon tip, applied bias voltage: −2.7 V, current scale: 0 pA (bright) to −5 pA (dark). (Reproduced with permission from [39]. Copyright American Institute of Physics 2008.)

in order to enhance current contrast between regions with different conductivities and to avoid electrical damage during the measurement. As-deposited samples only exhibit randomly distributed spots with higher leakage currents (Figure 4.13a). Clear ring or grain-shaped current structures with an increase in leakage current by a factor of approximately 5 compared to the surrounding areas could be observed for all annealed samples (Figure 4.13b,c) and the origin of the paths with higher conductivity is assigned to a larger conduction at the GBs [39].

Cross-section high resolution TEM (HR-TEM) analysis of samples confirms that $ZrO_2$ films become partially or fully nanocrystalline after the applied RTA conditions, whereas as-deposited samples are amorphous (Figure 4.14). It is also seen that the $ZrO_2$ surface and the underlying silicon surface of the samples (before and after RTA) are very smooth, which confirms the results from the AFM topography measurements. The crystallization of $ZrO_2$ layers during RTA does not increase surface roughness. Hence, grain structures could not be detected topographically even by AFM operated in tapping mode. All these facts provide evidence that larger leakage current densities emerge at the GBs of the $ZrO_2$ crystallites. For the as-deposited amorphous $ZrO_2$ films, however, only weak current spots are detected in the TUNA current maps. As the tunneling current is exponentially dependent on the thickness, the reason for the increased leakage current could be a smaller thickness at the GBs. However, the nanocrystalline $ZrO_2$ films and the underlying silicon show smooth surfaces, which implies that the main reason for the higher leakage currents is probably a different conduction mechanism in the GBs and the grains and not a variation in layer thickness. The difference in conduction mechanisms could be related to existence of increased densities of defects and electrically active centers at GBs. Increased leakage currents through the GBs have been also observed by C-AFM in other materials (e.g., $HfO_2$, $Al_2O_3$, $SrTiO_3$) [40–43].

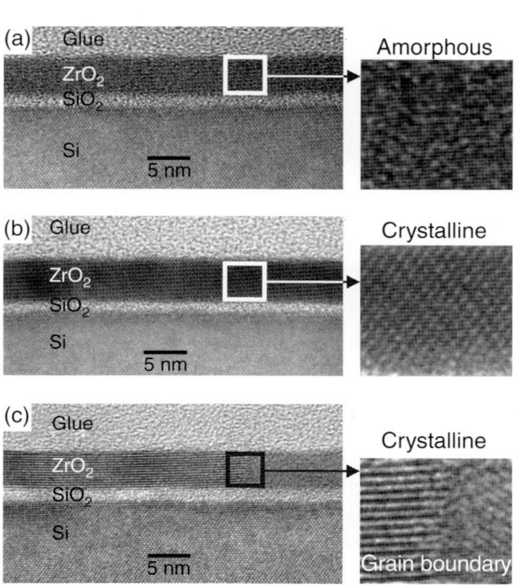


**Figure 4.14** HR-TEM images of: (a) as-deposited, (b) RTA at 450 °C, 30 s, and (c) RTA at 600 °C, 1 min samples. The images clearly show an increase in $SiO_2$ interfacial layer thickness and crystallization of $ZrO_2$ as a result of RTA. (Reproduced with permission from [22]. Copyright American Institute of Physics 2008.)

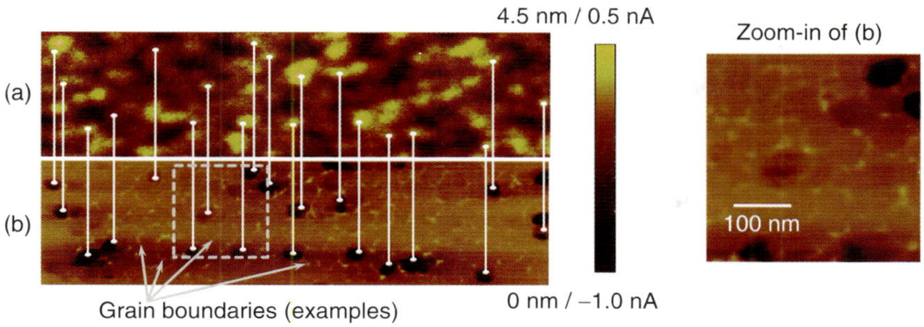


**Figure 4.15** Correlation between (a) topography and (b) leakage current distribution (applied bias voltage of −0.9 V, measured with 1 nA/V sensitivity) of $TiO_2(TiCl_4)$ sample. Scan area is $2 \times 0.5\ \mu m^2$. On the right, a zoom-in image of the leakage current map is presented (as indicated by the dashed frame in the original map). Note that the current scale in (b) includes positive currents to improve visibility. Actual currents do not exceed 0 A. The vertical lines point to corresponding areas in the topography and current maps [30].

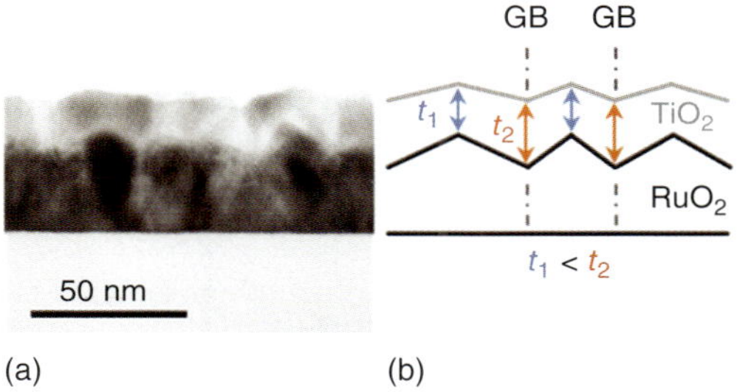


**Figure 4.16** (a) TEM cross-section image of sample $TiO_2(TiCl_4)$ on the polycrystalline $RuO_2$ layer grown by MOCVD. (b) Explanatory sketch: the $TiO_2$ is locally thicker over the $RuO_2$ grain boundaries and thinner over the $RuO_2$ grain tips with respect to the smoothening effect [30].

While the above studies demonstrate the increased leakage through the GBs there are also investigations in which the increased leakage currents through the bulk of grains have been observed [44, 45]. This is demonstrated in Figure 4.15 for $TiO_2$ films deposited on $RuO_2$, where the correlation between topography and leakage current distribution (Pt/Ir-coated tips are used here) is shown [30]. An increased leakage current mainly through the bulk of elevated $TiO_2$ grains and strongly reduced leakage current at the GBs can be clearly distinguished.

The transmission electron microscopy (TEM) image (Figure 4.16) reveals that the thickness of the $TiO_2$ layer is larger above $RuO_2$ GBs and smaller above $RuO_2$ grain tips. Higher conduction through grains rather than GBs can, thus, be partly explained by field enhancement effects due to smaller dielectric thickness, which is caused by the rough $RuO_2$ surface. The investigation of different types of $TiO_2$ layers deposited on identical $RuO_2$ layers [30] reveals that the grain sizes as well as grain orientations for various $TiO_2$ layers are different. Leakage currents are also quite different (Figure 4.1), which could be explained taking into consideration the fact that the tunneling current depends strongly not only on the thickness of the layer but also on its $k$-value, which, in turn, is closely related to the morphology of the film.

### 4.4.2 Assessment of the Interfacial $SiO_2$ Thickness

The C-AFM technique is also a powerful tool to correlate electrical with physical characteristics of the structures [22]. C-AFM has been used to determine the electrical oxide thickness of thin and ultrathin $SiO_2$ layers by applying a semi-analytical tunneling current model in both the direct and the Fowler–Nordheim regime regions of the $I$–$V$ characteristics [2, 46]. Here it is shown that by using C-AFM, it is possible to derive some physical properties not only of a single $SiO_2$ layer but also of high-$k$/$SiO_x$ stacks and, in particular, the thickness of the thin $SiO_x$ interfacial layer [22]. By performing a detailed analysis of the leakage current mechanisms, valuable qualitative (and, to some extent, even quantitative) information about the changes in the physical properties of the layers as a result of the annealing process could be obtained. The samples analyzed are the $ZrO_2$ layers subjected to RTA at various annealing conditions, which have already been discussed in Section 4.4.1.

In Figure 4.17a, the TUNA $I$–$V$ curves of different samples are shown. In these measurements highly p-type doped-diamond-coated tips were used. These tips are very suitable to measure electrical properties of dielectric films because of their large resistance to mechanical wear-out and chemical stability against oxidation. Although there are slight differences in the current level and the curve shape between the samples (Figure 4.17), this plot does not allow distinguishing a well-defined dependence on the annealing parameters.

As the analyzed layers are relatively thin, it is suggested that tunneling mechanisms govern the current through the layer stack. Therefore, $I$–$V$ curves are represented in Fowler–Nordheim coordinates (i.e., $\ln(I/V_g^{\,2})$ vs $1/V_g$ as shown in the inset of Figure 4.17a). First, a single-layer structure is assumed with the whole external applied voltage dropping across the dielectric. This was only done to roughly estimate from the slope of the curves whether a single conduction mechanism is operative in the measured voltage range or not. However, the existence of two regions with different slopes implies two different conduction mechanisms. Depending on the sample, the transition between the two mechanisms is observed in the range of 4.6-4.8 V. It was suggested [13, 25] that at lower fields, the tunneling is governed by the whole stack, whereas at higher fields, it is controlled by the thin $SiO_2$ interfacial layer (the two mechanisms are represented schematically in Figure 4.17a). Therefore, the current in the high voltage region (i.e., $V_g$ higher than 4.6-4.8 V) obeys the Fowler–Nordheim equation given by [47].

$$I = AV_s^2 \exp\left(-\frac{B_s d_s}{V_s}\right) \quad \text{with } B_s = \frac{8\pi}{3h}(2qm^*)^{\frac{1}{2}}\phi_b^{\frac{3}{2}} \tag{4.3}$$

where $V_s$ and $d_s$ are the voltage drop and thickness of $SiO_2$, $m^*$ is the effective electron mass in $SiO_2$, and $\phi_b$ is the barrier height at the Si/$SiO_2$ interface. The voltage drop across each layer of the double layer dielectric could be obtained from basic physical considerations:

$$V_z = \frac{V_g - V_{\text{off}}}{\frac{\varepsilon_z}{\varepsilon_s}\frac{d_s}{d_z} + 1} \quad \text{and} \quad V_s = \frac{V_g - V_{\text{off}}}{\frac{\varepsilon_s}{\varepsilon_z}\frac{d_z}{d_s} + 1} \tag{4.4}$$

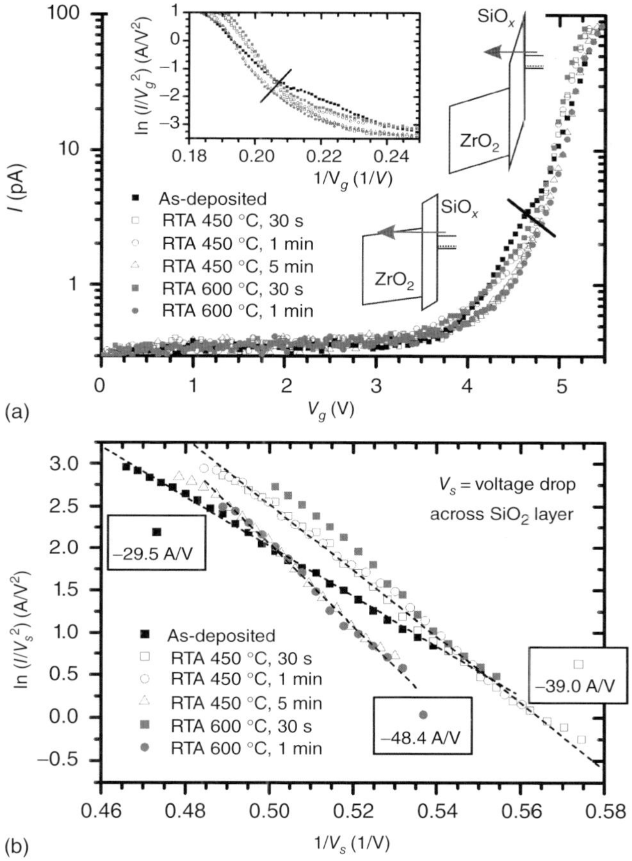


**Figure 4.17** (a) TUNA $I$–$V$ curves of as-deposited and at different conditions annealed $ZrO_2$ samples. In the inset, the $I$–$V$ curves of the samples are represented in Fowler–Nordheim coordinates. The Fowler–Nordheim plot reveals the existence of two conduction mechanisms, which are also schematically presented in the figure. (b) Fowler–Nordheim plot ($\ln(I/V_s^2)$ vs $1/V_s$) of $I$–$V$ curves in the higher voltage region ($V_g > 4.7$ V). Here, the voltage drop $V_s$ across $SiO_2$ is used to calculate the Fowler–Nordheim coordinates, as the current in this region is dominated by tunneling through the thin interfacial $SiO_2$ layer. (Reproduced with permission from [22]. Copyright American Institute of Physics 2008.)

$V_z$, $\varepsilon_z$, and $d_z$ are the voltage drop, permittivity, and thickness of $ZrO_2$, and $\varepsilon_s$ is the permittivity of $SiO_2$, that is, 3.9. For the calculation of the voltage drops $V_z$ and $V_s$, the layer thicknesses of the as-deposited stack measured by ellipsometry are used ($d_z = 4.8$ nm and $d_s = 1.1$ nm) and a permittivity of 20 for (amorphous) $ZrO_2$ is assumed [48]. $V_{\text{off}}$ is given by

$$V_{\text{off}} = \phi_{\text{MS}} + \phi_{\text{surf}} \tag{4.5}$$

where $\phi_{\text{MS}}$ is the work function difference between the tip and the substrate and $\phi_{\text{surf}}$ is the substrate surface potential. The determination of $V_{\text{off}}$ includes several inaccuracies due to the very limited availability of data for the tip work function

$\phi_{Tip}$, especially if highly p-type doped diamond coated tips were used whose parameters strongly depend on the fabrication method. In this work, values of 5.4 eV for the diamond band gap and 0 eV for its electron affinity [49] were used. As the doping level of the tip coating is very high, it can be assumed that the Fermi level is shifted into the diamond valence band and therefore, the work function of the diamond tip is 5.4 eV. Hence the work function difference between the tip and p-type Si is

$$\phi_{MS} = \phi_{Tip} - \phi_{p\text{-}Si} \approx 5.4\,\text{eV} - 4.9\,\text{eV} = 0.5\,\text{eV} \tag{4.6}$$

At measurement conditions used here, p-type Si is in strong inversion and the Si surface potential $\phi_{surf}$ is estimated to be about 0.9 eV. This leads to a $V_{off}$-value of

$$V_{off} = \phi_{MS} + \phi_{surf} \approx 0.5\,\text{eV} + 0.9\,\text{eV} = 1.4\,\text{eV} \tag{4.7}$$

To justify that the current in the high-voltage region is controlled by $SiO_2$, the $I$–$V$ curves are represented in Fowler–Nordheim coordinates (i.e., $\ln(I/V_s^2)$ vs $1/V_s$) by using the values of the voltage drop $V_s$ across $SiO_2$ Eq. (4.3) (Figure 4.17b). As can be seen, straight lines are obtained for all samples and the slope of the line increases with the thermal budget of the RTA process. Since the slope is proportional to the $SiO_2$ layer thickness $d_s$, this is an indication that with increasing thermal budget a thicker $SiO_2$ interfacial layer is formed. The calculation of $d_s$ from the slopes of the lines for all annealing conditions (Table 4.1) is performed by using an effective electron mass $m^*$ of 0.5 $m_0$ ($m_0$ is the free electron mass) in $SiO_2$ and a barrier height $\phi_b$ of 3.2 eV [10]. For the as-deposited sample, the calculated value of the $SiO_2$ thickness agrees well with the ellipsometrically measured $SiO_2$ interface thickness of 1.1 nm and also with the interfacial layer thickness obtained from HR-TEM analysis (Figure 4.14a). This proves the feasibility of the above considerations and gives rise to the assumption that the conduction at high voltages is indeed determined mainly by the thin interfacial $SiO_2$ layer. For annealed samples, however, the estimated electrical thickness $d_s$(TUNA) is 0.2–0.4 nm larger than the physical thickness $d_s$(TEM) (Figure 4.14 and Table 4.1). This difference may be due to the fundamental differences between the two methods. Another source of error,

**Table 4.1** Thickness of the interfacial $SiO_2$ layer as estimated from TUNA $I$–$V$ curves $d_s$ (TUNA), as determined from TEM image $d_s$ (TEM), and after correction of TUNA $I$–$V$ curves for process-induced positive oxide charge in $SiO_2$ $d_s^1$ (TUNA).

| Sample | $d_s$ (TUNA) (nm) | $d_s$ (TEM) (nm) | $d_s^1$ (TUNA) (nm) |
|---|---|---|---|
| As-deposited | 1.06 | 1.1 | 1.06 |
| RTA 450 °C, 30 s | 1.40 | 1.1–1.2 | 1.16 |
| RTA 450 °C, 60 s | 1.35 | — | 1.00 |
| RTA 450 °C, 300 s | 1.60 | — | 1.23 |
| RTA 600 °C, 30 s | 1.55 | — | 1.18 |
| RTA 600 °C, 60 s | 1.70 | 1.3–1.4 | 1.30 |

*Source:* Reproduced with permission from [22]. Copyright American Institute of Physics 2008.

especially for the annealed samples, is a possible change of the stack parameters (e.g., not only layer thicknesses but also permittivities, which are used to calculate the voltage drops) as a result of the thermal treatment. This seems to be the more likely reason having in mind the good agreement for the as-deposited sample for which the correct values of thickness and permittivity were used to calculate the voltage drops. Therefore, we recalculated the voltage drops using the layer thicknesses obtained by TEM. However, this did not result in significant changes of the calculated $d_s$(TUNA). Figure 4.18, which presents the forward and backward $I$–$V$ curves of as-deposited and annealed (RTA 450 °C, 1 min) samples, gives rise to the suggestion that the origin of the difference between the differently obtained thicknesses for the annealed samples could be electrically active defects created by the high-temperature annealing. The hysteresis in the low- and the high-voltage ranges (i.e., in the voltage ranges in which different conduction mechanisms are dominant) is also different. Generally, the hysteresis of the as-deposited sample is smaller in both bias regions and it progressively diminishes in the high voltage region where tunneling through the $SiO_2$ is the dominant conduction mechanism. Trapping of carriers, generation of charges which change the internal field, hence the current or generation of neutral traps which may affect the conduction mechanism could result in appearance of hysteresis. The disappearance of the hysteresis in the high-voltage region for the as-deposited sample reveals that tunneling through $SiO_2$ is only slightly affected independent of the hysteresis origin. Hence, the trapping and/or generation of defects take place in the high-$k$ dielectric layer. For the annealed sample (Figure 4.18), however, a large hysteresis is observed in both bias regions (i.e., also in the higher voltage range where tunneling through $SiO_2$ governs the current). In Table 4.2, the hysteresis of all samples at low and high voltages which is defined as the voltage difference between the forward and the backward curves at a current of 1 and 100 pA, respectively, are summarized. To account for the local inhomogeneity of the layers, each value represents an average of 10 $I$–$V$ curves measured

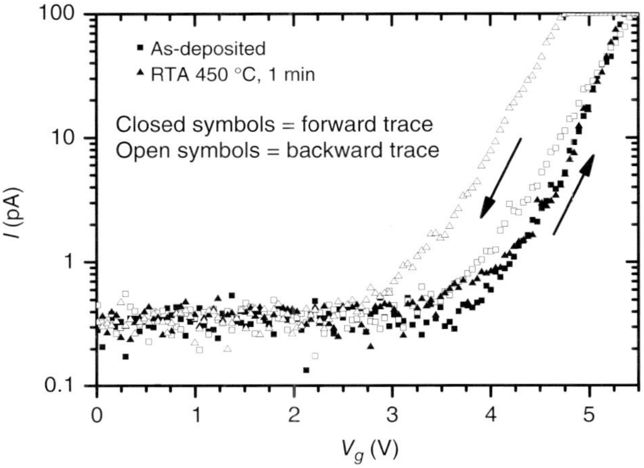


**Figure 4.18** $I$–$V$ hysteresis of as-deposited and annealed (450 °C, 1 min) samples. (Reproduced with permission from [22]. Copyright American Institute of Physics 2008.)

**Table 4.2** *I–V* hysteresis of different samples measured at 1 and 100 pA.

| Sample | *I–V* hysteresis (V) | |
|---|---|---|
| | At 1 pA | At 100 pA |
| As-deposited | 0.40 ± 0.1 | 0.09 ± 0.1 |
| RTA 450 °C, 30 s | 0.81 ± 0.1 | 0.30 ± 0.1 |
| RTA 450 °C, 60 s | 0.85 ± 0.1 | 0.50 ± 0.1 |
| RTA 450 °C, 300 s | 0.94 ± 0.1 | 0.48 ± 0.1 |
| RTA 600 °C, 30 s | 0.88 ± 0.1 | 0.50 ± 0.1 |
| RTA 600 °C, 60 s | 0.79 ± 0.1 | 0.47 ± 0.1 |

*Source:* Reproduced with permission from [22]. Copyright American Institute of Physics 2008.

at 10 different points. The results clearly show a distinct difference between the as-deposited and the annealed samples. The *I–V* hysteresis at 1 pA increases from 0.4 V for the as-deposited sample to 0.8-0.95 V for the annealed samples without a clear dependence on the RTA thermal budget. At 100 pA, the *I–V* hysteresis is as low as 0.1 V for the as-deposited sample and increases to 0.3 V for RTA 450 °C, 30 s sample and to about 0.5 V for the rest of the annealing conditions. Therefore, it can be concluded that the annealing process creates some defects in the stack that lead to a positive charge buildup in the $SiO_2$ layer which influences the exact extraction of $SiO_2$ thickness. So, the voltage drop $V_s$ across the $SiO_2$ was corrected with the voltage drop caused by this charge (i.e., 0.3 V for the RTA 450 °C, 30 s sample and 0.5 V for the rest of the samples). The corresponding estimated values $d_s^{\,1}$ (TUNA) for the $SiO_2$ thickness are given in Table 4.1. An excellent agreement between the TEM and TUNA values is obtained which reveals TUNA as a very suitable and sensitive technique to connect physical film properties and their modifications to their electrical behavior at nanoscale dimensions.

### 4.4.3 Trapping Phenomena and Degradation Mechanism in High-*k* Dielectric Stacks

As already mentioned, by using C-AFM techniques, it is possible to investigate also the different stages of the degradation process in high-*k* dielectric stacks starting from trapping phenomena, defect generation, electrical instabilities, soft and reversible BD phenomena up to a hard destructive BD and mechanical damage. Here it is shown how the TUNA *I–V* curves and their hysteresis could be used as a degradation monitor [22]. Several experiments which allow identification of possible reasons for the hysteresis and its relation to the degradation mechanism of high-*k* stacks are presented. As already mentioned, there are several phenomena which could be a reason for the appearance of hysteresis in TUNA *I–V* curves, for example, trapping of carriers, generation of charges which change the internal field, hence the current, or generation of neutral traps which may affect the conduction mechanism. By sweeping TUNA *I–V* curves to different end voltages $V_{end}$ it is possible to discriminate trapping in preexisting traps from defect generation. In Figure 4.19, the forward and backward $I$–$V$ curves of a $ZrO_2$ layer RTA at 600 °C for 1 min are shown. The slight shift of the backward curves to higher voltages when the end voltage is low enough (Figure 4.19a, $V_{end} = 3.7$ V) reveals a negative charge buildup in the stack. In

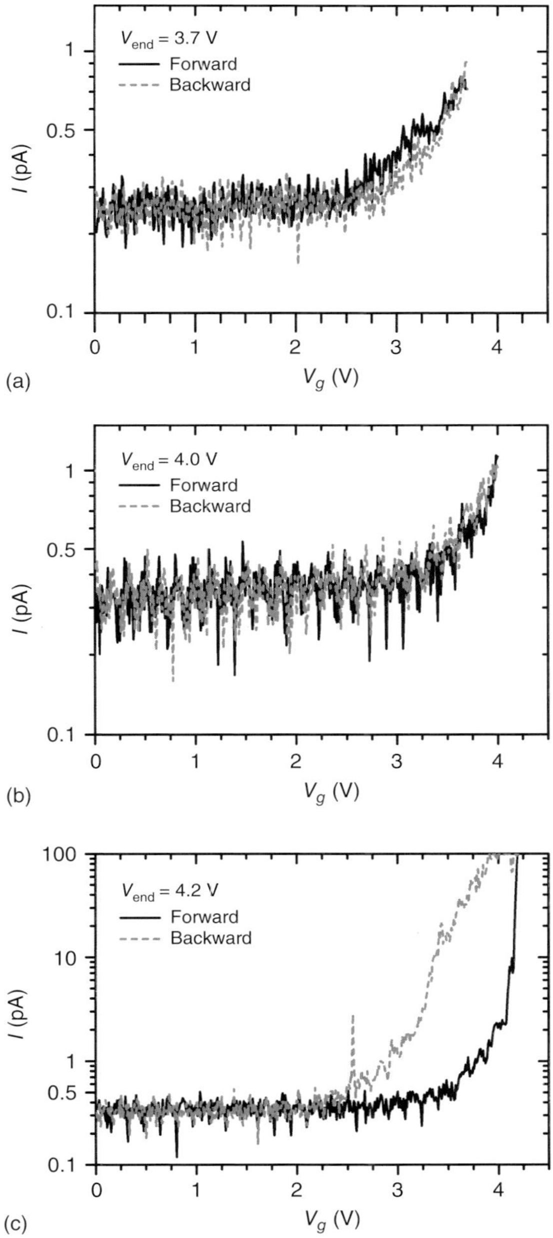


**Figure 4.19** $I$–$V$ hysteresis of the sample annealed at 600 °C, 1 min with different end voltages $V_{end}$: (a) 3.7, (b) 4.0, and (c) 4.2 V. The curves indicate different trapping phenomena resulting in different hysteresis (positive, zero, or negative). These measurements were performed with Pt/Ir tips. (Reproduced with permission from [22]. Copyright American Institute of Physics 2008.)

this case, the structure is not subjected to strong stressing conditions, therefore it is suggested that this negative charge buildup is rather due to trapping of electrons in preexisting traps than to trapping in newly generated traps. This is direct evidence for the existence of electron traps in the high-*k* dielectric which was previously found by the more complicated charge pumping technique and was identified as one of the significant reliability issues [50–52]. Here the high sensitivity of the method should be addressed. As already mentioned, the contact area between the tip and the sample is a few hundreds of square nanometers. This means that if a typical trap density of about $1 \times 10^{12}$ cm$^{-2}$ is assumed, there is only one trap located on an area of 100 nm$^2$. Therefore, each change in *I–V* curves is caused by one to several traps or defects. In other words, single defects can be studied by TUNA. On increasing the end voltage, the hysteresis first disappears (Figure 4.19b, $V_{end} = 4.0$ V) and then becomes negative (Figure 4.19c, $V_{end} = 4.2$ V), which indicates that another process such as stress generation of defects counteracts the negative charge trapping. This could be a stress-induced positive charge buildup that affects the internal electric field, hence the current. Another possibility is a stress-induced (neutral) trap generation which leads to a change of the conduction mechanism from tunneling (in the forward curve) to a trap-assisted tunneling (in the backward curve) through the stack, resulting in a significant increase of current [25]. These results demonstrate that the hysteresis is a complex phenomenon and more than one process may be responsible for it.

Measurements of successive *I–V* curves at one single spot can be used to get information about the degradation mechanisms. When measuring at a fixed location, the thickness of the layer remains intact, provided no hard BD or anodic oxidation has occurred. Anodic oxidation is prevented as the charge is injected from the substrate and the tip is grounded. Therefore, any change of the current level and the *I–V* curve hysteresis with repeated stress could be assigned to trapping of carriers and/or generation of defects which results in a change of tunneling barrier height, thus leading to a change of conductivity. Figure 4.20a,b represents the shifts of forward and backward *I–V* curves of the same sample presented in Figure 4.19 (i.e., $ZrO_2$, RTA 600 °C, 1 min). Figure 4.20c shows the comparison of hysteresis of the 1st and 100th *I–V* measurement. Several features are clearly visible:

1) The forward *I–V* curves shift progressively to lower voltages with the increase in the number of the measurements (Figure 4.20a). This shift could be due to (i) positive charge buildup which changes the internal field and, hence the current; (ii) stress-induced traps which serve as stepping sites for electrons, this way enabling their transport through the stack and increasing the current (trap-assisted tunneling). Here, the distinct difference from the behavior of $SiO_2$ layers should be mentioned. In $SiO_2$, negative charge trapping was observed for similar measurements [53], which implies a difference in the degradation mechanisms.
2) The backward curves are only slightly and irregularly affected by the stress (i.e., they are randomly shifted around the first backward curve, Figure 4.20b).
3) The hysteresis decreases with an increasing number of *I–V* measurements and it tends to diminish at a high enough number of measurements (Figure 4.20c).

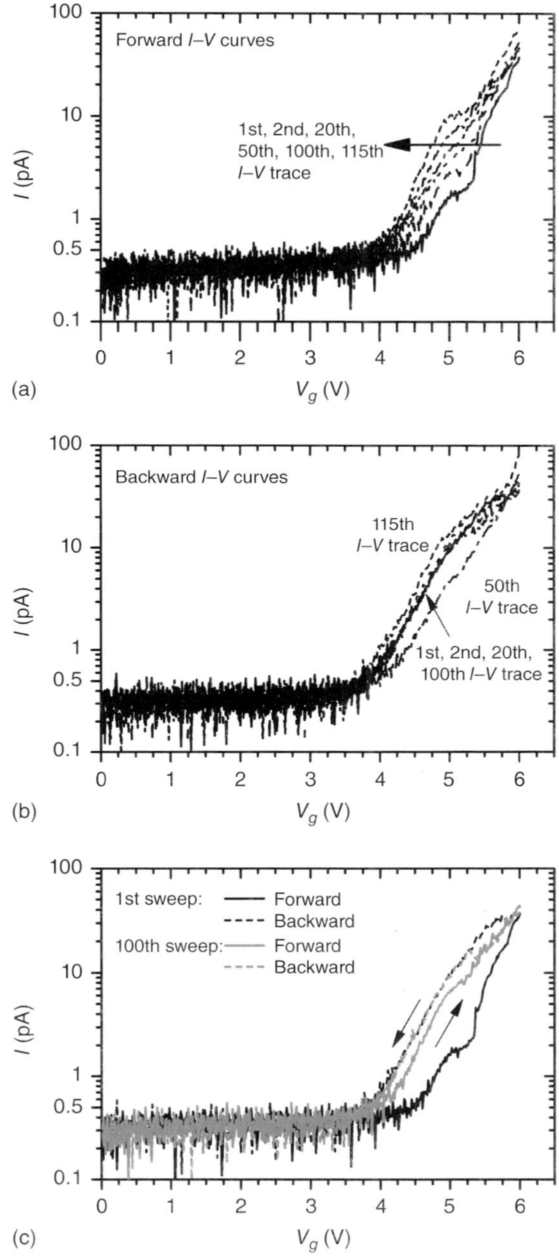


**Figure 4.20** Evolution of *I–V* curves of the sample annealed at 600 °C, 1 min with successive measurements at one point: (a) forward *I–V* curves, (b) backward *I–V* curves, and (c) comparison of hysteresis of the 1st and 100th *I–V* curves. (Reproduced with permission from [22]. Copyright American Institute of Physics 2008.)

On the basis of these observations, the following sequence of mechanisms leading to hysteresis and its progressive disappearance could be suggested. During the first forward $I-V$ measurement (provided the stress or $V_{\text{end}}$ is large enough), some kind of charges and/or traps are created which cause the hysteresis. These stress-induced defects should affect the second forward measurement. Indeed, it moves toward the backward curve, but nevertheless it does not follow it (i.e., the first backward and second forward $I$–$V$ curves do not coincide). This behavior is assigned to a partial recovery of the stress-induced defects which takes place at low fields. However, at the end of the second (and the following) forward measurement, these defects are again "born" and, therefore, the second (and the following) backward $I$–$V$ curve coincide with the first backward one. In other words, at the end of the second (and the following) forward measurement, the defect status of the stack is the same as at the end of the first forward measurement (i.e., the processes which take place during the second (and the following) forward measurement affect the backward $I$–$V$ curve similarly as the first forward measurement). This can be attributed to trapping/detrapping processes. A possible explanation is negative charge trapping in preexisting traps which occurs at low fields as demonstrated in Figure 4.19a. This negative charge trapping causes the shift of the second forward curve to higher voltages with respect to the backward one (i.e., partial recovery of defects which cause the hysteresis and hence, the second forward $I$–$V$ curve is located in between the first forward and backward $I$–$V$ curves). This charge may be detrapped at higher voltages. With increasing number of $I$–$V$ measurements, ever less damage is recovered (i.e., it transforms to a permanent damage), causing the forward $I$–$V$ curves to shift progressively toward the backward ones until hysteresis diminishes. This explanation, in fact, involves the preexisting electron traps as possible precursors for a stress-induced damage and their progressive transformation from "classical" traps to a permanent damage.

Another way to perform electrical stressing is to measure $I$–$V$ curves with different ramp rates – the slower the ramp rate, the longer the stressing, hence stronger damage could be anticipated. In Table 4.3, the hysteresis of the first $\Delta V_{\mathrm{I}}$ and second $\Delta V_{\mathrm{II}}$ $I$–$V$ curves as well as the voltage shift between the first and the second forward $\Delta V_{\mathrm{I-II}}(\mathrm{f})$ and backward $\Delta V_{\mathrm{I-II}}(\mathrm{b})$ curves analyzed at 1 pA are

**Table 4.3** Ramp rate dependence of the first $\Delta V_{\mathrm{I}}$ and the second $\Delta V_{\mathrm{II}}$ $I$–$V$ hysteresis as well as the voltage shift between the first and the second forward $\Delta V_{\mathrm{I-II}}(\mathrm{f})$ and backward $\Delta V_{\mathrm{I-II}}(\mathrm{b})$ $I$–$V$ curves.

| Ramp rate (V/s) | $\Delta V_{\mathrm{I}}$ (V) | $\Delta V_{\mathrm{II}}$ (V) | $\Delta V_{\mathrm{I-II}}$(f) (V) | $\Delta V_{\mathrm{I-II}}$(b) (V) |
|---|---|---|---|---|
| 0.05 | 0.68 | 0.4 | 0.32 | 0.04 |
| 0.10 | 1.06 | 0.42 | 0.57 | −0.07 |
| 0.14 | 0.69 | 0.55 | 0.15 | 0.01 |
| 0.50 | 0.82 | 0.34 | 0.48 | 0 |
| 1.00 | 1.30 | 0.42 | 0.94 | 0.06 |
| 2.00 | 0.73 | 0.24 | 0.44 | −0.05 |

These measurements were performed on the sample annealed at 600 °C for 1 min. All hysteresis values are analyzed at 1 pA.
*Source:* Reproduced with permission from [22]. Copyright American Institute of Physics 2008.

given as a function of the ramp rate. As seen, the hysteresis $\Delta V_{\mathrm{I}}$ changes in a very wide range (0.68-1.3 V) but without any relation to the ramp rate. In contrast, the hysteresis $\Delta V_{\mathrm{II}}$ varies only slightly around an average value of about 0.45 V. There is a significant shift between the first and second forward curves, but $\Delta V_{\mathrm{I-II}}(\mathrm{f})$ values are randomly scattered without any well-defined trend of change with the ramp rate. As seen in previous experiments, there is nearly no difference in the backward direction $\Delta V_{\mathrm{I-II}}(\mathrm{b})$. On the basis of these results, it may be concluded that the hysteresis does not depend on the ramp rate and, hence on the stressing time. The differences in $\Delta V_{\mathrm{I}}$ and $\Delta V_{\mathrm{I-II}}(\mathrm{f})$ values at different ramp rates could be assigned to differences in the local structural properties of the stack (this difference in the local properties is easy to suggest, having in mind the polycrystalline structure of the film). The independence on the stressing time needs more attention. In terms of stressing time, one single $I$–$V$ measurement performed at a ramp rate of 0.05 V/s should be equal to 40 $I$–$V$ measurements at 2 V/s. Therefore, this should result in a larger shift $\Delta V_{\mathrm{I-II}}$ of the second $I$–$V$ curve measuring at a ramp rate of 0.05 V/s. However, this is not the case. The difference between the two types of measurements (i.e., successive $I$–$V$ measurements at one location with a given ramp rate and one measurement with a slower ramp rate) is not the stressing time, but the way of stressing. In the case when the $I$–$V$ measurements are performed with different ramp rates, the structure is subjected only once to a degradation process (irrespective of the ramp rate). In the case of the successive $I$–$V$ measurements, however, the structure is forced multiple times through a given sequence of stressing processes. The absence of any clear trend of $\Delta V_{\mathrm{I-II}}(\mathrm{f})$ change with ramp rate actually reveals that the damage which causes the shift of $I$–$V$ curve depends rather on how many times the structure is subjected to a given stressing cycle than on how long it stays under stress. Hence, the obtained results give a support to the above proposed degradation mechanism which involves trapping/detrapping at already existing precursor sites and their progressive transformation to a permanent damage on increasing the number of stressing cycles (i.e., $I$–$V$ measurements). This mechanism can also explain the nearly constant value $\Delta V_{\mathrm{II}}$ as this hysteresis could be considered as a representative for the part of defects which are recovered after the first $I$–$V$ measurement, irrespective of the ramp rate.

The evolution of hysteresis was used to investigate the influence of $SiO_2$ doping on the trapping phenomena in $ZrO_2$ thin films (Figure 4.21) [54]. $ZrO_2$ films with various $SiO_2$ doping levels (pure $ZrO_2$ ($Z$), $SiO_2$-doped $ZrO_2$ ($ZS1$ ~1% $SiO_2$, $ZS2$ ~ 5% $SiO_2$) were deposited by atomic layer deposition (ALD) and subsequently annealed at 450 °C. The highest recovery (i.e., the highest hysteresis of the second sweep) is measured for $ZS2$ as this film can possess more preexisting traps due to the rather nanocrystalline structure. The hysteresis value of $Z$ stays constant at about 0.4 V after the second $I$–$V$ sweep (i.e., this electrically inhomogeneous layer shows a permanent high trapping level). In contrast, the hysteresis value of the $SiO_2$-doped layers decreases down to 0.2 V for $ZS1$ after the fifth sweep and even down to 0.1 V for $ZS2$ after the 20th sweep. The behavior of $ZS2$ samples which were annealed at higher temperatures (i.e., increase in the grade of crystallization) gives evidence that the trap concentration is related to the crystalline structure. Generally, a higher permanent trap density for $SiO_2$-doped

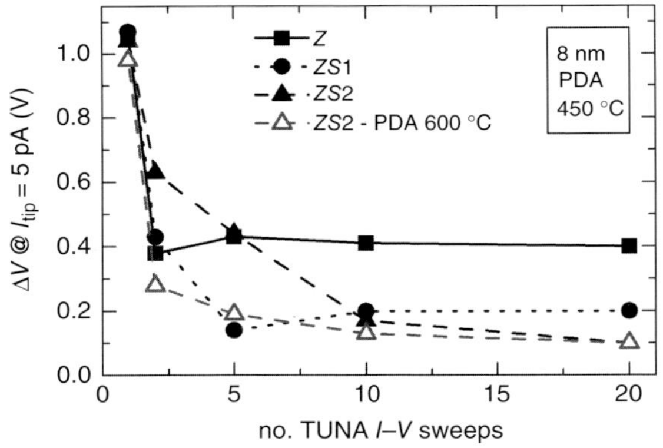


**Figure 4.21** Hysteresis (voltage shift between forward and backward TUNA *I–V* curve at 5 pA) after several measurement repetitions for 8 nm samples *Z*, *ZS*1, and *ZS*2. (Reproduced with permission from [54]. Copyright 2009, American Vacuum Society.)

layers compared to pure $ZrO_2$ was found. However, the $SiO_2$ doping also leads to lower hysteresis values after several measurement repetitions. Additionally, local TUNA *I–V*s can give suggestions on optimal $SiO_2$ concentrations in $ZrO_2$.

### 4.4.4 Reliability of High-*k* Dielectric Films

#### 4.4.4.1 Gate Oxide Reliability at the Nanoscale

Usually, reliability of dielectrics (gate oxides, in particular) is characterized by electrical measurements such as CVS or constant current stress (CCS) using conventional probe setups [55, 56]. For that, fabrication of gate electrodes is mandatory to analyze the integrity of the dielectrics on a reproducible basis. In this section, we show that C-AFM can also be efficiently applied to evaluate gate oxide reliability at the nanoscale for silicon dioxide layers as well as high-*k* dielectrics when using the CVS technique. For the evaluation of gate oxide integrity, time to BD CCS and CVS are used, which strongly depend on the size of the available gate electrode area [57]. Therefore, with decreasing gate area, BD at extrinsic defects becomes less evident. Time to BD ($t_{BD}$) is usually obtained by applying an accelerated electrical test to MIS devices. However, by using standard macroscopic measurements, it is not possible to recognize the local nature of BD and to correlate it to nanoscale features of the dielectric [58]. Next, the feasibility of using Weibull statistics for time-to-BD evaluation at the nanoscale is investigated.

For comparison, conventional electrical characterization (CVS) was performed using a Keithley K236 source measurement unit in combination with an automatic wafer prober on samples with metal electrodes. For characterization at the nanoscale (samples without metallization), a Veeco Dimension 5000 AFM system with a Nanoscope IV controller and a current amplifier with a maximum measurement range of 500 nA and a sensitivity of about 30 pA was used. This AFM system was also applied for measurements with the tip on metal gate structures of different sizes. Pt/Ir tips were used throughout all described experiments.

A comparison of conventional and C-AFM *I–V* measurements for a $Si_3N_4$ sample is depicted in Figure 4.22. Good coincidence is observed for measurements on gate electrodes with an area of 0.01 $mm^2$ (see also the results presented in Section 4.2). Additionally, an increase of average BD voltage for smaller gate areas can be deduced from the C-AFM data. This observation coincides with the weakest-link

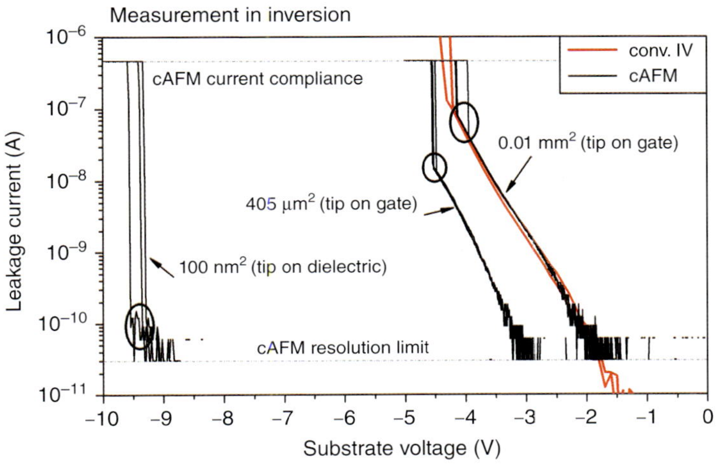


**Figure 4.22** Comparison of conventional and C-AFM-based *I–V* measurements for $Si_3N_4$-type sample; different gate contact areas and tip directly on dielectric were used – destructive dielectric breakdown is indicated by bold circles. (Reproduced with permission from [56]. Copyright 2011, American Vacuum Society.)

character of dielectric BD given by percolation theory, and can be described by the following power law [55, 59]:

$$t_{BD}(A_1) = t_{BD}(A_2) \cdot \left[\frac{A_2}{A_1}\right]^{\frac{1}{\beta}} \tag{4.8}$$

where $A_1$ and $A_2$ are the areas of the gate structures, $t_{BD}$ is the respective time to BD, and $\beta$ is the Weibull-plot parameter. The area of the MIS capacitors used for the conventional CVS measurements is $10^4$ μm$^2$ (0.01 mm$^2$), whereas the area of the tip was extracted from comparison of the *I–V* measurements at the nanoscale and the macroscopic measurements, and it equals approximately $10^{-4}$ μm$^2$ (100 nm$^2$). Hence, the gate areas of the samples differ by eight orders of magnitude (!). It should be noted that an additional shift of BD voltages is observed when probing the dielectric directly with the C-AFM tip. The reason for this additional shift is the work function difference between the C-AFM tip (Pt/Ir) and the gate electrode material (TiN) in contact with the dielectric, respectively. This difference in work function is considered in the following comparison between macroscopic and nanoscale CVS measurements.

A prediction of time to failure at lower voltages $V_{use}$ has been deduced for conventional CVS measurements by extrapolation from the measured time-to-failure values at higher substrate voltages $V_{sub}$, and a power-law dependency similar to Eq. (4.8) has been suggested as follows [57, 59]:

$$t_{BD}(V_{use}) = t_{BD}(V_{sub}) \cdot \left[\frac{V_{sub}}{V_{use}}\right]^{m} \tag{4.9}$$

The implications of Eqs (4.8) and (4.9) are also visible in the *I–V* sweeps shown in Figure 4.22. Here, time to failure occurs at a distinct substrate voltage within several hundred milliseconds, corresponding to the time delay between individual increments of the voltage ramp (staircase-like waveform). Owing to the small area used for the C-AFM measurements with the tip directly in contact, the probability for BD is significantly smaller than for the measurements on gate electrodes. Hence, a longer average time to failure is expected at comparable voltages. This means that a higher stress voltage has to be applied to yield equal time to failure.

In order to further investigate the BD characteristics of samples at the nanoscale, CVS tests were performed at different substrate voltages using the AFM system. A large number of measurements were conducted on each sample type to ensure statistical significance. For the measurement with the tip directly in contact with the dielectric, the current measured by the C-AFM sensor at the moment of BD increases from the sensitivity limit (before hard BD state) to current compliance (after BD state), that is, to the point where the preamplifier goes into saturation. The result of 100 individual CVS measurements with a stress voltage of −8.5 V (applied to the substrate) on $HfO_2$ samples is depicted in Figure 4.23.

It is evident from Figure 4.23 that the CVS measurements at the nanoscale can be adequately fitted by a Weibull plot, assuming a two-parameter Weibull distribution. The applicability of this commonly employed evaluation technique confirms that the BD mechanism at the nanoscale is identical to the BD mechanism resulting from macroscopic measurements. From the data, the Weibull slope parameter $\beta$ can be calculated, and the time to failure $t_{BD,63\%}$ for 63.2% of the samples under test can be extracted using the following equation:

$$\ln(-\ln(1 - F_{BD}(t))) = \beta \ln\left(\frac{t}{t_{BD,63\%}}\right) \tag{4.10}$$

where $F_{BD}$ is the accumulated failure rate.

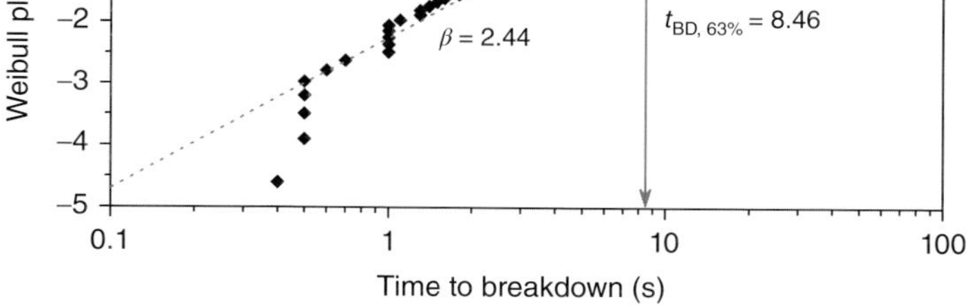


**Figure 4.23** Weibull plot for 100 measurements on a $SiO_2/HfO_2$ sample and extraction of time-to-failure 63rd percentile and Weibull slope parameter. (Reproduced with permission from [56]. Copyright 2011, American Vacuum Society.)

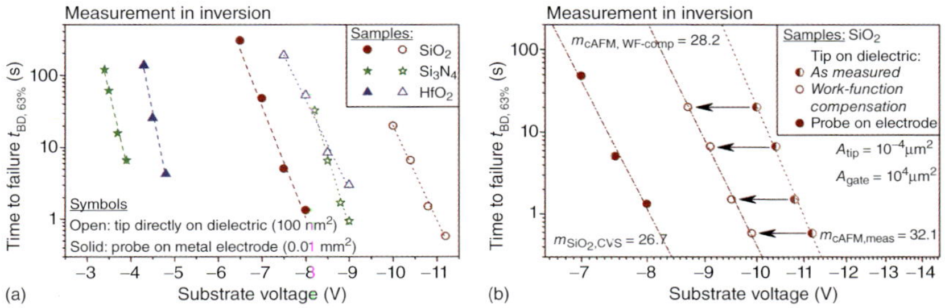


**Figure 4.24** (a) Time-to-failure plots (assuming Weibull distribution) of different samples for CVS measurement on gate contacts and directly on dielectric, respectively. (b) Measurement data for an 8 nm $SiO_2$ sample with compensation for different work-function materials between gate electrode and C-AFM tip; the respective slopes from voltage acceleration using Eq. (4.9) are listed. (Reproduced with permission from [56]. Copyright 2011, American Vacuum Society.)

Next, the CVS measurements were repeated at different substrate voltages and on different sample types both on the nanoscale using C-AFM and on planar capacitors with a gate area of 0.01 $mm^2$ using the conventional CVS setup. Figure 4.24a shows the evaluated time to failure (63rd percentile) for the different stress voltages and samples. Again, a significant shift between the measurements with C-AFM tip on the gate electrode and directly on the dielectric is evident. As previously mentioned, part of the shift is due to the different materials in contact with the dielectric. A work function of 5.6 and 4.3 eV is assumed for the Pt/Ir tip of the C-AFM probe and the TiN metallization of the gate electrodes, respectively. This work-function difference requires a higher gate voltage for the C-AFM measurement to achieve electric fields in the dielectric comparable to the conventional CVS measurements. Hence, compensation with respect to electric fields in the dielectric layers must be conducted for the C-AFM CVS measurements by reducing the substrate voltage by the work-function difference of 1.3 V as shown in Figure 4.24b for the 8 nm $SiO_2$ sample in order to directly compare the C-AFM and conventional CVS measurements.

The slope $m$ for voltage acceleration – see Eq. (4.9) – of the uncompensated (i.e., without correcting for the different work functions) CVS measurements at the nanoscale for the $SiO_2$ sample is 32.1. The slope of compensated C-AFM data is 28.2, and significantly closer to the slope (26.7) of the conventional CVS measurements on MIS capacitors. It should be noted, that the effect of charges in the dielectric layers and interface states which may be a source of additional voltage shifts has been neglected in the course of this study. To investigate whether the CVS measurements at the nanoscale compare to those using conventional MIS capacitors, the difference both in BD voltage and area between these sample types must be considered. So, in a first step the time to failure (63rd percentile) of the C-AFM measurements for compensated substrate voltages is extrapolated down to the substrate voltages used in the conventional $I$–$V$ setup following Eq. (4.9). In a second step, an area extrapolation using Eq. (4.8) has to be performed between the area of the gate electrodes and the area of the C-AFM tip. Here, the Weibull slope parameter $\beta_{CVS}$ determined from the Weibull plots (see Figure 4.23) is used

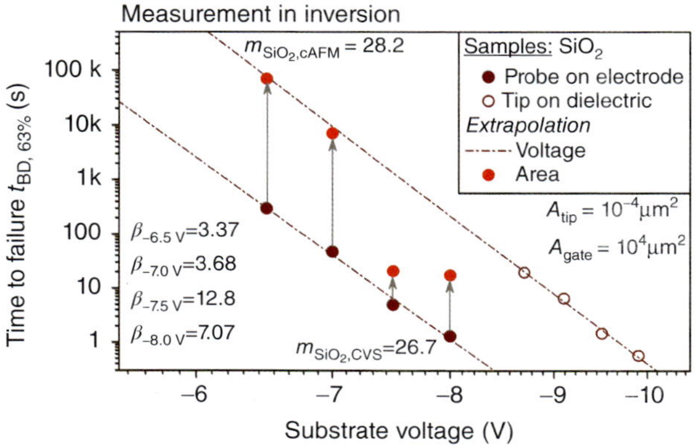


**Figure 4.25** Evaluation of measurement data for $SiO_2$ sample; voltage and area extrapolation were performed. (Reproduced with permission from [56]. Copyright 2011, American Vacuum Society.)

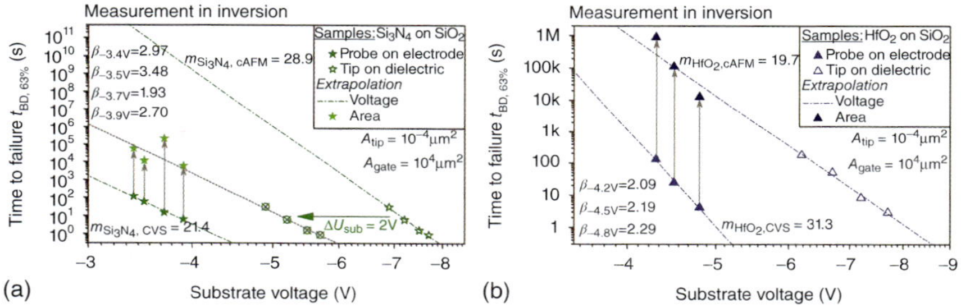


**Figure 4.26** Evaluation of measurement data for (a) $SiO_2/Si_3N_4$ and (b) $SiO_2/HfO_2$ sample; voltage and area extrapolation were performed. (Reproduced with permission from [56]. Copyright 2011, American Vacuum Society.)

as the area-scaling factor $\beta$ in Eq. (4.9). The required extrapolations for the $SiO_2$ sample are presented in Figure 4.25.

The extrapolations are in good agreement for the lower stress voltages used in the experiment. Similar evaluations were performed for the $HfO_2$ and the $Si_3N_4$ samples (see Figure 4.26). Results for the $HfO_2$ samples coincide with theory, whereas the C-AFM data for $Si_3N_4$ overestimates time to failure. One possible reason for the mismatch observed for the $Si_3N_4$ samples may be a shift in flatband voltage due to stronger charge trapping at higher stress voltages. Given that silicon nitride is known for its charge-trapping capability, there may be a tremendous difference between the flatband voltages at substrate voltages of about −4 V and of about −8 V. Assuming that this shift of flatband voltage for the C-AFM measurements is about 2 V higher than for the conventional CVS tests, a more decent fit can be obtained (see crossed circles and dashed line in Figure 4.26a). However, extraction of a nonlinear flatband voltage shift for different stress voltages has not yet been performed.

#### 4.4.4.2 In-Depth Analysis of Bimodal TDDB Distributions

Looking at it in more detail, the high-$k$ (HK) gate stack complicates the evaluation of the dielectric BD mechanism [60]. The different defect generation rates of high-$k$ dielectric and the $SiO_x$ interfacial layer (IL) may lead to bimodal BD statistical distributions [61]. Alternatively, one of the layers may control the overall HK/IL stack BD, specifically the IL, while the HK may have little impact on device lifetime [58, 62]. It is, therefore, critical to investigate the roles of the IL and HK dielectric on the overall degradation and BD of the gate stack. Here, it is demonstrated that C-AFM can be used to study the time-dependent dielectric breakdown (TDDB) in polycrystalline gate stacks and statistical analysis of $t_{BD}$ distributions measured on grains (Gs) and GBs is performed in a wide range of temperatures. The roles of the HK and IL layers, as well as the impact of the high-$k$ polycrystalline structure on the gate stack BD, are evaluated [60].

The gate dielectric stack under analysis consists of a 5 nm thick ALD $HfO_2$ film and an IL of about 1 nm $SiO_2$ grown on a Si epitaxial p-substrate. The gate stack was annealed at 1000 °C in $N_2$ ambient to simulate the thermal budget of the gate-first transistor fabrication process resulting in further crystallization (compared to as-deposited) of the HK film. The gate stack properties were investigated by C-AFM and Pt/Ir tips at temperatures of up to 240 °C. Having in mind the high lateral resolution of C-AFM, the contributions from G and GBs can be assessed independently. A CVS of −8.2 V was applied to the sample and $t_{BD}$ was extracted by measuring current-time characteristics. When the current reached 10 nA, which corresponds to the saturation of the setup electronics, BD was assumed to occur. To collect a statistically significant number of data points, an average of 60 random locations were measured at each $T$, and the corresponding Weibull distributions were analyzed.

The samples show a good topography-current correlation [60]. The current image demonstrates that the current flows preferentially through the GBs. In Figure 4.27a, the $t_{BD}$ Weibull distributions obtained at different temperatures are plotted. Note that such TDDB distributions show a bimodal behavior, exhibiting smaller slope values ($\beta$) for longer $t_{BD}$ (the dashed line is a guide to

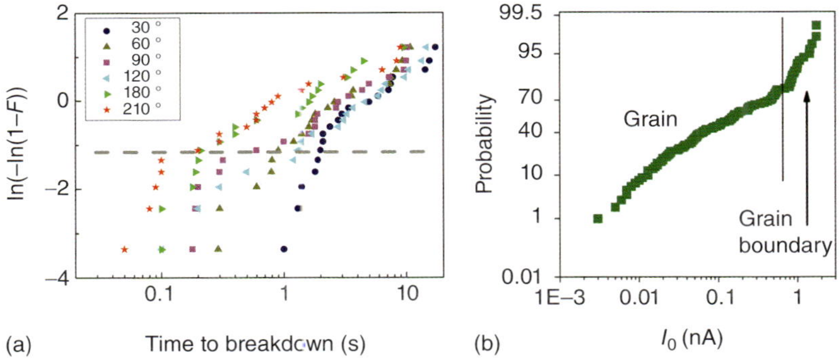


**Figure 4.27** (a) TDDB distributions obtained at different temperatures. A dashed line outlines the bimodal character of the distributions. (b) Cumulative distribution of the initial leakage currents, $I(t = 0\ s)$. Two types of locations with low (G) and high (GB) currents can be discerned. (Iglesias *et al.* 2013 [60]. Reproduced with permission of Elsevier.)

the eye for different modes). To investigate the origin of the bimodality, the $t_{BD}$ distributions measured on GBs and Gs were analyzed independently. To distinguish between G and GB locations, the magnitude of the initial current [$I(t = 0$ s)] was considered as criterion: as GBs have shown to be leakier than Gs, the locations with higher initial currents were assumed to correspond to GBs. As can be seen (Figure 4.27b), the cumulative probability distribution of $I(t = 0$ s) of all randomly selected sites stressed at 60 °C shows two distinguished slopes, which have been attributed to Gs (low currents) and GBs (high currents), respectively.

When the data for the G and GB sites are plotted separately (Figure 4.28), they are represented by different Weibull distributions. Note that, although each distribution in Figure 4.28 could be seen as a bimodal, it is not necessarily the case: the high slope tails of the $t_{BD}$ values which are very close to the setup resolution can also be affected by a possible mixture of the data from the G and GB sites.

The Weibull slope, $\beta$, and, scale factor, $\eta$ (time for a 63% probability of a BD occurrence), experimentally obtained at Gs (squares, filled symbols) and GBs (circles, filled symbols) for all $T$ are represented, respectively, in Figure 4.29a,b. Note that when G and GB TDDB data are analyzed independently (Figures 4.28 and 4.29), the Weibull distributions exhibit similar slopes. Therefore, the bimodal

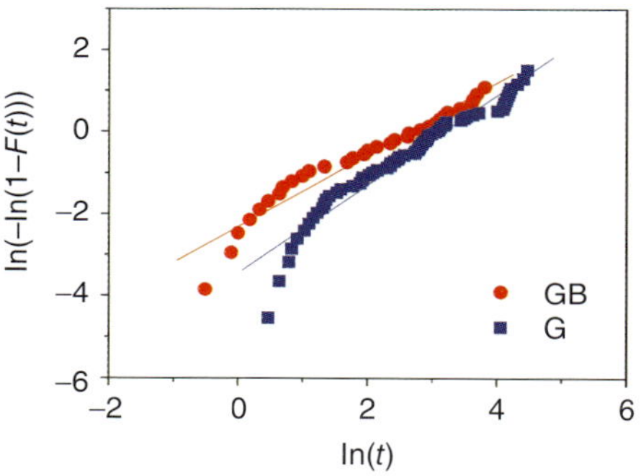


**Figure 4.28** Example of TDDB Weibull distributions obtained at 60 °C, at G (squares) and GB (circles) sites. (Iglesias *et al.* 2013 [60]. Reproduced with permission of Elsevier.)

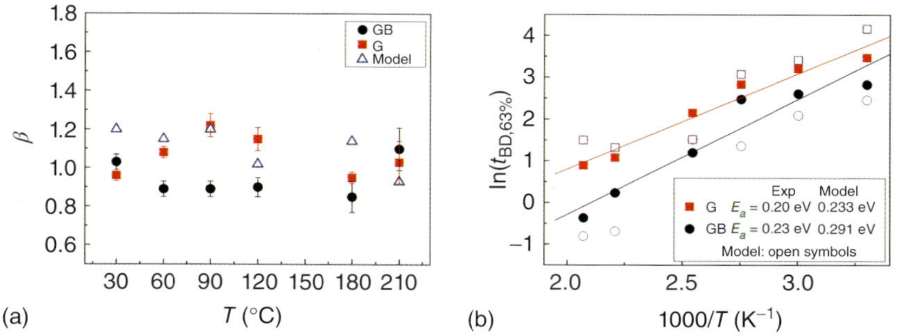


**Figure 4.29** Experimental (filled symbols) and modeled (open symbols) values of $\beta$ (a) and $\eta$ (b). Squares represent G and circles GB sites (in (a), $\beta$ was assumed to be the same at both site types). The lines in (b) show the Arrhenius dependency of the $\eta$ values. $E_a$ is the extracted activation energy for the G and GB data sets. (Iglesias *et al.* 2013 [60]. Reproduced with permission of Elsevier.)

feature arises only when G and GB data are combined within the same Weibull distribution (Figure 4.27), suggesting that differences in the electrical properties of each type of the sites in the HK layer can be the origin of such bimodal behavior [63]. This hypothesis is supported by the results of modeling the global TDDB distribution $F(t)$ as a combination of two different BD distributions containing either G or GB data (Eq. (4.11)):

$$F(t) = P_{\mathrm{GB}} \cdot F_{\mathrm{GB}}(t, \beta, \eta_{\mathrm{GB}}) + (1 - P_{\mathrm{GB}}) \cdot F_G(t, \beta, \eta_G) \tag{4.11}$$

where $P_{\mathrm{GB}}$ is the probability that the randomly positioned C-AFM tip was over a GB, and $F_G$ and $F_{\mathrm{GB}}$ are the Weibull distributions for Gs and GBs, respectively. In this model, the same $\beta$ value was considered in both cases, as suggested by the data in Figure 4.29a. In Figure 4.30, the TDDB distributions at 60 and 210 °C are fitted using Eq. (4.11). Note that the proposed Eq. (4.11) fits the global TDDB distributions well (continuous line, Figure 4.30), with $\beta$ and $\eta$ values close to those experimentally obtained over the entire range of temperatures (open symbols in Figure 4.29). The Weibull parameters extracted from G and GB distributions at different $T$ were analyzed in more detail. Figure 4.29a shows that $\beta$ is independent of the measurement positions (G or GB) and temperature. However, for a given $T$, $\eta$ is lower for GBs than for Gs. This means that, on average, the GB sites break faster, which is consistent with the observed preferential BD at the leakiest positions (GBs), likely due to an excess of oxygen vacancies, which tend to accumulate at GBs [26, 40, 64].

Moreover, $\eta$ decreases with $T$ [59], which could be related to a higher rate of trap generation at higher temperatures [65]. From the temperature dependence of $\eta$, the activation energies ($E_a$) of BD-related traps at Gs and GBs have been estimated assuming an Arrhenius dependence. In Figure 4.29b, the lines correspond to the Arrhenius fitting of $\eta$ for the data measured on Gs (filled circles) and GBs (filled squares) locations. Note that similar activation energies ($E_a$) of $0.20 \pm 0.04$ eV and $0.23 \pm 0.02$ eV were extracted for Gs and GBs, respectively. These values are also close to those obtained for the G and GB data predicted from Eq. (4.11) (Figure 4.29b, open symbols). Since $E_a$ reflects the nature of the defects, these results suggest that similar defects are likely responsible for the BD triggered at both G and GB sites. Similar $\beta$ and $E_a$ values for the Gs and GBs (with smaller $\eta$ for the latter) suggest that BD at both types of sites could be triggered by

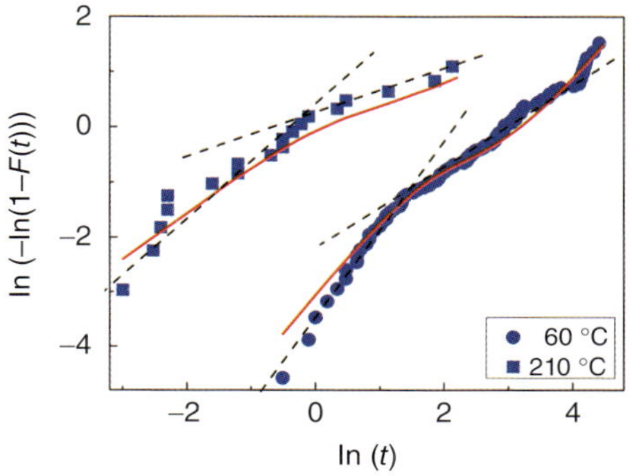


**Figure 4.30** TDDB distributions at 60 °C (circles) and 210 °C (squares) fitted using the Eq. (4.11) (continuous line). Broken lines outline the bimodal character of the distributions. (Iglesias *et al.* 2013 [60]. Reproduced with permission of Elsevier.)

the same process. Considering also that $\beta$ for $SiO_2$ does not depend on $T$ [26] and that an $E_a$ of ~0.2–0.25 eV has been extracted for the $SiO_2$ IL BD in the HK gate stack [63, 65], these results indicate that the stack BD could be triggered by the BD of the IL underlying the $HfO_2$ film, as was suggested earlier [63, 66]. Therefore, the formation of the BD path in HK gate stacks might be initiated by the degradation of the $SiO_2$ film. Then, the smaller $\eta$ values for the GB $t_{BD}$ data are associated with the faster degradation of the $SiO_2$ region below the GBs. Indeed, higher currents through the GBs indicate that the GBs represent less resistive $HfO_2$ regions [26] and, consequently, the voltage drop across those regions of the $SiO_2$ film is larger [63]. The IL beneath GBs, therefore, undergoes accelerated degradation resulting in smaller $t_{BD}$ values. Thus, the $SiO_2$ IL experiences two different degradation rates determined by the morphology of the overlaying HK dielectric, which could explain the global bimodal TDDB distributions.

## 4.5 Conclusion

The C-AFM technique has been demonstrated as a powerful tool to characterize high-$k$ dielectric stacks at nanoscale and to provide information which complements the results from conventional electrical and structural measurements. In order to take full advantage of this technique, dedicated efforts should be made to increase its sensitivity and to analyze correctly the data obtained. The sensitivity could be increased by using specially designed shielded probes which decrease parasitic capacitance. It is also shown that even without improved hardware solutions, C-AFM measurement sensitivity can significantly be improved by a two-step measurement procedure followed by an appropriate, simple numerical-data-correction procedure. The capability and potential of C-AFM have been demonstrated by performing a large spectrum of diverse studies on different high-$k$ dielectric stacks. The main implications of these studies could be summarized as follows:

1) Owing to the high sensitivity of the tunneling current on the thickness, and the $k$-value (hence crystal orientation), as well as the local nature of the C-AFM technique, this method proves to be extremely suitable to investigate film morphology and its relation to electrical properties of high-$k$ dielectric stacks. However, this relation is very sophisticated and the superposition and interplay of additional factors (e.g., surface flatness, defects, and electrically active centers) should be carefully accounted for in order to analyze the data correctly.
2) C-AFM has been demonstrated to be a very suitable and sensitive technique to connect physical properties of films to their electrical behavior at nanoscale dimensions. In particular, the thickness of the interfacial $SiO_2$ in high-$k$ stacks and its change under different annealing conditions have been evaluated from local TUNA $I$–$V$ curves.
3) Charge trapping and degradation phenomena are identified at nanoscale by employing local C-AFM $I$–$V$ curves and their hysteresis as a degradation monitor. A model for the degradation mechanism is inferred on the basis of the analysis of TUNA $I$–$V$ curve hysteresis phenomena.

4) C-AFM has been also successfully employed to investigate gate oxide reliability and integrity at the nanoscale using CVS. Area extrapolation over eight orders of magnitude assuming a Weibull distribution for C-AFM CVS measurements has been demonstrated. This allows for inline characterization of gate oxides with high spatial resolution and without the requirement of gate electrodes or back-end metallization. However, voltage-dependent effects, for example, charge trapping or generation of interface states, may impact electric field distribution and BD characteristics beyond the scope of basic Weibull statistics.
5) C-AFM can be very effective to analyze the impact of the high-*k* dielectric film morphology on the TDDB distribution in high-*k* gate stacks with a $SiO_2$ interfacial layer. The different electrical properties of grains and GBs in the polycrystalline dielectric layer have been clearly demonstrated by the obtained bimodal characteristics. The polycrystalline nature of the high-*k* dielectric layer may strongly influence the BD properties of the gate stacks. The results imply that the gate stack BD is triggered by the BD of the $SiO_2$ regions underlying the GBs.

## References


1 Blasco, X., Porti, M., Nafriía, M., Aymerich, X., Pétry, J., and Vandervorst, W. (2005) Electrical characterization of high-dielectric-constant/$SiO_2$ metal–oxide–semiconductor gate stacks by a conductive atomic force microscope. *Nanotechnology*, **16** (9), 1506–1511.

2 Frammelsberger, W., Benstetter, G., Kiely, J., and Stamp, R. (2006) Thickness determination of thin and ultra-thin $SiO_2$ films by C-AFM IV-spectroscopy. *Appl. Surf. Sci.*, **252** (6), 2375–2388.

3 Lanza, M., Porti, M., Nafría, M., Benstetter, G., Frammelsberger, W., Ranzinger, H., Lodermeier, E., and Jaschke, G. (2007) Influence of the manufacturing process on the electrical properties of thin (<4 nm) Hafnium based high-*k* stacks observed with CAFM. *Microelectron. Reliab.*, **47** (9-11), 1424–1428.

4 Rommel, M., Yanev, V., Paskaleva, A., Erlbacher, T., Lemberger, M., Bauer, A.J., and Frey, L. (2010) Electrical scanning probe microscopy techniques for the detailed characterization of high-k dielectric layers. *ECS Trans.*, **28**, 139–156.

5 Celano, U., Hantschel, T., Giammaria, G., Chintala, R.C., Conard, T., Bender, H., and Vandervorst, W. (2015) Evaluation of the electrical contact area in contact-mode scanning probe microscopy. *J. Appl. Phys.*, **117** (21), 214305-1–214305-5.

6 Olbrich, A., Ebersberger, B. and Boit, C. (1998) Nanoscale electrical characterization of thin oxides with conducting atomic force microscopy. IEEE 36th Annual International Reliability Physics Symposium, IEEE Proceedings, 98CH36173, pp. 163–168.

7 Pétry, J., Vandervorst, W., and Blasco, X. (2004) Effect of $N_2$ anneal on thin $HfO_2$ layers studied by conductive atomic force microscopy. *Microelectron. Eng.*, **72** (1-4), 174–179.

8 Porti, M., Nafría, M., and Aymerich, X. (2004) Nanometer-scale analysis of current limited stresses impact on $SiO_2$ gate oxide reliability using C-AFM. *IEEE Trans. Nanotechnol.*, **3** (1), 55–60.

9 De Wolf, P., Brazel, E., and Erickson, A. (2001) Electrical characterization of semiconductor materials and devices using scanning probe microscopy. *Mater. Sci. Semicond. Process.*, **4** (1-3), 71–76.

10 Porti, M., Nafría, M., Aymerich, X., Olbrich, A., and Ebersberger, B. (2002) Electrical characterization of stressed and broken down $SiO_2$ films at a nanometer scale using a conductive atomic force microscope. *J. Appl. Phys.*, **91** (4), 2071–2079.

11 Wu, Y.-L. and Lin, S.-T. (2006) Two-trap-assisted tunneling model for post-breakdown I-V characteristics in ultrathin silicon dioxide. *IEEE Trans. Device Mater. Reliab.*, **6** (1), 75–80.

12 Porti, M., Nafría, M., and Xavier Aymerich, X. (2003) Current limited stresses of $SiO_2$ gate oxides with conductive atomic force microscope. *IEEE Trans. Electron Devices*, **50** (4), 933–940.

13 Blasco, X., Pétry, J., Nafría, M., Aymerich, X., Richard, O., and Vandervorst, W. (2004) C-AFM characterization of the dependence of $HfAlO_x$ electrical behavior on post-deposition annealing temperature. *Microelectron. Eng.*, **72** (1-4), 191–196.

14 Bierwagen, O., Geelhaar, L., Gay, X., Piesins, M., Riechert, H., Jobst, B., and Rucki, A. (2003) Leakage currents at crystallites in $ZrAl_xO_y$ thin films measured by conductive atomic-force microscopy. *Appl. Phys. Lett.*, **90** (23), 232901-1–232901-3.

15 Efthymiou, E., Bernardini, S., Volkos, S.N., Hamilton, B., Zhang, J.F., Uppal, H.J., and Peaker, A.R. (2007) Reliability nano-characterization of thin $SiO_2$ and $HfSi_xO_y/SiO_2$ gate stacks. *Microelectron. Eng.*, **84** (9-10), 2290–2293.

16 Kremmer, S., Wurmbacher, H., Teichert, C., Tallarida, G., Spiga, S., Wiemer, C., and Fanciulli, M. (2005) Nanoscale morphological and electrical homogeneity of $HfO_2$ and $ZrO_2$ thin films studied by conducting atomic-force microscopy. *J. Appl. Phys.*, **97** (7), 074315-1–074315-7.

17 Lanza, M., Iglesias, V., Porti, M., Nafría, M., and Aymerich, X. (2011) Polycrystallization effects on the nanoscale electrical properties of high-k dielectrics. *Nanoscale Res. Lett.*, **6** (1), 108-1–108-9.

18 Martin, D., Grube, M., Weber, W.M., Rüstig, J., Bierwagen, O., Geelhaar, L., and Riechert, H. (2009) Local charge transport in nanoscale amorphous and crystalline regions of high-k $(Zr0_2)_{0.8}(Al_2O_3)_{0.2}$ thin films. *Appl. Phys. Lett.*, **95** (14), 142906-1–142906-3.

19 Porti, M., Aguilera, L., Blasco, X., Nafría, M., and Aymerich, X. (2007) Reliability of $SiO_2$ and high-k gate insulators: a nanoscale study with conductive atomic force microscopy. *Microelectron. Eng.*, **84** (3), 501–505.

20 Schwalke, U. and Stefanov, Y. (2005) Process integration and nanometer-scale electrical characterization of crystalline high-k gate dielectrics. *Microelectron. Reliab.*, **45** (5-6), 790–793.

21 Shubhakar, K., Pey, K.L., Bosman, M., Kushvaha, S.S., O'Shea, S.J., Kouda, M., Kakushima, K., and Iwai, H. (2014) Leakage current and structural analysis of annealed $HfO_2/La_2O_3$ and $CeO_2/La_2O_3$ dielectric stacks: a nanoscopic study. *J. Vac. Sci. Technol., B*, **32** (3), 03D125.

22 Paskaleva, A., Yanev, V., Rommel, M., Lemberger, M., and Bauer, A.J. (2008) Improved insight in charge trapping of high-k $ZrO_2/SiO_2$ stacks by use of tunneling atomic force microscopy. *J. Appl. Phys.*, **104** (2), 024108 1–024108 7.

23 Sun, Z.G., Kuramochi, H., Akinaga, H., Yu, H.H., and Gu, E.D. (2007) Conductive atomic force microscopy study of silica nanotrench structure. *Appl. Phys. Lett.*, **90** (4), 042106-1–042106-3.

24 Wu, Y.L., Huang, Ch.Y. and Lin, J.J. (2009) Comparison of charge trapping in silicon dioxide and hafnium dioxide at nanoscale. IEEE Proceedings of 16th IPFA, pp. 79–82.

25 Pétry, J., Vandervorst, W., Pantisano, L., and Degraeve, R. (2005) On the data interpretation of the C-AFM measurements in the characterization of thin insulating layers. *Microelectron. Reliab.*, **45** (5-6), 815–818.

26 Iglesias, V., Lanza, M., Zhang, K., Bayerl, A., Porti, M., Nafría, M., Aymerich, X., Benstetter, G., Shen, Z.Y., and Bersuker, G. (2011) Degradation of polycrystalline $HfO_2$-based gate dielectrics under nanoscale electrical stress. *Appl. Phys. Lett.*, **99** (10), 103510-1–103510-3.

27 Masuduzzaman, M., Xie, S., Chung, J., Varghese, D., Rodriguez, J., Krishnan, S., and Alam, M.A. (2012) The origin of broad distribution of breakdown times in polycrystalline thin film dielectrics. *Appl. Phys. Lett.*, **101** (15), 153511-1–153511-5.

28 Murakami, K., Rommel, M., Yanev, V., Bauer, A.J., and Frey, L. (2011) Current voltage characteristics through grains and grain boundaries of high-k dielectric thin films measured by tunneling atomic force microscopy. *AIP Conf. Proc.*, **1395**, 134–138.

29 Shubhakar, K., Raghavan, N., Kushvaha, S.S., Bosman, M., Wang, Z.R., O'Shea, S.J., and Pey, K.L. (2014) Impact of local structural and electrical properties of grain boundaries in polycrystalline $HfO_2$ on reliability of $SiO_x$ interfacial layer. *Microelectron. Reliab.*, **54** (9-10), 1712–1717.

30 Murakami, K., Rommel, M., Hudec, B., Rosová, A., Hušeková, K., Dobrocka, E., Rammula, R., Kasikov, A., Han, J.H., Lee, W., Song, S.J., Paskaleva, A., Bauer, A.J., Frey, L., Fröhlich, K., Aarik, J., and Hwang, C.S. (2014) Nanoscale characterization of $TiO_2$ films grown by atomic layer deposition on $RuO_2$ electrodes. *ACS Appl. Mater. Interfaces*, **6** (4), 2486–2492.

31 Rommel, M., Jambreck, J.D., Lemberger, M., Bauer, A.J., Frey, L., Murakami, K., Richter, C., and Weinzierl, P. (2013) Influence of parasitic capacitances on conductive AFM I-V measurements and approaches for its reduction. *J. Vac. Sci. Technol., B*, **31** (1), 01A108-1–01A108-7.

32 Yanev, V., Erlbacher, T., Rommel, M., Bauer, A.J., and Frey, L. (2009) Comparative study between conventional macroscopic IV techniques and advanced AFM based methods for electrical characterization of dielectrics at the nanoscale. *Microelectron. Eng.*, **86** (7-9), 1911–1914.

33 Garcia, R., Martinez, R.V., and Martinez, J. (2006) Nano-chemistry and scanning probe nanolithographies. *Chem. Soc. Rev.*, **35** (1), 29–38.

34 Frammelsberger, W., Benstetter, G., Kiely, J., and Stamp, R. (2007) C-AFM-based thickness determination of thin and ultra-thin $SiO_2$ films by use of different conductive-coated probe tips. *Appl. Surf. Sci.*, **253** (7), 3615–3626.

35 Erlbacher, T., Graf, T., DasGupta, N., Bauer, A.J., and Ryssel, H. (2009) Suppression of parasitic electron injection in silicon-oxide-nitride-oxide-silicon-type memory cells using high-k capping layers. *J. Vac. Sci. Technol., B*, **27** (1), 482–485.

36 Lányi, Š. (2005) Effect of tip shape on capacitance determination accuracy in scanning capacitance microscopy. *Ultramicroscopy*, **103** (3), 221–228.

37 Tran, T., Oliver, D., Thomson, D., and Bridges, G. (2001) "Zeptofarad" ($10^{-21}$F) resolution capacitance sensor for scanning capacitance microscopy. *Rev. Sci. Instrum.*, **72** (6), 2618–2623.

38 Buh, G.H., Tran, C., and Kopanski, J.J. (2004) PSPICE analysis of a scanning capacitance microscope sensor. *J. Vac. Sci. Technol., B*, **22** (1), 417–421.

39 Yanev, V., Rommel, M., Lemberger, M., Petersen, S., Amon, B., Erlbacher, T., Bauer, A.J., Ryssel, H., Paskaleva, A., Weinreich, W., Fachmann, C., Heitmann, J., and Schroeder, U. (2008) Tunneling atomic-force microscopy as a highly sensitive mapping tool for the characterization of film morphology in thin high-k dielectrics. *Appl. Phys. Lett.*, **92** (25), 252910-1–252910-3.

40 Iglesias, V., Porti, M., Nafría, M., Aymerich, X., Dudek, P., Schroeder, T., and Bersuker, G. (2010) Correlation between the nanoscale electrical and morphological properties of crystallized hafnium oxide-based metal oxide semiconductor structures. *Appl. Phys. Lett.*, **97** (26), 262906-1–262906-3.

41 Lanza, M., Porti, M., Nafría, M., Aymerich, X., Whittaker, E., and Hamilton, B. (2010) UHV CAFM characterization of high-k dielectrics: effect of the technique resolution on the pre- and post-breakdown electrical measurements. *Microelectron. Reliab.*, **50** (9-11), 1312–1315.

42 Lanza, M., Porti, M., Nafría, M., Aymerich, X., Benstetter, G., Lodermeier, E., Ranzinger, H., Jaschke, G., Teichert, S., Wilde, L., and Michalowski, P.P. (2011) Conductivity and charge trapping after electrical stress in amorphous and polycrystalline $Al_2O_3$-based devices studied with AFM-related techniques. *IEEE Trans. Nanotechnol.*, **10** (2), 344–351.

43 Menou, N., Popovici, M., Clima, S., Opsomer, K., Polspoel, W., Kaczer, B., Rampelberg, G., Tomida, K., Pawlak, M.A., Detavernier, C., Pierreux, D., Swerts, J., Maes, J.W., Manger, D., Badylevich, M., Afanasiev, V., Conard, T., Favia, P., Bender, H., Brijs, B., Vandervorst, W., Van Elshocht, S., Pourtois, G., Wouters, D.J., Biesemans, S., and Kittl, J.A. (2009) Composition influence on the physical and electrical properties of $Sr_xTi_{1-x}O_y$-based metal-insulator-metal capacitors prepared by atomic layer deposition using TiN bottom electrodes. *J. Appl. Phys.*, **106** (9), 094101-1–094101-7.

44 Ikeda, H., Goto, T., Sakashita, M., Sakai, A., Zaima, S., and Yasuda, Y. (2003) Local leakage current of $HfO_2$ thin films characterized by conducting atomic force microscopy. *Jpn. J. Appl. Phys., Part 1*, **42** (Pt.1 4B), 1949–1953.

45 Menou, N., Wang, X.P., Kaczer, B., Polspoel, W., Popovici, M., Opsomer, K., Pawlak, M.A., Knaepen, W., Detavernier, C., Blomberg, T., Pierreux, D., Swerts, J., Maes, J.W., Favia, P., Bender, H., Brijs, B., Vandervorst, W., Van Elshocht, S., Wouters, D.J., Biesemans, S., and Kittl, J.A. (2008) 0.5 nm EOT low leakage ALD $SrTiO_3$ on TiN MIM capacitors for DRAM applications in *Electron Devices Meeting, Proceedings of the IEEE International Conference*, San Francisco, CA, Dec 15–17, 2008 (ed. S. Deleonibus), Institute of Electrical and Electronics Engineers, Inc., Piscataway, NJ, pp. 1–4.

46 Frammelsberger, W., Benstetter, G., Schweinboeck, T., Stamp, R.J., and Kiely, J. (2003) Characterization of thin and ultra-thin $SiO_2$ films and $SiO_2$/Si interfaces with combined conducting and topographic atomic force microscopy. *Microelectron. Reliab.*, **43** (9-11), 1465–1470.

47 Lenzlinger, M. and Snow, E.H. (1969) Fowler-Nordheim tunneling into thermally grown $SiO_2$. *J. Appl. Phys.*, **40** (1), 278–283.

48 Zhao, X. and Vanderbilt, D. (2002) Phonons and lattice dielectric properties of zirconia. *Phys. Rev. B*, **65** (7), 75105-1–75105-10.

49 Schaadt, D.M., Yu, E.T., Vaithyanathan, V., and Schlom, D.G. (2004) Nanoscale current transport in epitaxial $SrTiO_3$ on $n^+$-Si investigated with conductive atomic force microscopy. *J. Vac. Sci. Technol., B*, **22** (4), 2030–2034.

50 Aoulaiche, M., Houssa, M., Degraeve, R., Groeseneken, G., De Gendt, S. and Heyns, M.M. (2005) Polarity dependence of bias temperature instabilities in $Hf_xSi_{1-x}ON$/TaN gate stacks. Proceedings of the ESSDERC Grenoble, France, pp. 197-200.

51 Lee, B.H., Choi, R., Sim, J.H., Krishnan, S.A., Peterson, J.J., Brown, G.A., and Bersuker, G. (2005) Validity of constant voltage stress based reliability assessment of high-k devices. *IEEE Trans. Device Mater. Reliab.*, **5** (1), 20–25.

52 Zafar, S., Kumar, A., Gusev, E., and Cartier, E. (2005) Threshold voltage instabilities in high-k gate dielectric stacks. *IEEE Trans. Device Mater. Reliab.*, **5** (1), 45–64.

53 Fiorenza, P., Polspoel, W., and Vandervorst, W. (2006) Conductive atomic force microscopy studies of thin $SiO_2$ layer degradation. *Appl. Phys. Lett.*, **88** (22), 222104-1–222104-3.

54 Weinreich, W., Wilde, L., Kücher, P., Lemberger, M., Yanev, V., Rommel, M., Bauer, A.J., Erben, E., Heitmann, J., Schröder, U., and Oberbeck, L. (2009) Correlation of microscopic and macroscopic electrical characteristics of high-k $ZrSi_xO_{2-x}$ thin films using tunneling atomic force microscopy. *J. Vac. Sci. Technol., B*, **27** (1), 364–368.

55 Degraeve, R., Groeseneken, G., Bellens, R., Ogier, J.L., Roussel, P.J., and Maes, H.E. (1998) New insights in the relation between electron trap generation and the statistical properties of oxide breakdown. *IEEE Trans. Electron Devices*, **45** (4), 904–911.

56 Erlbacher, T., Yanev, V., Rommel, M., Bauer, A.J., and Frey, L. (2011) Gate oxide reliability at the nanoscale evaluated by combining conductive atomic force microscopy and constant voltage stress. *J. Vac. Sci. Technol., B*, **29** (1), 01AB08-1–01AB08-4.

57 Stathis, J.H. (2001) Physical and predictive models of ultrathin oxide reliability in CMOS devices and circuits. *IEEE Trans. Device Mater. Reliab.*, **1** (1), 43–59.
58 Suñe, J. and Miranda, E. (2000) Soft and hard breakdown of ultrathin $SiO_2$ gate oxides. *Proc. Electrochem. Soc.*, **2000-2**, 333–344.
59 Wu, E.Y. and Suñe, J. (2005) Power-law voltage acceleration: a key element for ultra-thin gate oxide reliability. *Microelectron. Reliab.*, **45** (12), 1809–1834.
60 Iglesias, V., Martin-Martinez, J., Porti, M., Rodriguez, R., Nafría, M., Aymerich, X., Erlbacher, T., Rommel, M., Murakami, K., Bauer, A.J., Frey, L., and Bersuker, G. (2013) Bimodal CAFM TDDB distributions in polycrystalline $HfO_2$ gate stacks: the role of the interfacial layer and grain boundaries. *Microelectron. Eng.*, **109**, 129–132.
61 Nigam, T., Kerber, A. and Peumans, P. (2009) Accurate model for time-dependent dielectric breakdown of high-k metal gate stacks. IEEE International Reliability Physics Symposium Proceedings, pp. 523-530, 26–30.
62 Blasco, X., Nafría, M., Aymerich, X., Pétry, J., and Vandervorst, W. (2005) Nanoscale post-breakdown conduction of $HfO_2/SiO_2$ MOS gate stacks studied by enhanced-CAFM. *IEEE Trans. Electron Devices*, **52** (12), 2817–2819.
63 Bersuker, G., Heh, D., Young, C.D., Morassi, L., Padovani, A., Larcher, L., Yew, K.S., Ong, Y.C., Ang, D.S., Pey, K.L. and Taylor, W. (2010) Mechanism of high-k dielectric-induced breakdown of the interfacial SiO2 layer. IEEE International Reliability Physics Symposium Proceedings, pp. 373-378.
64 McKenna, K. and Shluger, A. (2009) The interaction of oxygen vacancies with grain boundaries in monoclinic $HfO_2$. *Appl. Phys. Lett.*, **95** (22), 222111-1–222111-3.
65 Chowdhury, M.H., Mannan, M.A., and Mahmood, S.A. (2009) Temperature dependent time-to-breakdown ($T_{BD}$) of TiN/$HfO_2$ n-channel MOS devices in inversion. *Microelectron. Reliab.*, **49** (5), 495–498.
66 Bersuker, G., Chowdhury, N., Young, C., Heh, D., Choi, R. and Misra, D. (2007) Progressive breakdown characteristics of high-k/metal gate stacks. IEEE International Reliability Physics Symposium Proceedings, pp. 49–54.

# 5

# Characterization of Grain Boundaries in Polycrystalline $HfO_2$ Dielectrics

*Shubhakar Kalya*[1], *Sean Joseph O'Shea*[2], *and Kin Leong Pey*[1]

[1] *Singapore University of Technology and Design (SUTD), Engineering Product Development (EPD) Pillar, 8 Somapah Road, Singapore 487372, Singapore*
[2] *A*STAR, Institute of Materials Research and Engineering (IMRE), Materials Processing and Characterization Department, 2 Fusionopolis Way, Singapore 138634, Singapore*

## 5.1 Introduction

Most of the electrical behavior and reliability studies of the $HfO_2$ high-$k$ (HK) dielectric-based metal oxide semiconductor (MOS) devices have been performed using standard electrical characterization methods at the device level [1–6]. These studies provide averaged information on the electrical and physical properties of a $HfO_2$ dielectric stack incorporating a thin $SiO_x$ ($x \leq 2$) interfacial layer (IL) on a silicon substrate. Conversely, scanning probe microscopy (SPM)-based techniques are able to characterize the morphological and local electrical properties of dielectrics at a nanometer scale [7–11]. In this chapter, conductive atomic force microscopy (CAFM), one of the important types of SPM, is used as a nanoscale-characterization tool to study the electrical and physical properties of grain boundaries (GBs) in polycrystalline $HfO_2$ dielectrics.

The degradation and breakdown (BD) phenomenon in a $HfO_2/SiO_x$ dielectric stack is different from that in ultrathin $SiO_2/SiO_xN_y$ dielectrics [6, 12, 13] and its analysis is complex owing to the polycrystalline nature of the $HfO_2$ dielectric and the presence of the $SiO_x$ IL. The different bonding strengths and defect generation mechanisms for both $SiO_x$ and $HfO_2$ layers within the stack further complicate investigations into the degradation and BD phenomenon [14–18]. In $HfO_2$-based advanced MOS devices, one of the major concerns is the variations in the local property of $HfO_2$ dielectric and reliability of the overall $HfO_2/SiO_x$ dielectric stack. The GBs in polycrystalline $HfO_2$ dielectrics, consisting of a high density of localized-process-induced traps, show complex current conduction mechanisms and serve as high-leakage paths [14–18]. Hence it is important to study the role of GB in the degradation and BD of $HfO_2/SiO_x$ dielectric stacks. The study of the degradation mechanism and the identity of the layer that drives

*Conductive Atomic Force Microscopy: Applications in Nanomaterials,* First Edition.
Edited by Mario Lanza.

the BD in the $HfO_2/SiO_x$ dielectric stack are very critical for understanding the reliability and failure mechanisms. As the GBs have a width ranging from 0.5 to 1 nm [15, 16], characterization of their electrical properties at these locations is challenging, and thus CAFM provides one of the few suitable tools for such characterization. In this chapter, the effect of polycrystallization of the $HfO_2$ dielectric and the role of the GB in the degradation and BD of $HfO_2/SiO_x$ dielectric stack based on CAFM measurements are described.

## 5.2 Experimental Details and Sample Specifications

Conductive AFM has been widely used to study local variations in the electrical characteristics of dielectric materials at a nanometer scale [8–11, 17–21]. In this work, CAFM characterization of polycrystalline $HfO_2$ dielectric was performed using a Pt-wire tip [22], which acts as a top gate. The use of an all-metal probe ensures electrical integrity of the tip during the experiments. The conducting Pt-wire tip collects and measures (via a current-voltage amplifier) the current flow through the dielectric at every pixel while scanning over the surface topography with a constant DC voltage bias applied to the sample [8]. Hence, the topographical and electrical properties of the dielectric can be collected simultaneously at a nanometer scale. The CAFM is operated in an ultrahigh vacuum environment to avoid artifacts which can strongly influence the electrical characterization and reliability analysis. Ultrahigh vacuum reduces oxidation and surface contamination [23, 24], and ensures no liquid meniscus forms at the tip-surface contact, both of which strongly modify the measured signals. In addition to the acquisition of topography and current images, current–voltage ($I–V$) spectroscopy is obtained at a fixed location on the surface by linearly ramping the bias voltage and measuring the current. Current–stress time ($I–t$) spectroscopy is also measured at a fixed location on the surface by measuring the current as a function of time under a constant applied stressing voltage.

Two samples of $HfO_2$ dielectric were fabricated using atomic layer deposition with thicknesses of ~4.5 nm and ~2.5 nm, and subjected to post-deposition annealing (PDA) at 600 and 1000 °C, respectively. A very thin layer of $SiO_x$ (~1 nm), IL was formed when the $HfO_2$ dielectric was deposited over the silicon substrate.

## 5.3 Formation of Grain Boundaries and Its Local Electrical Properties in $HfO_2$ Dielectric

As-deposited amorphous $HfO_2$ dielectric crystallizes to polycrystalline form upon standard PDA treatments [4, 25]. Polycrystallization of the $HfO_2$ dielectric could result in nonuniform morphological and local electrical characteristics. In turn, variations in the surface morphology and the local electrical properties of the $HfO_2$ dielectric could affect the performance and reliability of the

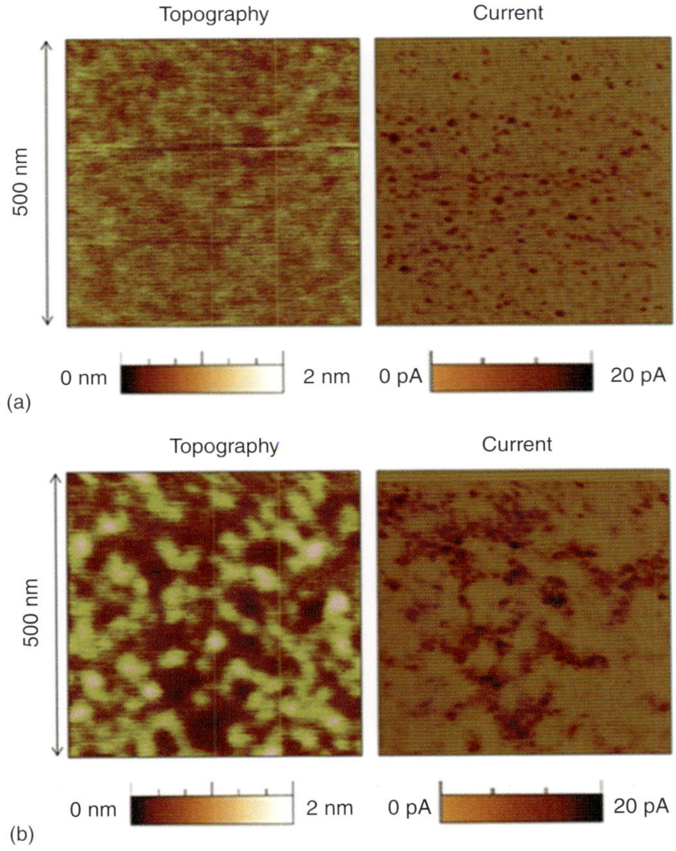


**Figure 5.1** Simultaneously collected topography-current images on the surface of (a) as-deposited and (b) 1000 °C annealed $HfO_2$ (2.5 nm)/$SiO_2$ (1 nm)/Si stack. The images clearly show changes in structural and electrical properties of the $HfO_2$ dielectric. (Reproduced with permission from [11], Copyright MDPI publishing group 2014.)

$HfO_2$-based MOS device. Figure 5.1a,b shows dielectric surface morphology and spatial leakage current profile of as-deposited and 1000 °C annealed $HfO_2$ dielectric films over a $500 \times 500\ nm^2$ area [11]. The as-deposited dielectric exhibits reasonably uniform topography and electrical conduction over the entire surface. In contrast, the annealed film shows an increase in surface roughness and an increase in the leakage current distribution and magnitude. The increase in surface roughness is believed to be due to the effect of polycrystallization in the $HfO_2$ dielectric, indicating microstructural change of the dielectric with larger inhomogeneity of topography. The annealed $HfO_2$ shows obvious granular structures (Figure 5.1b), thus also implying the formation of GBs, which are not distinct in the as-deposited films (Figure 5.1a). The increased leakage current and electrical inhomogeneity is due to the presence of the GBs in the polycrystalline structure formed during the PDA. In this regard, note that the more pronounced

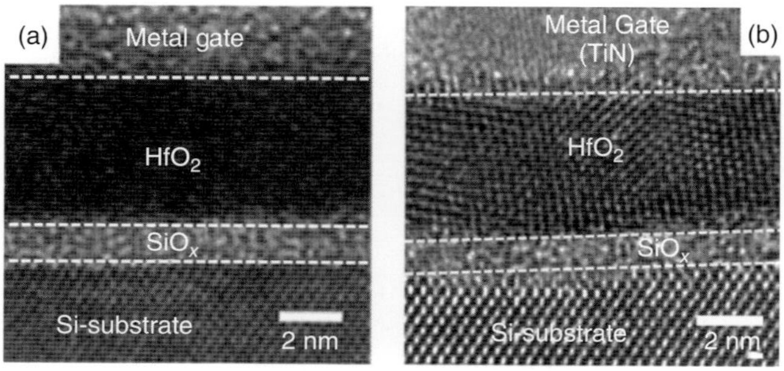


**Figure 5.2** TEM micrographs of (a) as-deposited $HfO_2$ and (b) 600 °C annealed $HfO_2$ on Si substrate with $SiO_x$ as IL. ((a) Reproduced with permission from [25]. Electrochemical Society. (b) Reproduced with permission from [8]. Elsevier B. V. 2013.)

leakage sites occur over the larger topographic regions (Figure 5.1b) and the high leakage current follows the topographic contours in these regions. In regions having more closely packed grain structure (e.g., lower left corner of Figure 5.1b), the current is small and appears uniform.

Figure 5.2a,b shows the transmission electron microscopy (TEM) cross section of a metal gate/$HfO_2$ (~4–5 nm)/$SiO_x$ (1 nm)/Si structure of as-deposited and 600 °C annealed samples, respectively. The TEM micrographs show the transformation from amorphous to polycrystalline microstructure upon annealing, depicting the change in the structural properties of $HfO_2$ after PDA.

The larger thickness fluctuation after annealing has implications for the performance and reliability of the $HfO_2/SiO_x$ dielectric stack. Variation of the surface roughness of the $HfO_2$ can affect the local electric field across the $HfO_2$ and thus also across the underlying $SiO_x$ IL. Hence, the nanoscale roughness may have an important effect on BD characteristics, as the BD is a local phenomenon that takes place by the formation of a percolation path under applied electric field stress.

In the following, the CAFM technique is applied to investigate conduction characteristics of the GBs in polycrystalline $HfO_2$ dielectrics. Figure 5.3a,b shows a CAFM topographical image and the corresponding current map, respectively, showing the grains and their GB contours. A higher leakage current, that is, bright spots on the current map at the GB locations, was observed corresponding to the depressed locations on the topographical image (Figure 5.3a), indicating the enhanced trap-assisted tunneling (TAT) [5] through the high concentration of pre-existing traps present at the GBs. Figure 5.3c shows a line profile of the topography and current map along the line $X_1$–$X_2$, clearly showing an increased leakage current at the GB location [8]. In general, higher conductivity is observed along the GBs with distinctly lower leakage current within the grain. A higher tunneling current at the GBs can arise from electrical defects (e.g., oxygen vacancy and crystal defects) and/or a smaller physical thickness. A higher gate tunneling leakage current was observed at the GBs attributed to enhance TAT through the dense chain of process-induced traps at the GB. In terms of thickness, regions of thinner film will have higher tunnel current at a given bias

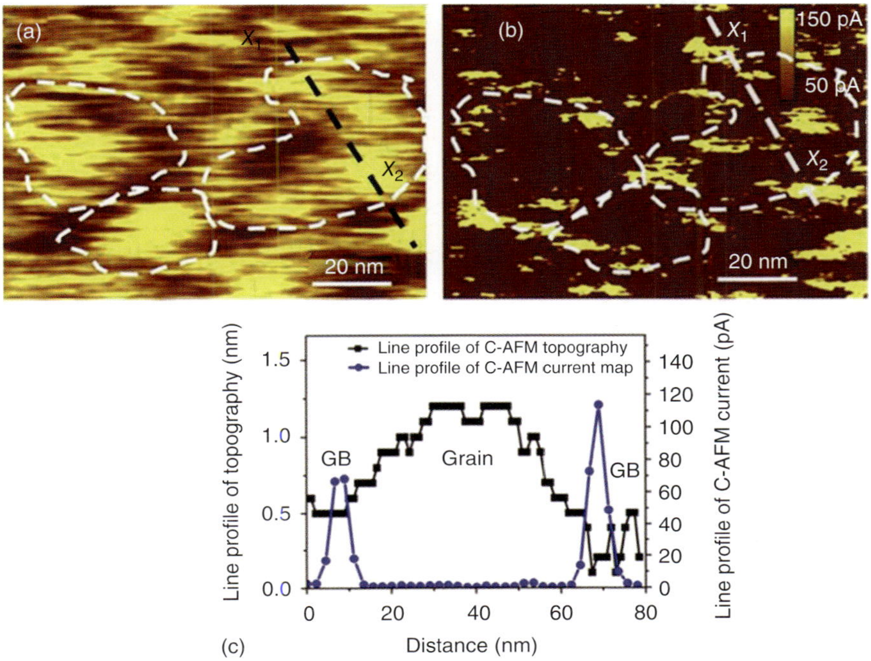


**Figure 5.3** (a) CAFM topography and (b) corresponding CAFM current map (at +5 V) of the $HfO_2$ gate dielectric depicting the grain and GB profile. (c) Line profile of topography and current map along line '$X_1$–$Y_2$' shown in (a) and (b). (Shubhakar *et al.* 2013 [8]. Reproduced with permission of Elsevier.)

voltage. It is observed in Figure 5.3,a,b that the thinner regions coincide with the GBs.

GBs, which arise from inter-grain orientational mismatch, serve as a source of localized high concentration of defects, which take the functional form of oxygen vacancies and contain a high density of favorable sites for oxygen vacancy segregation [14–16]. It is reported that grain size with an average lateral size ~30 nm was observed with 600 °C annealed $HfO_2$ [8]. Defects at the GBs may also be able to trap more electrons than those in the bulk owing to their deeper electronic states and tend to accumulate more at the GB than in the grain sites [15]. Thus the GBs, with a larger density of defects compared to the bulk grain, have a lower effective barrier height for electron transport from the CAFM tip to the substrate and this translates to a higher leakage current in CAFM.

A sequence of *I*–*V* spectroscopic measurements has been performed at different random locations of the $HfO_2$ dielectric (~2.5 nm annealed at 1000 °C) [18]. The results reveal significantly higher leakage currents through the GBs than through grain regions. A comparative analysis of conduction characteristics of the GBs and grains is carried out using *I*–*V* curves measured on fresh and electrically stressed sites. Figure 5.4 shows typical sets of four *I*–*V* curves measured sequentially on the grain (filled symbols) and GB (open symbols) regions. Comparison of the first measurements on the fresh (i.e., unstressed) locations (G_IV1

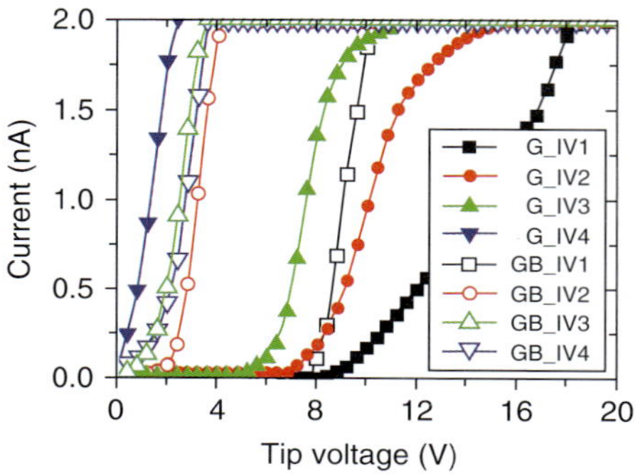


**Figure 5.4** Evolution of current through the grain (G_IV) and GB (GB_IV) during sequential RVS. The maximum current measured by CAFM is 2 nA. The first *I–V* curves (IV1) were taken on a fresh site. (Reproduced with permission from [18]. AIP Publishing 2011.)

and GB_IV1) shows that the grains are less conductive than the GBs, as expected [18]. Subsequent *I–V* measurements on the same GB or grain sites indicate that conduction through these regions is governed by different mechanisms. For the GBs, a dielectric BD was triggered by the first *I–V* sweep (GB_IV1). In contrast, at the grain (G_IV1-G_IV4), the *I–V* curves progressively shift to lower voltages via stress-induced leakage current [1], and BD is not triggered until several *I–V* sweeps have been applied.

## 5.4 RVS and CVS Stressing of $HfO_2/SiO_x$ Dielectric Stack

In a localized stressing of the $HfO_2/SiO_x$ dielectric stack, the tip was moved to the location of interest to perform ramped voltage stressing (RVS) or constant voltage stressing (CVS) with a bias applied to the sample. To analyze electrical conduction at these locations with RVS, the *I–V* curves were measured with the enhanced-conductive atomic force microscopy (E-CAFM) (having a larger dynamic range) on both GBs and grain regions as shown in Figure 5.5. E-CAFM uses a semiconductor parameter analyzer to apply the voltage and measure currents, which greatly increases the current range. Again, the fresh *I–V* curve at the GB location shows larger current with two conduction regimes, indicating that the BD generally occurs at lower voltages compared to grain regions. For grain regions, the initial current is lower, and the two conduction regimes are not seen. Moreover, the BD is triggered only after multiple *I–V* sweeps, indicating better robustness of grains to localized electrical stress.

Since the GBs are more conductive, most of the voltage applied to the $HfO_2/SiO_x$ dielectric stack drops across the interfacial $SiO_x$ layer beneath the GBs. This leads to faster degradation of the $SiO_x$ under the GB, resulting in a gradual increase in the current measured at the GBs as the stress proceeds, eventually resulting in the BD of the entire gate stack [18].

Figure 5.6 shows typical CVS stressing data (sample voltage +6 V) at different grain and GB regions. During a CVS test, the tip was maintained over a location for a specific period of time with a constant voltage applied between the tip and the sample. Different sets of current–time (*I–t*) readings were obtained. Over a number of locations, the dielectric degradation and BD occur significantly faster;

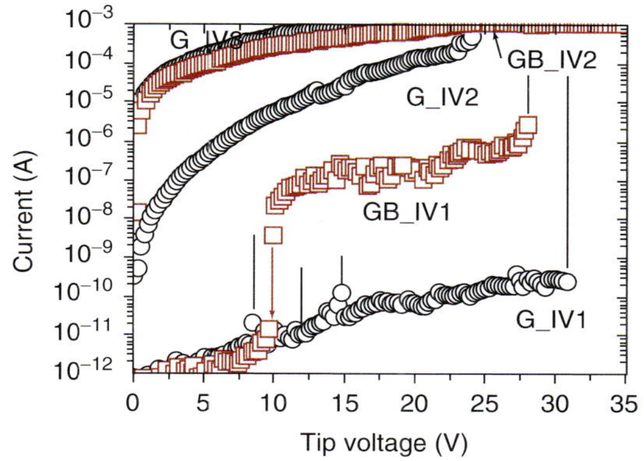


**Figure 5.5** Evolution of current through the grain (G_IV) and GB (GB_IV) during sequential RVS using E-CAFM. The maximum current measured by E-CAFM is 1 mA. The first *I–V* curves (IV1) were taken on a fresh site. (Reproduced with permission from [18]. AIP Publishing 2011.)

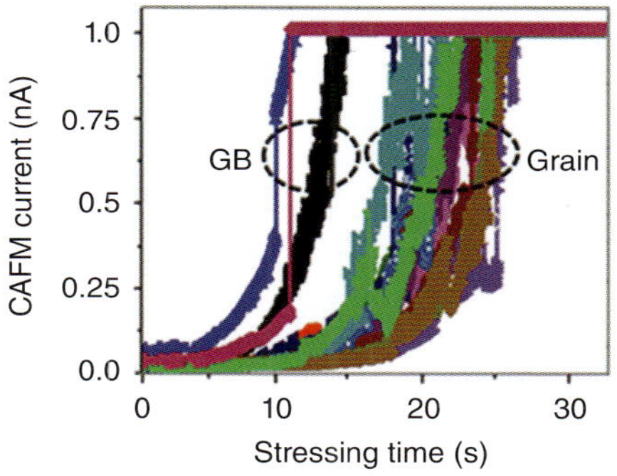


**Figure 5.6** CAFM leakage current evolution with stress time under constant voltage stress of +6 V at different locations on the dielectric. Note the CAFM current saturates at 1 nA in these experiments. (Shubhakar *et al.* 2013 [8]. Reproduced with permission of Elsevier.)

most probably because these stressed locations are GBs, or close to a GB, where the trap generation rate is higher. The pre-existing defects present in the GBs take part in an initial TAT leakage current. Under CVS stressing, additional defects are generated along the GB at a higher rate compared to grain regions, which leads to the formation of a percolation path, which in turn strongly enhances the local current and the related power dissipation [8, 26]. Hence, the time-dependent dielectric breakdown (TDDB) lifetime at the GBs is much lower than that at the grain locations as observed in Figure 5.6, supporting the hypothesis of GB-assisted degradation and BD of HK dielectrics [8, 26].

Figure 5.7 shows the Weibull probability plot of the TDDB distribution for CAFM-induced localized BD events of $HfO_2/SiO_x$ dielectric stack at ~20 different locations, shown in Figure 5.6. The results clearly show convexial distribution with steeper low-percentile and shallower high-percentile trends. This nonlinearity is due to the GB defects present in the $HfO_2$ dielectric. Hence GBs play a crucial role in determining the shape of the statistical distribution. The results obtained from CAFM data (Figure 5.6) are in good agreement with the simulation results [8], supporting faster degradation at the GB sites. So, nonuniform electrical characteristics throughout the polycrystalline $HfO_2$ dielectric film were observed by the *I–V* and *I–t* measurements.

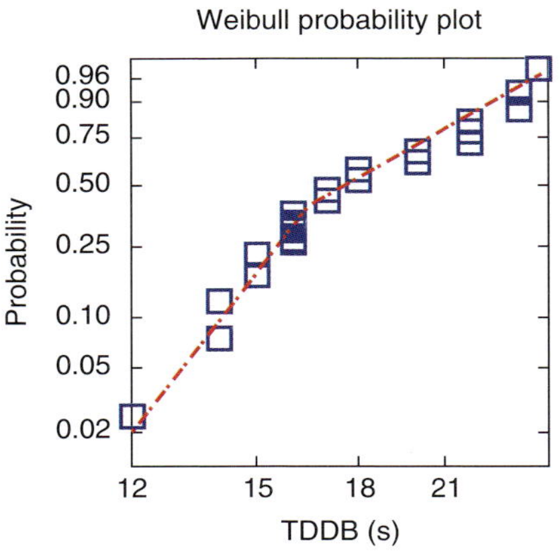


**Figure 5.7** Weibull plot of TDDB for CAFM-induced BD locations (~20) in the $HfO_2/SiO_x$ stack shown in Figure 5.6. (Shubhakar *et al.* 2013 [8]. Reproduced with permission of Elsevier.)

## 5.5 Uniform Stressing with Successive Scanning in CAFM Mode

In uniform stressing using CAFM, a fresh region of the $HfO_2$ surface was selected and successive images were taken (i.e., the tip "scans" the surface) over the area with a fixed applied bias voltage. The topography and current maps were collected simultaneously and compared with subsequent topography and current maps to observe the effect of uniform voltage stressing. The CAFM topography and corresponding current maps of a fresh $HfO_2/SiO_x$ surface obtained at sample bias of +6 V over a ~200 × 200 nm$^2$ area are shown in Figure 5.8a,c, respectively. Figure 5.8c shows the leakage spots (bright areas representing higher current) present in the $HfO_2/SiO_x$ dielectric stack. The topography and current map in Figure 5.8b,d, respectively, were obtained after the second scan at +6 V. The current map (Figure 5.8d) shows increased degradation (i.e., an increase in the leakage current and often the leakage spot area) of the dielectric stack in the second scan, indicating the +6 V sample bias used in the first scan (Figure 5.8b) was high enough to electrically stress the $HfO_2/SiO_x$ stack to create more leaky spots [27]. It should be noted that most of the leakage spots observed in the first scan were observed in the second scan but with increased current, indicating more severe degradation. Hence, the density of the leaky sites and their conductance increase with stressing. A line profile of the leakage current along the line $X_1$–$X_2$ (Figure 5.8c,d) after the first and second scan is shown in Figure 5.9. The line profiles allow for quantification of the increased leakage current and any growth of leakage spot area. These leaky sites are considered to be the precursors of dielectric BD [27]. Here, the specific degradation evolution under uniform stressing of either the grain or GB could not be identified. This observation highlights the need to scan a smaller area for GB studies in CAFM experiments. Hence, a smaller scanning area around 100 × 100 nm$^2$ area was selected to study the degradation at grain and GB regions with uniform stressing.

Figure 5.10 shows the CAFM topography and the corresponding current map (around 100 × 100 nm$^2$ area) obtained at a constant voltage over the same area of

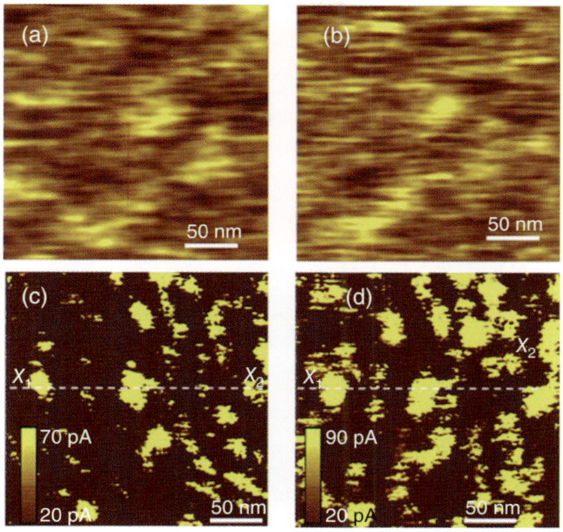


**Figure 5.8** CAFM topography map of a $HfO_2/SiO_x$ dielectric stack after (a) first scan and (b) second scan at +6 V sample bias. (c, d) Show the corresponding current maps after the first and second scans. (Reproduced with permission from [27]. IEEE Publishers 2012.)

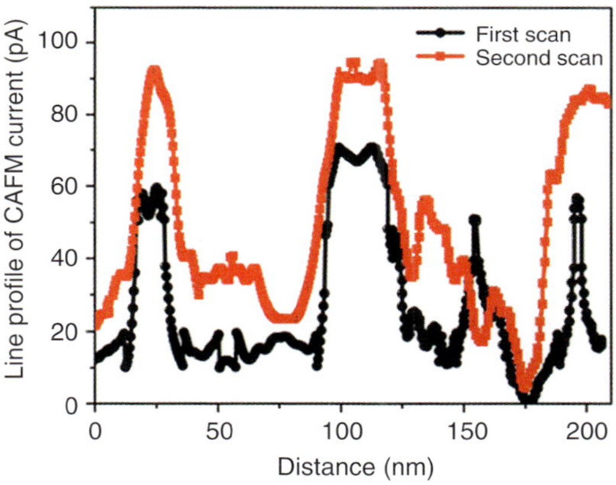


**Figure 5.9** Line profile of current map along the line "$X_1$–$X_2$" after the first and second scans shown in Figure 5.8c,d. (Reproduced with permission from [27]. IEEE Publishers 2012.)

the dielectric upon uniform stressing. Figure 5.10a shows the topography and, (b) and (c) show sequence of the current maps over the same area obtained at +6 V, respectively. These results show clearly the degradation under uniform stressing occurs primarily at the GB regions (the dashed lines in Figure 5.10). The magnitude of the leakage current and the number of high-intensity spots along the GB increases upon uniform stressing, indicating a higher rate of stress-induced trap generation compared to the grain regions. The line profiles of topography and successive current maps along the line $X$–$Y$ are shown in Figure 5.10d.

Higher leakage current at GBs is mainly due to TAT through the pre-existing and stress-induced traps in the $HfO_2$ [8, 27]. The magnitude of the leakage current and the number of leakage spots increases significantly along the GB upon successive stressing, indicating a higher rate of stress-induced trap generation compared to the grain regions. The faster degradation of the more conductive GB regions leads to earlier formation of percolation paths and BD spots compared to the grain regions [8, 17]. Underlying the GB, the thin $SiO_x$ IL region

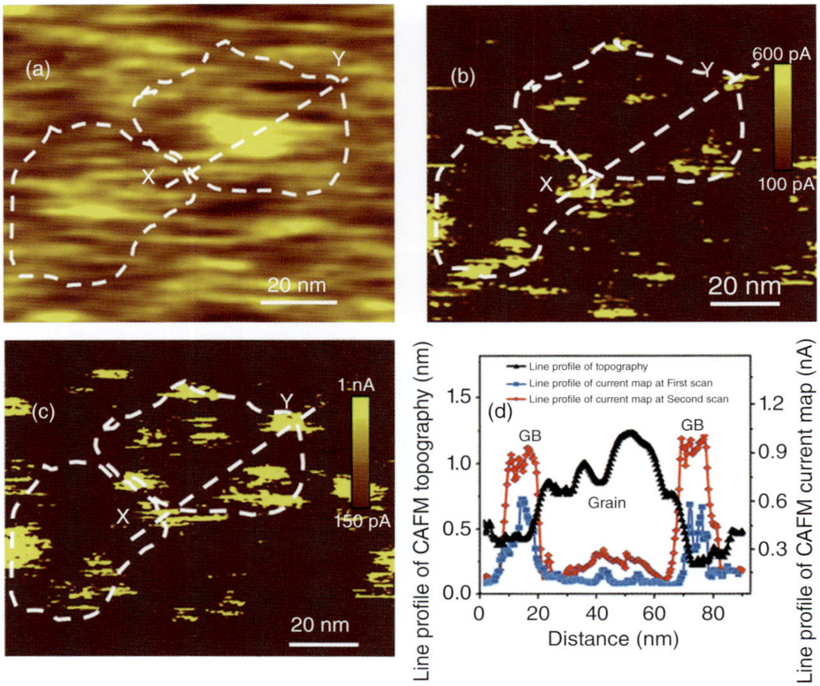


**Figure 5.10** (a) CAFM topography and leakage current profile at +6 V after the (b) first scan and (c) second scan. (d) Line profile of topography and current along the line *X–Y* shown in a, b and c. (Shubhakar *et al.* 2013 [8]. Reproduced with permission of Elsevier.)

experiences a higher electric field under the uniform stressing due to the lower electrical resistance at the GB, inducing stress-induced traps within the $SiO_x$ IL. Hence, degradation and BD of the $SiO_x$ IL can occur, which further enhances the degradation rate at the GB locations.

Modeling is very useful to gain microscopic understanding of the degradation process in uniform stressing of dielectrics. A Multi-physics-based percolation framework model was used to simulate the dynamics of the sequential and competitive evolution of soft-breakdown (SBD) spots during degradation of a $HfO_2/SiO_x$ dielectric stack [28]. This model leveraged on the combined use of kinetic Monte Carlo (KMC) routine to describe the microstructural variations and stochastics of defect nucleation and growth as well as the finite element model (FEM) to quantify the spatiotemporal evolution of the electric potential and electric field, and temperature distributions in the dielectric stack. The results point to $SiO_x$ IL being the first to BD and confirm that the preferential nucleation of SBD spots occurs beneath the grain boundary regions in the $HfO_2$; this is in good agreement with the above CAFM results.

Figure 5.11a–c shows the cross-sectional view of the electric-field (*E*-field) patterns of the $HfO_2/SiO_x$ dielectric stack for the three different phases of degradation. *Phase I* – For a fresh stack, the GB introduced in the $HfO_2$ contributes to

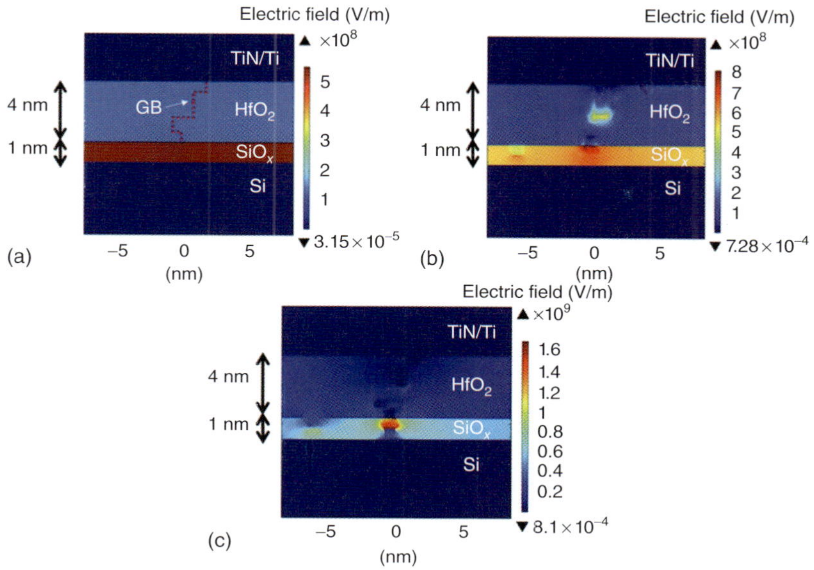


**Figure 5.11** Simulated cross-sectional electric field distribution in $HfO_2/SiO_x$ under different breakdown phases; (a) fresh device, (b) after $SiO_x$ IL SBD, and (c) subsequent $SiO_x$ IL and HK BD. (Reproduced with permission from [28]. IEEE Publishers 2016.)

a higher electric field across $SiO_x$ ($E_{SiO_x}$) in the $SiO_x$ IL beneath it (Figure 5.11a). This is because permittivity is a function of the vacancy density and it is locally higher at the GB regions in the $HfO_2$, which causes $E_{SiO_x}$ to increase, by Gauss's Law [29]. *Phase II* – Vacancies are generated with higher rate in the $SiO_x$ IL around the GB region owing to enhanced $E_{SiO_x}$. The vacancies tend to segregate toward the GB in the $HfO_2$, causing the value of dielectric constant at GB ($\kappa_{GB}$) to increase, tunneling current to go up, and resistivity to drop, resulting in further enhancement of $E_{SiO_x}$, initiating the first SBD event in the $SiO_x$ IL (Figure 5.11b). Following the percolation formation in the IL, the value of $E_{SiO_x}$ continues to increase owing to charge-trapping events at the SBD region [30, 31]. Some SBD events in $SiO_x$ IL below the grain are also generated in the dielectric, as there could be defect clusters located inside the grains of the $HfO_2$, either due to process-induced traps or stress-induced leakage current. *Phase III* – Eventually, the $HfO_2$ also undergoes SBD after the $SiO_x$ IL SBD event. At this stage, the percolation path is fully evolved and vertically bridges the gate and the substrate. After SBD formation in both the dielectric layers, the value of $E_{SiO_x} \gg$ electric field across $HfO_2$ ($E_{HfO_2}$), which drives the $SiO_x$ IL to undergo hard breakdown (HBD) when the field reaches the critical field-strength of $SiO_x$ [32, 33], eventually resulting overall in BD of the $HfO_2/SiO_x$ dielectric stack (Figure 5.11c). This approach is very helpful in explaining the enhanced field across the $SiO_x$ IL below the GB in the $HfO_2$ dielectric and provides insight into the physical mechanisms governing the defect generation and BD phenomenon.

## 5.6 Conclusions

We have shown how CAFM characterization at the nanometer scale of polycrystalline $HfO_2$ dielectrics can provide insight into the effect of GBs on the degradation and reliability of $HfO_2/SiO_x$ dielectric stacks. Nonuniform electrical characteristics throughout the polycrystalline $HfO_2$ film were observed by various stressing approaches (RVS, CVS, and uniform stressing) followed by subsequent current map imaging, $I$–$V$ and $I$–$t$ measurements. The GBs in polycrystalline $HfO_2$ are the dominant locations of high leakage current. The higher density of electronic traps at the GBs in $HfO_2$ is likely to be the limiting factor in the reliability of $HfO_2/SiO_x$ dielectric stacks. The increased concentration and preferential segregation of vacancies at the GB plays an important role in early percolation, supporting the hypothesis of GB-assisted degradation and BD of HK dielectrics. The $SiO_x$ IL below the GB regions is likely to be the BD location in $HfO_2/SiO_x$ dielectric stacks. Hence, the microstructure of the $HfO_2$ high-$k$ dielectric thin film plays a very significant role in determining the overall TDDB robustness.

## References


1 Wilk, G.D., Wallace, R.M., and Anthony, J.M. (2001) *J. Appl. Phys.*, **89** (10), 5243–5275.

2 Ribes, G., Mitard, J., Denais, M., Bruyere, S., Monsieur, F., Parthasarathy, C., Vincent, E., and Ghibaudo, G. (2005) *IEEE Trans. Device Mater. Reliab.*, **5**, 5–19.

3 Degraeve, R., Aoulaiche, M., Kaczer, B., Roussel, P., Kauerauf, T., Sahhaf, S., and Groeseneken, G. (2008) IEEE International Symposium on the Physical and Failure Analysis of Integrated Circuits (IPFA), pp. 1-6.

4 Migita, S. and Ota, H. (2011) IEEE International Conference on IC Design & Technology (ICICDT), pp.1-4.

5 Campera, A., Iannaccone, G. and Crupi, F. (2007) *IEEE Trans. Electron Dev.*, **54**, 83–89.

6 Pey, K.L., Ranjan, R., Tung, C.H., Tang, L.J., Lo, V.L., Lim, K.S., Selvarajoo, T.A.L., and Ang, D.S. (2005) *Microelectron. Eng.*, **80**, 353–361.

7 Ong, Y.C., Ang, D.S., Pey, K.L., O'Shea, S.J., Goh, K.E.J., Troadec, C., Tung, C.H., Kawanago, T., Kakushima, K., and Iwai, H. (2007) *Appl. Phys. Lett.*, **91**, 102905.

8 Shubhakar, K., Pey, K.L., Raghavan, N., Kushvaha, S.S., Bosman, M., Wang, Z., and O'Shea, S.J. (2013) *Microelectron. Eng.*, **109**, 364–369.

9 Chen, S.-H., Yu, C.F., Lin, Y.S., Xie, W.J., Hsu, T.W., and Tsai, D.P. (2008) *J. Appl. Phys.*, **104**, 114314.

10 Teichert, C. and Beinik, I. Chapter 23, (2011) Conductive atomic-force microscopy investigation of nanostructures in microelectronics, in *Scanning Probe Microscopy in Nanoscience and Nanotechnology*, vol. 2, Springer.

11 Lanza, M. (2014) *Materials*, **7**, 2155–2182.

12 Buchanan, D.A., Marwick, A.D., DiMaria, D.J., and Dori, L. (1994) *J. Appl. Phys.*, **76** (6), 3595.

13 DiMaria, D.J., Cartier, E., and Arnold, D. (1993) *J. Appl. Phys.*, **73** (7), 3367.
14 McKenna, K. and Shluger, A. (2009) *Appl. Phys. Lett.*, **95** (22), 222111-1–222111-3.
15 McKenna, K., Shluger, A., Iglesias, V., Porti, M., Nafría, M., Lanza, M., and Bersuker, G. (2011) *Microelectron. Eng.*, **88**, 1272–1275.
16 Bersuker, G., Yum, J., Iglesias, V., Porti, M., Nafría, M., McKenna, K., Shluger, A., Kirsch, P., and Jammy, R. (2010) Proceedings of the European Solid-State Device Research Conference (ESSDERC), pp. 333–336.
17 Shubhakar, K., Raghavan, N., Kushvaha, S.S., Bosman, M., Wang, Z.R., O'Shea, S.J., and Pey, K.L. (2014) *Microelectron. Reliab.*, **54**, 1712–1717.
18 Iglesias, V., Lanza, M., Zhang, K., Bayerl, A., Porti, M., Nafría, M., Aymerich, X., Benstteter, G., Shen, Z.Y., and Bersuker, G. (2011) *Appl. Phys. Lett.*, **99**, 103510.
19 Lanza, M., Iglesias, V., Porti, M., Nafria, M., and Aymerich, X. (2011) *Nanoscale Res. Lett.*, **6**, 108.
20 Porti, M., Gerardin, S., Nafría, M., Aymerich, X., Cester, A., Paccagnella, A., Schiavuta, P., and Pierobon, R. (2007) *Microelectron. Eng.*, **84**, 1956–1959.
21 Lanza, M., Porti, M., Nafria, M., Aymerich, X., Sebastiani, A., Ghidini, G., Vedda, A., and Fasoli, M. (2009) *IEEE Trans. Device Mater. Reliab.*, **9**, 529.
22 Bruker Pt-wire tip Bruker AFM probes: RMN-12PT300B, https://www.brukerafmprobes.com/Product.aspx?ProductID=3772
23 O'Shea, S.J., Atta, R.M., Murrell, M.P., and Welland, M.E. (1995) *J. Vac. Sci. Technol., B*, **13**, 1945.
24 Gordon, A.E., Fayfield, R.T., Litfin, D.D., and Higman, T.K. (1995) *J. Vac. Sci. Technol., B*, **13**, 2805.
25 Jung, H.-S., Jang, J.H., Cho, D.-Y., Jean, S.-H., Kim, H.K., Lee, S.Y., and Hwang, C.S. (2011) *Electrochem. Solid-State Lett.*, **14** (5), G17–G19.
26 Ranjan, R., Pey, K.L., Tung, C.H., Tang, L.J., Groeseneken, G., Bera, L.K., and De Gendt, S. (2004) IEEE International Electronic Devices Meeting (IEDM), Technical Digest, pp. 725–728.
27 Shubhakar, K., Pey, K. L., Bosman, M., Thamankar, R., Kushvaha, S. S., Loke, Y. C., Wang, Z.R., Raghavan, N., Wu, X., and O'Shea, S.J. (2012) IEEE International Symposium on the Physical and Failure Analysis of Integrated Circuits (IPFA), pp. 1–7.
28 Mei, S., Raghavan, N., Shubhakar, K., Bosman, M., and Pey, K.L. (2016) IEEE International Reliability Physics Symposium (IRPS), pp. 7A-2-1–7A-2-6.
29 Autran, J.L., Munteanu, D., and Houssa, M. (2003) *High-κ Gate Dielectrics* Chapter-3.4, Taylor and Francis.
30 Lombardo, S., Stathis, J.H., Linder, B.P., Pey, K.L., Palumbo, F., and Tung, C.H. (2005) *J. Appl. Phys.*, **98**, 121301.
31 Raghavan, N., Pey, K. L., Liu, W.H., and Li, X. (2010) IEEE International Reliability Physics Symposium (IRPS), pp. 778–786.
32 Padovani, A., Raghavan, N., Larcher, L., and Pey, K.L. (2013) *IEEE Electron Device Lett.*, **34**, 1289–1291.
33 Ribes, G., Mora, P., Monsieur, F., Rafik, M., Guarin, F., Yang, G., Roy, D., Chang, W.L., and Stathis, J. (2010) IEEE International Reliability Physics Symposium (IRPS), pp. 4A.2.1–4A.2.5.

# 6

# CAFM Studies on Individual GeSi Quantum Dots and Quantum Rings

*Rong Wu, Shengli Zhang, Yi Lv, Fei Xue, Yifei Zhang, and Xinju Yang*

*Department of Physics, Fudan University, Shanghai 200433, China*

## 6.1 Introduction

With the development of nanotechnology, owing to their unique properties compared to the bulk, various nanostructures have been fabricated and applied in almost all the fields of human life and scientific research. Among them, semiconductor nanostructures, particularly silicon-based nanostructures, have received considerable interest because of the novel electronic, electrical, and optical properties generated by their quantum confinement effects [1], as well as their great potential applications [2, 3].

Therefore, in recent decades, considerable efforts have been directed toward the fabrication of nanostructures as well as investigation of their properties. Despite the success achieved in the growth of nanostructures [4, 5], there is still a long way to go in the studies of their properties. This is mainly because the commonly used techniques to study properties are macroscopic. For example, nowadays, the optical and electrical properties of nanostructures are usually studied by photoluminescence (PL), optical absorbance, current-voltage spectroscopy, capacitance spectroscopy, deep-level transient spectroscopy, and admittance spectroscopy [6–8]. In these measurements, the obtained properties are averaged over a large number of nanostructures. Owing to the considerable size scattering in nanostructures, a lot of important quantum information could be flattened or even lost. Therefore, it is quite important to investigate the properties of single or individual nanostructures.

The invention of scanning probe microscopy (SPM) has provided scientists with a possible route to focus on single nanostructures. With the development of the SPM-based techniques, such as conductive atomic force microscopy (CAFM), electrostatic force microscopy (EFM), Kelvin potential force microscopy (KPFM), and scanning capacitance microscopy (SCM), the electrical properties of single nanostructures can be investigated in addition to topographic measurements [9, 10]. For example, CAFM has been applied to measure the local topographical and conductive properties on individual nanostructures simultaneously and independently, and exciting results have

*Conductive Atomic Force Microscopy: Applications in Nanomaterials*, First Edition.
Edited by Mario Lanza.

been achieved on various nanostructures, such as films, inorganic and organic heterostructures, graphene, and nanodots [11–19].

In this chapter, CAFM studies on individual GeSi quantum dots (QDs) and quantum rings (QRs) are reported. The influencing factors, including growth condition, native oxidation, normal force, bias voltage, and inter-dot coupling, on the conductive properties of the QDs are investigated. The conductive mechanism of QRs is suggested in combination with other conductive SPM measurements. With selective chemical etching, the conductance and composition distributions of GeSi QRs are obtained simultaneously on the same single QRs, and the dependence of the QRs' conductive property on their composition is investigated.

## 6.2 Conductive Properties of Individual GeSi QDs and QRs

In the last decade, self-assembled GeSi QDs have attracted much attention from researchers because of their great importance in the development of semiconductor technology as well as their promising potential for applications in electronics and optoelectronics [20–22]. Considerable progress has been made in their self-organized growth by changing the growth temperature, Ge layer thickness, deposition rate, substrate property, and other methods [23, 24]. In addition, many efforts have been made to clarify the growth mechanism of GeSi QDs [4, 25]. On the other hand, the physical properties of GeSi QDs have also attracted considerable attention owing to their importance and potential in electronic and optoelectronic applications. However, as mentioned earlier, the most commonly used experimental methods to investigate electrical properties, such as deep-level transient spectroscopy and capacitance-voltage measurements, are focused on ensembles of QDs [6, 7]. Therefore, the electrical properties of GeSi QDs are far from well understood. Fortunately, CAFM, which is available to investigate the conductive properties on individual QDs, offers an effective route to reveal the conductive properties of QDs. In recent years, CAFM has been employed to investigate the conductive properties of various QDs, including those of InAs, GeSi, GaN, InP, and $TiSi_2$ [26–31]. Using CAFM, the current distributions on individual QDs can be studied and their size-dependent current–voltage ($I$–$V$) characteristics obtained.

Another class of self-assembled nanostructures are QRs, which have, in recent years, attracted considerable attention for their unique properties and potential applications in nanoelectronic devices [32]. It has been reported that QRs hold especially interesting characteristics due to their ring-shaped geometry, such as persistent current and Aharonov-Bohm effects [33]. However, compared to the intensive theoretical studies on ideal or lithographed QRs [34, 35], the studies addressing self-assembled QRs are relatively scarce. Furthermore, most of the existing studies on self-assembled QRs deal with their growth mechanisms or electronic properties [36, 37], while some studies have been performed on their composition and strain distributions or atomic structures [38, 39]. The electrical characteristics of QRs, which are of vital importance to nanoelectronic applications, have been barely researched. Until now, very few experimental results have been reported on the conductive properties of individual self-assembled InAs and GeSi QRs [40, 41]. Therefore there is little understanding of the conductive mechanisms of the QRs or the main factors influencing their conductive properties.

This section discusses the electrical property studies on individual GeSi QDs and QRs. The effects of growth conditions on the current distribution of GeSi QDs are investigated. Results suggest that the growth temperature and capping process can significantly influence the conductive properties of QDs. Also, the conductive mechanism of QRs is discussed by using CAFM studies combined with SCM, and KPFM measurements. It has been found that the conductive distribution of GeSi QRs as well as their carrier density distributions are largely attributed to their topographic shapes rather than to the complete oxidation of the their central hole or their composition distributions.

### 6.2.1 Conductive Property Studies on Individual GeSi QDs

#### 6.2.1.1 Growth Temperature Dependence

Two kinds of QD samples grown at 550 and 640 °C have been investigated by CAFM, and the conductance distributions of these two types of QDs are significantly different [42]. The topographic and current images of Ge QDs grown at 550 and 640 °C are shown in Figures 6.1a,b and 6.2a,b respectively, with a bias voltage of 1.0 and 1.4 V applied to the sample. The line profiles along the marked lines are presented in Figures 6.1c and 6.2c for the two kinds of QDs. In the topographic images, it can be seen that both dome-shaped QDs have an average diameter of about 80 nm and an average height of about 12 nm. From the current images, it can be observed that for the QDs grown at 550 °C, the conductance at the center is higher than that at the periphery of the QD. On the contrary, a noticeable difference is observed in the conductance characteristics of the QDs grown at 640 °C, which show the periphery of the QD to be more conductive than at the center, resulting in a ring-shaped current distribution. These results show that the current distributions of the QDs with the same shape but grown at different temperatures are significantly different. Since the QDs have the same domelike shape, the different conductance distribution is attributed to the different diffusion levels of Si into the Ge dots at different growth temperatures.

To investigate the diffusion of Si into Ge, the topographic change with selective chemical etching is investigated by using 30% $H_2O_2$ for 30 min. This etchant can selectively remove GeSi material with Ge composition higher than 65 ± 5% [43]. As shown in Figure 6.3a, the dots grown at 550 °C are completely etched away, indicating that Ge composition in these dots is higher than 65%. However, for the QDs grown at 640 °C, as shown in Figure 6.3b, only a small part of the dots is etched away and a greater residual matter of the QDs is left on the surface. The result suggests that the most part of the QDs grown at 640 °C has Si content higher than 35% and this high Si concentration is definitely due to Si atoms diffusing into the Ge dots. Thus it can be demonstrated that the composition distribution of GeSi QDs is greatly influenced by their growth temperature, which would significantly change the current distribution.

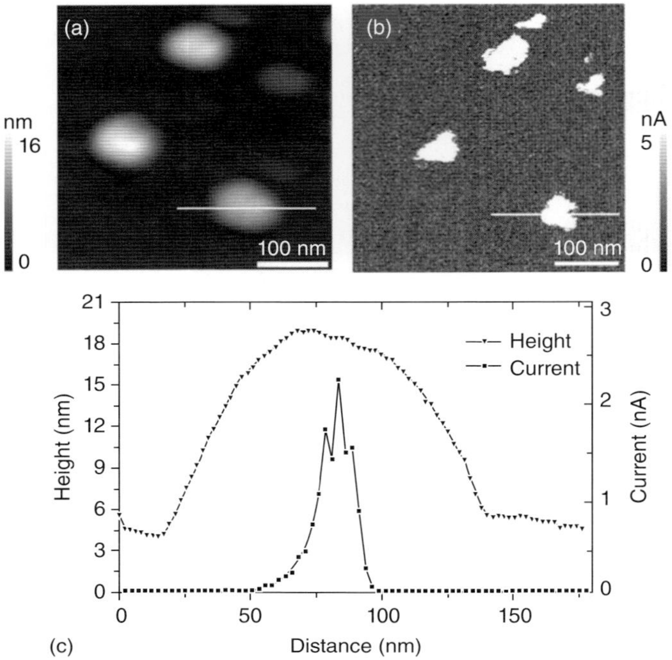


**Figure 6.1** Topographic (a) and current (b) images of Ge QDs grown at 550 °C measured at a sample bias voltage of 1.0 V. (c) Respective cross-sectional profiles along the solid lines marked in (a, b). (Xue *et al.* 2005 [42]. Reproduced with permission of Elsevier.)

To interpret the current distribution of different types of QDs, the total resistance measured by CAFM is considered as the contact resistance between the tip and the dot ($R_c$), the resistance of the dot itself ($R_d$), and the substrate resistance ($R_s$). As $R_s$ is much smaller than the other two parts because of the large contact area, it is not taken into account. The contact resistance $R_c$ should be related to the Schottky barrier between the tip and the surface. For QDs grown at 550 °C, the dot resistance $R_d$ is small due to the low resistivity of Ge, so the total resistance is mainly determined by the contact resistance $R_c$. Since Ge is enriched at the top center, the barrier height is smaller at the center of the QDs, resulting in higher current at the center. On the contrary, for the QDs grown at 640 °C and held at this temperature for 5 min before cooling down, a large amount of silicon is alloyed into the QDs. As a result, the dot resistance $R_d$ is largely increased owing to the much higher resistivity of Si compared with Ge, and this dominates the total resistance. Because $R_d$ is proportional to its current length and inversely proportional to the contact area between the dot and the tip, $R_d$ is smaller at its periphery than at its center owing to the smaller current path length (dot height) and larger contact area at the periphery. Meanwhile, because of the Si alloying, the surface composition variation is largely decreased and the resulting contact resistance difference between the center and periphery of the QDs is greatly reduced. Therefore, a ring-shaped current distribution is measured on the QDs grown at 640 °C.

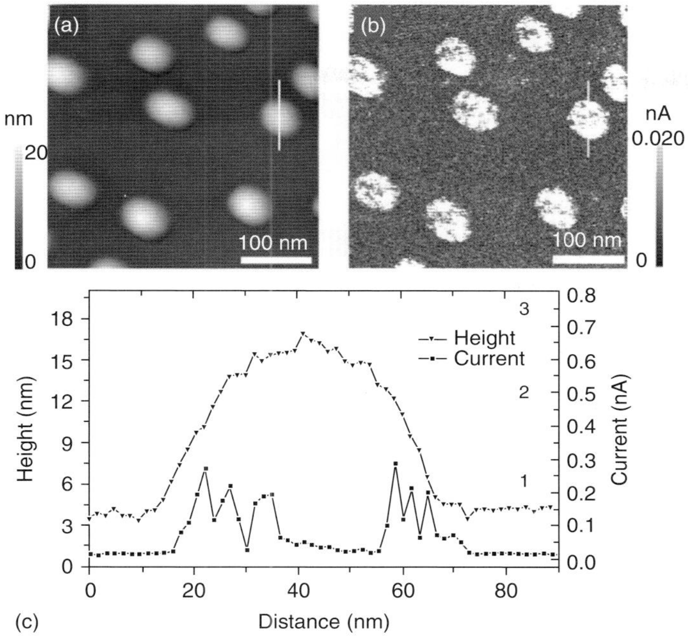


**Figure 6.2** Topographic (a) and current (b) images of Ge QDs grown at 640 °C obtained at 1.4 V. (c) Respective cross-sectional profiles along the solid lines marked in (a, b). (Xue *et al.* 2005 [42]. Reproduced with permission of Elsevier.)

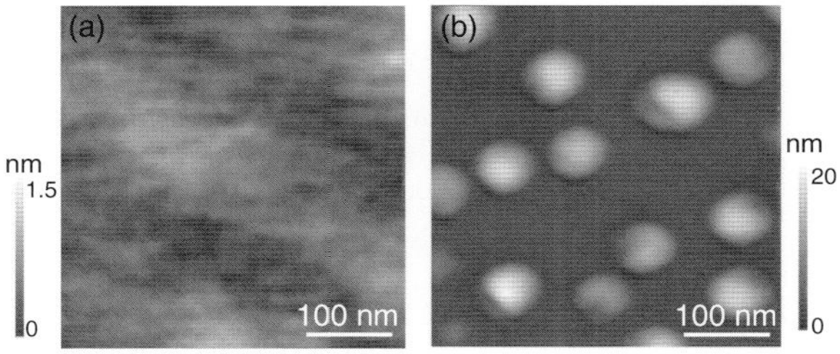


**Figure 6.3** Topographic images of Ge QDs grown at 550 °C (a) and 640 °C (b) after etching in 30% $H_2O_2$ solution for 30 min. (Xue *et al.* 2005 [42]. Reproduced with permission of Elsevier.)

#### 6.2.1.2 Electrical Property Changing with the Capping of Si Layer

Self-assembled GeSi QDs have promising applications in low-dimensional optical and electronic devices [44]. For this application, it is usually necessary to deposit a Si capping layer on GeSi QDs to avoid dot oxidation. However, the capping process could significantly change the topography as well as the physical properties of the QDs. For example, GeSi QRs were formed by capping a Si layer on GeSi QDs at a proper temperature [32]. A strain–relaxation-based model was supposed to explain the transformation of GeSi QRs [32]. This section describes, in addition to their topography and composition change, the electrical property variation with capping layer deposition on the QDs through studies by CAFM.

The topographic and current images of QDs grown at 640 °C before and after 0.32 and 2.0 nm Si capping layer deposition at the same temperature are

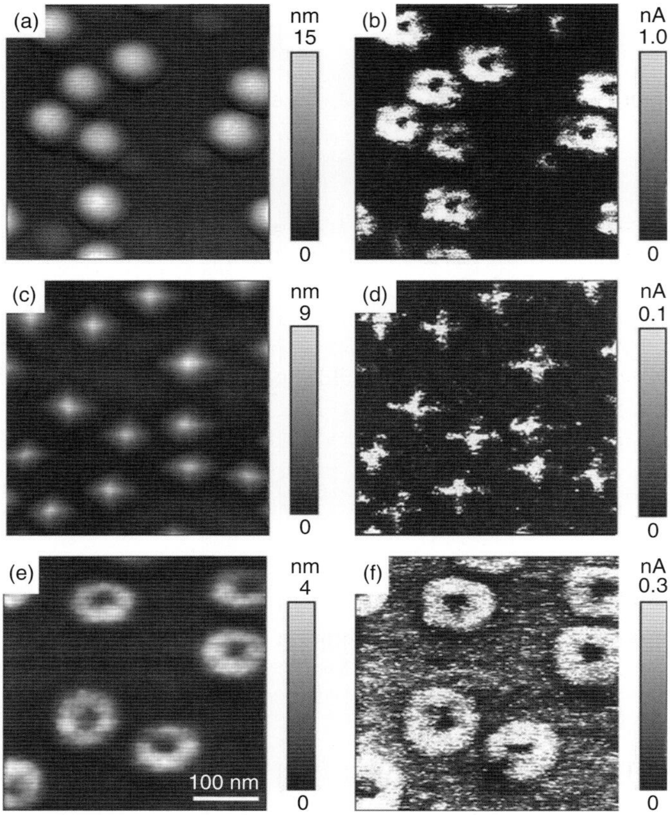


**Figure 6.4** Topographic and current images of GeSi QDs without Si capping (a, b), after 0.32 nm Si capping (c, d), and after 2 nm Si capping (e, f). (Adapted with permission from [44]. Copyright IOP Publishing 2009.)

shown in Figure 6.4a–f. From the topographic image (Figure 6.4a), before Si capping, the QDs are dome shaped with relatively uniform size of 70 nm in diameter and 12 nm in height. The corresponding current image shows that the conductance of individual QDs is higher at the periphery than that at the center, forming a ring-shaped current distribution as that shown in Figure 6.2b. After 0.32 nm Si capping, the shape of the QDs is totally changed to pyramidal, as shown in Figure 6.4c. Most of the QDs are about 80 nm in pyramid side length and 8 nm in height, and have four facets with {105} orientation. The current of the pyramid-shaped QDs is higher at the two diagonal lines than that at the facet regions, exhibiting a bright cross-shaped distribution, which is entirely different from that of the QDs without capping. With 2 nm Si capping, the QDs are changed to QRs, as shown in Figure 6.4e. Meanwhile, the current image measured at −3.5 V shows that the conductance of individual QRs is higher at the rim than that at the central hole, exhibiting a dark hole surrounded by a bright ring, as shown in Figure 6.4f. These results demonstrate that, with the deposition of the Si capping layer, the electrical properties of the QDs are significantly changed, accompanied by a change in the topography.

To interpret the topography and current distribution change with the capping layer, the composition distribution of individual QDs is measured by AFM combined with selective chemical etching as well as scanning Auger microscopy (SAM), as shown Figure 6.5. The selective etching was done in a 1 : 1 Vol. (28% $NH_4OH$)/(31% $H_2O_2$) solution (NHH) for 1 min. The NHH solution could selectively etch GeSi alloys with the etching rate exponentially proportional to the Ge concentration and it could remove GeSi alloys with Ge > 50% with 1 min etching [45]. The selective chemical etching result for QDs before capping (Figure 6.5a) shows that large parts of the QDs are etched away, which corresponds to Ge content greater than 50%. It can also be found that many QDs have some residue left at the side corners, suggesting these parts are enriched with Si and cannot be removed by NHH etching for 1 min. The typical Ge distribution of a single QD obtained by SAM is shown in Figure 6.5b, and the circle drawn there is derived from the corresponding

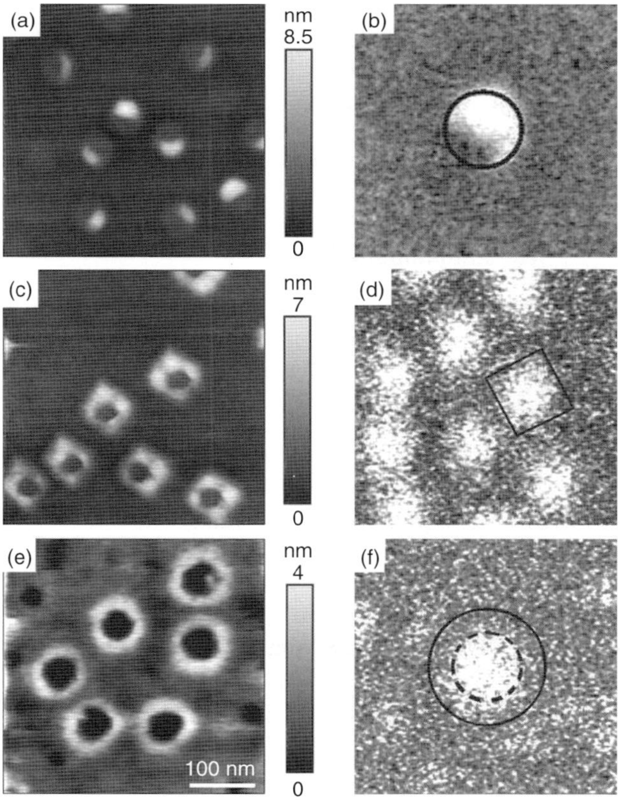


**Figure 6.5** Topographic and Ge content images of GeSi QDs before capping (a, b), after 0.32 nm Si capping (c, d), and after 2 nm capping (e, f), respectively. (Adapted with permission from [44]. Copyright IOP Publishing 2009.)

scanning electron microscopy (SEM) image, indicating the base of the QD. A basically uniform Ge distribution in the QD can be observed, except for the bottom left corner which is enriched with Si. Both results are consistent with the aforementioned assumption about the composition distribution of QDs for the ring-shaped current distribution and the nonsymmetrical current distribution of some QDs.

After 0.32 nm Si layer deposition, both the AFM and SAM images show that the Ge content is much higher at the center of the pyramids and that Si is abundant at the four corners and bottom edges. This composition distribution is consistent with the high currents measured at the center of the dot. However, the composition distribution is not consistent with higher current along the two diagonal lines, which may be due to the geometrical factor of the QDs, as the electrical field is higher on the edges than that on the facets. So it can be indicated that the composition distribution as well as the shape of the QDs determine the conductance distribution after 0.32 nm Si capping. In the case of the QRs formed after 2.0 nm Si capping, following selective chemical etching, their central holes become much deeper than the original ones, as shown in Figure 6.5e, indicating that the central hole of the QRs contains higher Ge concentration than the rim. The SAM results consistently present a bright region (high Ge content) at the center of the QR. Therefore both composition measurements suggest that central hole of QRs has higher Ge concentration, which however seems to be in conflict with the result that no current can be measured at the central hole. Details about the conductive mechanism of the GeSi QRs are discussed in the next section.

### 6.2.2 The Conductive Mechanism of GeSi QRs

As already mentioned, the GeSi QRs exhibit ring-shaped current distribution, similar to their topographic shape [46, 47]. Since the composition of the QRs shows that their central hole has higher Ge content than the rim, the ring-shaped current distribution cannot be interpreted using the composition distribution. In the following section, an attempt is made to reveal the conductive mechanism of the GeSi QRs.

Because the aspect ratio of the QRs' centers is very small (i.e., 1 nm depth vs 50 nm inner diameter), the electrical contact between the tip and the central hole should not be a problem. This type of ring-shaped current distribution was also observed by Mlakar *et al.* on InAs/GaAs QRs by CAFM [40], where the QRs' central holes also had low conductivity compared with the rim. In their paper, the low conductivity at the central hole was attributed to the complete oxidation of the central holes and the partial oxidation of the rims of the QRs. To investigate whether the current distribution of GeSi QRs is caused by the complete oxidation of their Ge-enriched central holes, the QR sample was etched in 10% hydrofluoric acid (HF) solution for 30 s to remove the native oxide layer from the surface. The topography and current images of the QRs measured immediately after the oxide layer removal are shown in Figure 6.6, together with the line profiles along the marked lines. The results exhibit a similar ring-shaped current distribution as observed before the HF etching, with the difference that the absolute current values have increased significantly. Therefore, these results indicate

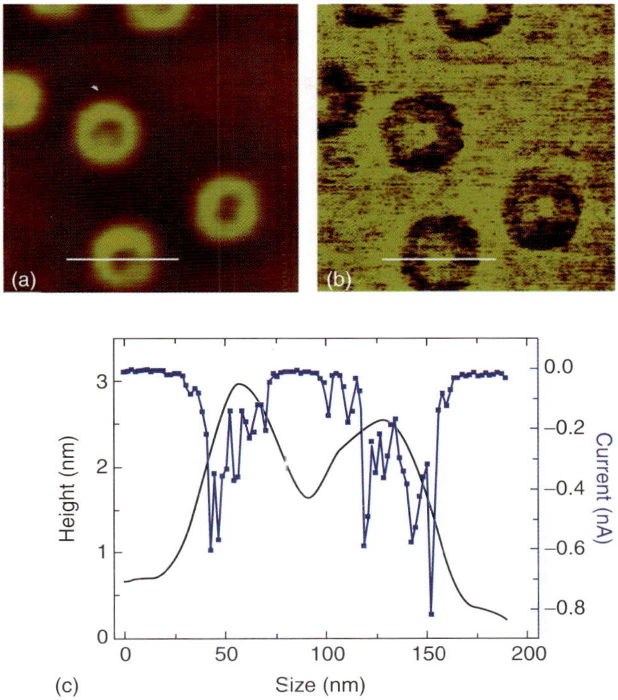


**Figure 6.6** Topography (a) and current (b) images of GeSi QRs after etching by HF for 30 s at a sample bias of −1 V. The topographic and current profiles of an individual QR as marked in (a) and (b) are shown in (c). (Reproduced with permission from [46]. Copyright American Institute of Physics 2011.)

that the ring-shaped current distribution of GeSi QRs could not be attributed to the complete oxidation of their central holes.

In addition, the current distributions of GeSi QRs with different surface composition distributions obtained by NHH etching for different times were investigated. As the etching rate of GeSi alloys in NHH solution was exponentially proportional to the Ge content [45], more time etching would remove more Ge components, resulting in less Ge/Si ratio on the surface. From the topography images as well as the topographic profiles, it is found that the etching rate and hence the Ge content at the central hole is greater than that of the rim, leaving a deeper and broader central hole as the etching time increases [46]. Although the fluctuation between different QRs could not be ignored, it could still be deduced that the central holes of the GeSi QRs have a higher Ge content than their rims, which is consistent with the reported results on GeSi QRs obtained by selective chemical etching [38] and those by transmission electron microscopy (TEM) [48]. However, from the current images, it can be found that the GeSi QRs with different surface composition distributions exhibit similar ring-shaped current distributions, that is, their rims are more conductive than the central hole for all cases. Therefore, these results confirm that the conductive distribution of GeSi QRs could not have resulted mainly from their composition distributions.

From the above experimental results, it appears that the ring-shaped current distribution is related to their ring-shaped topography; hence, we suggest that the QRs' topographic shape may be the main contributor to the ring-shaped current distribution of the GeSi QRs. If the topographic shape of the QRs has an important impact on their conductive properties, it should be expected that it will also contribute to their other electrical properties. Hence, KPFM, SCM, and CAFM measurements are performed to check the correlation between the electrical properties and topographic shapes of the QRs. The height, contact potential difference (CPD, by KPFM), current, and $dC/dV$ amplitude (by SCM) images of a typical QR before and after selective etching in BPA solution ($HF/H_2O_2/CH_3COOH = 1:2:3$) are shown in Figure 6.7. BPA etching can remove the GeSi alloys with Ge ratio more than 6% [49], leaving a Si-dominated surface, which can largely exclude the influence of the composition distribution. Similarly to the ring-shaped topography and current distribution, KPFM and SCM also present ring-shaped CPD and $dC/dV$ amplitude distributions, both before and after BPA etching.

From the KPFM images, it can be seen that the surface potential of the GeSi QRs exhibits a ring-shaped feature, showing that their rim has smaller CPD values than their central hole. In KPFM, the measured CPD is defined as $\Phi_{tip} - \Phi_{sample}$, where $\Phi_{tip}$ and $\Phi_{sample}$ are work functions of the tip and the sample, respectively. The conductive tip used in KPFM was a $W_2C$-coated Si tip. Its work function was calibrated with the highly oriented pyrolytic graphite (HOPG) sample [50] and obtained to be about 4.54 eV. Thus, the work functions of the rim, central hole, and the wetting layer (WL) of the QRs can be obtained, which are about 4.38, 4.31, and 4.34 eV before etching, and 4.48, 4.37, and 4.43 eV after BPA etching, respectively. The CPD difference (0.11 eV) between the rim and the central hole for the BPA-etched QRs is larger than that for the original QRs (0.07 eV), which may be due to the QRs' geometrical or compositional change after BPA etching. Nevertheless, the results present that for both QRs, the central hole has a smaller work function than the rim, even smaller than the WL.

From the SCM results, it can be seen that the $dC/dV$ amplitude obtained at the rim of the QRs is a little smaller than those obtained both at the central hole and the WL, also exhibiting a ring-shaped distribution. Since smaller $dC/dV$ amplitude corresponds to higher carrier density in the semiconductor for the tip/oxide/semiconductor structure [9, 10], the results suggest that the carrier density on the rim of the QRs is higher than that in the central hole. The averaged values of the $dC/dV$ amplitude for the QRs' rim, central hole, and WL are about 220, 260, and 250 mV before etching, and 130, 170, and 150 mV after etching, respectively. The results suggest that the QRs' central hole has the smallest carrier density.

All the above results indicate that the composition distribution, which mainly changes the absolute values, has little impact on the electrical property distributions (i.e., ring-shaped distributions) of QRs. In addition, as the BPA etching can remove most of the GeSi alloys with Ge content greater than 6%, the strain effect becomes considerably weak for BPA-etched QRs, which still present ring-shaped distributions. Therefore, it seems reasonable to assume that the topographic shape should have vital influence on electrical property distributions of the QRs,

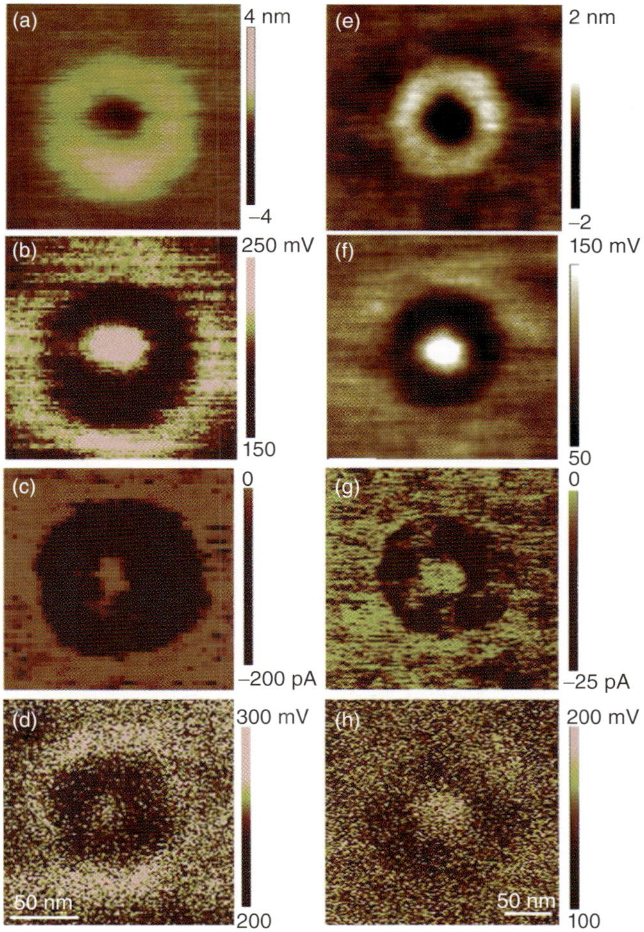


**Figure 6.7** Height image measured simultaneously with the CPD image before (a) and after BPA etching (e). CPD image obtained at an AC modulation of 2 V and a lift height of 10 nm before (b) and after BPA etching (f). Current image measured at a sample bias of −1 V before (c) and after BPA etching (g). d*C*/d*V* amplitude image obtained by applying 2 V AC modulation to the sample before (d) and after BPA etching (h). (Adapted with permission from [47]. Copyright Springer-Verlag 2012.)

while size, composition, and strain mainly influence the values. The reason for the topographic shape of the QRs to have a major impact on their electrical properties is not clear yet. In previous theoretical studies dealing with ideal QRs [35] or InAs/GaAs QRs [33, 51], both ring-shaped adiabatic potential and carrier probability density were achieved by considering the geometrical parameters of the QRs. Here also, we suggest that the ring-shaped surface potential and carrier density distributions of the QRs are attributed to their geometrical shapes. The correlation between the conductance distribution and carrier density distribution is direct, that is, higher carrier density results in larger conductivity. The correlation between the conductance distribution and the surface potential distribution could be described from the viewpoint of electron barrier height at the interface between the tip and the measured surface. From the results of KPFM, the electron barrier height for the rim ($\Phi_{B,\text{Rim}}$) of the QRs is found to be lower than that for their central hole ($\Phi_{B,\text{Center}}$) and WL ($\Phi_{B,\text{WL}}$). This is consistent with the CAFM results which show that the rim of the QRs is more conductive than their central hole and WL.

On the basis of the aforementioned considerations, a possible explanation for the electrical properties of the QRs has been suggested: the ring-shaped geometry determines that the rim of the QRs has a lower barrier height with the tip and a higher carrier density, resulting in a higher conductivity at the rim, compared to the central hole and the WL.

## 6.3 Modulating the Conductive Properties of GeSi QDs

Despite the above successive measurements on individual QDs and QRs, the electrical properties of these nanostructures could be influenced by factors other than their topography and composition. In this section, the influence of extrinsic factors (oxidation, normal force, and bias voltage) and intrinsic factor (inter-dot coupling) on the electrical properties of the QDs are discussed. The results find that the conductive properties of the QDs can be well modulated by bias voltage and inter-dot coupling.

### 6.3.1 Oxidation and Normal Force

CAFM has been proved to be a powerful tool to measure local topographical and electrical properties of individual QDs simultaneously and independently, and many exciting results have been obtained [52]. Nevertheless, several problems should be dealt with to make CAFM measurements more reliable, especially when the CAFM measurement is performed in atmosphere. In the latter case, an oxide layer will inevitably be formed on the QDs before and during the electrical property measurements. Thus, the oxidation effects on the electrical properties of QDs, which have previously been ignored, should be taken into consideration. Moreover, the selection of normal forces in experiments is optional; these forces may also affect the measured results as they can change the effective oxide layer thickness between the tip and the surface.

Typical $I–V$ curves measured on individual QDs grown at 640 °C after exposure to air for several days under a series of normal forces, labeled as $F=F_0$, $2F_0$, $3F_0$, and $4F_0$ ($F_0$ is the unit of force) are shown in Figure 6.8a. It can be seen that the $I–V$ characteristics change significantly with the normal forces. The measured current is greater at the larger force when the bias voltage is beyond a threshold value of about 1.5 V. To interpret the effects of the normal force, the thickness of the oxide layer between the metal-coated tip and surface of the QD is considered to bethe main factor. Therefore the Fowler–Nordheim tunneling (FNT) theory [53] which is often applied to describe the tunneling current in metal–oxide–semiconductor structures with oxide thicknesses of more than 3 nm is adopted to fit these $I–V$ curves, and the fitting curves are shown as solid

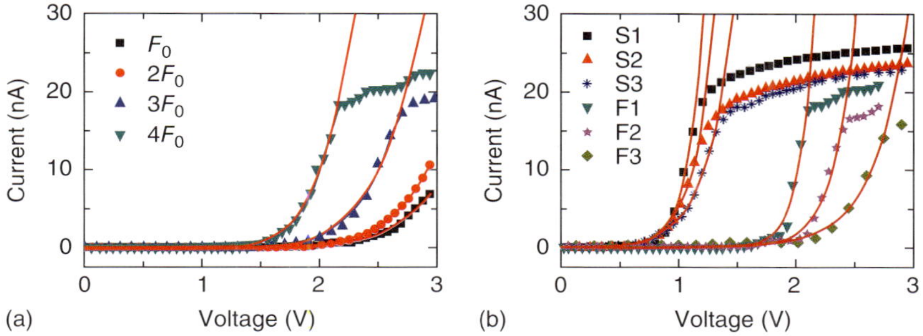


**Figure 6.8** (a) *I–V* characteristics of the QD under different normal forces: experimental data (symbols) and FNT fitting results (solid lines). (b) *I–V* curves obtained as a function of exposure time after etching with two fitted conductance models. The dotted lines are experimental data and the solid lines are fitted data. S1, S2, and S3 correspond to Schottky emission fitting at 30, 45, and 60 min after etching and F1, F2, and F3 correspond to FNT fitting at 150, 240, and 360 min after etching, respectively. (Adapted with permission from [52]. Copyright IOP Publishing 2006.)

lines in Figure 6.8a. It can be seen that all the simulated data agree well with the experimental results, except for the large voltage regions when using large forces ($F = 3F_0$ and $4F_0$) which are not in good agreement due to the saturated current limit of the instrument at these regions. From the fitting parameters, it can be found that the oxide thickness decreases with the normal force, indicating that a large normal force may help the tip to deeply penetrate the oxide layer.

In addition, the *I–V* curves measured on QDs after HF etching to remove the oxide layer as a function of exposure time are as shown in Figure 6.8b. After etching, the *I–V* characteristics show Schottky contact characteristics within 60 min. These *I–V* curves can be well fitted by the Schottky emission (SE) model [54], and the fitting results are shown as solid lines in Figure 6.8b. As the exposure time increases further, to more than 150 min, FNT features can be observed again because of an oxide layer forming on the Ge QDs, the same as those observed before etching (Figure 6.8a). Thus, the FNT model is used to fit the *I–V* curves, and the fitting data agree well with the experimental results, also shown as solid lines in Figure 6.8b. These results indicate that the Schottky contact is formed between the metal-coated tip and the Ge QD without an oxide layer. On exposure to air, the Schottky contact is destroyed owing to formation of the native oxide layer on the surface of the QD and the FNT model becomes available. Therefore, it is concluded that the *I–V* characteristics measured on GeSi QDs in ambient conditions are significantly influenced by the experimental conditions.

On the other hand, the influence of the oxide layer thickness and normal force on the current distribution of GeSi QDs has also been investigated. It is worth noting that the current distributions of all the Ge QDs exhibit almost the same ring-shaped characteristics for different exposure times and normal forces [52]. Although the absolute current values change obviously with the exposure time as well as normal force, the characteristics of the ring-shaped current distributions are almost the same.

From the above results, it is found that the current distribution of individual QDs remain almost the same for different thicknesses of the oxide layer or different normal forces applied, despite significant effects of the oxide layer on the current–voltage characteristics. Therefore, the effects of the oxide layer on the current distribution can be ignored, although the absolute current values are strongly influenced. Our results thus provide important evidence of the reliability of CAFM measurements of current distributions in ambient conditions.

### 6.3.2 Bias Voltage

In previous section, CAFM is applied to study the electrical properties of individual GeSi QDs or QRs by current imaging at a fixed bias voltage [55]. The influence of the applied bias on the current distribution has not been considered. Almost all the existing studies give current distributions of QDs at a fixed bias, and few reports have discussed the dependence of current distributions of the QDs on the applied bias [56]. In this work, the current distributions of individual self-assembled GeSi QDs are studied as a function of bias by using CAFM. By analyzing the bias-dependent current images of GeSi QDs and histograms of the current magnitude at different biases, the conductive mechanisms for QDs and for WLs are arrived at. The results suggest that the current distribution of GeSi QDs can be regulated by the applied bias.

The topographic and current images of self-assembled GeSi QDs measured at low biases from −0.5 to −2.0 V are shown in Figure 6.9. From the current images, a ring-shaped current distribution of individual QDs can be observed for all four biases. The conductivity at the periphery of the GeSi QDs is always higher than that at the center, and the current distribution does not show an obvious change with the bias. The typical current profiles across individual QDs at four biases are shown in the column to the right of the current images. They exhibit the same dual-peak feature, confirming that the characteristics of current distribution did not change with the bias. On the contrary, the current distributions of QDs obtained at high biases ranging from −2.5 to −4.0 V are shown in Figure 6.10. From the current images, it can be seen that all the current distributions of individual QDs have a disklike shape, which is totally different from the ringlike shape at low biases. The current intensity enhances sharply as the bias magnitude increases, and the central parts of QDs become more conductive than the other parts. The current profiles for four biases are also shown in the column to the right of the corresponding current images. All current profiles of GeSi QDs obtained at the high biases yield similar disklike shape characteristics, except for a little current magnitude increase and shape broadening with increasing bias magnitude.

To analyze the bias-dependent conductance distribution of QDs in detail, the statistical distributions of current magnitude have been calculated. The results show that at low biases, only one current peak is exhibited in the histograms of current magnitude, while at high biases, one or two additional peaks appear at the large-current region, except for the same small-current peak as obtained at low biases [55]. On the contrary, all current histograms of the WLs show a similar single-peak feature, where only one peak appears in the small-current

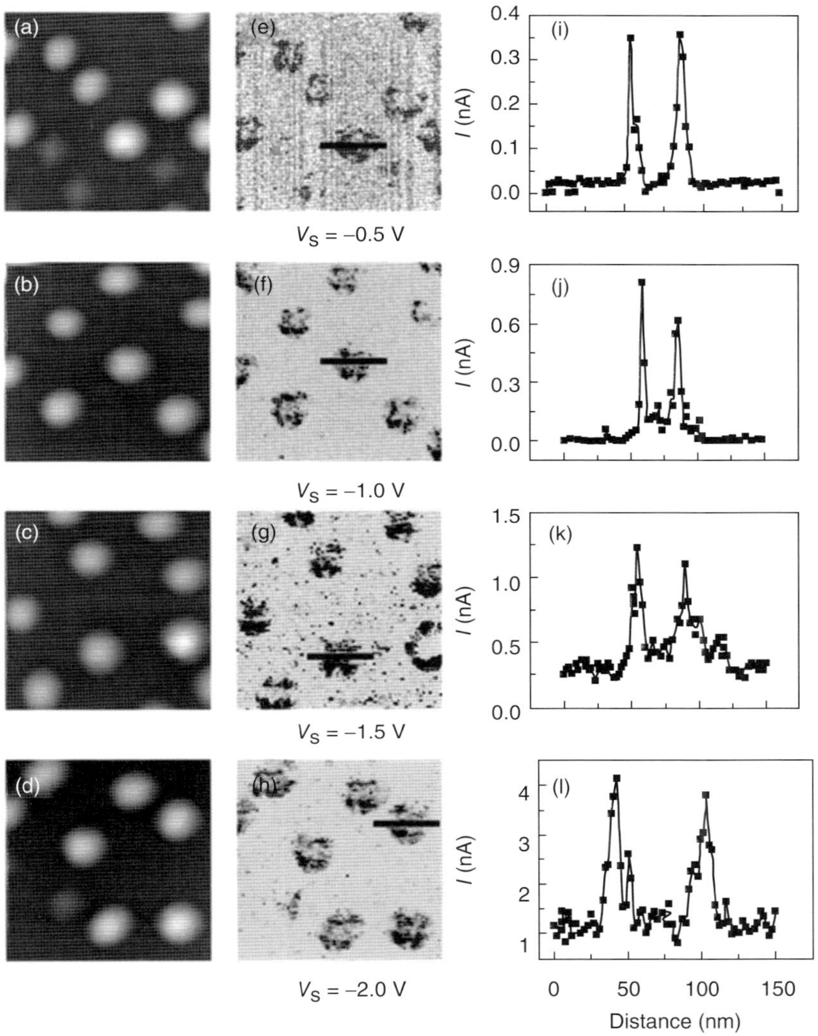


**Figure 6.9** Topographic (a–d), current (e–h) images of GeSi QDs, and current profiles along the marked lines (i–l) measured at different biases from −0.5 to −2 V, respectively. All image sizes are 0.5 μm × 0.5 μm. (Adapted from Ref. [55]). Reproduced with permission from [55]. Copyright IOP Publishing 2011.

region. These results indicate that the novel large-current peak(s) of QDs should be related to the quantum effects of QDs, that is, discrete energy levels in QDs. To understand the bias-dependent conductance, the dependence of the peak current on bias is investigated. It is found that the dependence of small-current peak $I_0$ on applied bias $V$ agrees with the FNT mode [53] under low biases and the SE model [54] under high biases. However, the dependence of the large-current peak on the bias does not obey the FNT or the SE model. This may be attributed to the direct transfer of the electrons and/or holes via the discrete energy levels in

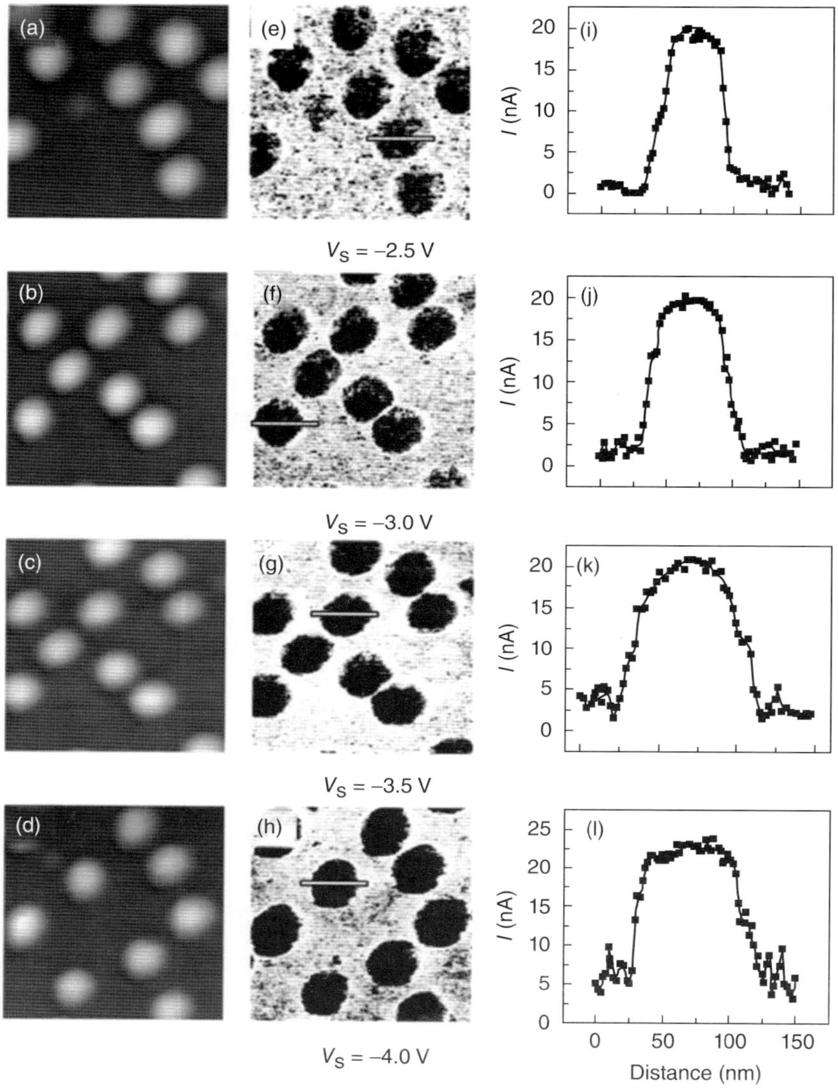


**Figure 6.10** The topographic (a–d), current (e–h) images of GeSi QDs, and current profiles along the marked lines (i–l) measured at different biases from −2.5 to −4 V. All image sizes are 0.5 μm × 0.5 μm. (Reproduced with permission from [55]. Copyright IOP Publishing 2011.)

QDs. As the bias increases above a certain value, the current at the center of the QDs increases much faster than at the periphery as more eigenstates localized at the QDs center contribute to the current. A similar viewpoint was put forth by Johal *et al.* to interpret the bias-dependent scanning tunneling microscopy (STM) tunneling-current distribution of InAs QDs [56]. Although a complete understanding of the mechanism is yet to be arrived at by including all contributions of the oxide layer, space charge, surface states, and so on, our results suggest that the conductive properties of GeSi QDs can be effectively modulated by the bias voltage.

### 6.3.3 Inter-Dot Coupling

In the above sections, the conductive properties have been considered in relation to individual QDs [57]. However, as the efficiency of single-layer (SL) QDs is relatively low, vertically aligned multilayer QDs are often adopted for practical applications. By repeating dot layers separated by spacer layers a few nanometers in thickness, a more homogeneous size distribution could be achieved, simultaneously with novel physical properties induced by vertical coupling [58, 59]. On the other hand, both high-density QDs and QD molecules have attracted a lot of interest for their potential applications, where lateral coupling between adjacent QDs significantly modify the propertiesof the QDs [60, 61]. The coupling effects have been investigated mainly by macroscopic techniques such as PL and admittance spectroscopy. Owing to large scattering in the size, separation, or composition distribution of the QDs [62–64], the quantum properties of coupled QDs obtained by the macroscopic methods would be greatly weakened or eliminated by averaging effects. Until now, there have been only a few microscopic studies performed by STM on InAs and PbSe QD clusters [65, 66]. Here, CAFM studies have been performed to investigate the conductive properties of individual SL and bilayer (BL) GeSi QDs with different dot densities. Both the vertical coupling between BL QDs and the lateral coupling between densely packed QDs are found to exist at room temperature, which significantly increase the QDs' conductance, suggesting that the electrical properties of QDs can be modified by the coupling effects.

The current images measured on sample A (SL) and sample B (BL) with low dot density are shown in Figure 6.11, together with their corresponding topographic images. It can be found that the average current of the BL QDs is much larger than that of the SL QDs of similar size. The typical current profiles of single small QDs for SL and BL samples (marked by circles) are also given on the upper right side, together with their corresponding height profiles. It can be seen that the sizes of the two QDs are similar (~50 nm in diameter and ~1.8 nm in height), but the current values of the BL QD are about twice that of the SL QD, and the same is true for large QDs. The same conclusion could be drawn by comparing the conductive results of high-density samples C (SL) with D (BL), which are shown in Figure 6.12. It could be found that the BL QDs are more conductive than the SL QDs, for both large and small QDs. So our results indicate that the BL QDs are more conductive than SL QDs of similar size for both low and high dot density samples.

In CAFM measurements, as the area of the current flow increases fast along the current path, the major contribution to the resistance is the fixed surface region which contacts with the tip. Hence without vertical coupling, the second-layer QDs will not influence the measured current. Therefore it can be suggested that enhanced conductance in BL QDs is attributed to the vertical-coupling effect of the QDs. Owing to the vertical coupling, the density of states which contribute to the electron tunneling would be larger than those of single QDs. Thus the QDs

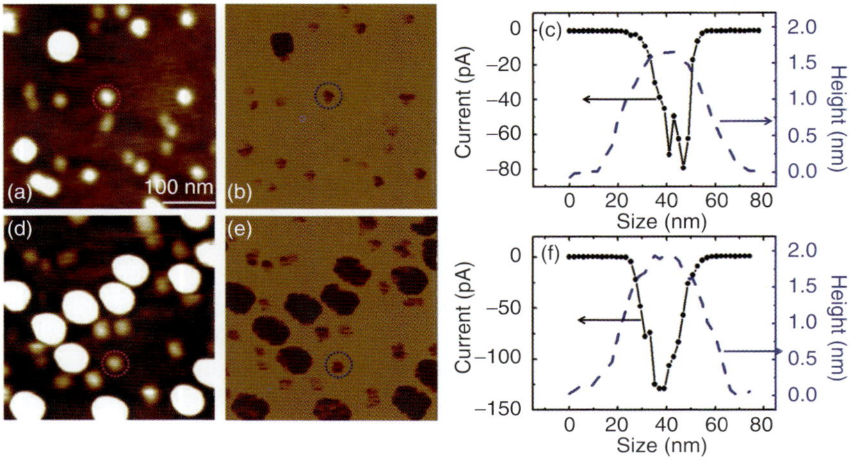


**Figure 6.11** Topography and current images of SL sample A (a, b) and BL sample B (d, e) with low dot densities. The height and current profiles of the marked small QDs of samples A and B are plotted in (c, f), respectively. (Reproduced with permission from [57]. Copyright Springer-Verlag 2012.)

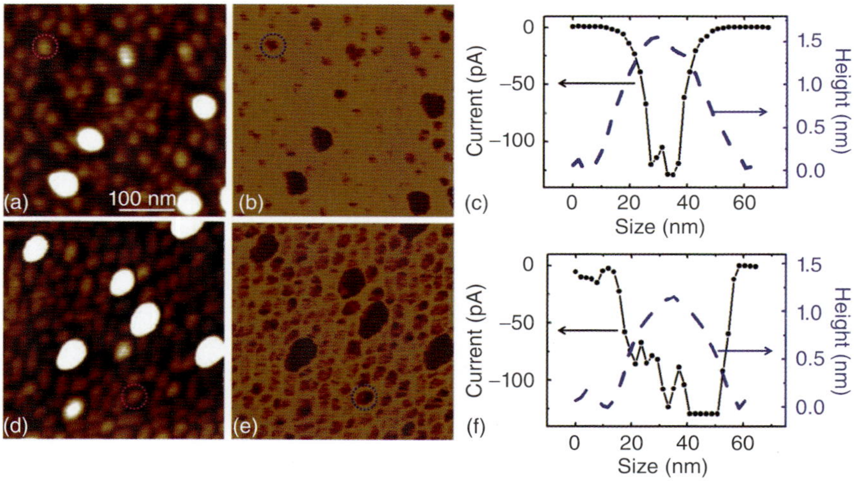


**Figure 6.12** Topography and current images of SL sample C (a, b) and BL sample D (d, e) with high dot densities. The height and current profiles of the marked small QDs of samples C and D are plotted in (c, f), respectively. (Reproduced with permission from [57]. Copyright Springer-Verlag 2012.)

in the second layer can contribute to the conductance through vertical coupling, which makes the QDs in the first layer more conductive.

On the other hand, the influence of the density of the QDs on their conductive properties is also considered. By comparing the current images of the SL and BL samples with different dot densities, it is found that the higher the dot density, the larger the average current of the QDs of similar sizes. As the current measured by

CAFM only comes from the contact area between the tip and the surface beneath it, and this area is smaller than that of a single QD, the dots nearby will not influence the measured current without lateral coupling. Thus our results indicate the existence of the lateral coupling between the closely packed QDs at room temperature, which increases the conductance of the QDs. Similar lateral-coupling effects have been observed on InAs and PbSe QD clusters by STM studies [65, 66], which found that the tunneling current increased with the dot number in the cluster. In their cases, the increase of current was interpreted by the increase of tunneling path when the QDs were closely packed. From the same viewpoint, the increased conductance of high-density QD samples may be explained as the conductive path between the tip and the sample increases with the dot density, that is, electron tunneling between laterally coupled QDs.

Furthermore, the influences of inter-dot coupling on the QDs' conductance are investigated as a function of bias voltage. The bias dependencies of the average current, which is deduced from the peak current in the histogram of current magnitude by calculating over a number of QDs of similar sizes, are shown in Figure 6.13. By comparing the results of the large and small QDs of sample A with those of sample B, it can be found that the average current values of BL QDs are larger than those of their corresponding SL QDs at all biases, for both large and small QDs. Additionally, the current enhanced ratio (CER) of BL QDs to SL QDs increases with the bias voltage, resulting in the significantly large conductance of BL QDs at high biases. Similar bias-dependent CER of BL QDs to SL QDs are

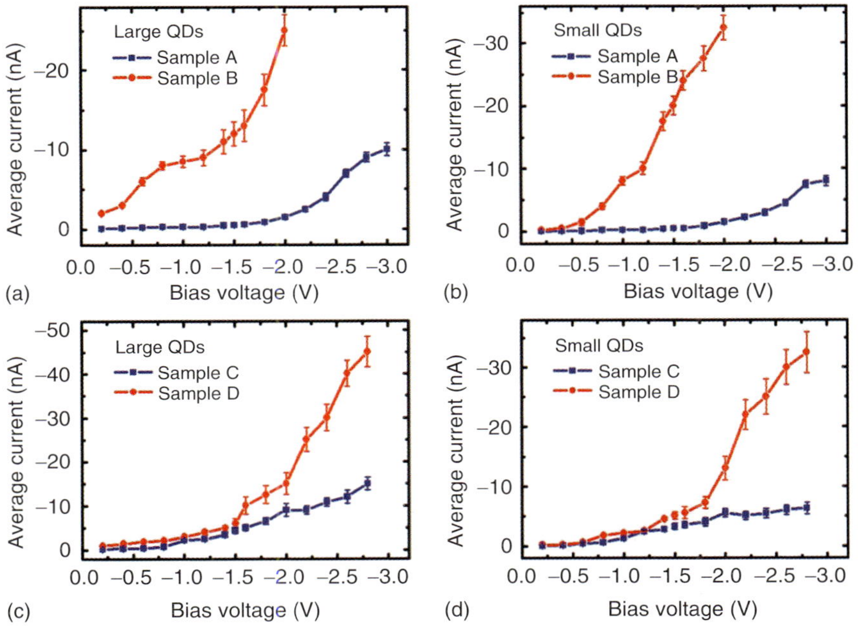


**Figure 6.13** The obtained average current as a function of bias voltage. (a) Large QDs of samples A/B, (b) small QDs of samples A/B, (c) large QDs of samples C/D, and (d) small QDs of samples C/D. (Reproduced with permission from [57]. Copyright Springer-Verlag 2012.)

observed for the high-density samples C/D, for both large and small QDs. On the other hand, the increase in the conductance with dot density can be observed on SL QDs of samples A/C. The CER of high-density QDs (sample C) to low-density QDs (sample A) also increases with the bias voltage, which are about six and four times at −2 V for large and small QDs respectively. For BL samples B/D, however, the increase in conductance with dot density could not be observed, which may be due to the already existing vertical coupling between the BL QDs.

The large increase in the conductance of QDs at large biases should be an exciting result, as it suggests that the conductive properties of the coupled QDs can be greatly regulated by bias voltage, which should be valuable for applications. The bias dependence of the conductance of individual QDs has been investigated in the previous section. It was found that the current of the QDs increases much faster with the bias than that of the WL; this was attributed to the discrete energy levels of QDs. Using a similar concept, the larger bias dependence of the conductance of the coupled QDs may be also attributed to the energy levels of the coupled QDs.

## 6.4 Simultaneous Measurements of Composition and Current Distributions of GeSi QRs

To realize the applications of self-assembled GeSi QDs or QRs in optoelectronic devices and quantum computers, it is essential to have a good understanding of their composition distributions and electrical properties [67]. Until now, various methods have been employed to investigate the composition distributions of GeSi quantum structures [68], such as anomalous X-ray diffraction (XRD) [69], cross-sectional TEM[70], tip-enhanced Raman spectroscopy [71], and atomic force microscopy (AFM) combined with selective chemical etching [70, 72, 73]. Many studies have reported on GeSi QDs, but much fewer on GeSi QRs. In addition, studies on the electrical properties of single QRs are considerablylacking and the electrical properties of QRs as well as the main factors influencing their conductive properties have not been well understood yet. So in the following, a method is developed that combines CAFM with selective chemical etching to investigate the composition and current distributions on single GeSi QRs simultaneously. Hence the correlation between the current and composition distributions of single QRs can be obtained. As the same QRs are investigated before and after etching, the fluctuation between different QRs is eliminated, and the exact distribution and correlation can be achieved. Using the same method, the composition and current distributions of GeSi QDs can also be obtained, but this is not discussed here.

To obtain the composition distribution of GeSi QRs, the etching rate (etched height versus etching time) should be obtained first. In previous studies applying a similar method, the etched height was usually measured by taking the WL as a benchmark. However, the WL would also be etched during the etching process, thus a more accurate height benchmark than the WL is required. Here, to act as the height benchmark, we utilize a nanoscale trench, which was scratched on the original surface of the QRs with a diamond tip scanning on the surface under a relatively large load. As the trench is about 20 nm in height, according to the growth conditions, the bottom of the trench is deep enough to reach the silicon buffer layer, thus it can hardly be etched by the NHH or BPA etchant. Therefore, the bottom of the trench can act as a benchmark to obtain the etched heights of the QRs as well as the WL, while the position of the trench can be used as a mark for AFM imaging on the same QRs after etching.

First, the large-scale topographic images including the trench before and after 10, 60, and 600 s NHH etching and 90 s BPA etching are measured. The quantitative analysis of the decline of the WL can be done with the height profiles along the same marked line across the trench before and after each etching process, which are vertically aligned according to the bottom of the trench. As a result, the etched heights of the WL are achieved to be about 0.8, 1.4, 1.9, and 2.2 nm for 10, 60, and 600 s NHH etching and 90 s BPA etching, respectively [67]. Therefore, our results suggest that the etching of the WL cannot be ignored. On the other hand, by this method, the absolute Ge content on the WL surface can be calculated from the etching rate or BPA etching limit, which is about 50% on the original WL surface, decreasing to 45%, 35%, 10%, and 6% after 10, 60, and 600 s NHH etching and 60 s BPA etching, respectively [67].

After the etched heights of the WL are obtained, the distributions of the etched height of the QRs could be achieved. The resulting topographic images of several QRs before and after different selective etching processes are presented in Figure 6.14a–e. It can be observed that the center of the QRs is etched much faster than their rim after 10 and 60 s NHH etching, while the rim as well as the WL is effectively etched after 600 s NHH etching and 90 s BPA etching. A quantitative analysis can be made with the height profiles along the same marked line, as shown in Figure 6.14f, which are vertically aligned by considering the etched height of the WL in each corresponding etching process. By comparing the two topographic images before and after each etching process as well as considering the etched height of the WL in the same etching process, the etched height at each point of the surface can be obtained. The topography and surface composition distribution of GeSi QRs before etching are as shown in Figure 6.15a,b. It can be observed that the surface Ge content at the central hole is higher than that at the rim; that is, the Ge content at the central hole is about 65% while that at the rim is about 45%. The composition distribution of the QRs is almost symmetric and ring shaped, except for the periphery of the rim, which has an asymmetric composition distribution. The composition distributions of different QRs are also deduced by the same method; these slightly differ from QR to QR, but the centers of the QRs always have higher Ge content than the rims.

The composition distribution perpendicular to the surface can also be obtained by a similar method. The cross-sectional topographic change of the same marked QR with the etching processes is shown in Figure 6.15c. By analyzing the topographic variation (etched height) with the etching process, the vertical composition distribution of a single, specific QR can be obtained. Figure 6.15d presents a schematic diagram of the vertical composition distribution of single GeSi QRs. The center of the QR is a Ge-rich core containing more than 65% Ge before selective etching. From top to bottom, and also from center to rim, the Ge content

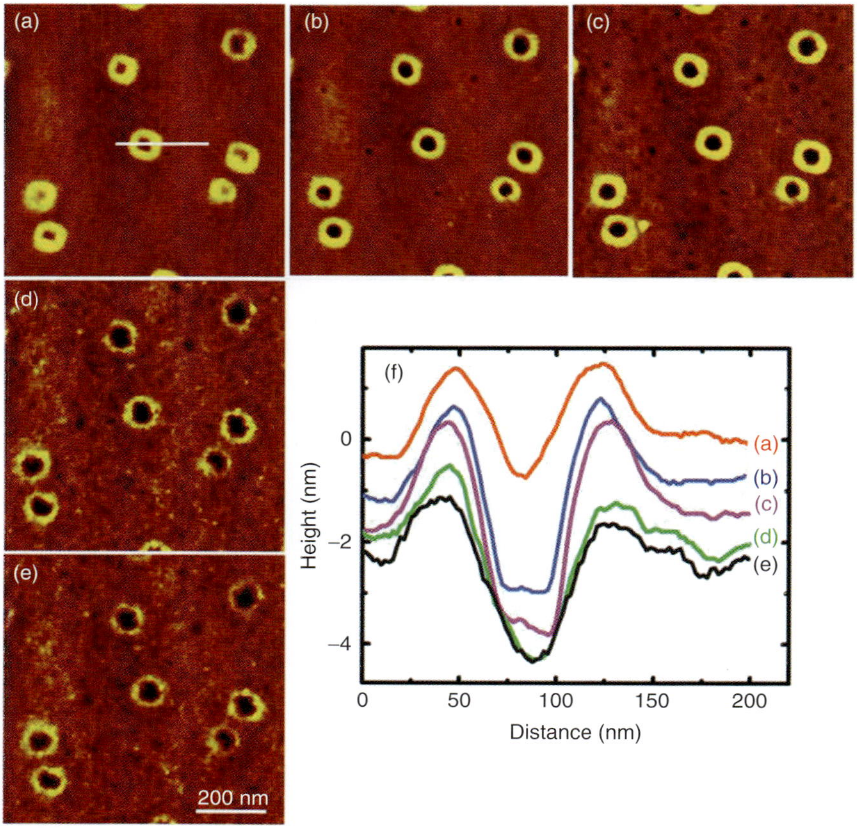


**Figure 6.14** Topography images of GeSi QRs before (a) and after etching in NHH for 10 s (b), 60 s (c) and 600 s (d) and in BPA for 90 s (e). (f) The height profiles of the same QR along the same marked line before and after each etching step. The curves are vertically offset by subtracting the etched height of the WL in the corresponding etching step. (Reproduced with permission from [67]. Copyright IOP Publishing 2013.)

decreases. Similar results of composition distribution of GeSi QRs have also been reported by using the same selective chemical etching combined with the AFM imaging method, finding that the Ge content on the surface was about 35% and 20% at the center and the ridges, respectively [38]. But in that research work, different QRs after NHH etching were imaged, and the etching of the WL was ignored. Therefore, our results present a more accurate and complete 3D quantitative composition distribution of single GeSi QRs.

Furthermore, the above *in situ* method is used in the CAFM measurements. The current images of the same GeSi QRs before and after selective chemical etching are measured simultaneously with their topographic images. The topographic and current images of the same QRs before and after 10, 60, and 600 s NHH etching and 90 s BPA etching measured at −1.0 V are shown in

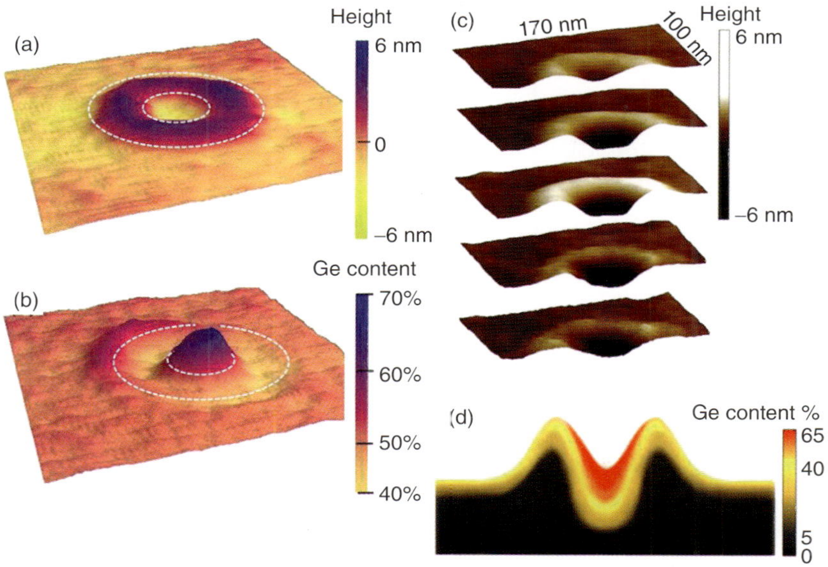


**Figure 6.15** Topography (a) and surface composition distribution (b) of the same area including a single QR. The two dashed circles indicate the inner and outer rings of the QR. (c) Cross-sectional topographic change before and after each etching process. (d) A schematic diagram of the vertical composition distribution of single GeSi QRs. (Reproduced with permission from [67]. Copyright IOP Publishing 2013.)

Figure 6.16a–j, respectively. From these images, ring-shaped geometric shape and ring-shaped current distributions can be clearly observed on all samples. The height profiles of the same QR along the same marked line before and after each etching process are given in Figure 6.16i. With the aforementioned method, the surface composition distributions of the single QRs before each etching process can be deduced. The composition profiles along the same line as in (I) are shown in (II), while the corresponding current profiles are plotted in (III). From these topographic, compositional, and current profiles of the same QR, it can be observed that the original GeSi QRs have higher Ge content at the central hole than that at the rim, but the measured current at the rim is larger than that measured at the central hole. This indicates that the current distribution of GeSi QRs is not consistent with their composition distribution. After selective etching, the Ge content in all areas of the surface decreases; however, owing to the higher Ge content at the hole than in other areas, the Ge content at the central hole decreased much faster than that at the rim, resulting in the surface composition distribution of GeSi QRs changing significantly. On the other hand, as the Ge content decreased, the absolute values of the measured current apparently decreased. However, the ring-shaped current distributions of QRs did not vary when there was a change in both the Ge content and GeSi composition distribution on the surface of the QRs. Even after BPA etching, the

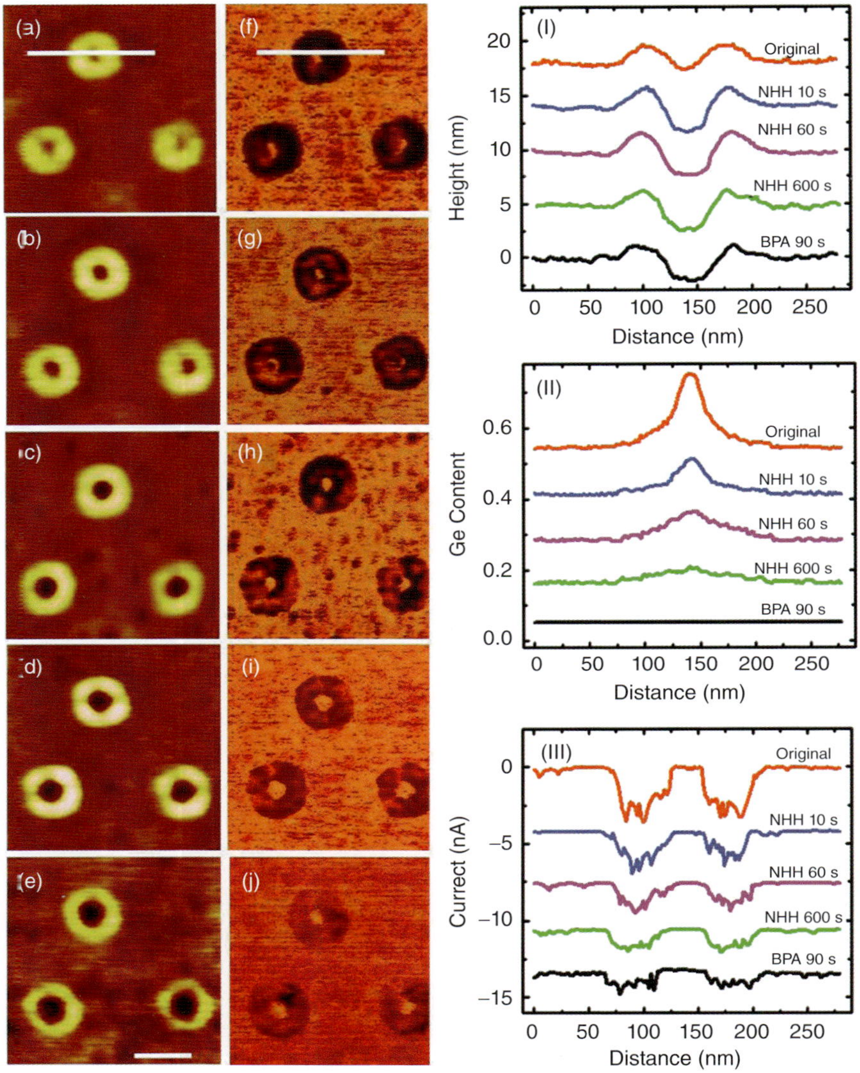


Figure 6.16 The topography and current distributions of GeSi QRs before (a, f) and after etching in NHH for 10 s (b, g), 60 s (c, h) and 600 s (d, i) and in BPA for 90 s (e, j). The sample was biased at −1.0 V and the tip was grounded. (I), (II), and (III) give the height, composition, and current profiles of the same QR along the marked line, respectively. The curves in (I), (II), and (III) are offset to various values for clarity. (Reproduced with permission from [67]. Copyright IOP Publishing 2013.)

Ge content distribution on the surface is flattened as it is dominated by Si atoms (>94%); still, it is found that the rim of these QRs is more conductive than the central hole. Therefore, we suggest that the conductance distribution of GeSi QRs is not consistent with their compositional distribution, just as discussed in Section 6.2.

## 6.5 Conclusions

In this chapter, the conductive properties of individual GeSi QDs and QRs have been studied by CAFM, combined with other conductive SPM studies as well as selective chemical etching. The results find that the current distribution of GeSi QDs changes significantly with the growth temperature as well as Si capping process, indicating that the growth condition can modify the conductive properties of the QDs. By combining with other conductive SPM measurements, the conductive mechanism of GeSi QRs is suggested: the ring-shaped geometry determines that the rim has a lower barrier height with the tip and a higher carrier density, resulting in a higher conductivity at the rim, compared to the central hole and the WL.

On the other hand, many factors which may influence the conductive characteristics of GeSi QDs are also investigated. It is found that their current values and $I$–$V$ characteristics vary obviously with native oxidation and normal force. The current distributions of GeSi QDs and QRs are not, however, influenced by native oxidation and normal force, making the measurement in atmospheric conditions reliable. Furthermore, the conductive properties of the QDs are found to be ably modulated by bias voltage and inter-dot coupling, which should have promising potential applications in nanoelectronics.

By combining with selective chemical etching, the conductance and composition distributions of GeSi QRs are obtained simultaneously on same single QRs, and the dependence of the conductive property of the QRs on their composition is established. The results indicate that the conductance distributions of GeSi QRs are not consistent with their composition distributions. Instead, the geometric shapes of the QRs may make the main contribution to the their current distributions.

## References

1 Barbagiovanni, E.G., Lockwood, D.J., Simpson, P.J., and Goncharova, L.V. (2014) Quantum confinement in Si and Ge nanostructures: theory and experiment. *Appl. Phys. Rev.*, **1** (1), 011302.

2 Priolo, F., Gregorkiewicz, T., Galli, M., and Krauss, T.F. (2014) Silicon nanostructures for photonics and photovoltaics. *Nat. Nanotechnol.*, **9** (1), 19–32.

3 Jia, B.H. (2015) Nanophotonics silicon solar cells: status and future challenges. *Nanotechnol. Rev.*, **4** (4), 337–346.

4 Aqua, J.N., Berbezier, I., Favre, L., Frisch, T., and Ronda, A. (2013) Growth and self-organization of SiGe nanostructures. *Phys. Rep.*, **522** (2), 59–189.

5 Li, X.L., Wang, C.X., and Yang, G.W. (2014) Thermodynamic theory of growth of nanostructures. *Prog. Mater Sci.*, **64**, 121–199.

6 Balberg, I. (2011) Electrical transport mechanisms in three dimensional ensembles of silicon quantum dots. *J. Appl. Phys.*, **110** (6), 061301.

7 Hamdaoui, N., Ajjel, R., Salem, B., Gendry, M., and Maaref, H. (2011) Coulomb charging effect of electrons in InAs/InAlAs quantum dots studied by capacitance techniques. *Physica B*, **406** (19), 3531–3533.

8 Koblmuller, G. and Abstreiter, G. (2014) Growth and properties of InGaAs nanowires on silicon. *Phys. Status Solidi RRL*, **8** (1), 11–30.

9 Oliver, R.A. (2008) Advances in AFM for the electrical characterization of semiconductors. *Rep. Prog. Phys.*, **71** (7), 076501.

10 Avila, A. and Bhushan, B. (2010) Electrical measurement techniques in atomic force microscopy. *Crit. Rev. Solid State Mater. Sci.*, **35** (1), 38–51.

11 Alvarez, J., Boutchich, M., Kleider, J.P., Teraji, T., and Koide, Y. (2014) Direct observation of the leakage current in epitaxial diamond Schottky barrier devices by conductive-probe atomic force microscopy and Raman imaging. *J. Phys. D: Appl. Phys.*, **47** (35), 355102.

12 Bi, J.H., Wei, G.D., Shang, M.H., Gao, F.M., Tang, B., and Yang, W.Y. (2014) Carrier transport in graphite/Si3N4-nanobelt/PtIr Schottky barrier diodes. *Appl. Phys. Lett.*, **105** (19), 191604.

13 Fisichella, G., Greco, G., Roccaforte, F., and Giannazzo, F. (2014) Current transport in graphene/AlGaN/GaN vertical heterostructures probed at nanoscale. *Nanoscale*, **6** (15), 8671–8680.

14 Lee, N., Jo, W., Liu, C., and Meny, C. (2014) Size dependent bipolar resistance switching of NiO nanodots for low-power and multi-state operation. *Nanotechnology*, **25** (41), 415302.

15 15. Villeneuve-Faure, C., Le Borgne, D., Bedel-Pereira, E., Chane-Ching, K.I.M., Hernandez-Maldonado, D., and Seguy, I. (2015) Cross Kelvin force microscopy and conductive atomic force microscopy studies of organic bulk heterojunction blends for local morphology and electrical behavior analysis. *J. Appl. Phys.*, **117** (5), 055501.

16 Lee, H., Qi, Y.B., Kwon, S., Salmeron, M., and Park, J.Y. (2015) Large changes of graphene conductance as a function of lattice orientation between stacked layers. *Nanotechnology*, **26** (1), 015702.

17 Gajewski, K., Kopiec, D., Moczala, M., Piotrowicz, A., Zielony, M., Wielgoszewski, G., Gorszalk, T., and Strupinski, W. (2015) Scanning probe microscopy investigations of the electrical properties of chemical vapor deposited graphene grown on a 6H-SiC substrate. *Micron*, **68**, 17–22.

18 Hentschel, C., Jiang, L., Ebeling, D., Zhang, J.C., Chen, X.D., and Chi, L.F. (2015) Conductance measurements of individual polypyrrole nanobelts. *Nanoscale*, **7** (6), 2301–2305.

19 Yin, J., D'Haese, C., and Nysten, B. (2015) Surface electrical properties of stainless steel fibres: An AFM-based study. *Appl. Surf. Sci.*, **330**, 65–73.

20 Bimberg, D. and Pohl, U.W. (2011) Quantum dots: promises and accomplishments. *Mater. Today*, **14** (9), 388–397.

21 Krasilnik, Z.F., Novikov, A.V., Lobanov, D.N., Kudryavtsev, K.E., Antonov, A.V., Obolenskiy, S.V., Zakharov, N.D., and Werner, P. (2011) SiGe nanostructures with self-assembled islands for Si-based optoelectronics. *Semicond. Sci. Technol.*, **26** (1), 014029.

22 Berbezier, I. and Ronda, A. (2009) SiGe nanostructures. *Surf. Sci. Rep.*, **64** (2), 47–98.

23 Baribeau, J.M., Wu, X., Rowell, N.L., and Lockwood, D.J. (2006) Ge dots and nanostructures grown epitaxially on Si. *J. Phys. Condens. Matter*, **18** (8), R139–R174.

24 Persichetti, L., Sgarlata, A., Fanfoni, M., and Balzarotti, A. (2015) Heteroepitaxy of Ge on singular and vicinal Si surfaces: elastic field symmetry and nanostructure growth. *J. Phys. Condens. Matter*, **27** (25), 253001.

25 Ye, H. and Yu, J.Z. (2014) Germanium epitaxy on silicon. *Sci. Technol. Adv. Mater.*, **15** (2), 024601.

26 Ichiro Tanaka, I.K. and Sakaki, H. (1999) Imaging and probing electronic properties of self-assembled InAs quantum dots by atomic force microscopy with conductive tip. *Appl. Phys. Lett.*, **74**, 844.

27 Tanaka, I., Tada, Y., Nakatani, S., Uno, K., Kamiya, I., and Sakaki, H. (2010) Resonant tunneling of electrons through single self-assembled InAs quantum dot studied by conductive atomic force microscopy. *Physica E*, **42** (10), 2606–2609.

28 Chung, H.-C., Chu, W.-H., and Liu, C.-P. (2006) Electron transport through individual Ge self-assembled quantum dots on Si. *Appl. Phys. Lett.*, **89** (8), 082105.

29 Smaali, K., Troyon, M., El Hdiy, A., Molinari, M., Saint-Girons, G., and Patriarche, G. (2006) Imaging the electric properties of InAs/InP(001) quantum dots capped with a thin InP layer by conductive atomic force microscopy: Evidence of memory effect. *Appl. Phys. Lett.*, **89** (11), 112115.

30 Nakayama, Y., Yamazaki, S., Okino, H., Hirahara, T., Matsuda, I., Hasegawa, S., and Ichikawa, M. (2007) Electrical conduction of Ge nanodot arrays formed on an oxidized Si surface. *Appl. Phys. Lett.*, **91** (12), 123104.

31 Toker, D., Balberg, I., Zelaya-Angel, O., Savir, E., and Millo, O. (2006) Size-dependent local conductance properties of CdSe nanocrystal ensembles. *Phys. Rev. B*, **73** (4), 045317.

32 Cui, J., He, Q., Jiang, X.M., Fan, Y.L., Yang, X.J., Xue, F., and Jiang, Z.M. (2003) Self-assembled SiGe quantum rings grown on Si(001) by molecular beam epitaxy. *Appl. Phys. Lett.*, **83** (14), 2907–2909.

33 Kleemans, N.A.J.M., Bominaar-Silkens, I.M.A., Fomin, V.M., Gladilin, V.N., Granados, D., Taboada, A.G., García, J.M., Offermans, P., Zeitler, U., Christianen, P.C.M., Maan, J.C., Devreese, J.T., and Koenraad, P.M. (2007) Oscillatory Persistent Currents in Self-Assembled Quantum Rings. *Phys. Rev. Lett.*, **99** (14), 146808.

34 Lucignano, P., Giuliano, D., and Tagliacozzo, A. (2007) Quantum rings with Rashba spin-orbit coupling: A path-integral approach. *Phys. Rev. B*, **76** (4), 045324.

35 Okunishi, T., Ohtsuka, Y., Muraguchi, M., and Takeda, K. (2007) Interstate interference of electron wave packet tunneling through a quantum ring. *Phys. Rev. B*, **75** (24), 245314.

36 Yu, L.W., Chen, K.J., Song, J., Xu, J., Li, W., Li, H.M., Wang, M., Li, X.F., and Huang, X.F. (2007) Self-assembled Si quantum-ring structures on a Si substrate by plasma-enhanced chemical vapor deposition based on a growth-etching competition mechanism. *Adv. Mater.*, **19** (12), 1577–1581.

37 Lei, W., Notthoff, C., Lorke, A., Reuter, D., and Wieck, A.D. (2010) Electronic structure of self-assembled InGaAs/GaAs quantum rings studied by capacitance-voltage spectroscopy. *Appl. Phys. Lett.*, **96** (3), 033111.

38 Stoffel, M., Malachias, A., Rastelli, A., Metzger, T.H., and Schmidt, O.G. (2009) Composition and strain in SiGe/Si(001) "nanorings" revealed by combined x-ray and selective wet chemical etching methods. *Appl. Phys. Lett.*, **94** (25), 253114.

39 Sztucki, M., Metzger, T.H., Chamard, V., Hesse, A., and Holý, V.C. (2006) Investigation of shape, strain, and interdiffusion in InGaAs quantum rings using grazing incidence x-ray diffraction. *J. Appl. Phys.*, **99** (3), 033519.

40 Mlakar, T., Biasiol, G., Heun, S., Sorba, L., Vijaykumar, T., Kulkarni, G.U., Spreafico, V., and Prato, S. (2008) Conductive atomic force microscopy of InAs/GaAs quantum rings. *Appl. Phys. Lett.*, **92** (19), 192105.

41 Filatov, D.O., Borodin, P.A., and Bukharaev, A.A. (2011) Study of local density of electron states in InGaAs/GaAs quantum rings by combined STM/AFM. *J. Surf. Invest.*, **5** (3), 547–553.

42 Xue, F., Qin, J., Cui, J., Fan, Y.L., Jiang, Z.M., and Yang, X.J. (2005) Studying the lateral composition in Ge quantum dots on Si (001) by conductive atomic force microscopy. *Surf. Sci.*, **592** (1-3), 65–71.

43 Denker, U., Stoffel, M., and Schmidt, O.G. (2003) Probing the Lateral Composition Profile of Self-Assembled Islands. *Phys. Rev. Lett.*, **90** (19), 196102.

44 Zhang, S.L., Xue, F., Wu, R., Cui, J., Jiang, Z.M., and Yang, X.J. (2009) Conductive atomic force microscopy studies on the transformation of GeSi quantum dots to quantum rings. *Nanotechnology*, **20** (13), 135703.

45 Katsaros, G., Rastelli, A., Stoffel, M., Isella, G., Känel, H.v., Bittner, A.M., Tersoff, J., Denker, U., Schmidt, O.G., Costantini, G., and Kern, K. (2006) Investigating the lateral motion of SiGe islands by selective chemical etching. *Surf. Sci.*, **600** (12), 2608–2613.

46 Zhang, S.L., Lv, Y., Jiang, Z.M., and Yang, X.J. (2011) Electrical properties of individual self-assembled GeSi quantum rings. *J. Appl. Phys.*, **110** (9), 094313.

47 Lv, Y., Cui, J., Jiang, Z.M.M., and Yang, X.J. (2012) Nanoscale electrical property studies of individual GeSi quantum rings by conductive scanning probe microscopy. *Nanoscale Res. Lett.*, **7**, 659.

48 Lee, S.W., Lee, C.H., Chang, H.T., Cheng, S.L., and Liu, C.W. (2009) Evolution of composition distribution of Si-capped Ge islands on Si(001). *Thin Solid Films*, **517** (17), 5029–5032.

49 Holländer, B., Buca, D., Mantl, S., and Hartmann, J.M. (2010) Wet chemical etching of Si, $Si_{1-x}Ge_x$, and Ge in $HF:H_2O_2:CH_3COOH$. *J. Electrochem. Soc.*, **157** (6), H643–H646.

50 Melitz, W., Shen, J., Lee, S., Lee, J.S., Kummel, A.C., Droopad, R., and Yu, E.T. (2010) Scanning tunneling spectroscopy and Kelvin probe force microscopy investigation of Fermi energy level pinning mechanism on InAs and InGaAs clean surfaces. *J. Appl. Phys.*, **108** (2), 023711.

51 Filikhin, I., Suslov, V.M., and Vlahovic, B. (2006) Electron spectral properties of the InAs/GaAs quantum ring. *Physica E*, **33** (2), 349–354.

52 Wu, R., Li, F.H., Jiang, Z.M., and Yang, X.J. (2006) Effects of a native oxide layer on the conductive atomic force microscopy measurements of self-assembled Ge quantum dots. *Nanotechnology*, **17** (20), 5111–5116.

53 Frammelsberger, W., Benstetter, G., Stamp, R., Kiely, J., and Schweinboeck, T. (2005) Simplified tunnelling current calculation for MOS structures with

ultra-thin oxides for conductive atomic force microscopy investigations. *Mater. Sci. Eng., B*, **116** (2), 168–174.
54 Oh, J. and Nemanich, R.J. (2002) Current-voltage and imaging of TiSi2 islands on Si(001) surfaces using conductive-tip atomic force microscopy. *J. Appl. Phys.*, **92** (6), 3326–3331.
55 Wu, R., Zhang, S.L., Lin, J.H., Jiang, Z.M., and Yang, X.J. (2011) Bias-dependent conductive characteristics of individual GeSi quantum dots studied by conductive atomic force microscopy. *Nanotechnology*, **22** (9), 095708.
56 Johal, T.K., Rinaldi, R., Passaseo, A., Cingolani, R., Vasanelli, A., Ferreira, R., and Bastard, G. (2002) Imaging of the electronic states of self-assembledInxGa1 – xAsquantum dots by scanning tunneling spectroscopy. *Phys. Rev. B*, **66** (7), 075336.
57 Zhang, Y.F., Ye, F.F., Lin, J.H., Jiang, Z.M., and Yang, X.J. (2012) Increased conductance of individual self-assembled GeSi quantum dots by inter-dot coupling studied by conductive atomic force microscopy. *Nanoscale Res. Lett.*, **7**, 278.
58 Koroknay, E., Schulz, W.M., Richter, D., Rengstl, U., Reischle, M., Bommer, M., Kessler, C.A., Rossbach, R., Schweizer, H., Jetter, M., and Michler, P. (2012) Vertically stacked and laterally ordered InP and In(Ga)As quantum dots for quantum gate applications. *Phys. Status Solidi B*, **249** (4), 737–746.
59 Chuang, K.Y., Tzeng, T.E., Liu, Y.C., Tzeng, K.D., and Lay, T.S. (2011) Photovoltaic response of coupled InGaAs quantum dots. *J. Cryst. Growth*, **323** (1), 508–510.
60 Wu, J., Hu, X., Lee, J., Kim, E.S., and Wang, Z.M.M. (2013) Epitaxially self-assemblied quantum dot pairs. *Adv. Opt. Mater.*, **1** (3), 201–214.
61 Heldmaier, M., Hermannstadter, C., Witzany, M., Wang, L.J., Peng, J., Rastelli, A., Bester, G., Schmidt, O.G., and Michler, P. (2012) Growth and spectroscopy of single lateral InGaAs/GaAs quantum dot molecules. *Phys. Status Solidi B*, **249** (4), 710–720.
62 Ghadi, H., Agarwal, A., Adhikary, S., Tongbram, B., Mandel, A., Chakrabarti, S., Pendyala, N.B., Prajapati, S., and Kumar, A. (2014) Effect of barrier thickness on structural, optical, and spectral behaviors of vertically strain coupled InAs/GaAs quantum dot infrared photodetectors. *J. Vac. Sci. Technol., B*, **32** (5), 051208.
63 Yuan, F., Jiang, Z., and Lu, F. (2006) Study of coupling effect in double-layer quantum dots by admittance spectroscopy. *Appl. Phys. Lett.*, **89** (7), 072112.
64 Wang, L., Rastelli, A., Kiravittaya, S., Benyoucef, M., and Schmidt, O.G. (2009) Self-assembled quantum dot molecules. *Adv. Mater.*, **21** (25-26), 2601–2618.
65 Ou, Y.C., Cheng, S.F., and Jian, W.B. (2009) Size dependence in tunneling spectra of PbSe quantum-dot arrays. *Nanotechnology*, **20** (28), 285401.
66 Steiner, D., Aharoni, A., Banin, U., and Millo, O. (2006) Level structure of InAs quantum dots in two-dimensional assemblies. *Nano Lett.*, **6** (10), 2201–2205.
67 Lv, Y., Cui, J., Jiang, Z.M., and Yang, X.J. (2013) Composition and conductance distributions of single GeSi quantum rings studied by conductive

atomic force microscopy combined with selective chemical etching. *Nanotechnology*, **24** (6), 065702.
68 Biasiol, G. and Heun, S. (2011) Compositional mapping of semiconductor quantum dots and rings. *Phys. Rep.*, **500** (4-5), 117–173.
69 Schülli, T.U., Vastola, G., Richard, M.I., Malachias, A., Renaud, G., Uhlík, F., Montalenti, F., Chen, G., Miglio, L., Schäffler, F., and Bauer, G. (2009) Enhanced relaxation and intermixing in Ge islands grown on PIT-patterned Si(001) substrates. *Phys. Rev. Lett.*, **102** (2), 025502.
70 Bollani, M., Chrastina, D., Montuori, V., Terziotti, D., Bonera, E., Vanacore, G.M., Tagliaferri, A., Sordan, R., Spinella, C., and Nicotra, G. (2012) Homogeneity of Ge-rich nanostructures as characterized by chemical etching and transmission electron microscopy. *Nanotechnology*, **23** (4), 045302.
71 Ogawa, Y., Toizumi, T., Minami, F., and Baranov, A.V. (2011) Nanometer-scale mapping of the strain and Ge content of Ge/Si quantum dots using enhanced Raman scattering by the tip of an atomic force microscope. *Phys. Rev. B*, **83** (8), 081302.
72 Katsaros, G., Stoffel, M., Rastelli, A., Schmidt, O.G., Kern, K., and Tersoff, J. (2007) Three-dimensional isocompositional profiles of buried SiGe/Si(001) islands. *Appl. Phys. Lett.*, **91** (1), 013112.
73 Lee, S.W., Chang, H.T., Lee, C.H., Cheng, S.L., and Liu, C.W. (2010) Composition redistribution of self-assembled Ge islands on Si (001) during annealing. *Thin Solid Films*, **518** (6), S196–S199.

# 7

# Conductive Atomic Force Microscopy of Two-Dimensional Electron Systems: From AlGaN/GaN Heterostructures to Graphene and $MoS_2$

*Filippo Giannazzo, Gabriele Fisichella, Giuseppe Greco, Patrick Fiorenza, and Fabrizio Roccaforte*

*Consiglio Nazionale delle Ricerche, Istituto per la Microelettronica e Microsistemi (CNR-IMM), Strada VIII, 5, 95121 Catania, Italy*

## 7.1 Introduction

Two-dimensional (2D) electron systems represent one of the most important developments in solid-state physics, both from the fundamental and the applications point of view. A *two-dimensional electron gas (2DEG)* is a system where carriers are free to move in two dimensions, but are tightly confined in the third. To date, 2DEGs have been practically realized in metal–insulator–semiconductor (MIS) devices biased above the inversion threshold ($V_{th}$), in semiconductor heterojunctions (e.g., AlGaAs/GaAs, AlGaN/GaN [1], etc.), in complex oxide interfaces (e.g., $LaAlO_3/SrTiO_3$) [2], and, more recently, in graphene and other two-dimensional materials [3], such as the transition metal dichalcogenides (TMDs) ($MoS_2$, $WS_2$, $MoSe_2$, $WSe_2$, etc.) and Xenes (silicene, phosporene, etc.). The general criterion to observe a 2DEG behavior is that the electrons system is confined within a thickness $t < \lambda_F$, where $\lambda_F = 2\pi/k_F \approx 2\pi^{1/2}/n^{1/2}$ is the 2D Fermi wavelength and n the carrier density. Clearly, in MIS devices and at semiconductor heterostructures, carrier confinement is induced by an external or a built-in electric field and, as such, the 2DEG in these systems has always a finite thickness (typically from ~1 to ~10 nm). In the case of monolayers of graphene (Gr) or TMDs, carrier confinement is "truly 2D," as these materials represent the thinnest unsupported crystalline solids (without dangling surface bonds) that can be realized in nature. Furthermore, few layers of Gr or TMDs show superior intralayer (versus interlayer) transport of carriers, due to the weak (Van der Waals) interlayer bonding with respect to the in-plane covalent bonding.

2DEG systems generally exhibit excellent carrier mobility, making them very suitable for high-frequency electronics applications, as well as for high-precision metrology (based on quantum Hall effect measurements). Furthermore, as 2DEG conductivity can be strongly affected by the interaction with the external environment (e.g., charge transfer by adsorbed molecules), these systems are also very useful for sensing applications.

*Conductive Atomic Force Microscopy: Applications in Nanomaterials,* First Edition.
Edited by Mario Lanza.

To meet the applications requirements, 2D electrons systems have been developed on large-area (wafer-scale) substrates by using advanced synthesis methods, including chemical vapor deposition (CVD), metal organic chemical vapor deposition (MOCVD), molecular beam epitaxy (MBE), and so on. As a matter of fact, the electrical quality of the 2DEG is affected by the presence of nanoscale structural/morphological defects or inhomogeneities, typically originating by the employed substrate and/or by the peculiar growth mechanism. In many cases, nanoscale resolution electrical characterization methods based on scanning probe microscopy, such as conductive atomic force microscopy (CAFM), scanning spreading resistance microscopy (SSRM), scanning capacitance microscopy (SCM), and Kelvin force probe microscopy (KFPM), proved to be fundamental tools to understand the nature and electrical activity of these defects/nonuniformities [4–11]. Furthermore, these techniques provided a guide to solve growth problems and to understand the phenomena limiting the performances of electronic devices based on these 2DEG systems. In particular, CAFM is the method of choice to investigate the mechanisms of current injection from contacts into the 2DEG and/or the lateral homogeneity of 2DEG electrical properties, as this technique allows performing high-resolution current mapping and local current–voltage ($I$–$V$) characterization at the nanoscale.

This chapter focuses on technologically relevant materials for electronics and optoelectronics, such as III–V nitrides (in particular, AlGaN/GaN heterostructures), and on emerging 2D materials currently the object of huge scientific interest, such as Gr and $MoS_2$. New concept hybrid Gr/AlGaN/GaN heterostructures are also considered. Examples of applications of the CAFM technique to these materials are presented, to illustrate how this characterization method can be a very important tool to understand some peculiar electrical properties of such 2D electron systems.

## 7.2 Nanoscale Electrical Characterization of AlGaN/GaN Heterostructures

Although III–V nitrides are typically key technological materials for optoelectronics devices, the continuous improvement of the material quality has recently opened new perspectives also in the fields of high-power and high-frequency applications. In particular, gallium nitride (GaN) and related Al-based alloys ($Al_xGa_{1-x}N$) are outstanding electronic materials for the next generation of high-frequency devices. In fact, due to the presence of spontaneous and piezoelectric polarization, a 2DEG is generated at the AlGaN/GaN interfaces, with typical sheet carrier densities $n_s$ in the order of $\sim 1 \times 10^{13}$ cm$^{-2}$, and high values of the carrier mobility (1000–1500 cm$^2$V$^{-1}$s$^{-1}$) [1, 12]. These unique features, combined with the high critical electric field (3.3 MV/cm) and electron saturation velocity ($2.5 \times 10^7$ cm/s) of GaN, enable the fabrication of high-electron mobility transistors (HEMTs) operating up to several tens of gigahertz and with a high-power handling capability [13].

The 2DEG at AlGaN/GaN interface is located at a depth typically ranging from 10 to 30 nm from the AlGaN surface. Hence, its electrical properties are considerably influenced by the surface properties of the AlGaN, such as donor states, roughness, or by any kind of structural inhomogeneity of GaN and of the AlGaN barrier layer.

One of the crucial issues to optimize the performances of GaN-based devices is the formation of low-resistance Ohmic contacts. Such contacts on AlGaN/GaN heterostructures are typically obtained using complex metal stacks, such as Ti/Al/Mo/Au, Ti/Al/Pt/Au, Ti/Al/Ti/Au, and Ti/Al/Ni/Au [14]. The possibility of using "Au-free" metal stacks, such as Ti/Al or Ta/Al bilayers [15], is presently under investigation, to facilitate the compatibility of GaN devices processing with Si complementary metal–oxide–semiconductor (CMOS) technology. To obtain an Ohmic behavior, these multilayers are typically subjected to rapid annealing processes in the temperature range of 700–900 °C, which cause significant metal intermixing and also interfacial reactions with the AlGaN barrier layer. In some cases, the transition from a rectifying to an Ohmic behavior has been correlated to the formation of an interfacial TiN layer upon annealing, as a consequence of the out-diffusion of N atoms from AlGaN. Nitrogen vacancies act as donors for AlGaN, thus resulting in a highly doped surface region and in the occurrence of carrier tunneling through the metal/AlGaN Schottky barrier. In other cases, the presence of structural defects in the AlGaN barrier layer allows a favorable situation for Ohmic contacts formation [16]. As an example, a "direct contact" between the 2DEG and metallic protrusions has been observed in the correspondence of threading dislocations of the AlGaN.

The current between source and drain contacts in HEMT devices is commonly modulated using a gate Schottky contact onto AlGaN. High work-function metals (such as Pt and Ni) are commonly used for Schottky contact formation onto GaN and AlGaN alloys, and moderate thermal budgets (~400 °C) are typically required to improve the electrical properties of the contact. In spite of this, the measured Schottky barrier heights (SBHs) are significantly lower than the expected theoretical values, due to the inhomogeneity of the metal/semiconductor interface [17] and/or to structural defect of the semiconductor [18]. Sometimes, a thin insulating film is inserted between AlGaN and the contact to reduce the gate leakage current in HEMT devices.

In this context, CAFM proved to be a powerful technique to address the complex surface and interface issues in III–V nitrides processing [19]. In the following subsections, some examples of applications of CAFM to relevant case studies are presented, that is, (i) the study of contacts to AlGaN/GaN heterostructures with different structural quality, (ii) the passivation of the AlGaN surface defects by rapid thermal oxidation (RTO), and (iii) the study of current transport through dielectrics for the gate insulated HEMT technology.

### 7.2.1 Contacts to AlGaN/GaN Heterostructures

CAFM has been employed to investigate at nanoscale the mechanisms of current injection from Ohmic or Schottky contacts to the 2DEG of AlGaN/GaN heterostructures. Roccaforte *et al.* [20] used CAFM to investigate the nanoscale carrier transport in Ti/Al/Ni/Au Ohmic contacts on AlGaN epilayers grown on Si(111). The cross-correlation with structural analyses (X-ray diffraction, transmission electron microscopy (TEM)) allowed to demonstrate that the structure and the electrical properties of the different phases formed inside the reacted layer upon annealing are crucial for the nanoscale current transport. CAFM was also employed to investigate the lateral homogeneity of Schottky contacts to III–V nitrides [21]. More recently, CAFM has been applied by Fontserè *et al.* [22] to study the properties of Ti/Al/Ni/Au Ohmic contacts and of Ni/Au Schottky contacts on AlGaN/GaN HEMT devices on Si wafers. A correlation between the atomic force microscopy (AFM) surface topography and the CAFM current maps on the Schottky and Ohmic contact regions was performed. It was shown that only a fraction (~5%) of the Ohmic contact area efficiently contributes to current injection. Analogously, the reverse leakage of the gate Schottky contact at the nanoscale was found to correlate with the topography.

Greco *et al.* [16] recently employed CAFM in combination with TEM analyses to investigate the impact of the AlGaN layer defectivity/microstructure on the electrical properties of Ti/Al Ohmic contacts fabricated on the AlGaN/GaN heterostructure. In particular, two $Al_{0.25}Ga_{0.75}N$/GaN heterostructures with a different crystalline quality, grown on Si(111) by MOCVD, were compared in this study. The difference between the two samples (from now on named LD and HD samples) was originated from different growth conditions and mainly consisted in the presence of characteristic defects ("V-defects") in the AlGaN layer of the sample HD [16]. CAFM morphology and current mapping on the AlGaN surface were preliminarily performed on both samples before metal deposition, in order to correlate the surface topography with the local current injection into the 2DEG. As schematically illustrated in Figure 7.1a, the AFM conductive tip (connected to a high-sensitivity current amplifier) scanned the AlGaN surface, while the bias was applied to a macroscopic contact fabricated on the sample front side. The surface morphologies of the LD and HD samples are shown in Figure 7.1b,e, respectively. While a flat surface is observed in the LD sample, the presence of a large number of surface defects, visible as depressions in the morphology, can be observed in the HD one. Representative cross-sectional TEM analyses of the LD and HD samples are also reported in Figure 7.1c,f, respectively. In particular, Figure 7.1f elucidates the nature of the depressions observed in the morphology of the HD sample, consisting of "V-shaped" defects in the AlGaN layer, localized at the end of dislocations passing through the AlGaN/GaN interface [23]. The CAFM current maps acquired on the two sample regions are reported in Figure 7.1d,g, respectively. In particular, for the HD sample, it is worth noting a one-to-one correlation between the surface defects and the regions with a higher conduction, visible in the C-AFM map as darker regions. This correlation is not visible in the LD sample, where the C-AFM current map shows an almost uniform current level.

Interestingly, after deposition of the metal (Ti/Al bilayer), a lower annealing temperature (500 °C) was required to obtain an Ohmic contact behavior on the HD sample with respect to the case of the LD one (800 °C). The specific contact resistance ($R_c$) of the annealed contacts onto the two different AlGaN/GaN

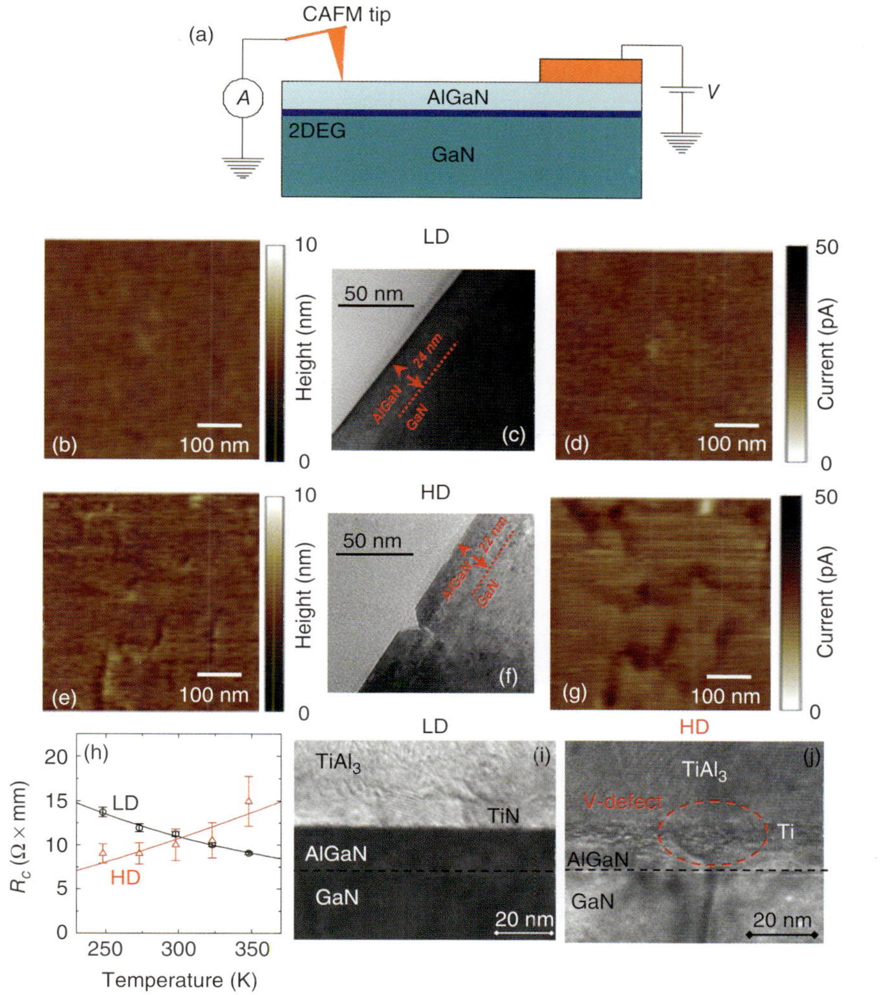


**Figure 7.1** (a) Schematic representation of the measurement setup for CAFM current mapping on the AlGaN surface. AFM topography (b) and the corresponding current map (d) on the LD sample. (c) Cross-section TEM image of the LD sample. AFM topography (e) and corresponding current map (g) on the HD sample. (c) Cross-section TEM image of the HD sample in which a V-shaped defect is visible. (h) Specific contact resistance $R_c$ as a function of the temperature $T$ for Ti/Al Ohmic contact formed on the HD sample (after annealing at 500 °C) and on the LD sample (after annealing at 800 °C). Cross-section TEM micrographs of annealed Ti/Al bilayers on (i) LD sample and (j) HD sample. (Greco *et al.* [16] 2014. Reproduced with permission of Elsevier.).

heterostructures was evaluated by measurements on transmission line model (TLM) test patterns [24]. Of note, by performing this analysis at different measurement temperatures (from 250 to 350 K), a temperature dependence opposite to that of $R_c$ has been observed in the two samples, with a decrease of $R_c$ as a function of $T$ for the annealed Ti/Al contact on the LD sample and an increase of $R_c$ versus $T$ for the contact on the HD one (see Figure 7.1h). While the contact behavior on the LD sample can be described by the thermionic-field emission (TFE) mechanism through the metal/semiconductor Schottky barrier (with a barrier height $\Phi_B = 0.63$ eV and a carrier concentration of $N_D = 1 \times 10^{19}$ cm$^{-3}$ determined from the fit of experimental data with the TFE model), the increase of $R_c$ with $T$ for the contact on the defective heterostructure of sample HD is similar to the behavior commonly observed in "metallic" systems. Figure 7.1i,j shows two representative cross-sectional TEM analyses of the Ti/Al contacts annealed at 800 and 500 °C on the LD and HD samples, respectively. In particular, in the HD sample (Figure 7.1j), an unreacted Ti layer is directly in contact with the AlGaN surface and penetrates inside the V-shaped defects present in AlGaN down to few nanometer distance from the 2DEG. On the other hand, in the LD sample (Figure 7.1i) annealed at 800 °C, the presence of a reacted TiN layer is observed at the interface with AlGaN. This interfacial reaction is expected to leave an excess of $N$ vacancies (donor centers) in the AlGaN surface region, consistently with the carrier density estimated from the fitting of $R_c$ versus $T$ dependence in Figure 7.1h.

### 7.2.2 Electrical Nanocharacterization of AlGaN Surface Passivated by a Rapid Thermal Oxidation

Clearly, structural defects of GaN and AlGaN barrier layer have an impact not only on the mechanisms of Ohmic contact formation but also on the leakage current of the gate Schottky contacts of GaN-based HEMT devices.

Recently, RTO has been proposed as a simple method to passivate the surface of AlGaN/GaN HEMTs and to reduce the leakage current. Of note, long oxidation processes have been used in the past also for the lateral isolation of AlGaN devices, that is, to locally suppress the 2DEG in the oxidized area [25, 26]. Instead, Greco *et al.* [27] employed a rapid process at 900 °C in oxygen ambient to achieve a controlled oxidation of the AlGaN surface and of conductive defects, such as dislocations and V-defects, while maintaining the 2DEG at the AlGaN/GaN interface. In that work, CAFM was employed together with structural and chemical characterization techniques, such as TEM and Energy Dispersive X-ray Analysis (EDX), to characterize this process.

Figure 7.2a shows a representative cross-sectional TEM of the $Al_{0.27}Ga_{0.73}N$/GaN heterostructure (with ~25 nm thick AlGaN barrier layer) grown on a Si(111) substrate. In the upper part of the AlGaN barrier layer, V-shaped defects are present. In spite of this, an atomically flat AlGaN surface is observed in the defect-free regions, as shown by the high-resolution TEM image in the insert of Figure 7.2a. A cross-sectional TEM micrograph of the heterostructure after the RTO at 900 °C is reported in Figure 7.2d. A uniform and conformal (~1.5 nm-thick) oxide layer (composed of both $Al_2O_3$ and $Ga_2O_3$ phases) is formed at the AlGaN surface, as shown in the insert in the upper part of the figure. Furthermore, the oxygen chemical map in the insert (lower part of the figure) indicates that oxidation occurs not only on the surface, but also on the walls of V-defects. The surface morphology and current maps measured by CAFM on the as-grown and oxidized samples are reported in Figure 7.2b,c,e,f,

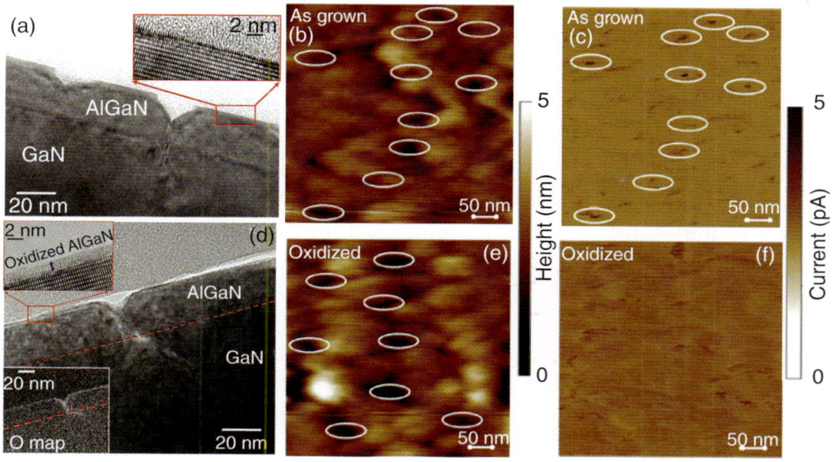


**Figure 7.2** (a) Cross-section TEM analysis of the as-grown AlGaN/GaN heterostructure and high-resolution image of the AlGaN layer (insert). Surface morphology (b) and CAFM current map (c) of the as-grown sample. The white circles indicate the morphological surface defects and the corresponding current spots. (d) Cross-section TEM analysis of the oxidized sample. A high-resolution image of the AlGaN layer is reported in the insert on the top and the oxygen chemical map in the proximity of a V-defect is reported in the bottom insert. Surface morphology (e) and CAFM current map (f) of the oxidized sample. (Panels (a–f) have been modified and reprinted with permission from [27], copyright from IOP Publishing 2014.)

respectively. The current maps were acquired in lateral configuration, by applying a forward bias of +1 V to the conductive tip with respect to an Ohmic contact fabricated on the AlGaN surface. The morphological images show the presence of defects (depressions) on the sample surface of both the as-grown and the oxidized sample (highlighted by the white circles). From the AFM images, the density of these surface defects was estimated to be in the order of $3\text{–}4 \times 10^9$ cm$^{-2}$, in both samples. In the as-grown sample, these morphological surface defects are clearly correlated with high current conductive spots (highlighted by white circles) observed in the current map. On the other hand, an overall reduction of the vertical conduction and no clear correlation between the topography and current map were found for the oxidized sample. Hence, it could be concluded by these CAFM analyses that near-surface defects acting as preferential conductive paths for the leakage current are almost completely passivated by the oxidation process. The results of this nanoscale electrical characterization are reflected in the reduction of the reverse leakage current at the gate Schottky diode fabricated on the passivated AlGaN surface.

### 7.2.3 CAFM on Dielectrics for Gate Insulated AlGaN/GaN Transistors

Different oxides ($SiO_2$, $Al_2O_3$, $Ga_2O_3$, $HfO_2$, $Sc_2O_3$, $La_2O_3$, NiO, etc.) have been studied as possible gate dielectrics in AlGaN/GaN transistors [28]. Understanding of the physical mechanisms involved in the electron conduction through gate dielectrics represents a crucial issue for this technology. CAFM has been extensively used in the past to investigate locally current transport in dielectrics on

Si and other semiconductors. Recently it proved to be a powerful technique to study conduction phenomena also in the case of dielectrics on III–V nitrides. As an example, Fiorenza *et al.* [29] applied CAFM to the study of epitaxial nickel oxide (NiO) thin films grown by MOCVD on AlGaN/GaN heterostructures. Due to its band gap (4 eV) and a relatively high permittivity (11.9), NiO is a promising candidate as insulating material for gate insulated HEMT devices. A significant reduction (more than two orders of magnitude) of the leakage current was observed in NiO(15 nm)/AlGaN/GaN MIS diodes compared to a simple Schottky diode. The transport mechanisms through the NiO have been investigated by temperature-dependent *I–V* measurements on the MIS diodes and nanoscale current mapping by CAFM.

The temperature dependence of forward *I–V* characteristics collected on the MIS device is shown in Figure 7.3a. A schematic representation of the device is also reported in the insert of Figure 7.3a. The current flowing through the MIS device was found to increase with the temperature. This behavior was well described by a Pool–Frenkel (PF) emission mechanism, as can be observed from the linear dependence of $J/E$ (with $J$ the current density and $E$ the electric field across the oxide) on $E^{1/2}$ for all the measurement temperatures (Figure 7.3b). From the fit of the data, the value of the relative dielectric constant ($\varepsilon_{\text{NiO}} = 10.9 \pm 0.1$) and of the emission barrier height ($E_c - \varphi_t = 0.20 \pm 0.02$ eV, with $E_c$ the NiO conduction band edge) of the deposited NiO film were extracted. The occurrence of the PF mechanism can be associated with material defects (like grain boundaries, twins, etc.) which result in localized states at 0.2 eV below

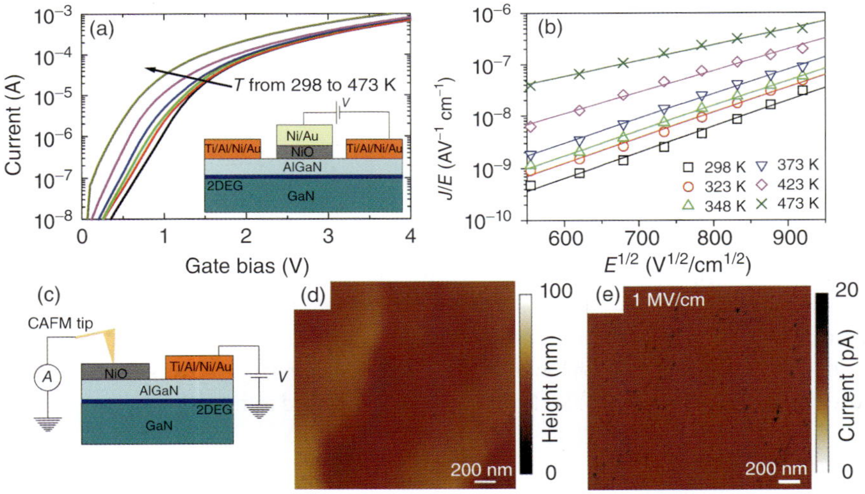


**Figure 7.3** (a) Forward *I–V* characteristics at different temperatures of the NiO/AlGaN/GaN MIS diode (see schematic in the insert). (b) Poole–Frenkel plot of the measured currents. (c) Schematic of the experimental setup for CAFM characterization of the NiO thin film. Surface morphology of the NiO film (d) and two-dimensional current map (e) collected for an electric field of 1 MV/cm applied to the dielectric. (Panels (a, b, d, e) have been modified and reprinted with permission from [29], copyright from American Institute of Physics 2012.)

the NiO conduction band edge, and are responsible for an enhanced current in a certain temperature range. In order to directly visualize the current conduction and have an idea of the spatial distribution and density inside NiO of the states responsible for the PF emission, two-dimensional C-AFM measurements were carried out using the experimental setup represented in Figure 7.3c. By this investigation, it was possible to correlate the morphological features on the bare NiO surface with the current map at a fixed electric field applied by the AFM tip. The surface morphology of NiO is shown in Figure 7.3d and clearly demonstrates that the film is free from macroscopic defects, that is, cracks, pits, and so on. The corresponding current map acquired in the same region at an applied electric field of 1 MV/cm across the NiO films is reported in Figure 7.3e. A density of conductive spots $\sim 6 \times 10^8$ cm$^{-2}$ could be estimated from the current map. No clear correlation between the spatial distribution of these conductive spots and the surface morphology shown in Figure 7.3d can be found. However, as their density is in the same order of magnitude of dislocations typically present in the underlying AlGaN layer, it cannot be ruled out that a correlation exists between conductive spots in the oxide and structural defects of the AlGaN/GaN heterostructure [18].

## 7.3 CAFM Characterization of Graphene and $MoS_2$

The isolation of graphene (Gr), the single atomic layer of sp$^2$ carbon, obtained by exfoliation of graphite represented the first experimental demonstration that two-dimensional (2D) materials can be stable under ambient conditions after separation from the bulk crystal [30]. This also opened the way to the investigation of an entire class of layered materials occurring in nature, which are composed by the vertical stacking of 2D sheets bonded by Van der Waals interaction [3]. Two-dimensional (2D) layered materials are currently the object of significant scientific interests, due to their unique electrical, optical, mechanical, and chemical properties, which make them attractive both from a basic and from a technological standpoint. To date, Gr has been the most widely studied 2D material, because of its rich physics and its excellent high carrier mobility, typically in the range from $10^3$ to $10^4$ cm$^2$V$^{-1}$s$^{-1}$. However, the lack of a band gap limits Gr application in electronics as a channel material for logic or switching field-effect transistors. On the other hand, other layered materials, such as 2D TMDs, are semiconductors with sizable band gaps [31]. This property has opened the way to their application both in thin-film transistors [32] and in novel device concepts based on vertical heterostructures of different 2D materials [3]. In particular, due to its high mechanical and chemical stability down to an ultimate single-layer thickness, $MoS_2$ is currently being considered as a candidate to replace Si for next-generation CMOS technology.

Exfoliation from the parent bulk crystals is not a suitable approach to obtain single- or few-layer 2D materials on large areas for electronics/optoelectronics applications. In the past decades, several scalable synthesis approaches have been developed to produce electronic quality Gr on wafer scale. In particular, the two main approaches currently used are (i) the CVD from hydrocarbon precursors on catalytic substrates (typically metals, such us Cu and Ni) [33] and (ii) the controlled graphitization of hexagonal silicon carbide (4H- and 6H-SiC) by high-temperature thermal treatments [34]. While using CVD-grown Gr on metals implies the transfer from the original substrate to an insulating/semiconducting one (with this manipulation typically introducing an unintentional doping and mobility degradation of Gr), thermal decomposition of SiC provides Gr directly on a semiconducting or semi-insulating substrate ready for electronics and/or optoelectronics applications. The structural and electrical properties of Gr on hexagonal SiC strongly depend on the substrate orientation, that is, the Si face (0001) [34, 35], the C face (000$\bar{1}$) [36, 37], or the nonpolar faces (11-20) and (1-100) [38]. In particular, highly uniform single- or few-layer Gr films with excellent thickness uniformity on wafer scale and very precise epitaxial orientation with respect to the substrate can be obtained in a controllable way on the (0001) face of SiC [34]. This is a result of the peculiar growth mechanism of Gr on this SiC orientation, that is, the formation of a $(6\sqrt{3}\times 6\sqrt{3})$R30 carbon reconstruction of the (0001) face, the so-called C buffer layer, which acts as the precursor of Gr formation [36]. In spite of its hexagonal structure similar to that of Gr, the C buffer layer is not electrically conductive, due to the partial covalent bonding with the SiC substrate, with the presence of a large density of Si dangling bonds. These charged dangling bonds are also responsible for the n-type doping (typically in the order of $10^{13}$ cm$^{-2}$) of Gr on SiC(0001) [39, 40].

As a matter of fact, Gr grown by the two above-mentioned approaches presents peculiar nanoscale inhomogeneities, which are closely related to the morphology of the substrate and specific growth mechanism. These local electrical and structural inhomogeneities are reflected in the macroscopic electrical behavior of Gr-based electronic devices. In Section 7.3.1, some examples of CAFM applications to Gr grown on the Si face of 4H-SiC are reported [41], to illustrate the huge potentiality of this characterization approach for 2D materials investigation.

In the case of $MoS_2$, the scalable growth by CVD methods is currently at its initial stages [42], and the best quality films (with electron mobility typically ranging from 10 to 200 cm$^2$V$^{-1}$s$^{-1}$) are still obtained by exfoliation from bulk molybdenite. Exfoliated $MoS_2$ thin films are typically unintentionally n-type doped [43]. Since methods for selective-area doping of $MoS_2$ are still lacking, source/drain contacts for a transistor structure are typically deposited directly on unintentionally doped $MoS_2$, resulting in the formation of a Schottky barrier. Experiments have shown that the SBH values on $MoS_2$ can range from ~25 to ~300 meV, going from low work-function metals (such as Sc or Ti) to high work-function ones (such as Ni or Pt) [44]. Such a behavior has been commonly ascribed to a Fermi level pinning in the upper part of the $MoS_2$ band gap. In Section 7.3.2, an investigation of the lateral homogeneity of the $MoS_2$ Schottky barrier and resistivity by the CAFM technique are reported.

Finally, the integration between Gr and AlGaN/GaN heterostructures is currently attracting increasing interest for potential applications in high-frequency electronics and optoelectronics. In Section 7.3.3, a CAFM-based investigation of current injection mechanisms at the Gr/AlGaN interface is presented.

### 7.3.1 Local Electrical Properties of Graphene 2DEG

Epitaxial Gr samples were grown on a semi-insulating "nominally" on-axis 4H-SiC (0001) wafer by a high-temperature (2000 °C) thermal treatment in Ar ambient at a pressure of ~1 atm performed in an inductively heated reactor [45]. A typical AFM morphological image of the sample is reported in Figure 7.4a, showing that Gr resides on a surface exposing micrometer-wide terraces separated by nanometer-high steps (see linescan in Figure 7.4b). This characteristic topography of the SiC substrate results from the evolution of atomic steps present in the virgin sample due to the unintentional miscut angle of the SiC wafer with respect to the growth axis (0001). A typical micro-Raman spectrum on the as-grown sample is reported in Figure 7.4c. The characteristic 2D peak at 2692 $cm^{-1}$ typical of graphitic carbon is clearly evident. This peak is symmetric and can be fitted by a single Lorentzian component with full-width-at-half-maximum (FWHM) of ~37 $cm^{-1}$ (as shown in the insert of Figure 7.4c), indicating that epitaxial Gr is mainly composed by a monolayer in the probed area [41]. Obviously, Raman spectra yield average information on a few micrometer area corresponding to the laser spot size.

The experimental configuration for CAFM measurements is illustrated in Figure 7.5a. Macroscopic Ohmic contacts (made by Ni/Au bilayers) have been deposited on Gr, leaving a circular bare Gr area (radius $r_2 = 200$ μm). AFM scans on ~1 μm × 1 μm regions were performed close to the center of that circular area by a Pt-coated Si tip. During the scan, 30 mV bias was applied to the macroscopic contact, and the lateral current flow inside Gr between this contact and the nanometric tip (apex curvature radius $r_{tip} \approx 20$ nm) was measured by a logarithmic current amplifier connected to the tip. All the measurements were carried out under ambient conditions. As illustrated schematically in Figure 7.5b, in this configuration the measured resistance R can be expressed as the series of several contributions, that is,

$$R = R_{tip} + R_c + R_{spr} + R_s + R_{c\_macro} \tag{7.1}$$

where $R_{tip}$ is the tip resistance, $R_c$ is the tip/Gr contact resistance, $R_{spr}$ is the spreading resistance encountered by the current to spread from the tip to Gr, $R_s$ is the series resistance, and $R_{c\_macro}$ is the macroscopic contact resistance.

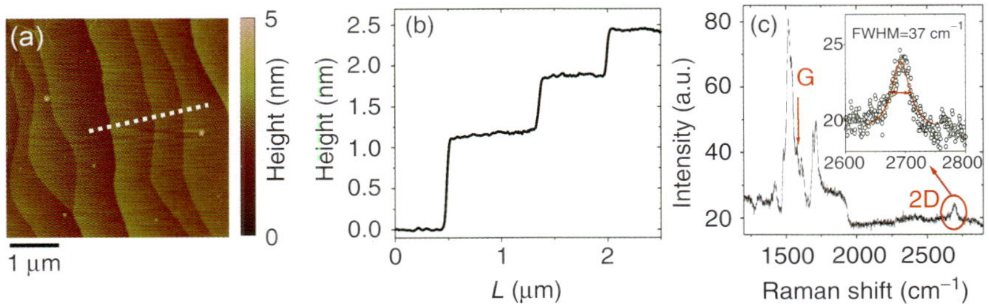


**Figure 7.4** (a) Typical AFM morphology and (b) height line profile of as-grown epitaxial Gr on nominally on-axis 4H-SiC(0001). (c) Typical micro-Raman spectrum on the same sample. (Panel (c), has been modified and reprinted with permission from [41], copyright from American Physical Society 2012.)

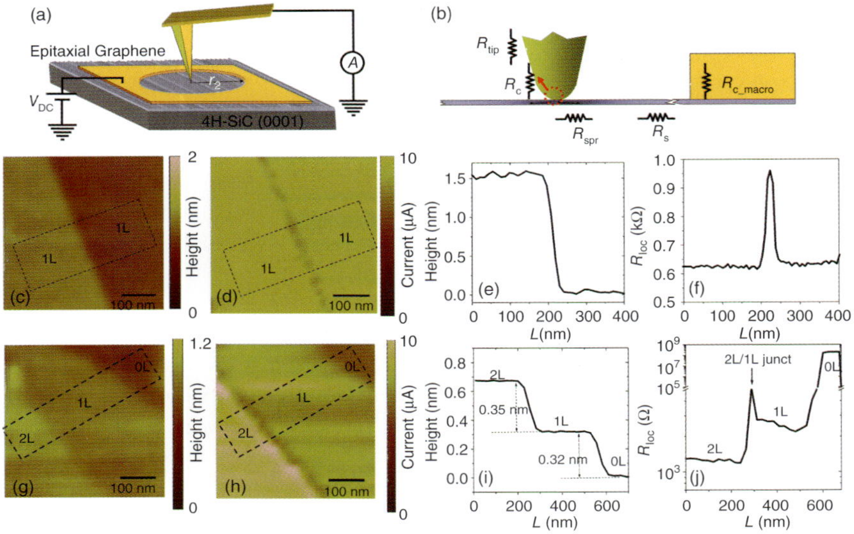


**Figure 7.5** (a) Schematics of the experimental setup for CAFM measurements on epitaxial Gr on 4H-SiC (0001). (b) Representation of the different contributions to the measured resistance *R*. Surface morphology (c), current map (d), height (e), and local resistance (f) line profiles for 1L Gr over ~1.5-nm-high step of SiC substrate. Surface morphology (g), current map (h), height (i), and local resistance (j) line profiles in a sample region with local inhomogeneity in the number of Gr layers. (Panels (a–j) have been modified and reprinted with permission from [41], copyright from American Physical Society 2012.)

The spreading resistance contribution $R_{spr}$ captures the information on the local resistance $R_{loc}$ of Gr with a lateral resolution comparable with the tip radius. Due to the 2D nature of epitaxial Gr, $R_{spr}$ can be expressed as $R_{spr} = R_{loc}/2\pi \ln(r_{tip}/r_0)$ [41], where $r_0$ is kept as the distance of Gr from the buffer layer ($r_0 \approx 3.3$ Å). To extract the $R_{spr}$ contribution and, hence, the local resistance of Gr, the other terms have been independently evaluated. In particular, $R_{tip}$ has been evaluated by CAFM measurements on metal structures, whereas the series resistance $R_s$ of epitaxial Gr and the macroscopic contact resistance $R_{c\_macro}$ have been measured using properly fabricated circular TLM test patterns [41]. The tip/Gr contact was found to be Ohmic from the linear *I–V* characteristics measured by the CAFM tip on Gr (not shown here). Furthermore, the tip/Gr contact resistance $R_c$ was minimized by properly setting the contact force between the tip and epitaxial Gr [46].

Figure 7.5c,d shows the AFM surface morphology and the corresponding current map measured in the case of a monolayer (1L) of Gr over a ~1.5-nm-high substrate step. The height and $R_{loc}$ line profiles in the selected areas are also reported in Figure 7.5e,f, respectively. The contrast in the current map clearly shows that 1L Gr is present at both sides of the step, and that a resistance peak is associated with the step itself. The extracted local sheet resistance ($R_{loc}$) profile of Gr shows ~55% enhancement of $R_{loc}$ in the step region, as compared to the surrounding 1L Gr areas. CAFM maps of Gr across steps of different heights were measured at different positions in the sample, and it was found that the local resistance of 1L Gr over substrate steps increases with the step height [41]. This behavior suggests that this effect is related to the interaction between Gr and the sidewall face of the SiC step. In particular, a lower electrostatic doping of epitaxial Gr over this face can result in an increase of the local resistance.

Figure 7.5g,h shows the AFM surface morphology and the corresponding current map measured on a region with submicrometer nonuniformities in the number of layers. In particular, three areas have been identified from the comparison of topography and current map that have been associated with the C-buffer layer (0L), a monolayer (1L), and a bilayer (2L) of Gr, as discussed in the following. The line profiles of the height and of the local sheet resistance $R_{loc}$ extracted from the measured current are reported in Figure 7.5i,j, respectively. The height profile shows a ~$0.32 \pm 0.1$ nm spacing between the 0L and the 1L areas, and a ~$0.35 \pm 0.1$ nm spacing between the 2L and the 1L ones, in good agreement with the values of 0L/1L and 1L/2L separation recently determined by atomic resolution scanning TEM on similar samples [47]. From the profile in Figure 7.5j, it is worth noting that the highest $R_{loc}$ values (~0.5 GΩ) are associated with the 0L region, whereas the 1L and 2L can be associated with the regions with ~0.62 and ~0.30 kΩ resistance values, respectively. The insulating character of the 0L area confirms that in this region the C buffer layer only is present, whereas the 1L and 2L regions can be assigned to monolayer and bilayer Gr, respectively. The most interesting feature of this profile is the peak at the junction between 1L and 2L of Gr. A remarkable increase of the resistance (~50%) is observed with respect to the value measured in the single-layer region. Furthermore, it is worth noting that this peak is very sharp and its width is just limited by the lateral resolution of the measurement, almost coincident with the tip contact radius (~20 nm), suggesting that the effect responsible for the enhanced resistance is extremely localized. Differently from the resistance enhancement of Gr on substrate steps, which is an electrostatic phenomenon due to a different charge transfer between the step edges and basal plane of SiC and Gr, the localized resistance enhancement at the 1L/2L junction is a purely quantum mechanical phenomenon. It can be ascribed to the destructive interference of electron wavefunctions of 1L and 2L Gr at the interface, as demonstrated by density functional theory quantum transport calculations [41].

### 7.3.2 Nanoscale Inhomogeneity of the Schottky Barrier and Resistivity in $MoS_2$

As previously discussed, the study of the metal/$MoS_2$ Schottky barrier currently represents an interesting topic for the development of prototypes of thin-film transistors based on this 2D semiconductor. In this context, CAFM has been employed to investigate the current injection from a nanoscale metal contact, that is, the AFM tip, to the surface of $MoS_2$ multilayers, with thicknesses ranging from ~30 to ~60 nm, exfoliated on a $SiO_2$ substrate [48]. Such relatively thick $MoS_2$ samples were chosen for this investigation since they guaranteed a better screening of the surface region (where the tip/$MoS_2$ Schottky contact is formed) from the effect of charges typically present at the interface with the substrate.

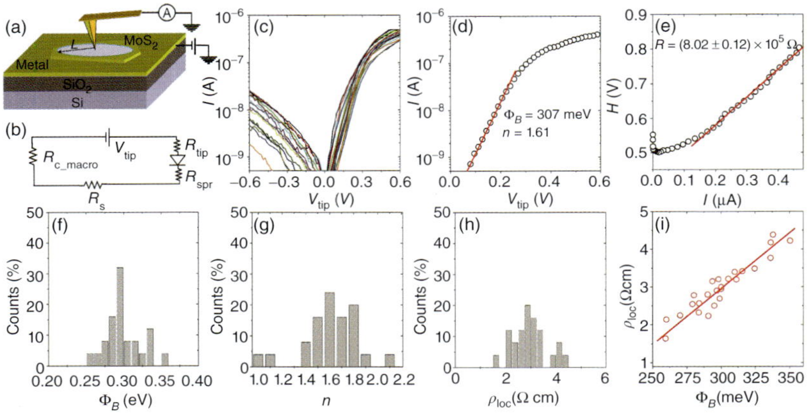

**Figure 7.6** (a) Schematics of the experimental setup for CAFM measurements on $MoS_2$ and (b) equivalent circuit. (c) Set of 25 $I$–$V_{tip}$ characteristics measured on a 500 nm × 500 nm array of tip positions with ~100 nm spacing on $MoS_2$. (d) Representative forward bias $I$–$V_{tip}$ characteristic from this set of measurements and fit with the thermionic emission law to extract the SBH and ideality factor. (e) $H$ function plot for the determination of the series resistance $R$. Histograms of the local SBH $\Phi_B$ (f), the ideality factors $n$ (g), and resistivities $\rho_{loc}$ (h) extracted from the full set of $I$–$V_{tip}$ characteristics reported in panel (c). (i) Plot of $\rho_{loc}$ versus $\Phi_B$. (Panels (a–i) have been modified and reprinted with permission from [48], copyright from American Physical Society 2015.)

Local $I$–$V$ measurements by CAFM were performed using Pt-coated Si tips with a curvature radius of $r_{tip} \approx 10$ nm. A contact geometry similar to that used for CAFM measurements on epitaxial Gr (consisting of a metal film with circular holes of radius $L \gg r_{tip}$) was adopted also for the $MoS_2$ flakes, as shown in the schematic of Figure 7.6a.

Figure 7.6c shows a set of 25 $I$–$V$ curves measured displacing the tip on a 500 nm × 500 nm array of tip positions (with ~100 nm spacing) in the central region of a ~50-nm-thick $MoS_2$ flake. For the sake of simplicity, the bias $V$ refers to the tip ($V_{tip}$). Clearly, all the curves exhibit an asymmetric behavior with respect to bias inversion, consistently with the expected Schottky behavior for the tip/$MoS_2$ contact. A representative forward bias $I$–$V_{tip}$ characteristic from this set of measurements is reported in Figure 7.6d. In the semilog plot, current exhibits a more than two-decade linear increase (from $1 \times 10^{-10}$ to $5 \times 10^{-8}$ A) with $V_{tip}$, followed by a saturation.

In order to extract the SBH $\Phi_B$ and the ideality factor $n$, the thermionic emission law was applied to fit the $I$–$V_{tip}$ curves in the forward bias regime,

$$I = AA^*T^2 \exp\left(-\frac{q\Phi_B}{kT}\right) \exp\left[\frac{q(V_{tip} - IR)}{nkT}\right] \tag{7.2}$$

where $q$ is the electron charge, $k$ is the Boltzmann constant, $T$ the absolute temperature ($T = 300$ K), $A = \pi r_{tip}{}^2$ the tip contact area, and $A^*$ the Richardson constant. At small forward biases the weight of the resistive term $R$ in the exponential factor of Eq. (7.2) is negligible and ln($I$) depends linearly on $V_{tip}$. By linear fitting of the forward bias characteristic in the low-voltage region, $\Phi_B = 307$ meV and $n = 1.61$ have been determined from the intercept and the slope of the fit, respectively. For higher bias values, $R$ causes a deviation of current from linearity and its saturation on the semilog scale.

Since the downward curvature in the high-voltage region of the $I$–$V_{tip}$ curves depends both on $n$ and $R$, Cheung's method [49] was applied to evaluate the $R$ contribution. In this method, the function $H$ is defined as $H = V_{tip} - nkTq \ln(IAA^*T^2)$, which depends on $I$ as $H = n\Phi_B + IR$. Figure 7.6e shows a plot of $H$ versus $I$ obtained from the forward bias $I$–$V_{tip}$ characteristic in Figure 7.6d. $R = (8.02 \pm 0.12) \times 10^5\ \Omega$ has been obtained by the slope of the linear fit for current values larger than 0.1 μA.

As illustrated in the equivalent circuit in Figure 7.6b, this resistance term $R$ in series with the tip/$MoS_2$ Schottky contact is the sum of several contributions, whose physical meaning has been already illustrated in Section 7.3.1, Eq. (7.1). In the present case, the two dominant contributions to $R$ are $R_{spr}$ and $R_s$, which both are associated with current transport within $MoS_2$ [48]. $R_{spr}$ is related to the local resistivity $\rho_{loc}$ of $MoS_2$ under the tip as

$$R_{spr} = \rho_{loc}/4r \tag{7.3}$$

$r$ being the contact radius. In the case of layered materials (such as graphite or $MoS_2$) with a strong anisotropy in the conductivity, $\rho_{loc}$ can be expressed as $\rho_{loc} = (\rho_\perp \rho_\parallel)^{1/2} = \gamma^{1/2}\rho_\parallel$, where $\rho_\perp$ and $\rho_\parallel$ are the out-of-plane and in-plane local resistivities, respectively, and $\gamma = \rho_\perp/\rho_\parallel$ is the anisotropy ratio [50]. Literature results on bulk single crystalline $MoS_2$ samples [51] indicate that $\gamma$ can range from $\sim 10^2$ to $\sim 10^4$, with typical values on the order of $\sim 10^3$. The $R_s$ term is mainly due to the lateral current transport within $MoS_2$ from the tip to the macroscopic contact and can be expressed in terms of the average value of the in-plane resistivity $\rho_\parallel$ as $R_s = \langle\rho_\parallel\rangle/2\pi t \ln(L/r_{tip})$, where $L$ is the radius of the circular hole in the metal contact and $t$ is the $MoS_2$ film thickness. Clearly, while the geometrical factors in the expressions of $R_{spr}$ and $R_s$ are in the same orders of magnitude, the local resistivity value $\rho_{loc}$ is expected to be significantly higher (30–100 times) than $\langle\rho_\parallel\rangle$ due to the high anisotropy ratio for $MoS_2$. As a result, $R_{spr}$ is the largest resistive contribution to $R$. Hence, $\rho_{loc}$ was estimated from the $R$ values measured at each tip position using Eq. (7.3), assuming $r \approx r_{tip}$.

By performing the same analysis on the full set of $I$–$V_{tip}$ characteristics of Figure 7.6c, the distributions of the local SBHs, ideality factors, and resistivity values at the different tip positions on $MoS_2$ have been determined. The histograms of the $\Phi_B$, $n$, and $\rho_{loc}$ values are reported in Figure 7.6f–h. An average SBH of 300 meV with a standard deviation of 24 meV has been estimated from the distribution in Figure 7.6f. Clearly, these SBH values are much lower than the ideal one expected according to the Schottky–Mott theory $\Phi_B = W_{Pt} - \chi_{MoS2} \approx 1.3$ eV (with $W_{Pt}$ the Pt work function and $\chi_{MoS2}$ the $MoS_2$ electron affinity), consistently with the commonly reported Fermi level pinning for most of the metals in the upper part of the $MoS_2$ band gap [44]. The histogram in Figure 7.6g shows that $n$ is close to unity only on $\sim$10% of the investigated $MoS_2$ area, whereas the average value of $n$ is 1.60 with a standard deviation of 0.23. Generally, the deviation of $n$ from unity indicates that current transport is not perfectly described by the thermionic emission theory. For a macroscopic Schottky contact, this can be ascribed to several reasons, including the presence of inhomogeneities within the contact area [52] and/or to interface states [53]. In the case of a nanometric size tip semiconductor Schottky diode, the SBH can be considered homogeneous within the contact area, whereas surface states can be responsible for $n > 1$. According to the theory of Schottky barriers with interface states [53], $n$ can be related to the density of surface states, whereas $\Phi_B$ can be related to their energy within the gap. The observed spread in the local $\Phi_B$ and $n$ values indicates a spatial distribution both of the energy and the density of the interface states. Finally, an average resistivity of 2.99 Ωcm with a standard deviation of 0.68 Ωcm has been estimated from the distribution in Figure 7.6h. The lateral variations of $\rho_{loc}$ can be ascribed to inhomogeneities in the carrier concentration and/or in the carrier mobility of $MoS_2$. To understand whether these local changes in the transport properties have an origin similar to that of the inhomogeneities in the SBH, a plot of the local $\rho_{loc}$ and $\Phi_B$ values measured at the different tip positions is shown in Figure 7.6i. A good correlation between these two quantities can be observed, with a linear increase of $\rho_{loc}$ versus $\Phi_B$ (the fit serves as a guide to the eye). This suggests that the source for the lateral variations of resistivity is the same as for the inhomogeneous SBH. Recent experimental investigations by scanning tunneling microscopy/scanning tunneling spectroscopy (STM/STS) showed the formation of energy states in the band gap of $MoS_2$ corresponding to the intentional removal of sulfur atoms from the surface by using the electric field of STM [54]. The presence of a high density of sulfur vacancies has been demonstrated even on as-exfoliated $MoS_2$ from molybdenite [55]. On the other hand, sulfur vacancies are also indicated as one of the sources of n-type doping in $MoS_2$ [43]. It is therefore reasonable that a local increase in the density of sulfur vacancies can lead both to a reduction of the SBH and to an increase of the carrier concentration, that is, to a reduction of $\rho_{loc}$.

### 7.3.3 Graphene Contacts to AlGaN/GaN Heterostructures

Recently, Gr transparent electrodes have been considered as alternative contacts for GaN and related alloys, especially for optoelectronics applications. As an example, it has been shown that Gr insertion between a metal and AlGaN gives rise to a Schottky contact with low barrier height, that is, ranging from ~0.4 to ~0.6 eV depending on the metal workfunction [56]. Another work indicates that a Gr interlayer between a metal (Cr) and AlGaN can provide an Ohmic contact to AlGaN/GaN heterostrucures even without performing any thermal treatment [57], which is a very interesting result for GaN devices processing. These differences in the electrical behavior can depend on several factors, such as the used metal layer and the structural properties of the AlGaN barrier layer.

Recently, Fisichella *et al.* [58] employed CAFM for the nanoscale electrical characterization of Gr contacts to AlGaN/GaN heterostructures, focusing on the role of the microstructure of the AlGaN barrier layer on the formation of rectifying or Ohmic contacts.

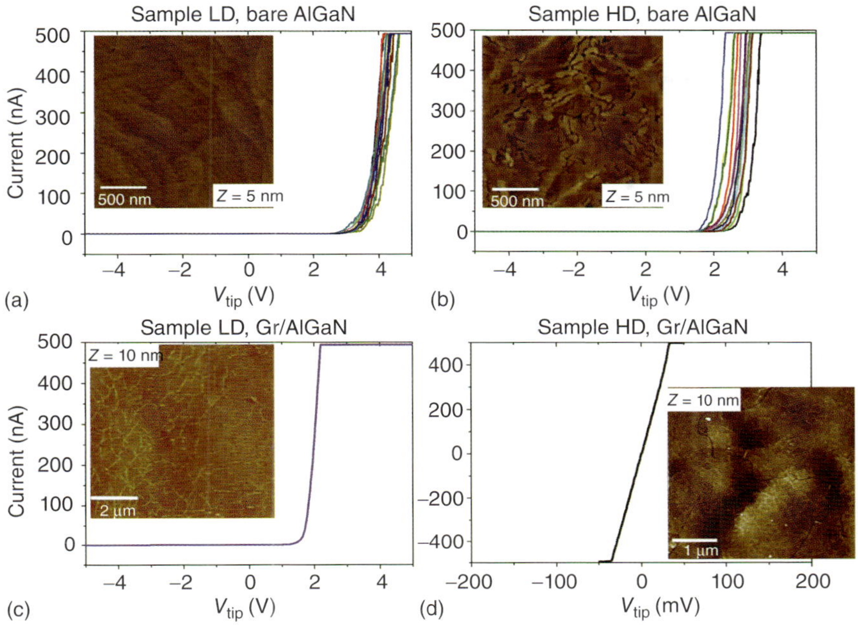


**Figure 7.7** *I–V* curves on the reference AlGaN/GaN samples LD (a) and HD (b) and on the Gr-coated samples LD (c) and HD (d). Representative AFM analyses of the topography of these samples are reported in the inserts. (Panels (a–d) have been modified and reprinted with permission from [58], copyright from American Institute of Physics 2014.)

Two AlGaN/GaN heterostructures of different quality, similar to those considered in Figure 7.1, were compared in this study. Also, in this case, the difference between the two samples (LD and HD) consisted in the presence of characteristic defects (V-defects) in the AlGaN layers, originating from different growth conditions [16, 27]. On these samples, Gr (grown on Cu by CVD) was transferred to the AlGaN surface using a procedure consisting in the electrochemical delamination from the Cu foil, followed by the thermocompression printing to the target substrate [59]. Two representative AFM analyses of the bare AlGaN surface of the LD and HD samples are reported in the inserts of Figure 7.7a,b, respectively. The LD sample exhibits an atomically smooth surface with low roughness (RMS = 0.85 nm), corresponding to a uniform (~24 nm thick) and defect-free AlGaN layer. Conversely, a much more structured AlGaN morphology can be observed in the HD sample (insert of Figure 7.7b), with the presence of a high density of characteristic surface depressions, which are responsible for the higher measured roughness value (RMS = 1.21 nm). Such morphological features are V-shaped defects in the AlGaN barrier layer, as already shown by TEM in Figure 7.1f, which cause a local reduction of the AlGaN layer thickness. Two typical AFM images of Gr transferred to the AlGaN surface of the LD and HD samples are reported in the inserts of Figure 7.7c,d, respectively. In both cases, a very uniform Gr coverage with a low level of cracks can be observed.

The enhanced roughness compared with the corresponding bare AlGaN samples (RMS = 1.50 nm for Gr on the LD sample and RMS = 2.10 nm for Gr on the HD sample) is due to typical corrugations (wrinkles) of the Gr membrane present since the CVD growth.

Local current–voltage ($I$–$V$) measurements by CAFM were carried out using an Au-coated tip at several positions on Gr-coated and on bare AlGaN regions of the two samples. The DC bias ramp was applied between the tip and a large-area contact on a bare AlGaN region.

In Figure 7.7a,b are reported the $I$–$V$ characteristics measured at the different positions on bare AlGaN in the LD and HD samples, respectively. For both samples all the $I$–$V$ curves exhibit a rectifying behavior, as typically found for not alloyed metal contacts to AlGaN. For each tip position, the Au/AlGaN SBH was calculated using the thermionic emission model [56]. The following average values and standard deviations have been obtained from SBH measured at different positions in LD and HD samples, that is, $\Phi_B = 0.95 \pm 0.12$ eV and $\Phi_B = 0.63 \pm 0.17$ eV. The lower average SBH value as well as the larger standard deviation in the case of the HD sample can be ascribed to the presence of the V-shaped depressions in the AlGaN barrier layer, acting as preferential current paths from the metal tip to the 2DEG, as already demonstrated in Figure 7.1e,g. The $I$–$V$ characteristics measured at different surface positions of the Gr-coated LD and HD samples are reported in Figure 7.7c,d, respectively. In the case of Gr on the LD sample, all the curves still exhibit a rectifying behavior, but with a significantly reduced SBH ($\Phi_B = 0.41 \pm 0.04$ eV) with respect to the contact to bare AlGaN. Interestingly, in the case of Gr on the HD sample (see Figure 7.7d), the $I$–$V$ characteristics reveal an Ohmic behavior. For both AlGaN surfaces, a very limited spread is found between the $I$–$V$ different curves, indicating an excellent lateral homogeneity of the Gr/AlGaN Schottky or Ohmic contacts. The excellent lateral uniformity of the Gr Schottky contact to AlGaN in comparison to a common metal contact can be ascribed to some peculiar properties of a Gr membrane with respect to a metal electrode, such as the conformability to the substrate surface and the higher electron mean free path $l$ ($l$~100 nm for Gr against $l = 1$–$10$ nm for common metals). It can be supposed that the Gr electrode has an averaging effect of the typical sources of Schottky barrier inhomogeneity (such as AlGaN surface potential fluctuations, defects, and dislocations) over a length scale in the order of the electron mean free path [8]. The Ohmic behavior of the Gr contacts to AlGaN with a high density of V-defects (HD sample) can be explained considering the combination of several effects. The local thinning of the AlGaN barrier layer at the V-defects positions, combined with the low values of the Gr/AlGaN SBH, give rise to many low-resistance conduction paths between Gr and the AlGaN/GaN 2DEG. In addition, the averaging effect by the Gr electrode can be supposed to connect several of these vertical conductive paths, since the V-defects separation (see the AFM image in Figure 7.7b) is similar or even lower than the typical Gr electron mean free path. This would also explain the highly homogeneous Gr Ohmic contact even in the presence of a high density of V defects in the AlGaN.

## 7.4 Conclusions

In conclusion, some relevant applications of the CAFM technique to the nanoscale electrical characterization of different 2D electron systems, from AlGaN/GaN heterostructures to graphene and $MoS_2$ thin films, have been reviewed.

In the case of AlGaN/GaN heterostructures, the impact of the AlGaN barrier layer structural defects on the current injection to the 2DEG has been elucidated by CAFM current mapping on the AlGaN surface. In particular, it has been shown how V-defects, that is, pits in the AlGaN layer formed under some growth conditions, can represent preferential paths for current injection to the 2DEG. In the presence of these structural features, the formation of Ohmic contacts to AlGaN/GaN by low temperature annealing of Ti/Al bilayer was achieved, and an unconventional temperature dependence of the contact resistance was observed, suggesting an "intimate" contact between the metal and the 2DEG. Furthermore, examples of CAFM-based studies on current transport through thin insulators thermally grown or deposited on the surface of AlGaN/GaN heterostructures have been reported.

CAFM was also employed to perform local resistance measurements in epitaxial Gr on SiC (0001). These analyses allowed to observe and quantify the local resistance enhancement due to characteristic features of epitaxial Gr, that is, the nanosteps of SiC substrate and the lateral variation of the number of Gr layers, in particular the monolayer/bilayer junction. The conductance degradation over SiC steps was related to a lower substrate-induced electrostatic doping of Gr over the step sidewall and is peculiar of epitaxial Gr/SiC interface. On the other hand, the effect on electronic transport of a junction between different numbers of layers was demonstrated to be a pure quantum mechanical effect.

CAFM analyses also allowed evaluation of the spatial inhomogeneities of the SBH and ideality factor of contacts on $MoS_2$, which have been ascribed to spatial variations in the density and energy of $MoS_2$ surface states. In addition, these analyses also provided information on the local resistivity of $MoS_2$, showing a nice correlation between the decrease of resistivity and that of SBH. These findings have been compared with recent literature results, which showed the role of sulfur vacancy clusters on as-exfoliated $MoS_2$ surfaces as preferential paths for current injection from metal contacts.

Finally, CAFM was employed to investigate the microscopic mechanisms of current transport at Gr/AlGAN interface, focusing on the role of the AlGaN layer quality. Gr contacts onto a uniform and defect-free AlGaN barrier layer exhibit a lower and much more uniform Schottky barrier with respect to a common metal (Au) on bare AlGaN. Interestingly, an Ohmic contact between Gr and AlGaN is obtained in the presence of a high density of V-defects which locally reduce the thickness of the AlGaN layer, creating preferential low-resistance current paths from Gr to the AlGaN/GaN 2DEG. Ohmic contact formation on AlGaN with this microstructure is peculiar to the Gr contact, since a lateral inhomogeneous Schottky barrier is observed using a common metal on the same AlGaN layer.

## Acknowledgments

A. La Magna and I. Deretzis (CNR-IMM) are acknowledged for the *ab initio* simulation activities on graphene. We thank G. Nicotra and C. Spinella (CNR-IMM) for the collaboration in atomic resolution TEM analyses on graphene. R. Lo Nigro (CNR-IMM) is acknowledged for the collaboration in dielectric growth on AlGaN/GaN heterostructures. A. Piazza (CNR-IMM) and S. Agnello (University of Palermo) are acknowledged for the collaboration with Raman measurements during the experiments on $MoS_2$. We thank F. Iucolano (STMicroelectronics) for the collaboration in the experiments on contacts to AlGaN/GaN heterostructures. S. Lo Verso, S. Di Marco, and S. Ravesi (STMicroelectronics) are acknowledged for the collaboration in the CVD graphene transfer. We thank R. Yakimova (University of Linkoping, Sweden) for the collaboration in epitaxial graphene growth. S. Di Franco and C. Bongiorno (CNR-IMM) are acknowledged for the precious technical support in devices fabrication and TEM analyses, respectively.

This work has been funded, in part, by MIUR in the framework of the FlagERA project "GraNitE: Graphene heterostructures with Nitrides for High-Frequency Electronics" (grant no. 3313).

## References

1 Ambacher, O., Smart, J., Shealy, J.R., Weimann, N.G., Chu, K., Murphy, M., Schaff, W.J., Eastman, L.F., Dimitrov, R., Wittmer, L., Stutzmann, M., Rieger, W., and Hilsenbeck, J. (1999) *J. Appl. Phys.*, **85**, 3222.

2 Ohtomo, A. and Hwang, H.Y. (2004) *Nature*, **427**, 423–426.

3 Geim, A.K. and Grigorieva, I.V. (2013) *Nature*, **499**, 419.

4 Giannazzo, F., Fiorenza, P., and Raineri, V. (2008) Carrier transport in advanced semiconductor materials, in *Applied Scanning Probe Methods*, vol. 8-10 (eds B. Bhushan, H. Fuchs, and M. Tomitori), Springer-Verlag, Heidelberg.

5 Giannazzo, F., Raineri, V., and Rimini, E. (2011) Transport properties of graphene with nanoscale lateral resolution, in *Scanning Probe Microscopy in Nanoscience and Nanotechnology 2* (ed. B. Bhushan), Springer-Verlag, Berlin Heidelberg.

6 Giannazzo, F., Sonde, S., Raineri, V., and Rimini, E. (2009) *Appl. Phys. Lett.*, **95**, 263109.

7 Giannazzo, F., Sonde, S., Raineri, V., and Rimini, E. (2009) *Nano Lett.*, **9**, 23.

8 Giannazzo, F., Sonde, S., Lo Nigro, R., Rimini, E., and Raineri, V. (2011) *Nano Lett.*, **11**, 4612–4618.

9 Sonde, S., Giannazzo, F., Vecchio, C., Yakimova, R., Rimini, E., and Raineri, V. (2010) *Appl. Phys. Lett.*, **97**, 132101.

10 Roccaforte, F., Giannazzo, F., and Raineri, V. (2010) *J. Phys. D*, **43**, 223001.

11 Eriksson, J., Pearce, R., Iakimov, T., Virojanadara, C., Gogova, D., Andersson, M., Syväjärvi, M., Lloyd Spetz, A., and Yakimova, R. (2012) *Appl. Phys. Lett.*, **100**, 241607.

12 Karmalkar, S., Shur, M.S., and Gaska, R. (2003) in *Wide Energy Bandgap Electronic Devices* (eds F. Ren and J.C. Zolper), World Scientific, Singapore.

13 Okumura, H. (2006) *Jpn. J. Appl. Phys., Part 1*, **45**, 7565.

14 Iucolano, F., Greco, G., and Roccaforte, F. (2013) *Appl. Phys. Lett.*, **103**, 201604.

15 Greco, G., Giannazzo, F., Iucolano, F., Lo Nigro, R., and Roccaforte, F. (2013) *J. Appl. Phys.*, **114**, 083717.

16 Greco, G., Iucolano, F., Bongiorno, C., Giannazzo, F., Krysko, M., Leszczynski, M., and Roccaforte, F. (2014) *Appl. Surf. Sci.*, **314**, 546–551.

17 Iucolano, F., Roccaforte, F., Giannazzo, F., and Raineri, V. (2007) *Appl. Phys. Lett.*, **90**, 092119.

18 Iucolano, F., Roccaforte, F., Giannazzo, F., and Raineri, V. (2007) *J. Appl. Phys.*, **102**, 113701.

19 Roccaforte, F., Giannazzo, F., Iucolano, F., Eriksson, J., Weng, M.H., and Raineri, V. (2010) *Appl. Surf. Sci.*, **256**, 5727–5735.

20 Roccaforte, F., Iucolano, F., Giannazzo, F., Alberti, A., and Raineri, V. (2006) *Appl. Phys. Lett.*, **89**, 022103.

21 Giannazzo, F., Roccaforte, F., Iucolano, F., Raineri, V., Ruffino, F., and Grimaldi, M.G. (2009) *J. Vac. Sci. Technol., B*, **27**, 789–794.

22 Fontserè, A., Pérez-Tomás, A., Placidi, M., Llobet, J., Baron, N., Chenot, S., Cordier, Y., Moreno, J.C., Jennings, M.R., Gammon, P.M., Fisher, C.A., Iglesias, V., Porti, M., Bayerl, A., Lanza, M., and Nafría, M. (2012) *Nanotechnology*, **23**, 395204.

23 Cheng, K., Leys, M., Degroote, S., Bender, H., Favia, P., Borghs, G., and Germain, M. (2012) *J. Cryst. Growth*, **353**, 88.

24 Schroder, D.K. (2006) *Semiconductors Materials and Device Characterization*, 3rd edn, Wiley, New York, p. 135.

25 Roccaforte, F., Giannazzo, F., Iucolano, F., Bongiorno, C., and Raineri, V. (2008) *Appl. Phys. Lett.*, **92**, 252101.

26 Roccaforte, F., Giannazzo, F., Iucolano, F., Bongiorno, C., and Raineri, V. (2009) *J. Appl. Phys.*, **106**, 023703.

27 Greco, G., Fiorenza, P., Giannazzo, F., Alberti, A., and Roccaforte, F. (2014) *Nanotechnology*, **25**, 025201.

28 Roccaforte, F., Fiorenza, P., Greco, G., Vivona, M., Lo Nigro, R., Giannazzo, F., Patti, A., and Saggio, M. (2014) *Appl. Surf. Sci.*, **301**, 9–18.

29 Fiorenza, P., Greco, G., Giannazzo, F., Lo Nigro, R., and Roccaforte, F. (2012) *Appl. Phys. Lett.*, **101**, 172901.

30 Novoselov, K.S., Geim, A.K., Morozov, S.V., Jiang, D., Zhang, Y., Dubonos, S.V., Grigorieva, I.V., and Firsov, A.A. (2004) *Science*, **306**, 666.

31 Wang, Q.H., Zadeh, K.K., Kis, A., Coleman, J.N., and Strano, M.S. (2012) *Nat. Nanotechnol.*, **7**, 699.

32 Radisavljevic, B., Radenovic, A., Brivio, J., Giacometti, V., and Kis, A. (2011) *Nat. Nanotechnol.*, **6**, 147.

33 Li, X., Cai, W., An, J., Kim, S., Nah, J., Yang, D., Piner, R., Velamakanni, A., Jung, I., Tutuc, E., Banerjee, S.K., Colombo, L., and Ruoff, R.S. (2009) *Science*, **324**, 1312–1314.

34 Emtsev, K.V., Bostwick, A., Horn, K., Jobst, J., Kellogg, G.L., Ley, L., McChesney, J.L., Ohta, T., Reshanov, S.A., Röhrl, J., Rotenberg, E., Schmid, A.K., Waldmann, D., Weber, H.B., and Seyller, T. (2009) *Nat. Mater.*, **8**, 203–207.

35 Vecchio, C., Sonde, S., Bongiorno, C., Rambach, M., Yakimova, R., Rimini, E., Raineri, V., and Giannazzo, F. (2011) *Nanoscale Res. Lett.*, **6**, 269.

36 Emtsev, K.V., Speck, F., Seyller, T., Ley, L., and Riley, J.D. (2008) *Phys. Rev. B*, **77**, 155303.

37 Nicotra, G., Deretzis, I., Scuderi, M., Spinella, C., Longo, P., Yakimova, R., Giannazzo, F., and La Magna, A. (2015) *Phys. Rev. B*, **91**, 155411.

38 Ostler, M., Deretzis, I., Mammadov, S., Giannazzo, F., Nicotra, G., Spinella, C., Seyller, T., and La Magna, A. (2013) *Phys. Rev. B*, **88**, 085408.

39 Varchon, F., Feng, R., Hass, J., Li, X., Ngoc Nguyen, B., Naud, C., Mallet, P., Veuillen, J.-Y., Berger, C., Conrad, E.H., and Magaud, L. (2007) *Phys. Rev. Lett.*, **99**, 126805.

40 Sonde, S., Giannazzo, F., Raineri, V., Yakimova, R., Huntzinger, J.-R., Tiberj, A., and Camassel, J. (2009) *Phys. Rev. B*, **80**, 241406(R).

41 Giannazzo, F., Deretzis, I., La Magna, A., Roccaforte, F., and Yakimova, R. (2012) *Phys. Rev. B*, **86**, 235422.

42 Zhan, Y., Liu, Z., Najmaei, S., Ajayan, P.M., and Lou, J. (2012) *Small*, **8**, 966.

43 Qiu, H., Xu, T., Wang, Z., Ren, W., Nan, H., Ni, Z., Chen, Q., Yuan, S., Miao, F., Song, F., Long, G., Shi, Y., Sun, L., Wang, J., and Wang, X. (2013) *Nat. Commun.*, **4**, 2642.

44 Das, S., Chen, H.-Y., Penumatcha, A.V., and Appenzeller, J. (2013) *Nano Lett.*, **13**, 100.

45 Virojanadara, C., Syväjarvi, M., Yakimova, R., Johansson, L.I., Zakharov, A.A., and Balasubramanian, T. (2008) *Phys. Rev. B*, **78**, 245403.

46 Nagase, M., Hibino, H., Kageshima, H., and Yamaguchi, H. (2009) *Nanotechnology*, **20**, 445704.

47 Nicotra, G., Ramasse, Q.M., Deretzis, I., La Magna, A., Spinella, C., and Giannazzo, F. (2013) *ACS Nano*, **7**, 3045–3052.

48 Giannazzo, F., Fisichella, G., Piazza, A., Agnello, S., and Roccaforte, F. (2015) *Phys. Rev. B*, **92**, 081307(R).

49 Cheung, S.K. and Cheung, N.W. (1986) *Appl. Phys. Lett.*, **49**, 85.

50 Koren, E., Knoll, A.W., Loertscher, E., and Duerig, U. (2014) *Appl. Phys. Lett.*, **105**, 123112.

51 Hermann, A., Samoano, R., Hadek, V., and Rembaum, A. (1973) *Solid State Commun.*, **13**, 1065.

52 Tung, R. (2014) *Appl. Phys. Rev.*, **1**, 011304.

53 Card, H. and Rhoderick, E.H. (1971) *J. Phys. D*, **4**, 1589.

54 Kodama, N., Hasegawa, T., Tsuruoka, T., Joachim, C., and Aono, M. (2012) *Jpn. J. Appl. Phys.*, **51**, 06FF07.

55 McDonnell, S., Addou, R., Buie, C., Wallace, R.M., and Hinkle, C.L. (2014) *ACS Nano*, **8**, 2880.

56 Fisichella, G., Greco, G., Roccaforte, F., and Giannazzo, F. (2014) *Nanoscale*, **6**, 8671.
57 Park, P.S., Reddy, K.M., Nath, D.N., Yang, Z., Padture, N.P., and Rajan, S. (2013) *Appl. Phys. Lett.*, **102**, 153501.
58 Fisichella, G., Greco, G., Roccaforte, F., and Giannazzo, F. (2014) *Appl. Phys. Lett.*, **105**, 063117.
59 Fisichella, G., Di Franco, S., Roccaforte, F., Ravesi, S., and Giannazzo, F. (2014) *Appl. Phys. Lett.*, **104**, 233105.

# 8

# Nanoscale Three-Dimensional Characterization with Scalpel SPM

*Umberto Celano*[1,2] *and Wilfried Vandervorst*[1,2]

[1] *KU Leuven, Department of Physics and Astronomy (IKS), Celestijnenlaan 200D, 3001 Leuven, Belgium*
[2] *IMEC, Kapeldreef 75, 3001 Heverlee (Leuven), Belgium*

## 8.1 Introduction

The recent transition of the semiconductor industry from planar to three-dimensional (3D) architectures is a technological evolution induced by the device scaling requirements which are contained in the so-called Moore's law. The latter describes the rate at which device dimensions decrease and device density increases. In the past, a scaling rate of 0.7×/year was achieved primarily by reducing the geometrical dimensions of the planar devices. However, the increasing importance of power reduction outranged the capabilities of planar short channel devices and led to the introduction of 3D architectures. Hence, modern transistors within the sub-20-nm technology nodes are almost exclusively based on 3D-finFET architecture [1, 2]. Note that since the fin width of a 14-nm finFET is ≈8–10 nm, state-of-the-art transistors are based on only a few tens of silicon atomic layers. While transistors were the first to move from 2D to 3D around 2011, nonvolatile memory devices followed in 2014 with the replacement of planar flash devices with 3D NAND using vertical cell stacking [3]. On a larger scale, chip stacking using silicon vias allowed 3D integration of chips for advanced volatile memory products such as dynamic random access memory (DRAM) [4]. Finally, in addition to the transition from 2D toward 3D, the future technology nodes foresee new materials such as III–V compound semiconductors to be fully integrated in 3D structures (trenches) embedded on regular Si substrates [5]. In essence, the 3D transition has already happened for most of the existing electronic devices (Figure 8.1). However, the introduction of 3D devices and stackable architectures poses unparalleled challenges to the semiconductor metrology, leading to the point that the cost of testing is becoming an important component of the cost of transistor fabrication. It is clear that conventional characterization methodologies start to be ineffective when applied to structures less than 100 nm. Certifying the control of the doping processes in 3D structures and providing chemical and electrical analysis with subnanometer spatial resolution in complex geometries are obviously now

*Conductive Atomic Force Microscopy: Applications in Nanomaterials*, First Edition.
Edited by Mario Lanza.

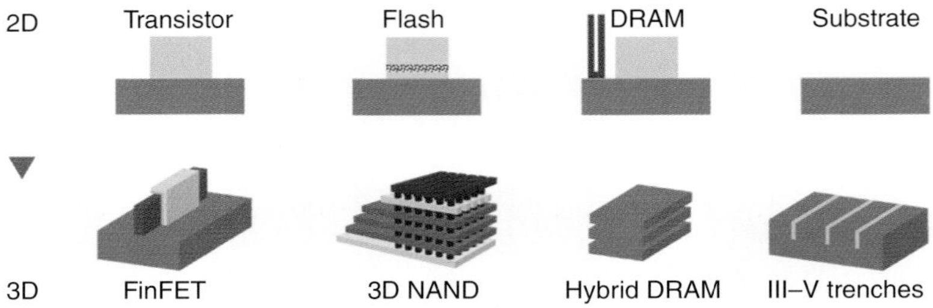


**Figure 8.1** Schematic of the progressive evolution of chip architecture from 2D to 3D (based on international technology roadmap for semiconductors ITRS) [56].

serious complications faced by the metrology community. Moreover, there is an increasing requirement to generate multiple types of information in the same analysis (and from the same device), necessitating a concurrent structural, compositional, and electrical sensing capability. Therefore, the preferred requirements for a valuable 3D characterization technique are as follows:

1) Nanoscale spatial resolution (e.g., sub-10-nm observation capability in 2D)
2) Multiple parameter sensitivity (e.g., electrical and structural)
3) Nanometer precision in probing the third dimension.

In this chapter we present an approach, termed scanning probe microscopy (SPM) *tomography* or *scalpel SPM*, which extends the nanometer scale 2D-analysis capabilities of SPMs toward 3D, thereby creating a valuable technique for the electrical 3D characterization of ultraconfined volumes. The potential applications of this technique range from fundamental material science, single device characterization, failure analysis, and quality control. This chapter describes the technique and its applications in detail. In the first section, scalpel SPM is described starting from a subnanometer tip-induced material removal method and ending in the 3D data treatment. In the second part, some practical applications of scalpel SPM are presented, in particular the 3D observation of the conductive filaments (CFs) in resistive switching (RS) devices.

## 8.2 SPM Metrology with Depth Information

SPM itself is a very versatile metrology concept providing information with 2D-nanometer resolution not only on shape (topography) and electrical properties such as local voltage, resistance, and capacitance but also on mechanical properties (hardness, adhesion) and chemical bonding structure. In particular, probing the electrical properties with concepts like Kelvin probe force microscopy (KPFM), scanning spreading resistance microscopy (SSRM), scanning capacitance microscopy (SCM), and conductive atomic force microscopy (C-AFM) have found numerous applications in semiconductor research and technology. Providing access to the in-depth distribution

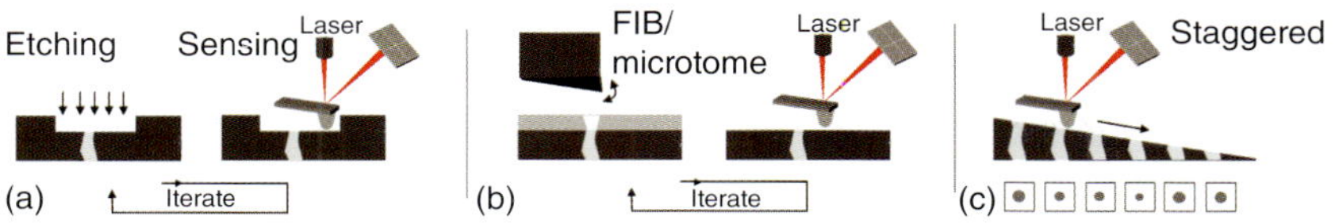


**Figure 8.2** SPM approaches adding the third dimension. (a) Alternating chemical etching and SPM sensing. (b) Physical removal by ion beam or microtome followed by SPM sensing. (c) Dedicated test structure creating a pseudo-3D volume (see text).

of these parameters (i.e., true 3D characterization) has already received a fair share of attention. A schematic overview of the proposed solutions can be seen in Figure 8.2. Magerle [6] used SPM in combination with subsequent calibrated chemical etching to probe ≈50 nm in depth, ultimately creating a data volume in 3D. The 3D analysis of semicrystalline polypropylene layers was demonstrated, revealing the differences between crystalline and amorphous regions [7]. The challenge in the approach is that the same area has to be repeatedly located in every etch cycle, leading to complex operations (Figure 8.2a). To overcome this problem, a dedicated liquid cell for *in situ* etching and sensing has been developed, providing an inlet and outlet of the etching solution without moving the sample [8]. In this configuration, a human bone with ≈10-nm resolution was measured using tapping mode AFM alternated with a hydrochloric acid etching. For application to the analysis of complex structures (e.g., integrated devices), the main limitation comes from the chemical etchants which can have undesired (chemical) interactions with the different parts (materials) of the sample. Alekseev *et al.* [9] proposed to combine C-AFM with an oscillating diamond knife (e.g., microtome) to cut extremely thin slices of the sample (Figure 8.2b). With a slice spacing of ≈12 nm, they could image a 3D volume of a conductive nanocomposite filled with multiwall carbon nanotubes (MWCNTs) and reconstruct the 3D conductive subnetworks by C-AFM. The clear conductance difference between the highly conductive MWCNTs and the polymer matrix provided a good contrast in the 3D visualization. By means of this approach they could quantify the critical filler concentration in the nanocomposite that determines the overall conductivity of the polymer. Bailey *et al.* [10] proposed a combination of AFM and ion milling to perform a morphological 3D observation (Figure 8.2b). However, the energetic ion–solid interaction can destroy the electrical properties in the near-surface regions, thus making high-resolution 3D electrical analysis very difficult with this approach. For the analysis of solid-state semiconductor devices, it is often possible to leverage the processing control and high quality of semiconductor processing to develop test structures which enable, based on the analysis of many identical structures, to probe the third dimension. For instance, one example is the fabrication of a dedicated sample with repetitive features staggered by a certain well-defined distance (Figure 8.2c). Using this approach the dopant diffusion under the gate of a finFET could be measured in 3D by placing side-by-side SSRM sections through multiple fins, whereby the gate edge of each device is staggered by a known nanometer increment relative to the next one. One single cross section through this set of fins then leads to multiple sections placed at known (different) distances relative to the gate edge.

Hence each of them represents a view on the carrier distribution at one specific (but different) location within the channel or source/drain region. Assuming that the repetitive structures are identical and by measuring the different 2D SSRM profiles across each cross section, it is now possible to collect multiple images leading to a 3D-carrier profile. More details on the practical sample preparation can be found elsewhere [11, 12]. Using this approach, Mody *et al.* [12] evaluated the gate underlap, conformality, and the source/drain doping in a finFET with nanometer-spatial resolution. Similarly, Schulze *et al.* [13] mapped the distributions of carriers in nanowire-based transistors. Although these approaches provided information not accessible otherwise, they suffer from the need to process dedicated test structures and, of course, from the inherent assumption that all devices have identical properties. It is also clear that the need for a repetitive test structure excludes the application for single-device or failure analysis.

## 8.3 Scalpel SPM: A Tip-Based Slice-and-View Methodology

### 8.3.1 General Description

In order to overcome these limitations, we developed a 3D-tomographic technique whereby we added a depth profiling capability to the 2D mapping of electrical AFM by a controlled tip-induced material removal leading to a slice-and-view approach. We named this approach *SPM tomography* or *scalpel SPM* [14]. Like the word suggests, we construct a 3D tomogram from a series of 2D sections of the sample. To accomplish the depth profiling with the tip-induced removal of sample material, we make use of conductive, wear-resistant diamond tips [15]. The hardness of diamond enables to physically remove (scrape) material quickly during the scan, without eroding the tip. The latter is important as the same tip is also used to perform electrical measurements (C-AFM, SSRM, etc.) on the freshly created 2D section. Hence, we slice in a controlled manner through the sample, collecting 2D maps at different depths. The consecutive planar SPM images are then combined into one stack whereby through interpolation in between the analysis planes, a complete 3D tomogram is created. In practice, it is possible with this SPM approach to achieve subnanometer vertical removal rates which provide ample depth resolution for the application of this technique in various scaled electronic devices. The possibility of inducing material removal with an AFM tip in contact mode has been known since the early days (and sometimes even perceived as a limitation since the sample gets damaged during the analysis). For instance, Hu *et al.* [16] observed that it is possible to form a 1-nm-deep hole on mica after topography measurements in contact mode. The latter could be induced either by one single scan at high loads or by multiple scans at low load, indicating for the first time a controllable tip-induced removal of material. Xu *et al.* [17] demonstrated a controlled material removal in InP with successive scans using a diamond tip at constant load force. By reducing the scan size continuously, the authors induced a staircase structure and probed in depth the dopant profile with SSRM. A similar approach has been used also

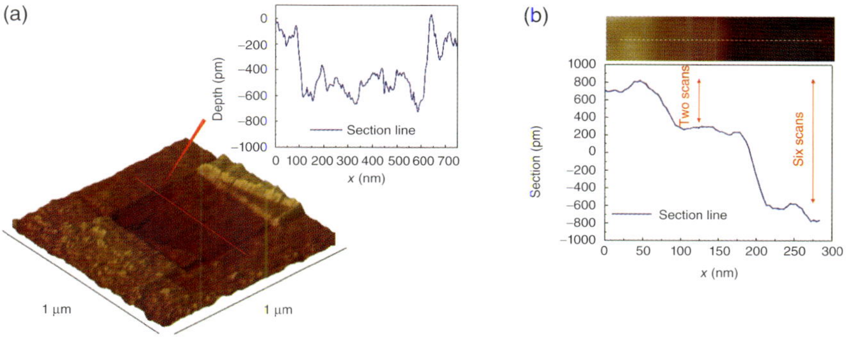


**Figure 8.3** Tip-induced material removal. (a) Crater formation by material removal on a blanket oxide sample (3-nm-thick $HfO_2$) using successive scans with a diamond tip. The diamond tip is scanned over $500 \times 500$ nm$^2$ at high force (≈μN). A topographic cross section (see insert) through the crater reveals a removal of ≈200 pm. (b) By repeating the number of scans progressively a higher amount of material is removed. This is seen in the cross section reported in (b); a removal rate of ≈200 pm/scan is found with a total removal of 1 nm $HfO_2$ after six scans.

by Ou *et al.* [18] for the 3D carrier profiling in Si nanowires. More recently, Schulze *et al.* [19] applied material removal with a diamond tip to characterize the resistance distribution within contact holes filled with carbon nanotubes and the carrier distribution in nanowire transistors [20]. At the foundation of all these observations lies the possibility of inducing a controlled removal of material from the sample's surface when scanning at relative high load force with an AFM tip. In our approach we exploit the hardness of the diamond tip, which allows maintaining a constant tip shape even after multiple scans, and the precise tip force control, to achieve a subnanometer tip-induced erosion capability. By adding a B-doping to the diamond film, a hard conductive probe is obtained which can also be used for electrical measurements (cfr also Section 8.2). For instance, Figure 8.3a shows an example similar to the experiment proposed by Hu and coworkers.

A $500 \times 500$ nm$^2$ region has been scanned on a thin oxide sample in contact AFM mode at relatively high load force (≈μN). A subsequent topographical analysis ($1 \times 1$ μm$^2$ analysis area) through the scanned region (Figure 8.3a) indicates that a subnanometer-deep crater is formed. The line scan in the insert indicates that a depression with a depth of ≈200 pm has been formed. By continuous scanning it is possible to increase the removed depth, as shown in Figure 8.3b. In this case, a thin $HfO_2$ layer was scanned with two- or six-tip passages, respectively. The resulting depth (cfr insert) indicates a constant removal rate of ≈200 pm/scan and a total depth removed, of 1 nm. Both examples demonstrate that one can explore the use of an SPM-tip scanning against a solid surface as a wear process, creating the possibility of inducing controlled removal of material.

Although, in principle, tip degradation and thus loss in lateral resolution may appear as a limiting factor favoring the application of the scalpel to soft materials only, the hardness of the diamond tip allows scalpel application to a wide range of materials such as Si, InP, TiN, Cu, Ti, TaN, Ru, $HfO_2$, $SiO_2$, $Al_2O_3$, and Au [19, 21–23]. Macroscopically the mechanical description of the processes involved the sliding of a sharp tip on a flat surface, and is contained in numerous publications and studies on wear, friction, and contact mechanics [24, 25]. However, at the nanoscale, wear is an atom-by-atom removal of material and is described as a stress-assisted chemical reaction (cfr Jacobs and Carpick [25]). In this case, the wear control arises from the precise load-force control enabled by the AFM feedback system. It goes without saying that creating access to deeper lying layers through the tip-induced material removal opens up a wide range of applications, spanning from fundamental material science to quality control, circuit editing, and reverse engineering.

Figure 8.4a shows a practical example where a 30-nm-thick Cu electrode is progressively eroded until the complete removal by continuous AFM scanning at constant load force [21]. In this case, the purpose was only to expose the TiN layer below the Cu electrode and open the related circuit path. This case is particularly interesting as the difference in hardness of the two materials also enables a precise end point control. Indeed, as Cu is much softer than TiN, it can be eroded with a small load force which does not trigger any removal in TiN. Hence, a complete removal of the Cu can be achieved (with automatic end point control) without affecting the integrity of the buried (TiN) line. Another example shown in Figure 8.4b illustrates the application of scalpel SPM for circuit editing and failure analysis. In this case, the tip-induced material removal is used to locally interrupt a 40-nm-thick metal line, constituted of 20 nm Cu and 20 nm Au. While in standard AFM measurements, material removal is avoided and considered as a measurement artifact, we achieve through a careful selection of the operational parameters a controlled subnanometer removal rate. This high degree of control now allows combining a high resolution in the third dimension with the nanometer lateral resolution and the probing specificity of various AFM techniques. For instance, in the examples discussed further on, we probe in each AFM scan (i.e., at each depth) the local variations in conductivity (using C-AFM or SSRM). By aligning and collating all the 2D AFM conductivity images, a 3D tomogram can be constructed. Figure 8.5 shows a schematic of the SPM tomography workflow.

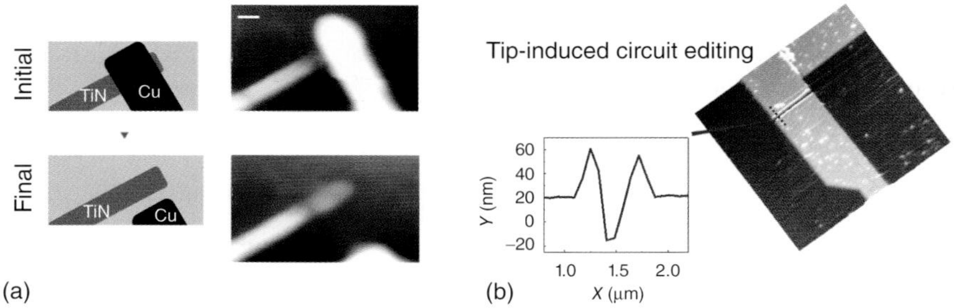


**Figure 8.4** Tip-induced material removal. (a) Removal of 30-nm-thick Cu electrode with end point control on the layer (TiN) underneath. (b) Circuit editing for failure analysis by interrupting an interconnect line. The insert shows a topographical scan across the line cut.

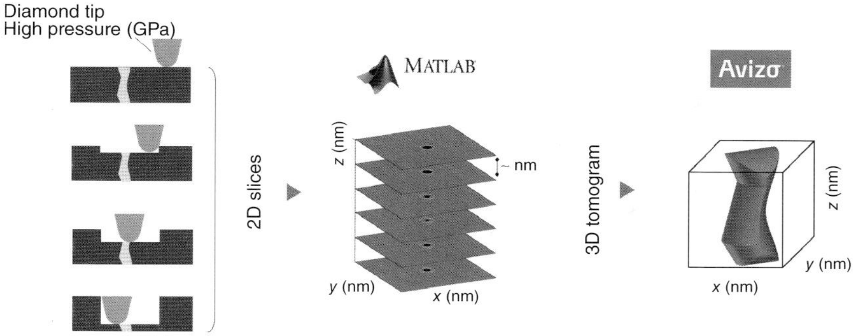


**Figure 8.5** SPM tomography procedure. The wear-resistant diamond tip is scanned on the surface at high pressure (≈GPa) inducing a controlled material removal. The acquired 2D images are aligned and interpolated by a dedicated software to generate a 3D tomogram.

### 8.3.2 Practical Implementation

The main conclusion from these general observations is that subnanometer removal rates can easily be achieved with tip-induced erosion. For analytical applications the material removal needs to be combined with the consecutive (or concurrent) measurement of a relevant property (electrical mechanical, chemical, etc.). However, targeting a quantitative 3D analysis requires not only a calibration of the removal rate but also of the response function relative to the measurement property. In each of these cases, having a tip with constant properties (tip shape, tip radius, and conductivity) is a stringent requirement influencing the choice of the tip material, geometry, and scanning parameters, to name a few. In principle, a large variety of tip materials and shapes (hardness control) and cantilever spring constants (force control) exist which can be used to create a wide range of interaction mechanics. However, a tip made of a soft material (such as a metal) will erode quickly, leading to a nonconstant tip shape and thus poor repeatability. Moreover, in case of tips with a metal coating, the latter will be removed very quickly prohibiting the electrical measurements. For those reasons we selected diamond as the material of preference for *scalpel SPM*. In general, two approaches exist, that is, either an Si tip coated with diamond or a full diamond tip (FDT). As the diamond-coated Si tips originate from tips for standard topography applications, they have a high aspect ratio and are very fragile. Indeed, despite the diamond coating, these tips can break close to the apex when a high load force is applied. This effect is mitigated in the FDTs which have a much more robust wide-base pyramidal design [26]. A comparison between FDT and a diamond-coated Si tip is shown in Figure 8.6a,b. Despite their reduced aspect ratio, FDT tips still provide excellent spatial resolution (below 5 nm) as their tip apex (inset in Figure 8.6a) may consist of a single diamond nanocrystal enhancing the outermost tip sharpness. Moreover, for applications like SSRM, the resolution is determined by the size of the pressure-induced β-Sn phase, which is much smaller than the actual tip radius [27]. Hence, through this confinement process, FDT tips do provide adequate spatial resolution. Since the tip is also used for electrical measurements (C-AFM,

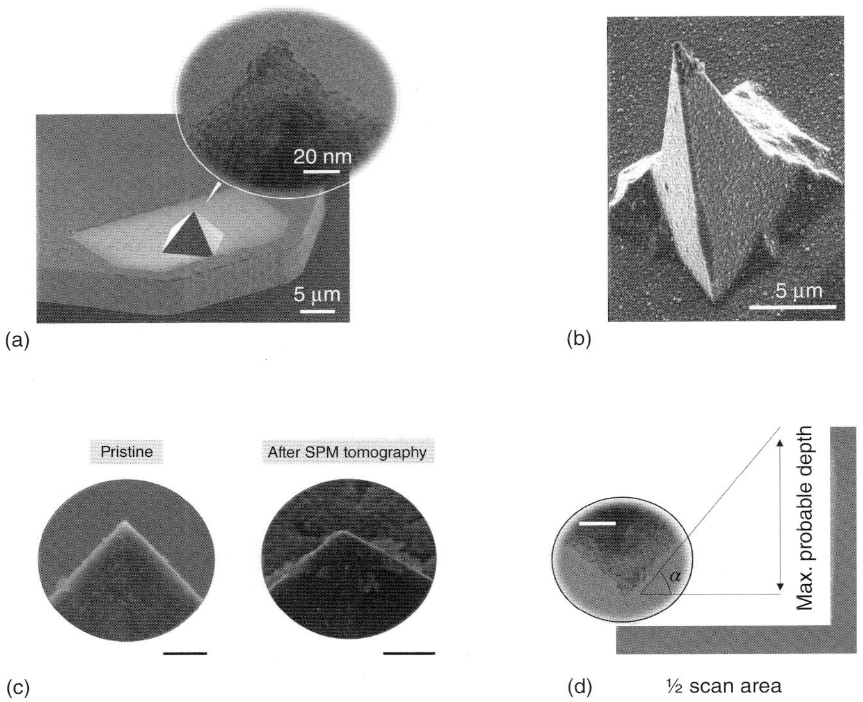


**Figure 8.6** Tip comparison and practical implementation. (a) SEM and TEM images of full diamond tip and (b) commercially available diamond-coated Si tip. (c) SEM observation of a pristine versus a used tip showing residues of the removed material on the tip body. (d) Maximum probing depth as determined by the angle formed between the tip apex and the sample surface (tip scale bar 20 nm).

SSRM, KPFM, etc.), it has to be conductive. For that purpose the diamond in both cases is B-doped (up to 1 at.%), leading to a bulk resistivity of $10^{-3}$ Ω-cm [15]. Based on the pressure (≈ few GPa) needed for the tip-induced material removal of metals and dielectrics, a cantilever spring constant greater than 10 N/m is generally required. For removal of organic material, typically a low stiffness (<1 N/m) is used, whereas for semiconductor materials the optimal spring constant ranges from 2 to 30 N/m. Different materials will show different removal rates (nm/scan) depending on their hardness. Note that the actual value for the removal rate depends on many parameters such as material type and deposition technique, layer thickness, and absorbed contaminants. For this reason a reliable evaluation of the removal rate for a specific material can only be obtained by a load force calibration on the material of interest (e.g., blanket sample). The tip can be scanned and the removal rate (nm/scan) versus the applied force is evaluated (similar to Figure 8.4b) [28]. By coupling the hardness of diamond and the tip load force, we can access material removal rates that span from zero to hundreds of nanometers per scan (depending on the sample). As SPM tomography is a destructive technique, a crater is formed in the region of interest whereby a large fraction of the removed material is accumulating on the side of the scan area and on the tip body (Figure 8.6c).

In combination with the intrinsic tip and material properties, the AFM scanning conditions play an important role in the removal rate. Tip velocity, the number of lines per scan, the tip–sample biasing, and the environment conditions (e.g., air, vacuum, or controlled humidity) are some of the tunable parameters. Here we provide some practical values used in the rest of the chapter and a brief description of their effect.

The *tip speed* is defined as the speed at which the tip scans the surface and it is reported in micro meter per second. Scanning the surface at lower speed translates in a longer time spent on the same location, exerting high pressure and leading to more stable tip–sample interactions. The usage of a small tip velocity not only has a dramatic impact on the time for each acquisition but also improves the quality of the removal and increases the removal rate. A commonly used value is in the range 0.5–1.5 μm/sec.

The number of *scan lines* acquired on each 2D AFM scan is generally associated with the graphical resolution of the final AFM image. This parameter plays a crucial role in the quality of the material removal as a higher number of lines translate into denser and less spaced passages of the tip in the same portion of area. In extreme cases, for example, very small scan sizes, the tip can pass many times on the same location within a single AFM scan, thus improving the effectiveness and smoothness of the removal process. We generally use 512 lines per scan, which increases the acquisition time and also provides high-resolution 2D AFM images. The latter is beneficial for the resolution of the final 3D interpolation. During the SPM electrical tomography, the DC bias between the tip and the sample has to fulfill two main requirements: (i) it has to be sufficiently high to induce a measurable current flow, thus creating a detectable contrast in the current map. (ii) It should not interfere with the measurement, for example, affecting the material removal rate. It is clear that this value is highly dependent on the conductivity properties of the material and will vary among different types of observation. In general, if the DC bias is not intentionally used as a mechanism for material removal during the scan, a low value is preferable; in this work, we use always (negative) voltage in the range of 10–100 mV with the tip grounded and the bias applied to the sample. However, thermal effects due to Joule heating can contribute to the eroding process since also at reduced voltages tens to hundreds of nanoamperes can flow (especially for conductors). Finally, the material removal depends on the environmental conditions during the analysis since it can induce structural and mechanical changes in the materials under investigation. In our measurements, we work generally at high vacuum ($\approx 10^{-5}$ mbar) although we were also able to obtain similar results when working in air under ambient conditions. With these typical scanning conditions, the entire analysis can take several hours; during this time any system-induced drift in the x–y plane needs to be compensated either by the operator or by dedicated control software. Due to the conical shape of the tip, a tradeoff exists between the achievable scan size and maximum depth of sampling. With simple trigonometric considerations, the maximum depth can be defined as $\tan\alpha$*(1/2 scan area), as schematically shown in Figure 8.6d. Using our FDT (tip angle 54°) the tip can penetrate at a maximum of 344 nm with a scan size of $500 \times 500$ nm$^2$; after this value, the sidewall of the tip will touch the edge of the crater. In essence, to obtain the analysis of a $500 \times 500$ nm$^2$ area, a safety margin as wide as 250 nm has to be considered to allow a maximum analysis depth of 344 nm.

*Tomogram formation and 3D reconstruction*: after the acquisition of the set of 2D images, the 3D reconstruction leads to a 3D dataset for the structural and electrical properties. This procedure starts with the displacement of the 2D images set in a 3D matrix by means of a dedicated MATLAB routine [19]. The slices are generally considered to be evenly spaced in the third dimension although nonequidistant spacing can be implemented if required. The newly generated 3D matrix is imported in a graphic rendering software (e.g., Avizo) where each slice is aligned. This step can be done using dedicated error-minimization algorithms between subsequent slices, or manually aligning consecutive slices by means of alignment markers. At this point the software can execute different interpolations, to reconstruct the missing information between slices. Generally, a linear interpolation based on the value of the nearby slices is preferred for our purpose. Recently, other groups have proposed the usage of dedicated software for 3D rendering, such as a 3D slicer [29].

## 8.4 Applications

### 8.4.1 Scalpel SPM for 3D Observation of Conductive Filaments in Resistive Memories

Now that the reader is familiar with the fundamental principles of scalpel SPM, we demonstrate its value in a few applications linked to the characterization of CFs in filamentary-based RS memories. Filamentary-based RS cells (resistive random access memory (ReRAM)) are emerging devices with extensive industrial interest due to their unique characteristic: their electrical resistance can be altered by applying an electrical voltage, reversibly up to a billion times [30–33]. In other words, they have an ON (0) and OFF (1) state which enables storing of digital information. The state of the cell remains unchanged even if power is removed, making it a nonvolatile memory device. The industrial interest arises from their appealing advantages, that is, they can be switched rapidly, consume little energy, and are nonvolatile [34]. Most of these properties find their origin in the (directional) migration of ions due to an electric field and the resulting formation/dissolution of a filamentary conduction path between two electrodes. The basic switching mechanism is the migration of metal ions (cations) or oxygen vacancies (anions), $V_o^{\bullet\bullet}$ (in the notation of Kröger and Vink) back and forth between two electrodes in a solid electrolyte (e.g., a 5-nm metal oxide). The interplay between ion migration and atomic scale redox reactions induces the formation and disruption of nanosized (few tens of cubic nanometer) CFs [33]. Devices based on the migration of cations are generally referred to as conductive-bridge random access memory (CBRAM) or electrochemical metallization cells (EMCs), [22] while devices based on the movement of oxygen vacancies are referred to as *valence change memory* (VCM) or *oxide resistive random access memory* (OxRRAM) [35]. Independently from the switching mechanisms, the fundamental understanding of the RS operations requires the observation of the CFs which is considered to be either a highly confined volume of sub-stoichiometric oxide or a nanoscopic metallic path (by accumulation of metal atoms) located in the solid electrolyte. The lateral dimensions of the CF, less than 10 nm, and the required specificity (sub-stoichiometric oxide) have made the CF observation a challenge for the existing metrology techniques. Although some transmission electron microscopy (TEM) observations have been reported, [36–38] the results are not unambiguous as TEM is still a projection technique averaging over the TEM lamella. As such, an exhaustive 3D observation of the CF in scaled bipolar devices is still missing. In particular, for the case of VCM, the CF observation is also hampered by the minimal compositional difference with the surrounding material as the RS is associated with the generation and migration of oxygen vacancies which are difficult to detect. Hence, although a theoretical framework based on *ab initio* calculations and various physical models [36–38] did describe the RS operations, the direct observation of the CFs for both types of memory is absent, creating a gap between physical understanding, theoretical modeling, and experimental device results. To overcome these limitations, we used scalpel SPM which combines the high lateral resolution of C-AFM with a subnanometer vertical resolution through the precise control of the material removal. A standard ReRAM cell is based on a cross-point structure (Figure 8.7a). The cross-section TEM of a Cu-based CBRAM cell is shown in Figure 8.7a (top inset), while an $HfO_2$-based VCM device is presented in Figure 8.7a (bottom inset). Both types of devices are, in essence, metal–insulator–metal structures in which the top electrodes have to be removed to observe the CFs. Once the top electrodes are removed, scalpel SPM is used to collect 2D slices of the electrolyte layer containing the CFs (i.e., oxide). The CBRAM device is a Cu/$Al_2O_3$/TiN-based memory which is placed at the cross-point between the bottom electrode (BE) and top electrode (TE). The memory element is based on a 5-nm amorphous $Al_2O_3$ sandwiched between 10-nm TiN (BE) and 40-nm Cu (TE), as shown in the inset of Figure 8.7a. The VCM device is based on a bilayer structure Hf (10 nm)/$HfO_2$ (amorphous, 5 nm) sandwiched between the Ru and TiN metal electrodes (bottom inset Figure 8.7a). The role of the 10-nm Hf metal layer is to induce at the interface, a sub-stoichiometric $HfO_x$ layer [39]. This interlayer increases the oxygen vacancy concentration and acts as a reservoir for mobile defects during the device operations.

The electrical switching characteristics of both devices are illustrated in Figure 8.7b,c. After an initial electroforming process (not shown), the devices show a bipolar RS behavior with two stable states, that is, a high-resistance state (HRS) and a low-resistance state (LRS) one. The change in resistance can be triggered using a voltage variation ($\pm 2$ V), whereby we limit the current through the device at 50 μA (CBRAM) in order to avoid the breakdown of the oxide and to limit the degradation. This is achieved by connecting the (TiN) BE of the memory element to the drain of a selector transistor in a one-transistor-one-resistor (1T1R) configuration [31]. The VCM characteristic can be triggered, as shown in Figure 8.7c, by using $\pm 1$ V swing. A simple comparison of the two *I–V*

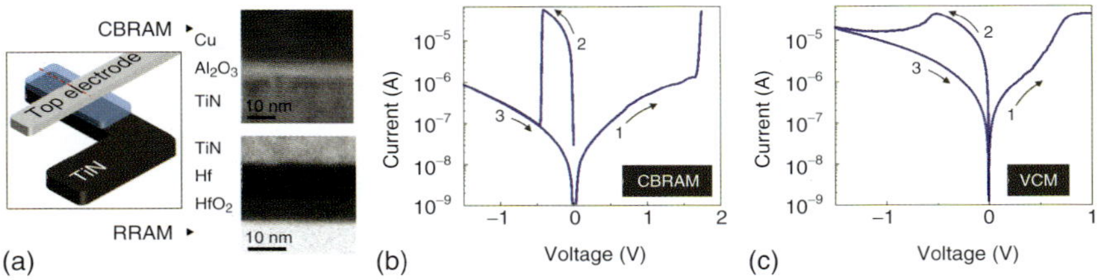


**Figure 8.7** Structure of the memory elements and their resistive switching characteristics. (a) Schematic of the crossbar device. High-resolution TEM images for CBRAM (top) and VCM (bottom) material systems. (b) Bipolar resistive switching characteristics, one representative cycle is reported for CBRAM. (c) VCM operated at 50 μA current compliance.

characteristics suggests a different physical switching mechanism for the two devices. This is visible, for example, from the different levels achieved by the HRS and LRS in the two material systems or by the very distinctive type of the reset transition (i.e., gradual and abrupt in the negative part of the *I–V*) [40]. CBRAM and VCM show similar values for the LRS when programmed at similar conditions; however, the resistance level achieved by the HRS shows two orders of magnitude difference. It goes without saying that the full characterization of the CF in real devices would be beneficial for three main reasons. First, it would dramatically enhance our understanding of the filament growth dynamics. Secondly, more insight would become available to locate the critical interface during filament cycling, which is essential in device optimization. Finally, the information about the shape and size of the CF would clarify the underlying physical mechanisms and contribute to advancing filamentary-based memory technology toward a reliable mature technology. As an example of the application of scalpel SPM (in this case, using C-AFM as probing concept), we performed the 3D observation of the CF for devices that had been subjected to a few tens of cycles before the analysis. Figure 8.8a–c presents the schematic workflow and data in case of a CBRAM cell.

The current map (i.e., using C-AFM) taken on the complete device after electrical programming (i.e., filament formation) is shown in Figure 8.8a. The presence of the TE shields the observation of the local conductivity variations linked to the CF, demonstrating that the traditional (2D) C-AFM is completely ineffective for the filament characterization. However, using scalpel C-AFM a complete characterization becomes possible. The starting point is the removal of the TE by the scalpel. Once the TE is removed, the oxide surface ($Al_2O_3$) is exposed and we can alternate between observing the local conductivity variations due to the CF and the progressive controlled erosion of the sample, leading to the two-dimensional (2D) current maps at different depths (Figure 8.8c). For the case of CBRAM we report the entire collection of C-AFM slices in Figure 8.8d(1–7). Using a removal rate of ≈0.5 nm/scan, the final collection constitutes 10 slices for the 5-nm-thick switching layer. The CF is clearly visible as a highly conductive spot in the middle of the active area, in contrast with the rest of the $Al_2O_3$ surface, which is still highly insulating (red background in Figure 8.8d). We stop the sectioning when the switching layer is completely removed and the highly

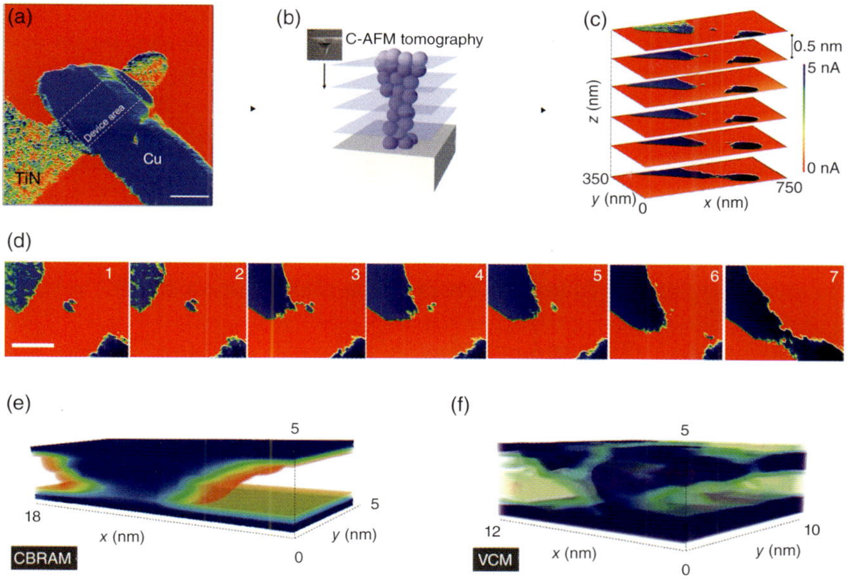


**Figure 8.8** (a) Planar 2D C-AFM, performed on a cross-point memory element in SET-state. The C-AFM is completely ineffective due to the presence of the TE shielding the local conductivity changes linked to the CF (scale bar 100 nm). (b) Schematic of the C-AFM tomography procedure: the scraping with the diamond tip creates multiple C-AFM maps taken at different heights of the CF after the removal of the TE. (c) 3D stacking of the collected 2D C-AFM slices. Note the average spacing between each slice is ≈0.5 nm. (d) Collection of 2D slices constituting the dataset for the 3D interpolation (scale bar 80 nm). The CF appears in the middle of the active area after top electrode removal. The highly conductive features on the top-left and bottom-right corners are the exposed parts of the TiN BE, which is progressively exposed during the removal of $Al_2O_3$. (e) Observation of the reconstructed 3D tomogram for the CF in a 5-nm-thick $Al_2O_3$ (Cu electrode was on top and TiN bottom). (f) CF observation (blue shape) by volume rendering within a 5-nm-thick $HfO_2$. The Hf electrode was on top and the TiN at the bottom. In both figures, the low current contribution in the tomogram is suppressed to enhance the contrast of the highly conductive features. (Reproduced with permission from [22]. Copyright American Chemical Society .)

conductive BE, for example, TiN, becomes visible. An analysis of the consecutive images clearly indicates that in contrast with the first C-AFM slice, the last one before the BE presents a CF with a much smaller size, indicating the shrinkage of the CF from the active electrode (Cu) to the inert counter electrode (TiN). The same procedure is repeated also for the VCM device (not shown, in our standard Hf/$HfO_2$/TiN test structure) after the cell has been programmed in LRS. The set of C-AFM slices representative of the CFs are aligned and interpolated to obtain a 3D tomogram for both CBRAM (Figure 8.8e) and VCM (Figure 8.8f). In both cases we suppress the low-current contribution in the 3D tomogram in order to highlight only the highly conductive features. The inert counter electrode was at the bottom. The 3D cross sections shown in Figure 8.8e,f clearly indicate the conical shape of the CFs in 3D. In Figure 8.8e, the CF (display threshold 5 nA, blue region) shrinks in dimensions when moving from TE to BE, that is, from the Cu/oxide interface toward the oxide/TiN interface. The cross-sectional area of the CF shrinks from ≈493 nm$^2$ on the Cu side down to ≈200 nm$^2$ on the side of the inert electrode. Figure 8.8f shows the CF of the VCM device (iso-surface at 50 pA, blue region). The area of the observed CF is smaller as compared to the CBRAM case, and shows a constriction of ≈40 nm$^2$ close to the (TiN) BE. The 3D tomograms unambiguously reveal that the CFs have the narrowest part close to the bottom (inert) electrode. It is also clear that the volume of the Cu filament can be approximated by a truncated cone, while the VCM filament shows a more elongated shape. The complete set of 2D C-AFM slices, the animated cross section and other images of the CFs are available elsewhere [22, 35].

### 8.4.2 Mechanisms for Filament Growth

The direct observation of the CFs in 3D is a key enabler for understanding the CF growth mechanisms. For instance, in case of CBRAM (Figure 8.8e), the shape observed by C-AFM tomography (constriction at inert electrode) is opposite to the conical shape predicted originally for chalcogenide-based electrochemical metallization memory (ECM) cells [41, 42], with the filament constriction close to the active electrode. It is clear that our 3D tomogram unambiguously reveals that the CF has the narrow part close to the inert electrode instead of at the active electrode. These observations are also in agreement with TEM observations by Yang *et al.* [43] in amorphous silicon and by Liu *et al.* in $ZrO_2$ [44]. The observation of these apparent discrepancies have raised a strong debate in the community about the possible mechanisms for (reversed) filament growth [45], which was recently concluded with an unified cation-transport-centric theory [38] whereby the rate limiting step determines the shape of the CF. On the other hand for VCM cells, the CF has the predicted conical shape with the filament constriction close to the active electrode interface. Our observation indicates that the switching mechanism in bipolar oxide-based VCM relies on the formation/cycling of a local conductive region (filament) created during the electroforming process (in the $HfO_2$) between the Hf and the TiN electrodes [35].

1) *Filament growth in CBRAM:* To describe the CF filament formation in oxide-electrolyte-based CBRAMs, we consider a field-assisted ion transport inside the $Al_2O_3$ (Figure 8.9a). Under the effect of a positive bias, Cu atoms are ionized and diffuse into the switching layer. At the same time, the high electric field, ≈$10^9$ V/m, might generate oxygen vacancies within the $Al_2O_3$. These can contribute to a leakage current throughout the electrolyte and also facilitate the Cu migration as ionic transport along chains of O-sites of the Al–O bonding in the amorphous matrix [46]. However, given the low cation mobility in the $Al_2O_3$, the Cu cations may only travel a short distance inside the dielectric before they are reduced by capturing electrons injected in the electrolyte under the high electric field. Hence, the $Cu^+$ ions are reduced back to Cu near the active electrode and become an extension of it. The $Cu^+$ ions that are subsequently injected will preferentially become reduced at the end of the already existing filament, as its end constitutes the point with the highest electric field. Therefore, the filament growth can be viewed as a repetition of ionization and injection, slow (limited) migration, and reduction processes.

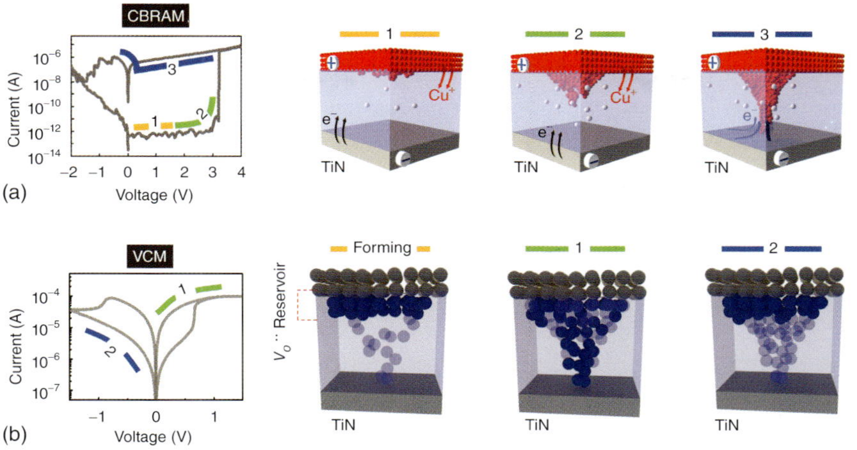


**Figure 8.9** Illustration of the electrochemical processes during resistive switching. (a) CBRAM: (1) The Cu oxidizes and $Cu^+$ ions are injected in the $Al_2O_3$. The high electric field might also lead to the formation of oxygen vacancies in the dielectric layers (white balls in the cartoon). (2) The slow migration of $Cu^+$ ions in the switching layer implies that a reduction reaction occurs before the $Cu^+$ reaches the inert electrode. (3) The filament growth continues and the CF eventually shorts the two electrodes, thereby creating the low resistive state. (b) VCM: the oxygen atoms start to leave their lattice position and drift toward the anode, leaving behind a local conductive path that joins the sub-stoichiometric $V_o^{\cdot\cdot}$ reservoir. On each subsequent memory cycle, the number of $V_o^{\cdot\cdot}$ will be reshuffled at the constriction of the CF inducing the resistance change and flipping of the bit-state. (Reproduced with permission from [22]. Copyright American Chemical Society 2014.)

Under these assumptions, the filament grows from the active electrode, for example, Cu, toward the inert electrode. Obviously, in this case, the cation transport is the rate-limiting process for the CF formation. Note that our C-AFM tomography only reveals the shape of the filament and does not provide any compositional information on the filament, leaving the debate on its chemical composition still open. The final CF can thus be constituted of different conductive species such as $Cu^{n+}$ ions, $Cu_{0,}$ or Al–Cu–O phases. For chalcogenides, sulphites, or ionic-conductors, the cation mobility is much higher as compared to the case of amorphous dielectrics [47]. Therefore, the cations can easily reach the counter electrode, before they are reduced, thereby initiating the nucleation of the CF at the counter electrode. In this scenario, the reduction and nucleation of the filament are the rate-limiting processes for the filament growth, and the CF starts to grow from the inert electrode to the active electrode.

2) *Filament formation in VCM:* Our 3D tomography, as shown in Figure 8.8f, clearly indicates that the wider part of the CF appears at the top interface. In this case, the conductive path (or CF) is thought to be formed by the creation/dissolution of a chain of oxygen vacancies ($V_o^{\cdot\cdot}$). Indeed, through the creation of the high electric field inside the oxide during the electroforming ($\approx 10^9$ V/m), oxygen atoms start to leave their lattice position and drift toward the anode, leaving behind an oxygen vacancy. The accumulation of $V_o^{\cdot\cdot}$ into a

chain of oxygen vacancies leads to the formation of a conductive path shorting the BE with the $Hf/HfO_2$ interface. This stack acts as a $V_o^{\cdot\cdot}$ reservoir and locally provides an exchange layer to sustain the successive CF rearrangements [48, 49]. Its functionality is created (Figure 8.9b) by adding a thin Hf layer on top of the $HfO_2$, thereby scavenging some oxygen atoms from the oxide layer. On each subsequent memory cycle, a number of $V_o^{\cdot\cdot}$ will be added/removed at the constriction of the CF, inducing the resistance change and flipping the bit state. Since the CF-formation mechanism is more linked to defect creation, its formation process resembles the percolation statistics for oxide breakdown and, as such, leads to a CF with a reduced lateral growth as compared to the case of CBRAM (cfr Figure 8.8e,f).

Although the observable electrical characteristics of both types of VCM are similar, our observations unambiguously reveal that their intimate physical switching mechanism is different. In one case (CBRAM), a foreign species (metal cations) is introduced into a solid electrolyte to short the two electrodes. In the other case (VCM), an intrinsic type of defect (oxygen vacancies) is generated in the oxide layer in order to short the electrodes by a conductive path (metallic-rich phase).

### 8.4.3 Chemical Nature of the Filament

Although our technique is not capable of any chemical analysis, it is interesting to note that the different chemical nature of the CFs in CBRAM and VCM can be deduced from the CF observations using scalpel SPM. This is evident by comparing the conductive spots of CFs induced with the same programming conditions and showing comparable values for the LRS (Figure 8.10a). After removing the TE with scalpel SPM, a relatively high current (a few nA) is flowing in the CF when probed with a few tens of millivolt in the case of CBRAM. On the contrary, for VCM a strongly reduced current (a few pA) is detected. This difference originates from the different chemical nature of the CFs. Indeed, in CBRAM the Cu in the electrolyte is fairly stable against the ambient and its LRS value remains fairly unaffected after the TE removal. On the other hand, the intimate nature of the defects (oxygen vacancies) constituting the CFs in VCM, and in particular their much higher reactivity with the environmental oxygen, will lead to a reoxidation

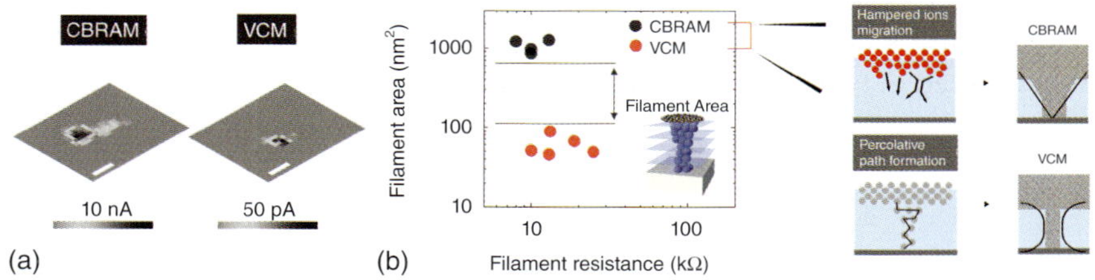


**Figure 8.10** (a) Current maps of two representative CFs; note the different sizes/conductivity as measured with the same conditions (scale bar 20 nm). (b) Comparison between CFs' sizes observed at the top electrode interface, for cation-versus anion-based devices. In the inset we show the schematic of the final shape difference induced by the different switching mechanisms in CBRAM and VCM.

of the top part of the filament such that in the top layer the conductivity is lost. The detected conductivity is then a combination of the tunneling resistance through this top layer and the filament conductance in the underlying part. The correct assessment of the LRS devices thus requires analysis of VCM devices in high vacuum ($<10^{-5}$ mbar), whereas CBRAM devices can be measured in air. Although this process does affect the apparent conductivity of the CF, it does not affect the observation of its size (below the reoxidized layer). This is shown in Figure 8.10b, where the sizes of experimentally observed CFs are compared again for the case of filaments showing the same LRS. The difference in filament sizes is consistent with differences in the filament formation process for cation- versus anion-based devices. Indeed, the growth model proposed for CBRAM in Figure 8.9a leads to a wide-base conical shape for the CF in CBRAM [22, 50]. Conversely in VCM, the CF formation arises from the generation of defects in the bulk of the oxide following the statistics of the percolation theory [51] for dielectric breakdown. When sufficient defects are generated to connect the reservoir with the BE, the CF is established, thereby limiting the lateral growth of the CF and prohibiting the wide-base conical shape as compared to the case of CBRAM (Figure 8.10b).

### 8.4.4 Scalpel SPM for Failure Analysis

Due to its site selective analysis capability, scalpel SPM represents an appropriate methodology for the analysis of a single device and for failure analysis [52]. In a first example we used the scalpel to investigate the appearance of multiple (LRS) resistance states in consecutive cycles in the case of CBRAM. In general, such a device shows, following multiple resent events (Figure 8.11a), the same LRS level and reset transition shape ($CF_1$), but then suddenly evolves for consecutive events into another LRS resistance ($CF_2$) that also shows a different type of reset transition. From the standpoint of the device operation, this is a source of undesired variability. The changes in the LRS are reflected also in the HRS and thus on the entire resistive window (RW) of the memory element, leading to potential corruption of the stored data. The observed variability suggests that multiple CFs might alternate in controlling the LRS/HRS states. However, it is widely accepted that the filament formation is a winner-takes-all process where, after the first CF formation, all the subsequent cycles will occur at the same location. To investigate this phenomenon, we use C-AFM tomography to study the morphology of the CF for a device showing this specific electrical behavior as shown in Figure 8.11a. Here it is clear that the reset characteristics present LRS values clustered in two resistance levels ($CF_{1,2}$). $CF_1$ shows for a few cycles a stable LRS $\approx$30 k$\Omega$ with abrupt reset transitions. Afterwards the CF evolves toward a different resistance state $CF_2 \approx 20$ k$\Omega$ (LRS) for a few more cycles and the reset transition becomes progressive. When applying the scalpel to such a device (after it is put in the LRS), a 3D tomogram is obtained (Figure 8.11b). The sections taken from this tomogram show that the CF is not as a single conductive path but rather exists as two branches located closely together. At the surface one can observe two spots whereby one (left) is much smaller than the other. In cross section 1, two filaments are visible whereby one (left) is only partially formed and the other runs completely through the $Al_2O_3$. The 3D cross section of this filament clearly

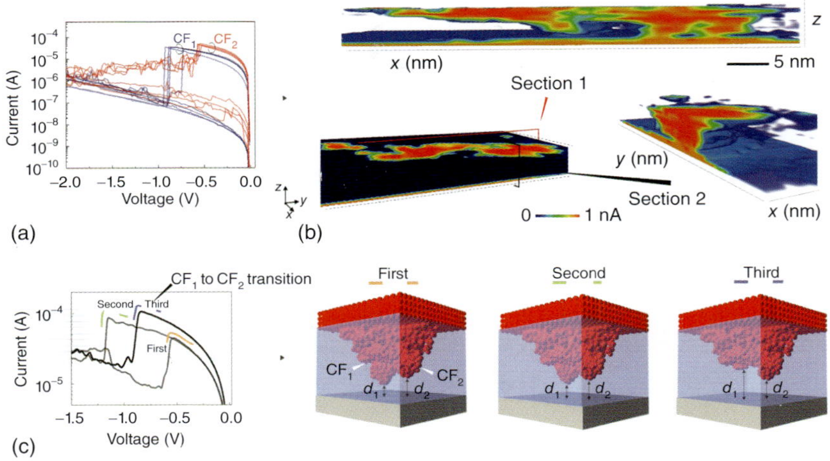


**Figure 8.11** Analysis of a device with variable switching. (a) Resistance state fluctuation observed in consecutive reset traces of CBRAM. Note that the multiple LRS and reset transitions observed suggest that different CFs are alternating. (b) Cross sections extracted from 3D tomogram, (top) two cross-section showing two filaments. (middle) CF-size at the top-electrode interface (right). 3D section through CF shortening the electrodes. (c) Proposed model for the two-filament instability. During the forming process, $CF_1$ and $CF_2$ attempted to grow in the $Al_2O_3$. Finally, $CF_1$ connects and its morphology changes during cycles. Due to the structural rearrangements in the $CF_1$ driven by the subsequent reset, on the third cycle $CF_2$ becomes the favored CF for the next formation, leading to a resistance state variability associated with the potential fluctuation between the two CFs. (Reproduced with permission from [52]. Copyright Royal Society of Chemistry 2016.)

shows it shorting the two electrodes. Although only one CF is fully connected to the BE, the presence of the second filament represents a potential reliability threat. Indeed, the electrochemistry describing the filament formation does not exclude the possibility of multiple attempts of filament formation and the nucleation of multiple filaments has been often reported by means of TEM [38, 43]. In principle, one would expect that the CF that undergoes a complete formation (i.e., shorting the two electrodes in LRS), dominates the subsequent cycles and becomes the only one responsible for the RS. As such, the variability would also be limited. While this description can be applicable for large (planar) structures, [43] it is no longer justified at reduced dimensions. In fact, when the solid electrolyte is only a few nanometers thick, there is a real risk of resistance variability due to the competition in CF formations, as schematically presented in Figure 8.11c.

Indeed, in the presence of two competing CFs, one needs to realize that the filament that was completely formed ($CF_1$), undergoes continuous structural rearrangements during the reset process. Following each reset event it will become disconnected with a gap which will have a certain thickness, $d_1$, while the partially formed CF will be separated from the BE by a gap with thickness $d_2$. In the next formation cycle, $CF_1$ will be favored for the formation only if $d_1 < d_2$. This condition is not always satisfied, especially considering the operation of the device for millions of cycles. In fact, the reset transition is a thermally assisted chemically activated process triggered by the energy dissipated in the reset, that is, it requires a formed CF and thus a current flow to occur. On the other hand, an originally incomplete filament ($CF_2$) is less sensitive to the reset process (as minimal current will flow through it) and thus will undergo less structural changes. With continuous resets taking place on $CF_1$ its properties might degrade such that formation of $CF_2$ becomes more favorable, for instance, when the distance between $CF_2$ and the BE becomes lower than the one of (the reset) $CF_1$. Figure 8.11c schematically shows this effect over three consecutive cycles, whereby the third cycle shows a change in the LRS level. For the first and second cycles, the rupture of the CF does not change the role of $CF_1$, which is formed again on the subsequent set-step (this is also associated with a relatively similar $R_{LRS} \approx 20$ kΩ during the ON state). In the third cycle, the reset is taking place at higher $V_{reset} \approx -1{,}4$ V and it involves more power, leading to a more pronounced reset of $CF_1$. This results in a favored situation for $CF_2$ to complete its formation on the next set-transition that will show a new LRS value and potentially a different type of reset transition. This type of reliability threat is more pronounced in case of a thin switching layer and a prolonged endurance cycling. In fact, the CF formation is a self-accelerated process due to the concentration of the $E$ field lines on the CF, which is more rapidly growing toward the BE. This leads to a large space between the CF connected and the others in case of a thick switching layer, as also reported by TEM [43]. The space dramatically reduces for integrated devices, and in combination with millions of cycles the mechanism modeled in Figure 8.11c can become relevant. As a second application of scalpel SPM for failure analysis, we investigated a stuck in low resistive state (stuck in LRS) or a stuck-set as failure mode [53]. This failure affects the endurance of the device, reducing the number of possible switching cycles and therefore the device lifetime. Endurance failures are considered a major concern for any RS device, as they degrade the RW and reduce the device lifetime [54, 55]. In general, the endurance depends on many factors such as amplitude of the pulse, switching speed, read-out conditions, and temperature [53]. One common type of endurance failure is the stuck in LRS, often referred to as *stuck-set*. In this case, the HRS resistance tends to decrease during cycles until it reaches the same value of the LRS, thus closing the RW (Figure 8.12a). This is often referred to as *resistive window* collapse. In this section we provide some understanding on this failure and its physical mechanism by selecting a device stuck in LRS after cycling and using C-AFM tomography to study the morphology of the CF. Figure 8.12c shows the dataset of C-AFM current maps, recorded in the cross-point memory element. It is immediately evident that the CF is no longer a conductive spot occupying a fraction of the active electrode. Instead the CF now appears as a long conductive platelet that runs through the entire device (cell size $200 \times 200$ nm$^2$). By observing the shape evolution of the CF within the oxide, three spots are visible in the proximity of the BE (Figure 8.12c). The 3D tomogram (Figure 8.12b) suggests that the CF is constituted of three branches that have merged into a long and unrecoverable conductive platelet. The extension of this platelet is ≈150 nm and the three spots have an equivalent radius ≈12 nm. It is clear that in this configuration the reset of the CF is impossible due to the absence of a single filament constriction or localized point for the CF to

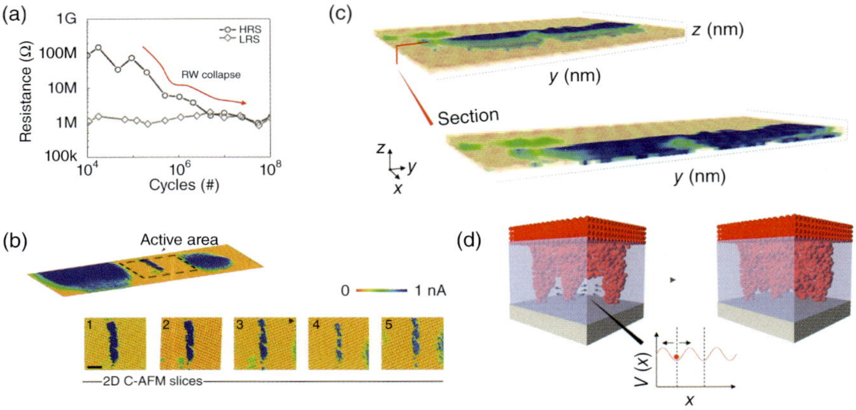


**Figure 8.12** Failure analysis of a device stuck in LRS. (a) Endurance failure with resistive window collapse due to stuck in LRS. (b) 2D current maps of the device under test. Note that after the top electrode removal the CF appears as a large conductive platelet (scale bar 50 nm). (c) Surface section and cross section extracted 3D tomogram of a failed device, showing the CF degenerated into a conductive wall. The latter shorts the bottom and top electrodes and runs through the entire switching layer. (d) Schematic of the proposed model for the stuck in LRS. Three original CFs join together into one platelet during cycling. (Reproduced with permission from [52]. Copyright Royal Society of Chemistry 2016.)

break. Actually the increased cross-section size of the narrowest point along this platelet requires a larger reset power to effectively induce the reset switching. In agreement with this platelet formation, we observed (not shown) that increasing the reset voltage without limiting the current often results in successful reset. These additional tests confirm that the stuck-in-LRS failure is not a hard failure but relates to the gradual formation of a wider filament that requires larger switching power during the cycling. In Figure 8.12d we schematically describe a possible scenario for such filament degeneration. From the previous section we know that multiple CF formation is not excluded and that during prolonged cycling those CFs could alternate as connected CF. In this scenario, over prolonged cycling an over injection of Cu inside the solid electrolyte could lead to an agglomeration of the CFs, leading to the formation of a platelet.

## 8.5 Conclusions and Outlook

This chapter provided an overview of the development of SPM tomography. The basic principles of scalpel SPM were introduced in the first part of the chapter, while its application in a broad study on the physical mechanisms of filamentary-based RS was presented as a real case study. Although scalpel SPM represents an exploratory field of research, the widespread use of 3D structures in the field of semiconductor devices has triggered a growing interest in the technique as it can provide a unique methodology for the characterization of confined volumes. Its unique 3D-imaging capability opens new possibilities for the 3D investigation in the area of material science, quality control, reverse engineering, and failure analysis. Nevertheless, the current state of scalpel SPM presents some limitations which need to be further improved:

- Data acquisition and serial sectioning are currently achieved using a contact-mode AFM technique. To limit the tip degradation, the load force and tip velocity have to be reduced, thus leading to relatively long acquisition time even for small volumes. In addition, this long acquisition is prone to image distortion induced by the piezo drift which has to be manually compensated for during the analysis.
- Being based on the collection of multiple 2D slices of the sample at different heights, scalpel SPM is a destructive technique. After the analysis, a crater of the same size of the scan area is formed. The major limitation is the impossibility to observe dynamic phenomena, for example, repeated RS events inside the same device by means of this approach.
- Scalpel SPM uses a tip-induced, pressure-assisted material removal and therefore it requires an accurate selection of the load force, tip shape, and tip materials for the removal process. Since each tip–sample combination will require a slightly different pressure for material removal, it is very common to overestimate the load force and destroy the sample during the slicing phase. Furthermore, as the tip will apply the same pressure in each point on the scan area, in case of anisotropies such as roughness or voids in the material, the exerted pressure can be too high and thus destroy the layer during the slicing. For this reason, in the current design our scalpel SPM has a poor yield, often requiring multiple attempts for a successful characterization.
- Physical contact between tip and sample is required to remove material, implying that we are restricted to contact-based AFM modes. Noncontact or oscillating modes, for example, tapping or peak force are not possible, although they would allow magnetic force microscopy, KPFM, or intermitting modes, thus broadening the range of observable material properties.

In summary, scalpel SPM is an extremely valuable technology in view of the growing need for 3D metrology on 3D devices and confined volumes. Satisfying the request for advanced metrology on such devices, advancing the state of the art in the field of 3D SPM will be extremely rewarding.

## References

1 Hisamoto, D., Kedzierski, J., Anderson, E., Takeuchi, H., Asano, K., King, T., Bok, J., and Hu, C. (1998) *IEDM Tech. Dig.*, **38**, 1032.

2 Natarajan, S., Agostinelli, M., Akbar, S., Bost, M., Bowonder, A., Chikarmane, V., Chouksey, S., Dasgupta, A., Fischer, K., Fu, Q., Ghani, T., Giles, M., Govindaraju, S., Grover, R., Han, W., Hanken, D., Haralson, E., Haran, M., Heckscher, M., Heussner, R., Jain, P, James, R., Jhaveri, R., Jin, I., Kam, H., Karl, E., Kenyon, C., Liu, M., Luo, Y., Mehandru, R., Morarka, S., Neiberg, L., Packan, P., Paliwal, A., Parker, C., Patel, P., Patel, R., Pelto, C., Pipes, L., Plekhanov, P., Prince, M., Rajamani, S., Sandford, J., Sell, B., Sivakumar, S., Smith, Song, B., Tone, K., Troeger, T., Wiedemer, J., Yang, M., and

Zhang, K. (2014) A 14 nm logic technology featuring 2nd-generation FinFET, air-gapped interconnects, self-aligned double patterning and a 0.0588 μm2 SRAM cell size, in Electron Devices Meeting (IEDM) Technical Digest, IEEE, pp. 3.7.1–3.7.3.
3 Chandrasekaran, N. (2013) Challenges in 3D Me emory Manufacturing and Process Inte egration, *IEDM Tech. Dig.*, 344.
4 Kawano, M., Uchiyama, S., Egawa, Y., Takahashi, N., Kurita, Y., Soejima, K., Komuro, M., Matsui, S., Shibata, K., Yamada, J., Ishino, M., Ikeda, H., Saeki, Y., Kato, O., Kikuchi, H., and Mitsuhashi, T. (2006) A 3D Packaging Technology for 4 Gbit Stacked DRAM with 3 Gbps Data Transfer, IEDM Technical Digest, IEEE, pp. 1–4.
5 Merckling, C., Waldron, N., Jiang, S., Guo, W., Collaert, N., Caymax, M., Vancoille, E., Barla, K., Thean, A., Heyns, M., and Vandervorst, W. (2014) *J. Appl. Phys.*, **115**, 023710.
6 Magerle, R. (2000) *Phys. Rev. Lett.*, **85** 2749.
7 Dietz, C., Zerson, M., Riesch, C., Gigler, A.M., Stark, R.W., Rehse, N., and Magerle, R. (2008) *Appl. Phys. Lett.*, **92**, 143107.
8 Dietz, C., Röper, S., Scherdel, S., Bernstein, A., Rehse, N., and Magerle, R. (2007) *Rev. Sci. Instrum.*, **78**, 053703.
9 Alekseev, A., Efimov, A., Lu, K., and Loos, J. (2009) *Adv. Mater.*, **21**, 4915.
10 Bailey, R.J., Geurts, R., Stokes, D.J., de Jong, F., and Barber, A.H. (2012) *MRS Proc.*, **1421**, 07.
11 Schulze, A. 2013 Two- and three-dimensional dopant and conductivity profiling in confined volumes, KU Leuven, PhD thesis.
12 Mody, J., Zschätzsch, G., Kölling, S., De Keersgieter, A., Eneman, G., Kambham, A.K., Schulze, A., Chiarella, T., Horiguchi, N., Hoffmann, T., Eyben, P., Vandervorst, W., Leuven, K.U., and Arenberg, K. (2011) (2006) IEDM Technical Digest, IEEE, 119.
13 Schulze, A., Florakis, A., Hantschel, T., Eyben, P., Verhulst, A.S., Rooyackers, R., Vandooren, A., and Vandervorst, W. (2013) *Appl. Phys. Lett.*, **102**, 052108.
14 Celano, U., Goux, L., Belmonte, A., Schulze, A., Opsomer, K., Detavernier, C., Richard, O., Bender, H., Jurczak, M., and Vandervorst, W. (2013) *IEDM Tech. Dig.*, 21.6.1.
15 Hantschel, T., Niedermann, P., Trenkler, T., and Vandervorst, W. (2000) *Appl. Phys. Lett.*, **76**, 1603.
16 Hu, J., Xiao, X.-D., Ogletree, D.F., and Salmeron, M. (1995) *Surf. Sci.*, **327**, 358.
17 Xu, M.W., Hantschel, T., and Vandervorst, W. (2002) *Appl. Phys. Lett.*, **81**, 177.
18 Ou, X., Das Kanungo, P., Kögler, R., Werner, P., Gösele, U., Skorupa, W., and Wang, X. (2010) *Adv. Mater.*, **22**, 4020.
19 Schulze, A., Hantschel, T., Dathe, A., Eyben, P., Ke, X., and Vandervorst, W. (2012) *Nanotechnology*, **23**, 305707.
20 Schulze, A., Hantschel, T., Eyben, P., Verhulst, A.S., Rooyackers, R., Vandooren, A., Mody, J., Nazir, A., Leonelli, D., and Vandervorst, W. (2011) *Nanotechnology*, **22**, 185701.

21 Celano, U., Goux, L., Opsomer, K., Iapichino, M., Belmonte, A., Franquet, A., Hoflijk, I., Detavernier, C., Jurczak, M., and Vandervorst, W. (2014) *Microelectron. Eng.*, **120**, 67.

22 Celano, U., Goux, L., Belmonte, A., Opsomer, K., Franquet, A., Schulze, A., Detavernier, C., Richard, O., Bender, H., Jurczak, M., and Vandervorst, W. (2014) *Nano Lett.*, **14**, 2401.

23 Celano, U., Goux, L., Belmonte, A., Giammaria, G., Opsomer,K., Detavernier, C., Richard, O., Bender, H., Irrera, F., Jurczak, M., and Vandervorst, W. (2014) Progressive vs. abrupt reset behavior in conductive bridging devices: a C-AFM tomography study, in IEDM Technical Digest, IEEE, pp. 14.1.1–14.1.4.

24 Carpick, R.W. and Salmeron, M. (1997) *Chem. Rev.*, **97**, 1163.

25 Jacobs, T.D.B. and Carpick, R.W. (2013) *Nat. Nanotechnol.*, **8**, 108.

26 Hantschel, T., Demeulemeester, C., Eyben, P., Schulz, V., Richard, O., Bender, H., and Vandervorst, W. (2009) *Phys. Status Solidi A*, **206**, 2077.

27 Mylvaganam, K., Zhang, L.C., Eyben, P., Mody, J., and Vandervorst, W. (2009) *Nanotechnology*, **20**, 305705.

28 Celano, U. and Vandervorst, W. (2014) Integrated Reliability Workshop Final Report (IIRW), 2014 IEEE International, 1.

29 Buckwell, M., Montesi, L., Hudziak, S., Mehonic, A., and Kenyon, A.J. (2015) *Nanoscale*, **7**, 18030.

30 Lee, M.-J., Lee, C.B., Lee, D., Lee, S.R., Chang, M., Hur, J.H., Kim, Y.-B., Kim, C.-J., Seo, D.H., Seo, S., Chung, U.-I., Yoo, I.-K., and Kim, K. (2011) *Nat. Mater.*, **10**, 625.

31 Govoreanu, B., Kar, G. S., Chen, Y., Paraschiv, V., Kubicek, S., Fantini, A., Radu, I. P., Goux, L., Clima, S., Degraeve, R., Jossart, N., Richard, O., Vandeweyer, T., Seo, K., Hendrickx, P., Pourtois, G., Bender, H., Altimime, L., Wouters, D. J., Kittl, J. A., and Jurczak, M. (2011) $10 \times 10\,nm^2$ Hf/HfO × Crossbar Resistive RAM with Excellent Performance, Reliability and Low-Energy Operation, IEDM Technical Digest, Washington, DC, pp. 31.6.1 – 31.6.4.

32 Wong, H.-S.P., Lee, H.-Y., Yu, S., Chen, Y.-S., Wu, Y., Chen, P.-S., Lee, B., Chen, F.T., and Tsai, M.-J. (2012) *Proc. IEEE*, **100**, 1951.

33 Waser, R. and Aono, M. (2007) *Nat. Mater.*, **6**, 833.

34 Yang, J.J., Strukov, D.B., and Stewart, D.R. (2012) *Nat. Nanotechnol.*, **8**, 13.

35 Celano, U., Goux, L., Degraeve, R., Fantini, A., Richard, O., Bender, H., Jurczak, M., and Vandervorst, W. (2015) *Nano Lett.*, **15**, 7970.

36 Kamiya, K., Yang, M., Nagata, T., Park, S.-G., Magyari-Köpe, B., Chikyow, T., Yamada, K., Niwa, M., Nishi, Y., and Shiraishi, K. (2013) *Phys. Rev. B*, **87**, 155201.

37 Park, G.-S., Kim, Y.B., Park, S.Y., Li, X.S., Heo, S., Lee, M.-J., Chang, M., Kwon, J.H., Kim, M., Chung, U.-I., Dittmann, R., Waser, R., and Kim, K. (2013) *Nat. Commun.*, **4**, 2382.

38 Yang, Y., Gao, P., Li, L., Pan, X., Tappertzhofen, S., Choi, S., Waser, R., Valov, I., and Lu, W.D. (2014) *Nat. Commun.*, **5**, 4232.

39 Chen, Y.Y., Komura, M., Degraeve, R., Govoreanu, B., Goux, L., Fantini, A., Raghavan, N., Clima, S., Zhang, L., Belmonte, A., Redolfi, A., Kar, G.S.,

Groeseneken, G., Wouters, D.J., and Jurczak, M. (2013) *IEDM Tech. Dig.*, 10.1.1.

40 Celano, U., Goux, L., Belmonte, A., Opsomer, K., Degraeve, R., Detavernier, C., Jurczak, M., and Vandervorst, W. (1919) *J. Phys. Chem. Lett.*, **2015**, 1919.

41 Guo, X., Schindler, C., Menzel, S., and Waser, R. (2007) *Appl. Phys. Lett.*, **91**, 133513.

42 Valov, I., Waser, R., Jameson, J.R., and Kozicki, M.N. (2011) *Nanotechnology*, **22**, 254003.

43 Yang, Y., Gao, P., Gaba, S., Chang, T., Pan, X., and Lu, W. (2012) *Nat. Commun.*, **3**, 732.

44 Liu, Q., Sun, J., Lv, H., Long, S., Yin, K., Wan, N., Li, Y., Sun, L., and Liu, M. (2012) *Adv. Mater.*, **24**, 1844.

45 Valov, I. and Waser, R. (2013) *Adv. Mater.*, **25**, 162.

46 Sankaran, K., Goux, L., Clima, S., Mees, M., Kittl, J., Jurczak, M., Altimime, L., Rignanese, G., and Pourtois, G. (2012) *ECS Trans.*, **45**, 317.

47 Yang, Y. and Lu, W. (2013) *Nanoscale*, **5**, 10076.

48 Degraeve, R., Fantini, A., Clima, S., Govoreanu, B., Goux, L., Chen, Y. Y., Wouters, D. J., Roussel, P., Kar, G. S., Pourtois, G., Cosemans, S., Kittl, J. A., Groeseneken, J. A., Jurczak, M., and Altimime, L. (2012) Symposium on VLSI Technology, Honolulu, HI pp. 75–76.

49 Goux, L., Fantini, A., Govoreanu, B., Kar, G., Clima, S., Chen, Y.-Y., Degraeve, R., Wouters, D.J., Pourtois, G., and Jurczak, M. (2012) *ECS Solid State Lett.*, **1**, P63.

50 Celano, U., Goux, L., Opsomer, K., Belmonte, A., Iapichino, M., Detavernier, C., Jurczak, M., and Vandervorst, W. (2013) *Nanoscale*, **5**, 11187.

51 Degraeve, R., Groeseneken, G., Bellens, R., Ogier, J.L., Depas, M., Roussel, P.J., and Maes, H.E. (1998) *IEEE Trans. Electron Devices*, **45**, 904.

52 Celano, U., Giammaria, G., Goux, L., Belmonte, A., Jurczak, M., and Vandervorst, W. (2016) *Nanoscale*. **8**, 13915 doi: 10.1039/C5NR08735J

53 Chen, Y.Y., Degraeve, R., Clima S., Govoreanu, B., Goux, L., Fantini, A., Kar, G. S., Pourtois, G., Groeseneken, G., Wouters, D. J., and Jurczak, M. (2012) International Electron Devices Meeting, IEEE, 2012, pp. 20.3.1–20.3.4.

54 Chen, B., Lu, Y., Gao, B., Fu, Y.H., Zhang, F.F., Huang, P., Chen, Y.S., Liu, L.F., Liu, X.Y., Kang, J.F., Wang, Y.Y., Fang, Z., Yu, H.Y., Li, X., Wang, X.P., Singh, N., Lo, G.Q., and Kwong, D.L. (2011) *IEDM Tech. Dig.*, **12.3**, 283.

55 Lv, H., Xu, X., Liu, H., Liu, R., Liu, Q., Banerjee, W., Sun, H., Long, S., Li, L., and Liu, M. (2015) *Sci. Rep.*, **5**, 7764.

56 ITRS Consortium 2014 International Technology Roadmap for Semiconductors, Metrology Reports, www.itrs.net.

# 9

# Conductive Atomic Force Microscopy for Nanolithography Based on Local Anodic Oxidation

*Matteo Lorenzoni and Francesc Pérez-Murano*

*Microelectronics Institute of Barcelona (IMB-CNM, CSIC), NEMS & Nanofabrication group, Campus de la Universitat Autònoma de Barcelona, E-08193 Bellaterra, Spain*

## 9.1 Introduction to AFM Nanolithography

Since the first attempts made in the 1990s [1–3], the use of scanning probe microscopy as a lithographic tool has broadly developed and nowadays a variety of tools/applications are grouped under the name of scanning probe lithography (SPL) techniques. While in the past SPL was confined to the academic field, the upcoming challenges to conventional lithography are multiplying the efforts toward robust, efficient, and large throughput SPL technologies. In very general terms we can say that the working principle always relies on the interaction between a sharp probe and a substrate. The typical radius of curvature of the tip apex is 10–50 nm and the tip–substrate separation might vary from the close contact condition (<1 nm) to some nanometers during dynamic approaches, when an atomic force microscope (AFM) is operated in dynamic modes. If appropriate forces, heating, and/or electric fields are applied, the tip can induce various physical and chemical modifications on the substrate surface. Consequently, localized nanostructures are generated through physical modifications and/or chemical reactions on the material surface. AFM nanolithography possesses the versatility to pattern a wide range of materials, including metals, semiconductors, polymers, and biological molecules in different media. Numerous AFM-based lithographic techniques have been developed [4–6], which can be classified according to their operation principles into two groups: force-assisted or bias-assisted AFM nanolithography. Another way of classifying the techniques can be based on the fabrication process: one can (i) remove material, (ii) modify the material, or (iii) add some material. In force-assisted AFM nanolithography [7], a large force (≥100 nN) is applied via the tip, and its interaction with the substrate is mainly mechanical. Typical methods include indentation and plowing, thermomechanical writing, nanomanipulation [3], and dip-pen nanolithography [8, 9] (DPN). In the recent past, remarkable results have been achieved by means of thermal lithography [10, 11] in which a specially designed probe is able to deliver local heating with nanometric resolution. Remarkable examples of SPL experiments are given in Figure 9.1, with panels

*Conductive Atomic Force Microscopy: Applications in Nanomaterials*, First Edition.
Edited by Mario Lanza.

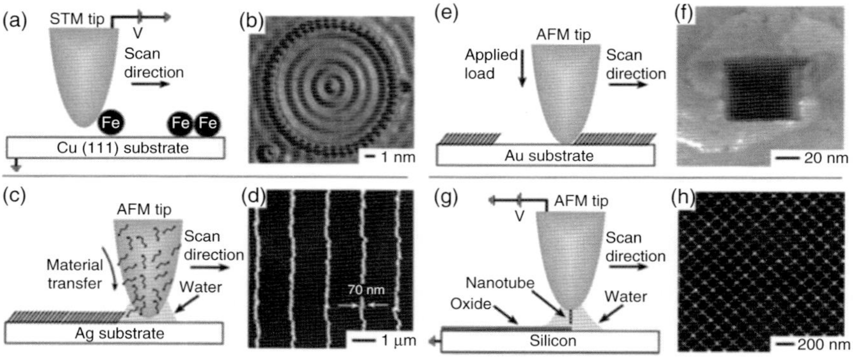


**Figure 9.1** Four possible approaches to scanning probe lithography and respective examples of patterns: (a) STM displacement of atoms on a surface, which can generate patterns such as (b) a "quantum corral" ring of 48 Fe atoms on Cu. (Reprinted with permission from Ref. [3]). (c) Dip-pen nanolithography, which can direct the deposition of SAMs such as (d) 70-nm-wide features of 16-mercaptohexadecanoic acid on Ag, to be used as an etch resist. (e) Nanoshaving removal of SAM regions, which can pattern features such as (f) a square hole within octadecanethiolate SAMs on Au. (g) Local anodic oxidation with a carbon-nanotube-modified AFM tip, which can selectively oxidize a surface, making patterns such as (h) 10-nm-wide (2-nm-tall) silicon oxide lines spaced by 100 nm [12]. Adapted from Ref. [4]. Copyright 2005 American Chemical Society.

A–B depicting the famous scanning tunneling microscopy (STM)-based atomic manipulation experiment with the creation of a "quantum corral" dated 1993. The scope of this chapter is an exhaustive description of the physics and chemistry behind one specific bias-assisted technique, local anodic oxidation (LAO); if the reader is interested in other SPL techniques, we recommend the exhaustive review of Ref. [6].

## 9.2 Local Anodic Oxidation

LAO has received several names such as field-induced oxidation, nano-oxidation, or scanning probe oxidation. By means of LAO integrated with other fabrication steps it is possible to produce a wide variety of electronic and mechanical nanodevices (see Figure 9.2). The elements that can be directly produced by LAO are, basically, dielectric barriers, masks for selective etching, and templates [13]. Dagata *et al.* firstly reported the modification of a hydrogen-passivated Si surface with an STM in 1990 [1], followed a few years later by the use of an AFM in contact mode to produce sub-100-nm oxide lines [14]. The schematic of contact LAO experiments is reported in Figure 9.3. As depicted, a conductive AFM tip is brought into contact with the semiconductor surface at a certain force set point. In these early attempts, it was already possible to recognize that the features produced were removed by a mild hydrofluoric acid (HF)-based wet etching, producing trenches where the protrusions were placed, similar to what is depicted in the panels b–c of Figure 9.3, pointing to the conclusion that the features produced were made of oxide. From this experiment it is possible to calculate the volume

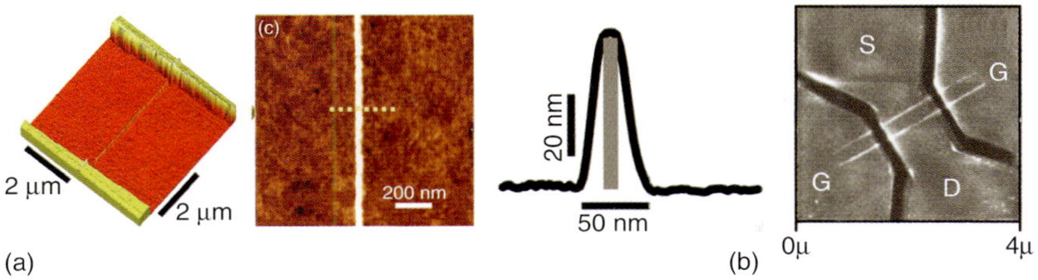


**Figure 9.2** (a) AFM images of a silicon nanowire connected to platinum electrodes; the height profile of the nanowire is also reported. Adapted from Ref. [15]. Copyright 2010 IOP Publishing, LDT. (b) The AFM height image of a single-electron transistor with oxide tunneling barriers. Between source (S) and drain (D) an island is defined by two oxide lines (light lines). Side gates (G) have been mechanically machined at the sides of the island (dark lines). Adapted from Ref. [16]. Copyright 2000 AIP Publishing.

mismatch between the pristine surface and the newly formed oxide; that is, for 1 nm of oxide above Si surface, there is approximately 0.4 nm of $SiO_2$ underneath the baseline. To obtain further confirmation the chemical composition of such features was later investigated by employing time-resolved photoemission spectroscopy of nanostructures grown by LAO [17]. Apart from silicon, LAO has been used to modify different surfaces such as silicon nitride, III–V compound semiconductors, metals (both bulk and films) [18, 19], perovskite oxides [20], and Cerium oxide [21]. The technique also showed its versatility when employed on molecularly functionalized/passivated surfaces such as self-assembled monolayers (SAMs) [22] and methyl-terminated silane monolayers [23]. If we consider the reaction volume underneath the tip as an electrochemical cell, the anodic semireaction that involves silicon is the following:

$$Si + 4h^+ + 4(OH^-) \rightarrow SiO_2 + 2H_2O \tag{9.1}$$

Hydrogen is generated at the cathode (meniscus/tip interface):

$$2H^+ + 2e^- \rightarrow H_2 \tag{9.2}$$

Clearly, the above reaction brings about the formation of stoichiometric $SiO_2$. When different substrates are involved, the incorporation of oxygen leads to oxides with different stoichiometry; for example, LAO on GaAs produces oxides poor in As [24]. In the case of silicon carbide, the direct anodic reaction becomes:

$$SiC + 8h^+ + 4(OH^-) \rightarrow SiO_2 + 2H_2O + CO_2 \tag{9.3}$$

where carbon atoms are completely eliminated in the form of carbon dioxide. Carbon inclusion is also likely to happen, leaving residuals with a SiOxCy stoichiometry. The water meniscus responsible for the reaction is formed due to capillary condensation when tip and sample are brought into contact in a humid environment. Usually when LAO is performed in air, no oxide growth is observed for relative humidity below 20%. Even if the mechanism of water meniscus formation was not yet elucidated, Pérez-Murano *et al.* firstly performed LAO in tapping mode in 1995 [2]. Following that, García *et al.*

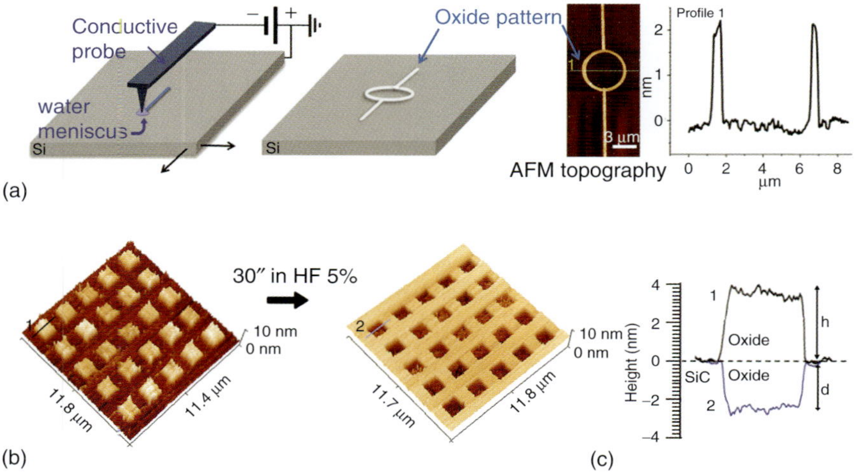


Figure 9.3 (a) Schematics of the LAO process on Si with an AFM operating in contact mode. The feature fabricated on Si(100) is the oxide pattern, the corresponding AFM topography and height profile are also reported. (b) The AFM topography of oxide squares fabricated on silicon carbide before and after a wet etching in HF. The oxide protrusions are readily etched, as expected for $SiO_2$. (c) The height profile before and after the etching; from this kind of experiment it is possible to calculate the volume mismatch between the pristine material (dotted line) and the newly formed oxide.

were able to further develop LAO with the AFM operating in dynamic mode [25]. The meniscus formation is induced by means of a pulsed electric field and is maintained during the oxidation process, being able to stretch to a few tens of nanometers. By amplitude modulation LAO the lateral resolution has been considerably improved, allowing, for example, the fabrication of SiNW transistors with sub-10-nm channel widths [15] (see Figure 9.2a) or the creation of chemical guiding patterns for the directed self-assembly of block copolymers [26]. It is worth mentioning that several devices have been fabricated integrating LAO with conventional fabrication methods. As a first example, Keyser showed in the year 2000 the fabrication of a single-electron transistor taking advantage of the ultrathin oxide tunneling barriers defined by LAO [16] (see Figure 9.2b). By manipulating organic liquid menisci it was also possible to produce sub-10-nm carbonaceous features [27], showing how moving from ambient-based oxidation to alternative media/substrates the potentiality of tip-based fabrication could greatly expand.

## 9.3 Kinetics of LAO

The mechanism and kinetics of Si oxidation surfaces by AFM has been extensively studied [28–30] since the very first experiments. LAO works efficiently on those materials (semiconductors and metals) that form a passivation layer in the presence of oxygen, and the native layer growth stops at a certain thickness, a few Angstrom in the case of Si. The parallelism with LAO is straightforward. During

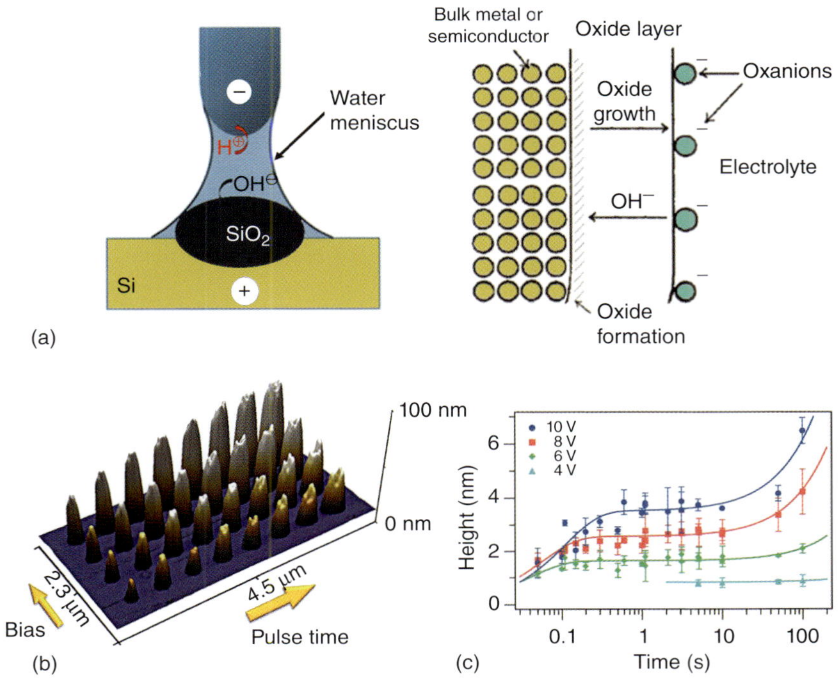


**Figure 9.4** (a) The LAO mechanism scheme in detail. (b) 3D AFM topography of oxide dots fabricated on silicon carbide at various pulse times/bias. Note that the maximum oxide height on such substrates is much higher than the few nanometers reported for silicon. In (c) we report the fitting of LAO data (oxide height vs log of exposition time) to the "two-path" model as reported in Ref. [31]. (Panel (c) adapted from Ref. [31]. Copyright 2006 WILEY-VCH Verlag GmbH & Co.)

tip-driven oxidation of Si the reaction occurs at the $Si/SiO_2$ interface, the factor playing a key role is the transport of ionic oxidizing species across the growing silicon oxide layer, resulting in a self-limiting reaction (see Figure 9.4), with a critical oxide height $h_L$ at which oxide growth stops. Low-temperature oxidation theory (Cabrera-Mott theory) was used early on to describe LAO kinetics and oxide growth mechanism. The theory *is based on the assumption that cation migration occurs under the influence of a potential built up across the growing oxide film* [32]. In the absence of space charge the so-called Mott potential $V_M$ creates an electric field $E = V_M/h$ that is the driving force allowing the growth of the first oxide layers ($h$ is the oxide height). For very thin oxide layers and strong electric fields, the rate of oxidation is determined only by the rate at which ions are formed at the interface. If we consider the anodic formation of an oxide film in an electrolyte containing oxide ions, $V$ is no more the Mott potential but the built-in voltage across the film. This and other assumptions led to an inverse logarithmic rate law [33], with the following form:

$$\frac{h}{h_1} = \ln\left(\frac{h_1 ut}{h_L^2}\right) \tag{9.4}$$

where $h \gg h_1$ and $h_1 \approx 10^{-5}$ cm. The term $h_L$ is the limiting height (approximately 10 nm for LAO) and $u$ is the drifting velocity of ions through the oxide. The Cabrera-Mott theory alone was able to account for experimental evidence; therefore, different models were proposed to explain the experimental data. It is also worth mentioning how some authors successfully attempted to fit experimental data to empirical kinetic expressions; Teuschler proposed a power rate law [30], expressed as follows:

$$h = \alpha \left( \frac{v_0}{v_{ox}} \right)^{\gamma} \tag{9.5}$$

with $\alpha$ and $\gamma$ as parameters, $\upsilon_{ox}$ the writing speed, and $\upsilon_0$ an arbitrary constant. The deviation from conventional growth kinetics is explained by the cooperative effect of water content into the system combined with a very high electrical field below the tip. More specifically, some experimental evidences could not be explained by the Cabrera-Mott theory: (i) it was possible by applying the same bias pulse to grow similar oxide features both on thin native oxide layers and thicker thermal oxide; (ii) when a modulated bias was used, the limiting thickness was increased; and (iii) the very high initial growth rates (several tens of nanometers per second) observed indicates such a high carrier concentration that would generate a space charge no more negligible. According to those observations, Dagata and coworkers [34, 35] introduced the effect of space charges generated by the extremely high electric field conditions ($E > 10^6$ V/cm) and fast initial growth. In this model, anionic species (i.e., $OH^-$ or other rate limiting species), initially formed from water dissociation, produce an accumulation of negative charges $Q_{max}$ at the tip/oxide interface that decays in time. At the same time, a certain percentage of oxyanions, drifting through the growing oxide, is trapped producing additional space-charge effects. The space charge region generated is responsible for the rapid reduction of the growth rate as the oxidation proceeds. Considering what is stated, different kinetics laws were proposed. Dubois and Bubendorff [36], trying to account for space-charge limited growth, proposed a power-of-time law in the form:

$$h(t, V) = h_0 \left( \frac{t}{t_0} \right)^{\frac{1}{\delta+1}} \tag{9.6}$$

where $h$, $V$, and $t$ are the oxide thickness, the applied voltage, and the pulse time, respectively, $t_0$ is a time constant depending on voltage, and $h_0$ and $\delta$ are fitting parameters. Dagata proposed a two-path explanation [34] for the anodic oxidation: initially, the direct oxidation process (Eq. (9.1)) is predominant, but over time a second indirect path of the oxidation occurs in which the charged defects are treated as reaction intermediates. The two models were compared by Kinser [31], showing that the model based on a two-path reaction better reproduces the data in case of LAO of Si in inert organic solvents; the resulting fit is shown in Figure 9.4c. A comprehensive explanation of LAO kinetics is not yet available, but the amount of experimental data is abundant and tentative kinetics model have been proposed for alternative substrates [37, 38] or alternative electrolytes [39]. In the following section we better describe the observed faradaic (ionic) and electronic currents flowing during LAO experiments.

## 9.4 Measurement of Electrical Current During LAO

Understanding and controlling the oxidation process during LAO experiments can be enhanced by detecting the electrochemical current. The oxidation process occurs through an electrochemical reaction. According to Eq. (9.1), in the case of silicon oxidation, each $SiO_2$ molecule requires the circulation of four electrical charges. Therefore, measuring the electrical current provides an essential insight about the oxide growth kinetics.

The electrical current originated from the electrochemical reaction is typically of the order of pico-amperes. Its measurement is complex not only because of the small magnitude of the current but also because it competes with other mechanisms of charge transport like electron tunneling, capacitive transients, and other types of faradaic currents. Because of that, we do not find many examples of current measurement during LAO experiments.

Depending on the condition in which we perform the LAO, the relation between electrochemical current and the volume of oxide growth can be estimated from the following relations:

1. In case the tip is scanned over the surface at a constant speed $v$, a steady-state electrochemical current, $I$, flows. That can be calculated from

$$I = ne\frac{N_A \rho_{ox} w_{ox}}{M_{ox}} h_{ox} v \tag{9.7}$$

2. In case of a voltage pulse of duration $t$ while the tip is kept static over the surface, the average electrochemical current, $\hat{I}$ that flows is

$$\hat{I} = ne\frac{N_A \rho_{ox}}{M_{ox}} \frac{V_{ox}}{t} \tag{9.8}$$

Equation (9.8) is obtained from the integration with respect to $t$ of the instantaneous current, $I(t)$:

$$I(t) = ne\frac{N_A \rho_{ox}}{M_{ox}} \frac{dV_{ox}}{dt} = ne\frac{N_A \rho_{ox}}{M_{ox}} \left( h_{ox} \frac{dA_{ox}}{dt} + A_{ox} \frac{dh_{ox}}{dt} \right) \tag{9.9}$$

where $n$ is the number of electrical charges per oxide molecule grown, $e$ is the electron charge, $N_A$ is the Avogrado number, $\rho_{ox}$ is the density of the oxide grown, $M_{ox}$ is the molecular mass of the oxide, $V_{ox}$ the volume of oxide grown, $A_{ox}$ the effective oxidized surface area, and $w_{ox}$ and $h_{ox}$ are the width and the average thickness of the oxide line created when performing the oxidation at constant speed $v$.

It is important to note that, according to Eqs (9.7) and (9.9), the steady-state current can be almost directly related to the oxide thickness, while the instantaneous current is a signature of the instantaneous increase of the oxide volume $dV_{ox}/dt$, which includes both contributions from the increase of oxide thickness and effective oxide area (i.e., vertical and lateral increase of the oxide dimensions). Studying LAO kinetics during the oxidation process, we can encounter different regimes which can be identified with different oxide growth rates for the vertical and lateral oxidation.

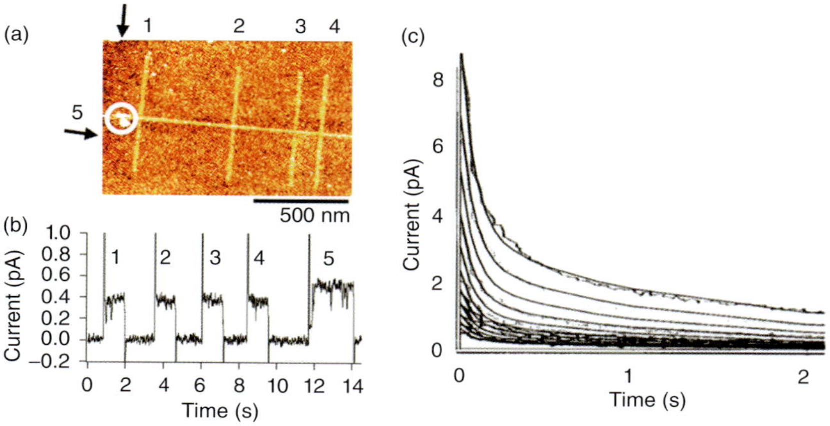


**Figure 9.5** (a) AFM height image of oxide lines fabricated at a constant speed of 1.0 μm $s^{-1}$ and voltage $V_{ox} = 7$ and 8 V for the vertical and horizontal lines, respectively. (b) Detected current measured during the writing process; during LAO in contact mode at constant speed, current is stable and constant. Peaks/spikes in current are produced at the beginning of each line, while instant drop in current happens when the tip crosses an oxide line previously fabricated. (Adapted from Ref. [41] Copyright 2004 IOP Publishing, LDT.) (c) Time evolution of the total instant current during a scanning probe oxidation experiment (see Ref. [42]).

Early experiments were reported by Ruskell *et al.* [40], showing a good correlation between the steady-state electrical current and the volume of silicon oxide growth, indicating that most of the detected current was related with the electrochemical reaction, except an initial current spike due to the onset of the voltage application (see Figure 9.5a,b). Avouris *et al.* [43] showed a good correlation between oxide growth and instantaneous current, but they had to assume a charge efficiency of 50%, indicating that additional electronic current was present in the measurements.

To avoid the presence of displacement current during the detection of instantaneous current, Pérez-Murano *et al.* [44] developed a procedure consisting of applying the voltage with the AFM tip separated from the surface, and then approaching the tip to the surface until the water meniscus is formed and so the oxidation process starts (see Figure 9.5c). In this way, the initial moments of the oxidation process can be captured and the correlation between silicon oxide growth kinetics and electrical current evolution investigated. A similar method to detect the instantaneous current, developed by Kim *et al.* [45], was in this case applied to $TiO_2$. In both works it is shown that the detection of the electrochemical current reveals the phenomenology behind LAO and, specifically, the effect of the space-charge buildup in the kinetics of the oxidation and the influence of the meniscus geometry. A deep investigation of the knowledge that can be extracted from the current measurements has been presented by Dagata *et al.* [42, 46], revealing how the oxidation proceeds differently when it is performed in contact mode or in noncontact mode.

Several works have employed electrical current detection during the oxidation process in order to understand the actual reactions occurring when oxidizing thin layers of materials. Fernandez-Cuesta *et al.* [47] showed how the detection of the steady-state current allowed them to distinguish two different processes during the application of LAO on thin silicon nitride films deposited on silicon: the conversion of silicon nitride into silicon oxide and the incorporation of silicon into the bottom of the silicon nitride film. Martín *et al.* [48] applied the procedure previously developed [44] to investigate the mechanism of LAO on thin poly(methyl methacrylate) (PMMA) layers deposited on silicon. The experiments point toward the presence of ionic transport through PMMA during the voltage application, occurring even without the local removal of PMMA. If higher voltage is applied, PMMA is eliminated by an electrochemical oxidation process.

Kuramochi and coworkers [46, 49–51] have also investigated different aspects of the LAO process by detecting the electrochemical current. They showed that current detection during LAO in dynamic mode allows a better monitoring of the oxidation process. In addition, using AFM probes with carbon nanotubes as apex, they could obtain information about the variation of the water meniscus as they detected changes in the electrochemical current when passing over previously oxidized lines [41]. They also showed that an enhanced spatial resolution can be obtained from the current measurements if compared to simple topographical measurements.

The difficulty in measuring and discriminating the current that results from the oxidation reactions is the reason LAO experiments are performed on a voltage-controlled mode rather than a current-controlled mode (feedback on current). Only very few examples of current-controlled LAO patterns have been reported. Among these, Pellegrino *et al.* [52] have shown an increase in the pattern uniformity in current-controlled oxidation compared to voltage-controlled oxidation.

## 9.5 Conclusions

While current-controlled nanolithography using scanning probe microscopes has been successfully applied to other systems, as in the case of current-induced removal of thin molecular glass resist [53], it is more difficult to ensure its applicability in LAO, where the presence of a water meniscus and the existence of high charge density may produce complex variation in the conditions under which the oxidation proceeds. However, an accurate control of the water meniscus and of the chemical properties of the tip may ensure more stable conditions for current-controlled LAO.

## Acknowledgments

Financial support from the European Union's Seventh Framework Programme FP7/2007-2013, under Grant Agreement No 318804 (SNM) is acknowledged.

## References


1 Dagata, J.a., Schneir, J., Harary, H.H., Evans, C.J., Postek, M.T., and Bennett, J. (1990) Modification of hydrogen-passivated silicon by a scanning tunneling microscope operating in air. *Appl. Phys. Lett.*, **56** (20), 2001–2003.

2 Pérez-Murano, F., Abadal, G., Barniol, N., Aymerich, X., Servat, J., Gorostiza, P., and Sanz, F. (1995) Nanometer-scale oxidation of Si(100) surfaces by tapping mode atomic force microscopy. *J. Appl. Phys.*, **78** (11), 6797.

3 Crommie, M. F.; Lutz, C. P.; Eigler, D. M. Confinement of electrons to quantum corrals on a metal surface. *Science.* 1993, **262** (5131 ), 218–220.

4 Gates, B.D., Xu, Q., Stewart, M., Ryan, D., Willson, C.G., and Whitesides, G.M. (2005) New approaches to nanofabrication: molding, printing, and other techniques. *Chem. Rev.*, **105** (4), 1171–1196.

5 Tseng A A, Pellegrino L, Shirakashi J. I. In *Encyclopedia of Nanoscience and Nanotechnology*; H S, Nalwa., Ed.; American Scientific Publishers: Valencia, 2012; pp. 171–207.

6 Garcia, R., Knoll, A.W., and Riedo, E. (2014) Advanced scanning probe lithography. *Nat. Nanotechnol.*, **9** (8), 577–587.

7 Rosa, L.G. and Liang, J. (2009) Atomic force microscope nanolithography: Dip-Pen, nanoshaving, nanografting, tapping mode, electrochemical and thermal nanolithography. *J. Phys. Condens. Matter*, **21** (48), 483001.

8 Piner, R.D., Zhu, J., Xu, F., Hong, S., and Mirkin, C.A. (1999) "Dip-Pen" nanolithography. *Science*, **283** (5402), 661–663.

9 Zhang, H. and Mirkin, C.A. (2004) DPN-generated nanostructures made of gold, silver, and palladium. *Chem. Mater.*, **16** (8), 1480–1484.

10 Mamin, H.J. and Rugar, D. (1992) Thermomechanical writing with an atomic force microscope tip. *Appl. Phys. Lett.*, **61** (8) 1003.

11 Pires, D., Hedrick, J.L., De Silva, A., Frommer, J., Gotsmann, B., Wolf, H., Despont, M., Duerig, U., and Knoll, A.W. (2010) Nanoscale three-dimensional patterning of molecular resists by scanning probes. *Science*, **328** (5979), 732–735.

12 Dai, H., Franklin, N., and Han, J. (1998) Exploiting the properties of carbon nanotubes for nanolithography. *Appl. Phys. Lett.*, **73** (September), 1508–1510.

13 Garcia, R., Martinez, R.V., and Martinez, J. (2006) Nano-chemistry and scanning probe nanolithographies. *Chem. Soc. Rev.*, **35** (1), 29–38.

14 Day, H.C. and Allee, D.R. (1993) Selective area oxidation with a scanning force microscope. *Appl. Phys. Lett.*, **1993** (62), 2691–2693.

15 Martínez, R.V., Martínez, J., and Garcia, R. (2010) Silicon nanowire circuits fabricated by AFM oxidation nanolithography. *Nanotechnology*, **21** (24), 245301.

16 Keyser, U.F., Schumacher, H.W., Zeitler, U., Haug, R.J., and Eberl, K. (2000) Fabrication of a single-electron transistor by current-controlled local oxidation of a two-dimensional electron system. *Appl. Phys. Lett.*, **76** (4), 457.

17 Lazzarino, M., Heun, S., Ressel, B., Prince, K.C., Pingue, P., and Ascoli, C. (2002) Atomic force microscope anodic oxidation studied by spectroscopic microscopy. *Appl. Phys. Lett.*, **81** (15), 2842.

18 Xie, X.N., Chung, H.J., Sow, C.H., and Wee, a.T.S. (2006) Nanoscale materials patterning and engineering by atomic force microscopy nanolithography. *Mater. Sci. Eng., R*, **54** (1-2), 1–48.

19 Rolandi, M., Quate, C.F., and Dai, H. (2002) A new scanning probe lithography scheme with a novel metal resist. *Adv. Mater.*, **14** (3), 191–194.

20 Pellegrino, L., Pallecchi, I., Bellingeri, E., Canu, G., Siri, A.S., Marré, D., Yanagisawa, Y., Ishikawa, M., Matsumoto, T., Tanaka, H. *et al.* (2010) AFM nanopatterning of transition metal oxide thin films. *J. Nanosci. Nanotechnol.*, **10** (7), 4471–4476.

21 Yang, N., Doria, S., Kumar, A., Jang, J.H., Arruda, T.M., Tebano, A., Jesse, S., Ivanov, I.N., Baddorf, A.P., Strelcov, E. *et al.* (2014) Water-mediated electrochemical nano-writing on thin ceria films. *Nanotechnology*, **25** (7), 075701.

22 Krämer, S., Fuierer, R.R., and Gorman, C.B. (2003) Scanning probe lithography using self-assembled monolayers. *Chem. Rev.*, **103** (11), 4367–4418.

23 Maoz, R., Frydman, E., Cohen, S.R., and Sagiv, J. (2000) Constructive nanolithography": inert monolayers as patternable templates for in-situ nanofabrication of metal-semiconductor-organic surface structures—a generic approach. *Adv. Mater.*, **12** (10), 725–731.

24 Mori, G., Lazzarino, M., Ercolani, D., Sorba, L., and Heun, S. (2005) Locatelli, A. Desorption dynamics of oxide nanostructures fabricated by local anodic oxidation nanolithography. *J. Appl. Phys.*, **97** (11), 1–8.

25 García, R., Calleja, M., and Rohrer, H. (1999) Patterning of silicon surfaces with noncontact atomic force microscopy: field-induced formation of nanometer-size water bridges. *J. Appl. Phys.*, **86** (4), 1898.

26 Fernández-Regúlez, M., Evangelio, L., Lorenzoni, M., Fraxedas, J., and Pérez-Murano, F. (2014) Sub-10 Nm resistless nanolithography for directed self-assembly of block copolymers. *ACS Appl. Mater. Interfaces*, **6** (23), 21596–21602.

27 Martínez, R.V., Losilla, N.S., Martinez, J., Huttel, Y., and Garcia, R. (2007) Patterning polymeric structures with 2 Nm resolution at 3 Nm half pitch in ambient conditions. *Nano Lett.*, **7** (7), 1846–1850.

28 Dagata, J.A., Perez-Murano, F., Abadal, G., Morimoto, K., Inoue, T., Itoh, J., and Yokoyama, H. (2000) Predictive model for scanned probe oxidation kinetics. *Appl. Phys. Lett.*, **76** (19), 2710.

29 Snow, E.S., Jernigan, G.G., and Campbell, P.M. (2000) The kinetics and mechanism of scanned probe oxidation of Si. *Appl. Phys. Lett.*, **76** (13), 1782.

30 Teuschler, T., Mahr, K., Miyazaki, S., Hundhausen, M., and Ley, L. (1995) Nanometer-scale field-induced oxidation of Si(111):H by A conducting-probe scanning force microscope: doping dependence and kinetics. *Appl. Phys. Lett.*, **67** (21), 3144.

31 Kinser, C.R., Schmitz, M.J., and Hersam, M.C. (2006) Kinetics and mechanism of atomic force microscope local oxidation on hydrogen-passivated silicon in inert organic solvents. *Adv. Mater.*, **18** (11), 1377–1380.

32 Fehlner, F.P. and Mott, N.F. (1970) Low-temperature oxidation. *Oxid. Met.*, **2** (1), 59–99.

33 Cabrera, N. and Mott, N. (1949) Theory of the oxidation of metals. *Rep. Prog. Phys.*, **12**, 163–184.

**34** Dagata, J.a., Inoue, T., Itoh, J., Matsumoto, K., and Yokoyama, H. (1998) Role of space charge in scanned probe oxidation. *J. Appl. Phys.*, **84** (12), 6891.

**35** Morimoto, K., Pérez-Murano, F., and Dagata, J. (2000) Density variations in scanned probe oxidation. *Appl. Surf. Sci.*, **158**, 205–216.

**36** Dubois, E. and Bubendorff, J.-L. (2000) Kinetics of scanned probe oxidation: space-charge limited growth. *J. Appl. Phys.*, **87** (11), 8148.

**37** Lorenzoni, M. and Torre, B. (2013) Scanning probe oxidation of SiC, fabrication possibilities and kinetics considerations. *Appl. Phys. Lett.*, **103** (16), 163109.

**38** Buyukkose, S., Okur, S., and Aygun, G. (2009) Local oxidation nanolithography on Hf thin films using atomic force microscopy (AFM). *J. Phys. D: Appl. Phys.*, **42** (10), 105302.

**39** Kim, Y., Kang, S.K., Choi, I., Lee, J., and Yi, J. (2005) In situ negative patterning of P-silicon via scanning probe lithography in HF/EtOH liquid bridges. *J. Am. Chem. Soc.*, **127** (26), 9380–9381.

**40** Ruskell, T.G., Pyle, J.L., Workman, R.K., Yao, X., and Sarid, D. (1996) Current-dependent silicon oxide growth during scanned probe lithography. *Electron. Lett*, **32** (15), 1411.

**41** Kuramochi, H., Ando, K., Shikakura, Y., Yasutake, M., Tokizaki, T., and Yokoyama, H. (2004) Nano-oxidation and in situ faradaic current detection using dynamic carbon nanotube probes. *Nanotechnology*, **15** (9), 1126–1130.

**42** Dagata, J.a., Perez-Murano, F., Martin, C., Kuramochi, H., and Yokoyama, H. (2004) Current, charge, and capacitance during scanning probe oxidation of silicon. I. Maximum charge density and lateral diffusion. *J. Appl. Phys.*, **96** (4), 2386.

**43** Avouris, P., Hertel, T., and Martel, R. (1997) Atomic force microscope tip-induced local oxidation of silicon: kinetics, mechanism, and nanofabrication. *Appl. Phys. Lett.*, **71** (April), 285–287.

**44** Pérez-Murano, F., Martín, C., Barniol, N., Kuramochi, H., Yokoyama, H., and Dagata, J.a. (2003) Measuring electrical current during scanning probe oxidation. *Appl. Phys. Lett.*, **82** (18), 3086–3088.

**45** Kim, T.Y., Ricci, D., Zitti, E.D., and Cincotti, S. (2007) A study of the transient current during the formation of titanium oxide nanodots by AFM anodic oxidation. *Surf. Sci.*, **601** (21), 4910–4914.

**46** Dagata, J.A., Perez-Murano, F., Martin, C., Kuramochi, H., and Yokoyama, H. (2004) Current, charge, and capacitance during scanning probe oxidation of silicon. II. Electrostatic and meniscus forces acting on cantilever bending. *J. Appl. Phys.*, **96** (4), 2393.

**47** Fernandez-Cuesta, I., Borrisé, X., and Pérez-Murano, F. (2005) Atomic force microscopy local oxidation of silicon nitride thin films for mask fabrication. *Nanotechnology*, **16** (11), 2731–2737.

**48** Martín, C., Rius, G., Borrisé, X., and Pérez-Murano, F. (2005) Nanolithography on thin layers of PMMA using atomic force microscopy. *Nanotechnology*, **16** (8), 1016–1022.

49 Kuramochi, H., Pérez-Murano, F., Dagata, J.a., and Yokoyama, H. (2004) Faradaic current detection during anodic oxidation of the H-passivated P-Si(001) surface with controlled relative humidity. *Nanotechnology*, **15** (3), 297–302.

50 Kuramochi, H., Ando, K., Tokizaki, T., and Yokoyama, H. (2004) In situ detection of faradaic current in probe oxidation using a dynamic force microscope. *Appl. Phys. Lett.*, **84** (20), 4005.

51 Kuramochi, H., Tokizaki, T., Yokoyama, H., and Dagata, J.a. (2007) Why nano-oxidation with carbon nanotube probes is so stable: I. Linkage between hydrophobicity and stability. *Nanotechnology*, **18** (13), 135703.

52 Pellegrino, L., Bellingeri, E., Siri, a.S., and Marré, D. (2005) Current-controlled lithography on conducting SrTiO[sub 3−$\delta$] thin films by atomic force microscopy. *Appl. Phys. Lett.*, **87** (6), 064102.

53 Kaestner, M. and Rangelow, I.W. (2011) Scanning proximal probe lithography for Sub-10 Nm resolution on calix[4]resorcinarene. *J. Vac. Sci. Technol., B*, **29** (6), 06FD02.

# 10

# Combination of Semiconductor Parameter Analyzer and Conductive Atomic Force Microscope for Advanced Nanoelectronic Characterization

*Vanessa Iglesias*[1], *Xu Jing*[2], *and Mario Lanza*[2]

[1] *International Iberian Nanotechnology Laboratory, Av. Mestre José Veiga, 4715-330 Braga, Portugal*
[2] *Soochow University, Collaborative Innovation Center of Suzhou Nanoscience and Technology, Institute of Functional Nano & Soft Materials (FUNSOM), 199 Ren-Ai Road, Suzhou 215123, China*

## 10.1 Introduction

Silicon-based integrated circuits (ICs) are the backbone of today's semiconductor world, the metal oxide semiconductor field-effect transistor (MOSFET) being its main [1] and more common device. Since the 1960s the number of transistors per unit area has been doubling every 1.5 years [2], leading to an important shrinking in the transistor size, as well as a reduction in the electronic device dimensions and speed increment. However, this continuous miniaturization has been slowing down since a physical limit has been reached. The thickness reduction of $SiO_2$ dielectric layer, which currently shows sizes at nanometric scale, is being affected by quantum phenomena (i.e., tunneling current) that could threaten the proper functioning of the entire device [3]. In addition, different aging mechanisms can be triggered due to the persistent increase of the electric fields in nanoscaled devices. Among them, channel hot carrier (CHC) degradation and bias temperature instabilities (BTI), or time-dependent dielectric breakdown (TDDB), have been identified as the most relevant degradation mechanisms in MOSFETs [4]. Despite this accelerated degradation, the expected lifetime of metal oxide semiconductor (MOS) devices under normal operating conditions could be several years. Therefore, the time required to deteriorate the gate oxide under these conditions is too large to be investigated in feasible time scales. For this reason, accelerated electrical tests are commonly used to degrade materials, which are placed in test structures (generally MOS capacitors or transistors) [3, 5]. The aim of this procedure is to emulate the electrical conditions to which the devices are subjected in the real ICs, but increasing drastically the factors that could activate the different degradation mechanisms. This is an important issue to extrapolate the expected lifetime in a reasonable time as well as to study the device reliability. There are different approaches to perform these tests, which usually consist of the application of voltages or injection of currents through the structures and the

*Conductive Atomic Force Microscopy: Applications in Nanomaterials*, First Edition.
Edited by Mario Lanza.

measurement of the evolution of their electrical properties. The most common tests used in dielectric yield and reliability assessment are

- *Constant voltage stress* (CVS). It consists of the application of a constant voltage to the stack and the measurement of the gate current over time until the dielectric breakdown (BD) is triggered. In this test, BD is detected as an increase of the leakage current through the gate. This type of stress is used to analyze the TDDB by measuring the time-to-breakdown ($t_{BD}$).
- *Constant current stress* (CCS). In this case, a constant current is injected until BD. The voltage evolution over time is monitored through a voltage versus time curve. BD is detected when a sudden drop in voltage occurs. This stress is used to measure the injected charge to BD and, in some cases, $t_{BD}$, as well.
- *Ramped voltage stress* (RVS). A continuous increasing voltage is applied to the structure until BD. During this test, gate current is measured and plotted in current–voltage ($I$–$V$) characteristics. BD is detected as a sudden increase at the gate current.

More information regarding these types of stresses can be found in Ref. [5]. The aforementioned analyses are most of the time applied to structures with a gate electrode (typically of hundreds of micrometer square), using a semiconductor parameter analyzer (SPA) and a probe station. The sample is located and fixed on the metallic chuck using a vacuum system, and four micromanipulators are used to position macroscopic tips at the different terminals of every device. The macroscopic tip of the probe station has a typical radius at the apex of several micrometers and it is placed on the metallic contact. Ultralow noise triaxial cables are used to connect the tips to the source-monitor units (SMUs) of the SPA. The SMU can precisely force the voltage or current signal, and simultaneously measure the other magnitude (current or voltage). Moreover, the SMUs are designed with low-noise logarithmic amplifiers that permit measuring current down to some femtoamperes, almost three orders of magnitude below the noise level of the conductive atomic force microscope (CAFM). The whole setup is introduced in a Faraday cage to protect the signals collected from electromagnetic interferences. A picture of the setup is shown in Figure 10.1.

The information obtained at the device level is spatially averaged since the whole gate area is affected. Nevertheless, local information at the nanometric scale is missing. Considering that most of the problems associated with the materials (and also some of the failure mechanisms that affect the gate oxide) have their origin at nanoscale, this lack of information leads to a partial understanding. Therefore, in order to provide accurate information, it seems reasonable to study them also at the nanoscale [6]. To this purpose, different tools, such as scanning probe microscopy (SPM)-based techniques, have been used to perform such nanoscale analyses. Among them, especially when studying the electrical properties, the CAFM takes up a stand-out position.

CAFM is a very effective tool that allows performing analyses in small localized areas through the measurement of current maps, obtained simultaneously with topographical images, or $I$–$V$ curves at fixed locations of the sample [7, 8]. Current maps allow analyzing the conductivity changes and homogeneity on a

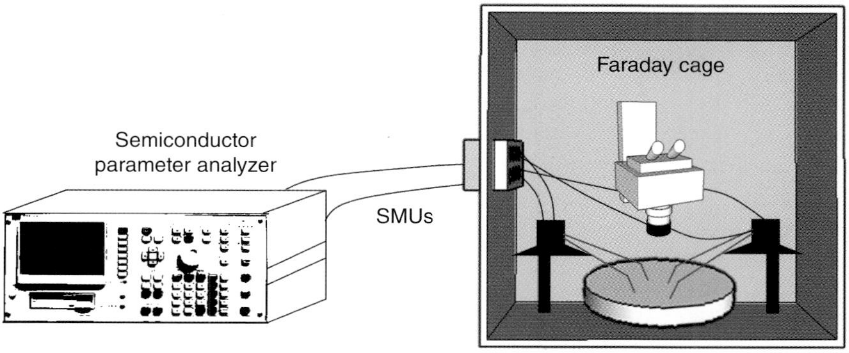


**Figure 10.1** Schematic of the probe station setup used for standard device level characterization.

selected area. During a current map, the CAFM tip scans the surface of the sample under a constant potential difference. *I–V* curves allow analyzing the different conduction mechanisms that can appear at different voltage ranges. In this second case, the test is performed by applying an RVS while the tip remains in a fixed position. Usually, CAFM experiments are carried out directly on a bare oxide, that is, no gate electrode on the structure. This configuration could be understood as a nano-MOS structure, where the tip plays the role of the gate electrode. Hence, a *nano-capacitor* with an area corresponding to the contact area between tip and sample ($\sim$10–100 nm$^2$) is defined [9, 10].

The combination of both device level and nanoscale characterization to analyze the reliability of thin dielectrics has shown to be very useful for a better understanding of the processes that affect the device [11, 12]. Nevertheless, a priori, this combination is not an easy task. The reason is that after the device is stressed in the probe station, the electrode covering the dielectric needs to be selectively etched in order to expose the gate oxide, so that it can be scanned with the CAFM tip. Furthermore, depending on the CAFM characteristics, localization of specific areas could be complicated and take a long time. To minimize some of these experimental difficulties that could appear, the structures under study should allow an easy gate electrode removal and also have the proper dimensions for the scanning of the whole gate area of the device [11].

## 10.2 Combination of SPA and CAFM for Local Channel Hot Carrier Degradation Analysis

The combination of SPA and CAFM can be very advantageous because, on one hand, the SPA can induce a homogeneous stress over the sample (imitating the operation conditions), and, on the other, the CAFM can help detect the local effects of such a stress. These capabilities would allow conducing advanced studies about many device phenomena, such as, for example, CHC degradation in MOSFET transistors.

CHC is a reliability issue affecting MOSFET transistors. When a voltage above the threshold is applied to the gate of a MOSFET and at the same time a potential difference is applied between the drain and source, electrical current between these last two terminals can flow [1]. Basically the gate potential controls the formation of a conductive channel between source and drain, along which charge carriers can flow. Under normal conditions, all charge carriers should follow this direction but it is known that, due to inhomogeneities in the electrical field along the channel, some electrons and holes may acquire a very high kinetic energy, which can make them deviate from this path. In many cases, these highly energetic (or hot) charge carriers can flow through the gate oxide, arriving at the gate electrode [13]. This oxide penetration, which is usually precluded by impact ionization, can degrade the reliability of the oxide, enhancing the leakage current and contributing to the failure of the whole device. Therefore, studying the impact of CHC degradation is of utmost importance for MOSFET technology. Several device-level analyses have corroborated that CHC within the channel appears in a very inhomogeneous way [14], but the real visualization of leakage spots and their single and statistical analysis is something more challenging. In this context, the CAFM is an ideal tool to provide this information.

A good example of SPA + CAFM combination to study CHC degradation was reported in 2014 by Bayerl *et al.* [11]. He analyzed pMOSFET transistors with 1.4-nm SiON gate dielectric and a polysilicion/NiSi gate electrode (40- and 60-nm thick, respectively). The size of the channels was $L = 1$ μm and $W = 0.5$ μm. First of all, the transistors were stressed at the device level with the SPA by applying −2.6 V to the drain and gate during 200 s, while source and bulk were grounded. After that, the transistors were characterized by measuring their typical characteristics, namely, $I_D - V_D$, $I_D - V_G$, and $I_G - V_G$. Unstressed pMOSFET transistors were also characterized, and they were considered as reference. The plots obtained (see Figure 10.2) corroborate the damage introduced with the SPA during the 200 s stress. This can be observed from the clear reduction of $I_D$ versus $V_D$ (Figure 10.2a), larger $V_G$ required to achieve the same $I_D$ (Figure 10.2b), and larger $I_G$ at similar $V_G$ of stressed transistors (Figure 10.2c).

After the device-level degradation, the interesting point is to use the CAFM to monitor the number of weak spots induced in the oxide. To do so, the

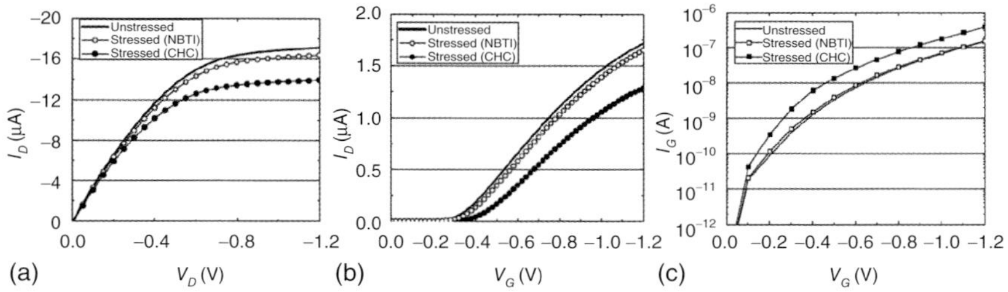


**Figure 10.2** Typical (a) $I_D - V_D$, (b) $I_D - V_G$, and (c) $I_G - V_G$ characteristics of a MOSFET recorded before (continuous line) and after NBTI (open symbols) and CHC (full symbols) stress. (d) Breakdown statistics from which $V_{BD}$ (63%) was estimated. (Reproduced with permission from [11]. Copyright IEEE 2014.)

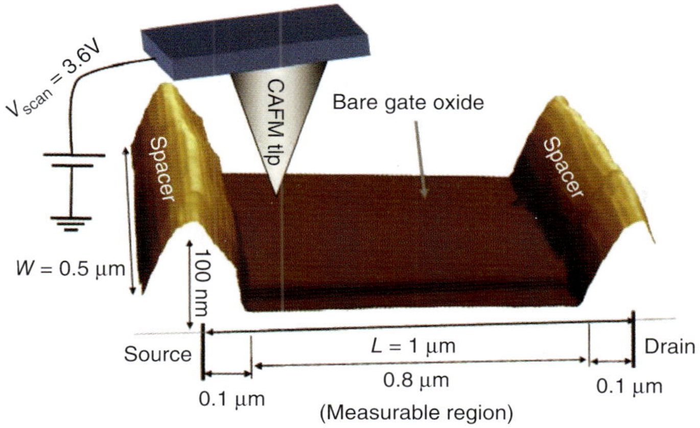


**Figure 10.3** AFM 3-D topographic image of a pMOSFET after the removal of the gate electrode, showing a very flat surface (RMS < 0.2 nm) between the two spacers. Placing the CAFM tip on the gate oxide region, the conduction through the oxide can be studied with nanometer lateral resolution. The dimensions of the area under study and the applied polarization voltage are indicated. (Reproduced with permission from [11]. Copyright IEEE 2014.)

polysilicon/NiSi electrode needs to be very carefully removed. Bayerl and coworkers used a highly selective wet etching [11]. After that, the surface of the gate oxide is exposed and the gate oxide reliability can be studied by means of CAFM (see Figure 10.3). It should be highlighted that this methodology could be also used to study thin-film transistors and single-back gate transistors.

It should be highlighted that finding the right transistors with the tip of the CAFM may be a challenging task for beginners, but with the right marks in the wafers experienced users should not have any problem. Figure 10.4 shows

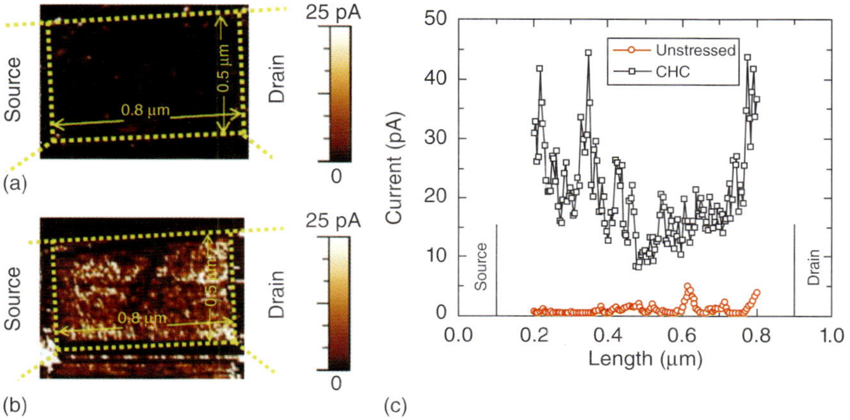


**Figure 10.4** Typical current images obtained at a gate voltage of 3.6 V in (a) nonstressed and (b) CHC-stressed MOSFETs. (c) Average current measured with the CAFM tip along the channel for the nonstressed (circles) and CHC (squares) stressed MOSFETs. (Modified and reprinted with permission from [11]. Copyright IEEE 2014.)

**Table 10.1** Average and dispersion of the current measured in $W \times 0.05\ \mu m^2$ regions taken in Figure 10.4b, close to the source, drain, and at the center of the channel.

| | Source | | Center | | Drain | |
|---|---|---|---|---|---|---|
| | $I$ (pA) | $\sigma_1$ | $I$ (pA) | $\sigma_1$ | $I$ (pA) | $\sigma_1$ |
| Reference | 0.4 | 0.1 | 1.1 | 0.2 | 1.1 | 0.9 |
| CHC | 24.9 | 38.9 | 13.6 | 18.4 | 25.4 | 75.6 |



the typical current maps obtained on the gate oxide when biasing the structure at 3.6 V. The dashed areas indicate the part that corresponds to the channel. While unstressed devices (Figure 10.4a) show almost no current spots, the devices subjected to the 200 s test (Figure 10.4b) show many conductive spots along the channel. Interestingly, it can be easily observed that the density of the spots is lower at the center and larger at the source and drain, meaning that the CHC degradation was more severe at these last two regions. The differences in the current levels can be better observed from the superimposed cross sections (Figure 10.4c). Table 10.1 further corroborates these differences by displaying the averaged current and dispersion at each region of the channel.

## 10.3 Combination of CAFM and SPA for Resistive Switching Analyses

Similar studies can be developed to analyze other device phenomena, such as resistive switching (RS) in nonvolatile memory devices. For a different device, the type of stresses with both SPA and CAFM will be different, but again the large resolution of the CAFM can be definitive to gain very valuable information about the samples and electronic phenomena.

### 10.3.1 Device-Level Stress with SPA Followed by CAFM Characterization

Resistive random access memories (RRAMs) are considered one of the most promising technologies for a variety of information storage applications, due to the easy structure and superior performance [15–18]. The RRAM unit cell is based on metal–insulator–metal (MIM) or metal–insulator–semiconductor (MIS) nanocomposites, and its working principle is based on tuning the electronic resistance of the insulator by applying electrical pulses. This phenomenon, called RS, allows differentiating two different logic states, the low resistive state (LRS) and the high resistive state (HRS), which allow simulating the ones and zeros of the binary code, making it possible to store information [19]. The LRS/HRS are characterized by the flow of large/low currents through the dielectric, namely, $I_{LRS}/I_{HRS}$. In some cases, an initial "forming process" is needed to get a stable switching between both states. This "forming process" is, in many cases, related to the formation of a conductive filament through the dielectric that afterwards can be partially destroyed (reset) and recovered (set) again. To get this initial process, a current limit needs to be usually applied; otherwise, the dramatic effects related to the irreversible dielectric breakdown (BD) could take place, including avalanche current, BD spot propagation, thermal heat, insulator contamination by metal migration, and dielectric-breakdown-induced epitaxy (DBIE) [20–23]. The RS phenomenon can be classified into two types: unipolar and bipolar. The RS is unipolar when the electrical pulses applied to induce the HRS-to-LRS (set) and LRS-to-HRS (reset) transitions are of the same polarity (does not matter that it is positive or negative). If the polarities of the electrical pulses needed to induce the set and reset processes are opposed, then it is bipolar. These two different types of RS can be easily distinguished by the shape of the $I$–$V$ curves [15].

The RS phenomenon has been known since the early 1960s [24], but it was not until the middle of the 2000s that it started to become popular and widely investigated to overcome several technical limits of memory technology. During the past decade, RS has been observed in a large variety of binary and ternary thin oxide films, including Cr-doped $SrZrO_3$, $SrTiO_3$, NiO, $TiO_2$, $Cu_2O$, and $HfO_2$-based devices (among others) [25–29]. Thanks to this capability, as well as to their compatibility with the current semiconductor industry, these materials have been proposed as the new dielectric film for RRAM. Nevertheless, the microscopic physical mechanisms driving the RS are still not fully understood. That is, among other reasons, because scaling is strongly linked to the question of whether the switching current is distributed homogeneously across the device area or localized to one or a few conducting filaments (CFs). In the most advanced RRAM devices, the RS seems to be a local phenomenon that takes place at very small areas [30], showing strong similarities with the well-known current-limited reversible dielectric BD. In this context, the CAFM has been shown to be a very powerful technique to clarify the nature and physics behind the RS phenomenon.

Multiple studies reported the use of CAFM to study RS with nanometric resolution [19]. One of the first works that used the CAFM to analyze the electronic properties of RRAM memories at the device level was reported by Son and Shin [31]. He used Hg nanoparticles on a NiO film and, after inducing the HRS and LRS, removed the particle and monitored the number of filaments and their properties, such as CF size and currents driven (Figure 10.5a). The CAFM was also used by Shubhakar *et al.* [32] to study the forming process of CFs from sequences of spectroscopic $I$–$V$ curves (Figure 10.5b). The locations of the samples promoting the formation of CFs can be clearly distinguished in CAFM current maps (Figure 10.5c), given the large lateral resolution of the tip [36]. The 3D shape of the filament can be assessed using tip-induced tomography after the SPA test [33] (see also Chapter 8). The RS phenomenon can also be induced with the tip of the CAFM. One method is to do it through current maps (Figure 10.5e) [34]. This allows the observation of many different CFs, allowing the statistical analysis of their properties (Figure 10.5f) [35]. Moreover, as this changes the properties of a considerable area, chemical tools can also be used to find out the changes in the material (Figure 10.5g) [34]. Another option is the application of sequences

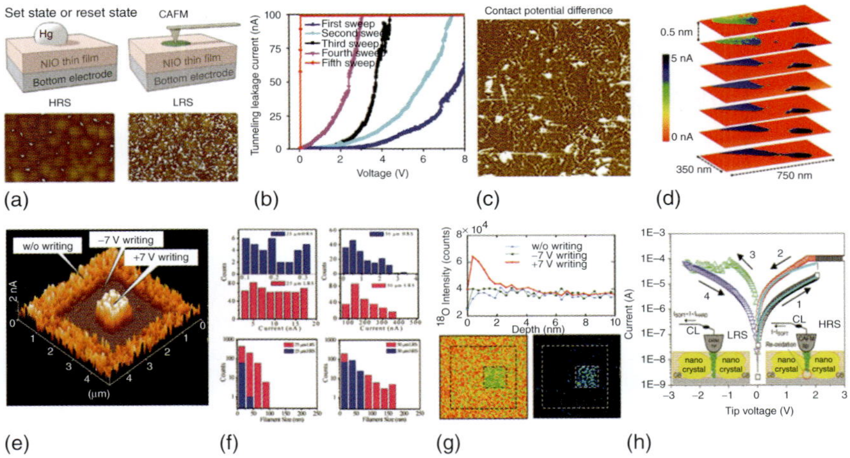


**Figure 10.5** (a) Current maps of Hg/NiO/Pt nanocapacitors that have been fixed at HRS and LRS (after the Hg electrode removal). (Reproduced with permission from [31]. Copyright 2008 AIP Publishing LLC.) (b) High BD hardness. (Reproduced with permission from [32]. Copyright IEEE 2012.) (c) Current map of a polycrystalline $HfO_2$ stack annealed at 100 °C showing leakage current only at the grain boundaries. (Modified and reproduced with permission from [36]. Copyright 2010 AIP Publishing LLC.) (d) Three-dimensional characterization of CFs in RRAM cells using CAFM tomography. (Reproduced with permission from [33]. Copyright IEEE 2013.) (e) Current image of Rh-tip/$Ni_{1+\delta}O$/Pt film after set and reset process at specific locations. (Reproduced with permission from [34]. Copyright 2008 AIP Publishing LLC.) (f) Statistical analysis of the current spots measured with CAFM in RRAM cells. (Reprinted from [35]. Copyright 2012 IOP Publishing.) (g) Time-of-flight secondary ion mass spectroscopy (TOF-SIMS) depth profile and maps of $^{18}O$ ions in areas in HRS and LRS. The ability of the CAFM to scan large areas through current maps allows chemical characterization with tools of lower spatial resolution. (Reproduced with permission from [34]. Copyright 2008 AIP Publishing LLC.) (h) Example of RS behavior observed only at grain boundaries of polycrystalline $HfO_2$ films. (Reproduced with permission from [12]. Copyright 2012 AIP Publishing LLC.)

of $I$–$V$ curves with enhanced current dynamic range (Figure 10.5h), which allows monitoring the set and reset processes [12]. More information about the use of RS can be found in Ref. [19].

As an example, here we describe the combination of a probe station and the CAFM tests applied to study the RS phenomena in MIM structures with polycrystalline $HfO_2$ layers as dielectric [37]. The samples analyzed consist of a 10.6-nm-thick Hf layer deposited by sputtering on a Pt/Ti/$SiO_2$/Si substrate and followed by a 600 °C annealing in $O_2$ ambient for 30 min to form an $HfO_2$ layer. After that, 100-nm-thick TiN top electrodes were deposited at room temperature. Devices were patterned by lithography technique to form isolated squared memory cells with sizes ranging from $2 \times 2\ \mu m^2$ to $100 \times 100\ \mu m^2$. A schematic of the structure is shown in Figure 10.6. Note that in this schematic dimensions are not scaled (to facilitate its understanding).

Initially, two different TiN/$HfO_2$/Pt MIM structures were stressed at device level by applying sequences of RVS with the SPA at the probe station (see Figure 10.1). As previously mentioned, a common feature of most switching materials is the necessity of an initial electroforming step, before stable RS can

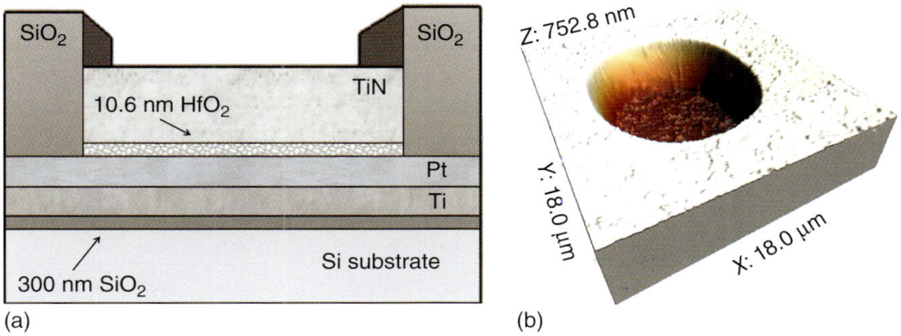


**Figure 10.6** (a) Schematic structure of the sample under study. MIM structures with an $HfO_2$ layer as dielectric. Areas ranging from $2 \times 2\ \mu m^2$ to $100 \times 100\ \mu m^2$. (b) Topographic map collected with the CAFM on the capacitor displayed in (a).

be achieved. In this case, an external voltage sweep from 0 up to 20 V (positive forming) was applied for the electroforming even as a current limitation of 100 $\mu$A was forced. Figure 10.7a shows a typical forming $I$–$V$ curve. The current gradually increases until BD takes place, which is observed as a sudden jump up to higher currents. In this case it happens around 7 V. Note that the maximum current for voltages larger than 7 V corresponds to the current compliance, set by the user. Once the controlled BD takes place, the device current increases by several orders of magnitude remaining in an LRS state (Figure 10.7b). After that, when an opposite voltage sweep (negative) is applied (Figure 10.7c), the device partially recovers its resistance (in the present case at −2.5 V), achieving an HRS state where it will remain (Figure 10.7d) until a new positive ramped voltage is applied. A complete set/reset cycle is represented in Figure 10.7e,f. After five cycles, one of the structures was kept at the HRS (Figure 10.7e, circles) and the other at the LRS (Figure 10.7f, triangles). The repeatable switching between the LRS and HRS when successive bipolar ramped voltage cycles are applied (not shown) indicates the observation of the RS phenomenon.

Once both structures were set to HRS and LRS, the top electrode (TiN layer) of the capacitors was scanned with the tip of the CAFM. Ideally, the top electrode should be removed to achieve a better resolution, but in this preliminary study the stack was kept. Obviously, a point-to-point correlation between the topography of the $HfO_2$ layer and its electrical properties cannot be performed and information about the origin of RS cannot be extracted with this procedure. However, a preliminary nanoscale analysis can also be performed, as already demonstrated in [38]. Figure 10.8 shows the current maps measured at the HRS (a) and LRS (b) capacitors when applying a voltage of 1.5 V between the top and bottom electrodes, which simulates a read process. Note that when a current scale of 700 pA was used, multiple locations drive current in both cases. However, a uniform current distribution is not observed, demonstrating that the current mainly flows through CFs. When the scale is increased up to 33 nA, no conductive spots were observed in the HRS structure (not shown). On the contrary, in the LRS capacitor, even at this scale (shown in the inset), one CF with current larger than the rest was observed. These results suggest that although the current in the HRS

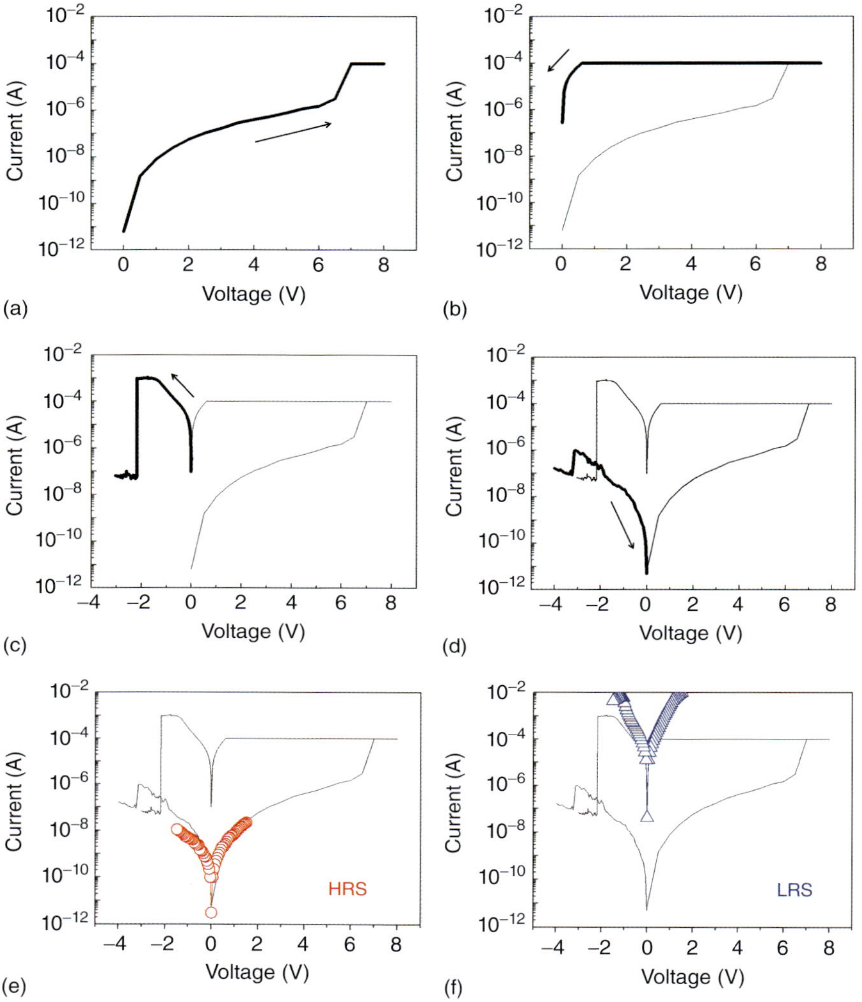


**Figure 10.7** Sequences of *I–V* curves (a–f) collected in a TiN/$HfO_2$/Pt device. The forming process and the currents before set and reset in HRS and LRS can be clearly observed.

can flow through many nanosized partially formed conductive filaments of similar properties (probably related to soft BD events), the high currents measured in the LRS is mainly driven by only one completely formed CF (hard BD, which is a unique event [39]) with larger conductivity.

These CAFM results agree with the previous probe station observations of area dependence only on HRS, as shown in Figure 10.9. In LRS, although many conductive filaments could also be formed (as for the HRS), the main part of the current flows through a dominant filament, masking any area dependence and current flowing through the rest of the partially formed CF.

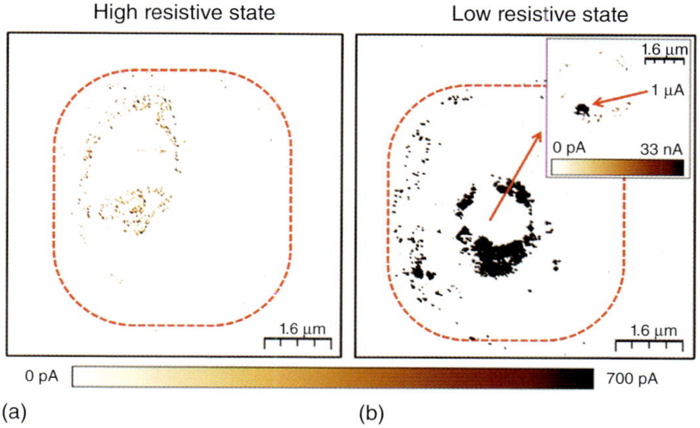


**Figure 10.8** Current images of (a) HRS and (b) LRS (with different scales) when scanning the capacitor with the tip of the CAFM by applying 1.5 V. The red dotted line outlines the capacitor. (Iglesias *et al.* 2012 [37]. Reproduced with permission of Elsevier.)

**Figure 10.9** Resistance measured in HRS and LRS for devices with different areas. (Iglesias *et al.* 2012 [37]. Reproduced with permission of Elsevier.)

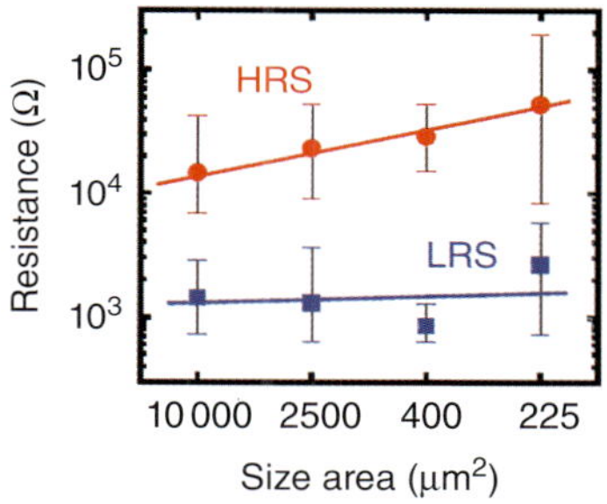


### 10.3.2 Direct Connection of SPA to the CAFM

Previous topography images do not reflect the morphological properties of the high-*k* dielectric, but of the TiN electrode. To determine which features in the insulator stack promote the formation of reversible CFs, blanket $HfO_2$ layers (before top TiN electrode deposition) were analyzed at nanoscale with the CAFM by measuring sequences of $I$–$V$ curves at different locations of the sample. Note that a standard CAFM current window is typically of three orders of magnitude (from pA to nA) and that would not be enough to measure the complete $I$–$V$ characteristics (i.e., observe the set and reset processes). To overcome these limitations and to be able to observe the set/reset transitions, a modification was carried out on a standard CAFM. In order to expand the dynamic range for current measurement of a standard CAFM from 1 pA to 1 mA, the linear $I$–$V$ converter of the AFM was substituted by a SPA [40, 41]. Since this equipment is based on SMUs, not only currents (when a voltage is applied) but also voltages (when a current is forced) can be measured. The use of SMUs leads to an improvement in the electrical performance of the CAFM and to an increase in the current dynamic range (from noise level up to milliamperes) while keeping the nanometer resolution. Moreover, the use of this setup allows forcing a current limitation, which is

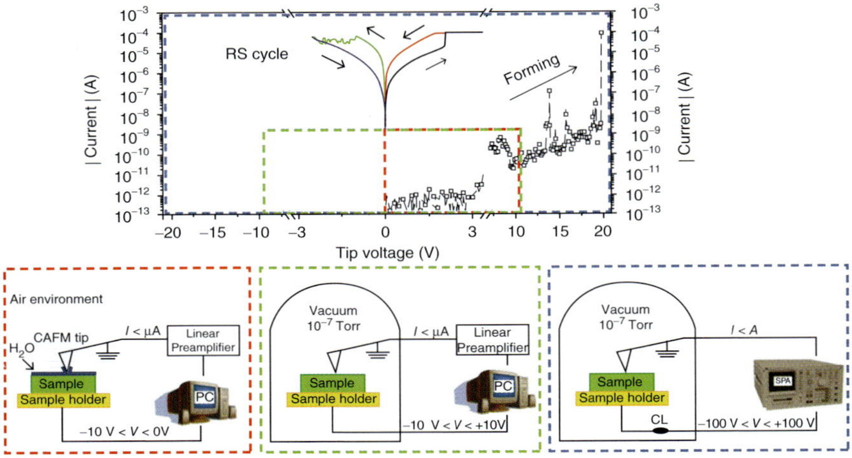


**Figure 10.10** Voltage/current ranges that can be applied/measured with three different CAFM setups: (i) a standard setup (red dashed schematic), (ii) a standard setup working in vacuum, and (iii) CAFM working in vacuum and connected to an SPA. The part of the graph that each setup can measure has been highlighted with dashed lines of different colors. (Modified and reprinted from [42]. Copyright ECS, 2014.)

necessary for limiting the damage during the forming/set processes. It should be also highlighted that if bipolar RS needs to be measured, an environmental chamber is necessary to avoid the apparition of local anodic oxidation when electrons are injected from the tip [42] (see Chapter 9). A schematic of the connection between the CAFM and SPA is displayed in Figure 10.10.

In spite of the improvement in the electrical measurement by using the CAFM + SPA setup, the fact that AFM controls by its side the scanning parameters and SPA controls the sample biasing and electrical measurement implies some drawbacks. The use of two different instruments to control the tip position and to measure the electrical variables forces a synchronization between both to correlate topography and electrical measurements. As a consequence, the measurement of current maps is relatively slow, which is a handicap for obtaining reliable current maps due to the drifts of the piezos and the wear out of the tip. Therefore, using this setup, we only collected spectroscopic $I$–$V$ curves (for more information about wide current range maps, see Chapters 10 and 11).

Sequences of RVS were applied by using the CAFM tip at 24 different locations of the $HfO_2$ layer [12]. The positions were previously determined through topographical maps. To enable a bipolar RS, the CAFM measurements were performed in a $N_2$ environment to avoid anodic oxidation when the electrons were injected from the tip (see Chapter 9). Since high current densities are measured during the electrical measurements and CAFM works in contact mode, conductive (B-doped) diamond-coated silicon tips were used to improve their resistance to wear out. Two types of $I$–$V$ curves were observed (see representative examples in Figure 10.11): most of the curves (21 of 24 locations) show forming of voltages above 11 V, while the others (at only 3 of 24 locations) exhibit a lower BD voltage ($V_{BD} \sim 4$ V), which is comparable to that measured at the device level. After

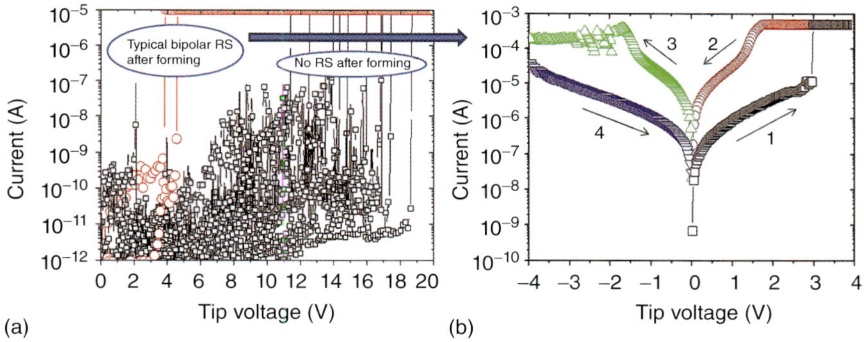


**Figure 10.11** (a) Nanoscale forming processes at 12 different positions of the $HfO_2$ layer of sample A. For clarity in this figure we only show 12 of the 24 locations analyzed. (b) Typical bipolar RS behavior obtained only at those locations where the $V_{BD}$ during the forming process was remarkable lower (red circles in panel a). (Reproduced with permission from [43]. Copyright American Institute of Physics 2012.)

the CF formation, an additional set and reset cycle was performed to determine the electrical properties of each of the analyzed locations. Only those sites which exhibit lower forming voltages (see Figure 10.11a) demonstrate a typical bipolar RS (Figure 10.11b) similar to that in the MIM structures (see Figure 10.6a), with comparable magnitudes of the LRS and HRS currents. The sites with larger forming voltages do not show the RS effect, meaning that an irreversible BD is caused by the forming process. Therefore, from these results it was possible to elucidate that the RS observed at device level is not only controlled by a very reduced area, as demonstrated in the previous section, but also that the RS phenomenon takes place preferentially at those positions in the device with a weaker *dielectric strength*. Since it is widely known that in polycrystalline $HfO_2$ gate dielectrics the leakiest positions correspond to GBs (see Chapter 5), the results in Figure 10.11 lead to the conclusion that RS phenomenon is preferentially registered at these positions.

The conclusions extracted from Figures 10.8 and 10.11 have been only possible thanks to the genuine combination of the CAFM + SPA. The only difficulty is that the cost of the SPA is high and its mobility is reduced (compared to standard sourcemeters). In this sense, we would like to highlight that the use of CAFM + sourcemeter is also valid. For example, this setup has been used to measure sequences of $I$–$V$ curves on $HfO_2$, and currents between 10 pA and 1 mA have been observed [44]. Hui *et al.* [45] used this setup to collect current versus time ($I$–$t$) curves on Cu foils, extending the range of experiments that can be performed.

## 10.4 Conclusions

In conclusion, CAFM studies can be complemented with the SPA to perform an advanced characterization of materials and devices. One option is to perform a device-level test using the SPA and later conduct CAFM tests to monitor the effect of the SPA test. In this chapter, we have explained how to study both CHC and RS following this procedure. This methodology brings the associated problem of having to remove the metallic electrode (which is necessary for the SPA tests). After a careful electrode removal, the surface of the underlying dielectric is exposed, and it can be reliably scanned with the tip of the CAFM. Another possibility is to combine simultaneously the electronic capabilities of the SPA and the lateral resolution of the CAFM technique. To do so, the CAFM tip can be directly connected to the SPA. This methodology allows applying/measuring large voltages and currents, as well as using current limitation. The only limitation of this methodology is that the synchronization of the tip movement and the SPA is complex, and for this reason it is only recommendable for spectroscopic measurements. In Chapter 11, a solution to this limitation is presented. Nevertheless, the combination of SPA + CAFM simultaneously allowed, for the first time, the visualization of set and reset processes at a single CF.

## References


1 Edgar, L. J. (1925) Method and apparatus for controlling electric currents. US Patent 1, 745, 175, October 22.

2 Moore, G.E. (1965) Cramming more components onto integrated circuits. *Proc. IEEE*, **86** (1), 82–85.

3 Stathis, J. H. and DiMaria, D. J. (1998) Reliability projection for ultra-thin oxides at low voltage. IEEE International Electron Devices Meeting, pp. 167–170.

4 Huard, V., Parthasarathy, C. R., Bravaix, A., Guerin, C., and Pion, E. (2009) CMOS device design-in reliability approach in advanced nodes in Proceedings of IEEE International Reliability Physics Symposium (IRPS), pp. 624–633.

5 Rodríguez, R. (2000) Analisis de la degradación y ruptura dieléctrica de capas finas de $SiO_2$ en dispositivos MOS sometidos a estresses estáticos y dinámicos. PhD thesis. Electronic Engineering Department of the Autonomous University of Barcelona.

6 Lanza, M. (2010) CAFM Nanoscale electrical characterization of gate stacks for advanced MOS devices. PhD thesis. Electronic Engineering Department of the Autonomous University of Barcelona.

7 Murrell, M.P., Welland, M.E., O'Shea, S.J., Wong, T.M.H., Barnes, J.R., and McKinnon, I.A.W. (1993) Spatially resolved electrical measurements of $SiO_2$ gate oxides using Atomic Force Microscope. *Appl. Phys. Lett.*, **62** (7), 786–788.

8 O'Shea, S.J., Atta, R.M., Murrell, M.P., and Welland, I.M.E. (1995) Conducting atomic-force microscopy study of silicon dioxide breakdown. *J. Vac. Sci. Technol., B*, **13** (5), 1945–1952.

9 Frammelsberger, W., Benstetter, G., Kiely, J., and Stamp, R. (2006) Thickness determination of thin and ultra-thin $SiO_2$ films by CAFM IV-spectroscopy. *Appl. Surf. Sci.*, **252** (6), 2375–2388.

10 Frammelsberger, W., Benstetter, G., Kiely, J., and Stamp, R. (2007) CAFM-based thickness determination of thin and ultra-thin $SiO_2$ films by use of different conductive-coated probe tips. *Appl. Surf. Sci.*, **253** (7), 3615–3626.

11 Bayerl, A., Wu, Q., Porti, M., Martin-Martínez, J., Lanza, M., Rodriguez, R., Velayudhan, V., Nafria, M., Aymerich, X., Gonzalez, M.B., and Simoen, E. (2014) A Conductive AFM nanoscale analysis of NBTI and Channel hot-carriers degradation in MOSFETs. *IEEE Trans. Electron Devices*, **61** (9), 3118–3124.

12 Lanza, M., Bersuker, G., Porti, M., Miranda, E., Nafría, M., and Aymerich, X. (2012) Resistive switching in hafnium dioxide layers: Local phenomenon at grain boundaries. *Appl. Phys. Lett.*, **101**, 193502.

13 grasser, T., Tang, T.W., Kosina, H., and Selberherr, S. (2003) A review of hydrodynamic and energy-transport models for semiconductor device simulation. *IEEE*, **91** (2), 251–274.

14 Amat, E. *et al.* (2009) Competing degradation mechanisms in short-channel transistors under channel hot-carrier stress at elevated temperatures. *IEEE Trans. Device Mater. Reliab.*, **9** (3), 454–458.

15 Waser, R., Dittmann, R., Staikov, G., and Szot, K. (2009) *Adv. Mater.*, **21**, 2632–2663.

16 Yu, S., Chen, H.- Y., Gao, B., Kang, J., and Wong, P.H.- S. (2013) *ACS Nano*, 7, 2320–2325.

17 Govoreanu, B., Kar, G. S., Chen, Y. Y., Paraschiv, V., Fantini, A., Radu, I. P., Goux, L., Clima, S., Degraeve, R., Jossart, N., Richard, O. (2011) $10 \times 10\,nm^2$ Hf/HfOx crossbar resistive RAM with excellent performance, reliability and low-energy operation IEEE International Electron Devices Meeting, Washington, DC, December 5–7, 2011.

18 Liu, Q., Long, S.B., Wang, W., Zuo, Q.Y., Zhang, S., Chen, J.N., and Liu, M. (2009) *IEEE Electron Device Lett.*, **30**, 1335.

19 Mario Lanza (2014) A review on resistive switching in high-k dielectrics: a nanoscale point of view using conductive atomic force microscope. *Materials*, 7, 2155–2182.

20 Obreja, V.V.N., Codreanu, C., Poenar, D., and Buiu, O. (2011) *Microelectron. Reliab.*, **51**, 536.

21 Lanza, M., Porti, M., Nafría, M., Aymerich, X., Whittaker, E., and Hamilton, B. (2010) *Microelectron. Reliab.*, **50**, 1312.

22 Seo, S.H., Hwang, J.S., Yang, J.M., Hwang, W.J., Song, J.Y., and Lee, W.J. (2013) *Thin Solid Films*, **546**, 14.

23 Raghavan, N., Pey, K.L., Shubhakar, K., and Bosman, M. (2011) *IEEE Electron Device Lett.*, **32**, 78–80.

24 Hickmott, T.W. (1962) *J. Appl. Phys.* USA,, **33**, 2669–2682.

25 Lin, C.C., Lin, C.Y., Lin, M.H., Lin, C.H., and Tseng, T.Y. (2007) Voltage-polarity-independent and high-speed resistive switching properties of v-doped $SrZrO_3$ thin films. *IEEE Electron Devices Soc.*, **53** (12), 3146–3151.

26 Szot, K., Dittmann, R., Speier, W., and Waser, R. (2007) Nanoscale resistive switching in SrTiO3 thin Flim. *Phys. Status Solidi RRL*, **1** (2), R86–R87.

27 Russo, U., Ielmini, D., Cagli, C., and Lacaita, A.L. (2009) Filament conduction and reset mechanism in NiO-based resistive-switching memory (RRAM) devices. *IEEE Trans. Electron Devices*, **56** (2), 186–192.
28 Huang, J.J., Kuo, C.W., Chang, W.C., and Hou, T.H. (2010) Transition of stable rectification to resistive-switching in Ti/$TiO_2$/Pt oxide diode. *Appl. Phys. Lett.*, **96**, 262901.
29 Chen, A. and Lin, M. R. (2011) Variability of resistive switching memories and its impact on crossbar array performance. Reliability Physics Symposium (IRPS), 2011 IEEE International, Monterey, CA.
30 Govoreanu, B., Kar, G. S., Chen, Y. Y., Paraschiv, V., Fantini, A., Radu, I. P., Goux, L., Clima, S., Degraeve, R., Jossart, N., Richard, O., and Kubicek, S. (2011) 10 nm × 10 nm Hf/HfOx crossbar resistive RAM with excellent performance, reliability and low-energy operation, in Proceedings of the IEEE International Electron Devices Meeting, Washington, DC, December 2011, pp. 31.6.1–31.6.4
31 Son, J.Y. and Shin, Y.H. (2008) Direct observation of conducting filaments on resistive switching of NiO thin films. *Appl. Phys. Lett.*, **92**, 222106. 10.1063/1.2931087.
32 Shubhakar, K., Pey, K. L., Bosman, M., and Thamankar, R. (2012) Nanoscale physical analysis of localized breakdown events in HfO2/SiOX dielectric stacks: a correlation study of STM induced BD with C-AFM and TEM, 19th IEEE International Symposium, 1–7.
33 Celano, U., Goux, L., Belmonte, A., Schulze, A., Opsomer, K., Detavernier, C., Richard, O., Bender, H., Jurczak, M., and Vandervorst, W. (2013) Conductive-AFM tomography for 3D filament observation in resistive switching devices. IEEE International Electron Device Meeting, Washington, December 10, 2013.
34 Yoshida, C., Kentaro, K., Takahiro, Y., and Yoshihiro, S. (2008) Direct observation of oxygen movement during resistance switching in NiO/Pt film. *Appl. Phys. Lett.*, **93**, 042106. 10.1063/1.2966141.
35 Singh, B., Mehta, B.R., Varandani, D., Savu, A.V., and Brugger, J. (2012) CAFM investigations of filamentary conduction in $Cu_2O$ ReRAM devices fabricated using stencil lithography technique. *Nanotechnology*, **23**, 495707. 10.1088/0957-4484/23/49/495707.
36 Iglesias, V., Porti, M., Nafría, M., Aymerich, X., Dudek, P., Schroeder, T., and Bersuker, G. (2010) Correlation between the nanoscale electrical and morphological properties of crystallized hafnium oxide-based metal oxide semiconductor structures. *Appl. Phys. Lett.*, **97**, 262906. 10.1063/1.3533257.
37 Iglesias, V., Lanza, M., Porti, M., Nafría, M., Aymerich, X., Liu, L.F., Kang, J.F., Bersuker, G., Zhang, K., and Shen, Z.Y. (2012) Nanoscale observations of Resistive Switching High and Low conductivity states on TiN/HfO2/Pt structures. *Microelectron. Reliab.*, **52**, 2110–2114.
38 Kim, D.-I., Yoon, J., Park, J.-B., Hwang, H., Kim, Y.-M., Kwon, S.-H., and Kim, H.-K. (2011) Nonlinear current-voltage behavior of the isolated resistive switching filamentary channels in CuC nanolayer. *Appl. Phys. Lett.*, **98**, 152107.

39 Suñé, J., Placencia, I., Barniol, N., Farrés, E., Martín, F., and Aymerich, X. (1990) On the breakdown statistics of very thin $SiO_2$ films. *Thin Solid Films*, **185** (2), 347–362.

40 Blasco, X., Nafria, M., and Aymerich, X. (2005) Electrical characteristic measurement instrument has atomic force microscope with conductive tip, comprising scanner, sample support, voltage generator and current generator. Patent no. ES2224890-A1; ES2224890-B1; WO2005119205-A1.

41 Aguilera, L., Lanza, M., Porti, M., Grifoll, J., Nafría, M., and Aymerich, X. (2008) Improving the electrical performance of a conductive atomic force microscope with a logarithmic current-to-voltage converter. *Rev. Sci. Instrum.*, **79** Art. no.073701, 073701.

42 Shi, Y., Ji, Y., Hui, F., Iglesias, V., Porti, M., Nafria, M., Miranda, E., Bersuker, G., and Lanza, M. (2014) Elucidating the origin of resistive switching in ultra-thin hafnium oxides through high spatial resolution tools. *ECS Trans.*, **64** (14), 19–28.

43 Lanza, M., Zhang, K., Porti, M., Nafria, M., Shen, Z.Y., Liu, L.F., Kang, J.F., Gilmer, D., and Bersuker, G. (2012) Grain boundaries as preferential sites for resistive switching in the HfO2 resistive random access memory structures. *Appl. Phys. Lett.*, **100**, 123508.

44 Lanza, M., Bayerl, A., Gao, T., Porti, M., Nafria, M., Jing, G., Zhang, Y., Liu, Z., and Duan, H. (2013) Graphene-coated Atomic Force Microscope tips for reliable nanoscale electrical characterization. *Adv. Mater.*, **25**, 1440–1444.

45 Hui, F., Vajha, P., Shi, Y., Ji, Y., Duan, H., Padovani, A., Larcher, L., Li, X.-R., Xu, J.-J., and Mario Lanza (2016) Moving graphene devices from lab to market: advanced graphene-coated nanoprobes. *Nanoscale*, **8**, 8466–8473.

# 11

# Design and Fabrication of a Logarithmic Amplifier for Scanning Probe Microscopes to Allow Wide-Range Current Measurements

*Lidia Aguilera and Joan Grifoll-Soriano*

*Technology Transfer Company, Knowledge Innovation Market, Pallars St., 179-185 08005 Barcelona, Spain*

## 11.1 Introduction

The use of conductive atomic force microscopes (CAFMs) has become essential in many fields of nanotechnology, as it allows to characterize and to modify materials and particles with very large spatial resolution. It measures the current flowing from a very sharp tip to a sample (and vice versa), and in most experiments CAFMs are typically between 1 pA (the noise level of the CAFM) and some nanoamperes (the saturation level of the CAFM). The currents collected by the tip are sent to a computer for building up current-related plots and images. But in between them, a current–voltage ($I$–$V$) converter is necessary because the computer cannot read currents, only voltages. Most of the $I$–$V$ converters in the market make a linear conversion, finding a compromise between the accuracy they want to obtain and the number of bits of the converter. Despite this being a good solution in most cases (especially from the accuracy point of view), this methodology has one associated limitation: the current range covered by the preamplifier is limited to three or four orders of magnitude (covering larger ranges would prohibitively increase the number of bits of the converter). For example, many works showed saturation currents collected with the CAFM at large voltages [1–14], masking some phenomena that may take place at larger currents, such as the dielectric breakdown (BD) of thin dielectrics. One possible solution (which is now commercially available) would be to use a converter with a variable gain, which in total would allow processing larger current ranges. This methodology allows an effective observation of current phenomena with large precision at low or high current ranges, but not both at the same time [15, 16].

In this chapter we present the design, fabrication, and characterization of a logarithmic preamplifier [17] capable of processing currents between 10 pA and 1 mA simultaneously, without the need for adjusting the data scale. The developed device provides repetitive response and it was modeled empirically. The low signal-to-noise ratio (SNR) of the system provides high reliability to measurements. The calibration of the unit is also described, and the results of the verification, validation, and integration in the CAFM system (carrying out real

*Conductive Atomic Force Microscopy: Applications in Nanomaterials,* First Edition.
Edited by Mario Lanza.

measurements) are discussed. As an example, the setup has been tested by analyzing BD spots induced in thin (<10 nm) $SiO_2$ and $HfO_2$ layers.

## 11.2 Fabrication of a Logarithmic Preamplifier for CAFMS

The design of the logarithmic preamplifier proposed in this work [17] is based on the block diagram of Figure 11.1 and, as will be shown later, it allows relative current measurements in the range of 10 pA to 1 mA. The preamplifier is based on a main logarithmic converter block (LCB), and other auxiliary blocks, including the (i) reference current, (ii) input current, (iii) low-pass filter, (iv) offset adjustment, (v) adder, and (vi) notch filter.

### 11.2.1 Design

The LCB provides in the output an amplified voltage value that is logarithmically proportional to the input current $I_{IN}$ and the reference current $I_{REF}$, which are the currents that flow through the structure under test and a reference current, respectively. The logarithmic current-to-voltage ($I$–$V$) converter is based on a virtual ground-type configuration, using a bipolar junction transistor (BJT) as a nonlinear device in the feedback loop. The most popular implementations use the logarithmic $I$–$V$ transfer characteristic inherent in PN junctions of a BJT when a voltage is applied across the base–emitter junction ($V_{BE}$) [18, 19]. Once the input voltage of the transistor has assumed an appropriate threshold value (about 0.6 V in silicon transistors), the current of the collector always increases exponentially with soft increments in input voltage, as $V_{BE}$ increases logarithmically depending on the collector current. This allows obtaining a wide-range signal observing a finite range of voltages by introducing the obtained signal into the feedback loop of an operational amplifier (OPAMP). In Figure 11.2, the output voltage of the OPAMP ($V_{OUT}$) is the $V_{BE}$ provided by the transistor placed in the feedback loop. Using an NPN transistor, $V_{OUT}$ is defined by Eq. (11.1).

$$V_{OUT} = -\frac{k \cdot T}{q} \cdot \ln\left(\frac{I_C}{I_S}\right) \approx -\frac{k \cdot T}{q} \cdot \ln\left(\frac{I_{IN}}{I_S}\right) \tag{11.1}$$

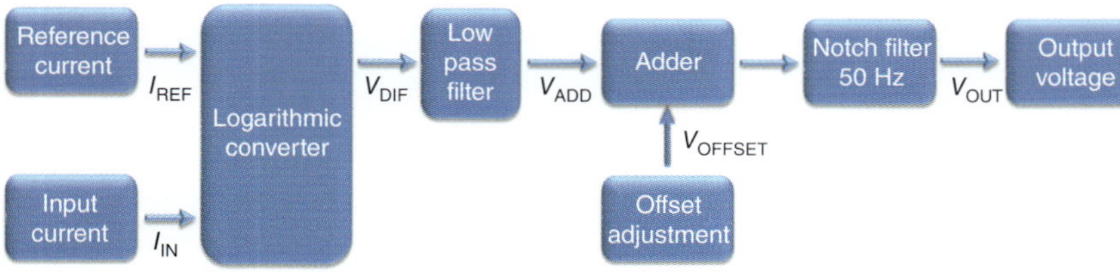


**Figure 11.1** Block diagram of the logarithmic current-to-voltage preamplifier.

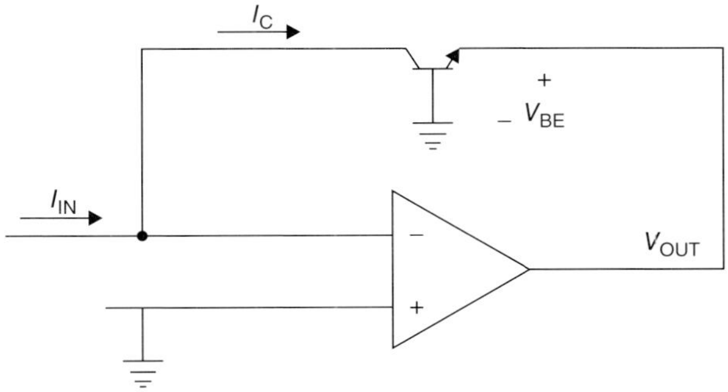


**Figure 11.2** Simplified circuit for describing the working principle of the logarithmic preamplifier.

where

$K$ = $1.381 \times 10^{-23}$ J/K (Boltzmann constant)
$T$ = absolute temperature (degree Kelvin)
$Q$ = $1.602 \times 10^{-19}$ C (elementary positive charge)
$I_C$ = collector current (mA or same units as $I_{IN}$ and $I_S$)
$I_{IN}$ = log-amp input current (mA or same units as $I_C$ and $I_S$)
$I_S$ = reverse saturation current (mA or same units as $I_{IN}$ and $I_C$)

To compensate for the large temperature dependence of the output voltage (due to the saturation currents of the bipolar transistors), a differential configuration has been chosen. Two basic logarithmic $I$–$V$ amplifiers make the conversion of currents applied at the inputs, $I_{IN}$ and $I_{REF}$, into logarithmic voltages ($V_{IN}$ and $V_{REF}$). After this conversion, their difference is amplified with a gain ($G_{DIF}$) chosen to ensure that the voltage applied to the atomic force microscopy (AFM) electronics does not exceed the limits (in this case, ±3 V). Figure 11.3 illustrates the circuit diagram implemented for the LCB. The OPAMP used is the OPA129 from Burr-Brown, which provides ultralow bias current features. The device providing the logarithmic transfer function is the Dual Match Pair Transistor HFA3134 from Intersil Corporation. SMD formats are used to eliminate the inductances originated by the pads of the components. Guarding and shielding techniques are used to minimize leakage current and noise pickup that could exceed the input bias current of the amplifier [20, 21].

$$V_{O\_REF} = -\frac{K \cdot T}{q} \cdot \ln\left(\frac{I_{REF}}{I_{S\_REF}}\right) \tag{11.2}$$

$$V_{O\_IN} = -\frac{K \cdot T}{q} \cdot \ln\left(\frac{I_{IN}}{I_{S\_IN}}\right) \tag{11.3}$$

where $I_{S\_IN}$ and $I_{S\_REF}$ are the saturation currents of the bipolar transistors in the two log $I$–$V$ amplifiers where the input and reference currents are connected, respectively. Those voltages are driven to a differential amplifier circuit providing

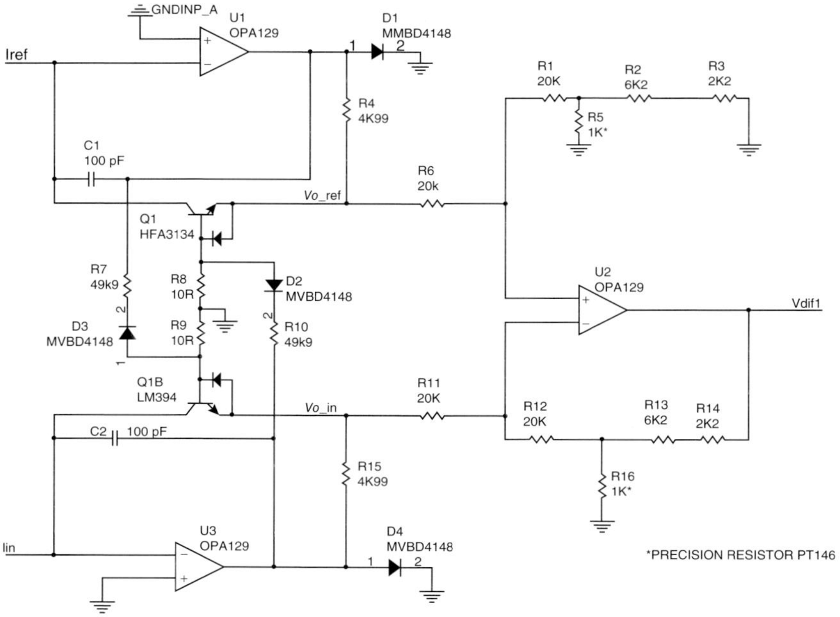


**Figure 11.3** Circuit diagram of the logarithmic converter block.

the difference between $V_{O\_REF}$ and $V_{O\_IN}$ and adding a $G_{DIF}$ gain as shown in Eq. (11.4).

$$V_{DIF1} = (V_{O\_REF} - V_{O\_IN}) \cdot G_{DIF} \tag{11.4}$$

By replacing Eqs (11.2) and (11.3) into (11.4), Eq. (11.5) is obtained:

$$V_{DIF1} = \frac{K \cdot T}{q} \cdot G_{DIF} \cdot \left( \ln \left( \frac{I_{IN} \cdot I_{S\_REF}}{I_{S\_IN} \cdot I_{REF}} \right) \right) \tag{11.5}$$

The saturation current ($I_S$) of the transistors is not independent of temperature variations; in fact, it changes abruptly with it. Those variations could produce variations in sensitivity about 2 mV/°C; this means that changes about 18 mV could appear in base–emitter voltage ($V_{BE}$). The temperature variation of $V_{DIF}$ on the saturation current was compensated for using a matched paired transistor so that the saturation current is identical:

$$I_{S\_REF} = I_{S\_IN} = I_S \tag{11.6}$$

Then, we can rewrite Eq. (11.5) using Eq. (11.6):

$$V_{DIF1} = \frac{K \cdot T}{q} \cdot G_{DIF} \cdot \left( \ln \left( \frac{I_{IN}}{I_{REF}} \right) \right) \tag{11.7}$$

And, rewriting Eq. (11.7) using logarithm with base 10 instead of the natural logarithm, we obtain Eq. (11.8):

$$V_{DIF1} = \frac{K \cdot T}{q} \cdot G_{DIF} \cdot \ln(10) \cdot \log \left( \frac{I_{IN}}{I_{REF}} \right) \tag{11.8}$$

On the other hand, the differential gain ($G_{DIF}$) is given by Eq. (11.9):

$$G_{DIF} = \frac{R_{13} + R_{14}}{R_{16}} \tag{11.9}$$

Finally, the output voltage $V_{DIF}$ is represented in Eq. (11.10):

$$V_{DIF} = \frac{K \cdot T}{q} \cdot \frac{R_{13} + R_{14}}{R_{16}} \cdot \ln(10) \cdot \log\left(\frac{I_{IN}}{I_{REF}}\right) \tag{11.10}$$

The amplifier gain $G_{DIF}$ has been calculated taking into account the input voltage range of the AFM electronics. In our system, the voltage provided by the logarithmic preamplifier is sent to an acquisition card that allows an input range voltage of ±3 V. Finally, the resistor-capacitor (R15-C2, R4-C1) pairs provide frequency compensation to the amplifiers and keep stability within 300 kHz bandwidth. When the input current is greater than 2 mA, the emitter resistance could contribute to a significant deviation of the theoretical transistor curve; and the use of R8 and R9 at the transistor base allows an increase of the base voltage of the transistor when the input current increases, compensating for the emitter voltage drop. It has to be mentioned that the effects of the emitter resistance of Q1 and Q2 transistors can degrade the circuit's accuracy at currents above 100 μA. The networks formed by R10-D2-R8 and R7-D3-R9 compensate for those errors. The R5 and R16 resistances have a positive temperature coefficient of 3500 ppm/°C and are used to keep the differential amplifier gain ($G_{DIF}$) for positive and negative values paired to temperature changes. The constant value (gain factor, FG) (multiplying the logarithm in Eq. (11.10)) will determine the volts per decade provided by the LCB output. The appropriate FG has to be calculated then to obtain the desired voltage range at the output (not exceeding ±3 V, as mentioned before).

$$F_G = \frac{K \cdot T}{q} \cdot \frac{R_{13} + R_{14}}{R_{16}} \cdot \ln(10) \tag{11.11}$$

The preamplifier gain has been calculated for a range of 4 V between the minimum and maximum value measured, allowing the possibility of measuring lower and higher currents to the minimum and maximum measured values, respectively. Supposing a unitary FG of 1 V per decade, we obtain an output voltage ($V_{DIF1}$) of −7 V when the measured input current is 10 pA (Eq. (11.10)) and 1 V when the measured current is 1 mA, giving a voltage range of 8 V. To obtain a 4 V range, we need to use an FG of 0.5, and using Eq. (11.12) we can obtain the value of $G_{DIF}$.

$$F_G = \frac{K \cdot T}{q} \cdot G_{DIF} \cdot \ln(10) = 0.5 \tag{11.12}$$

Developing Eq. (11.12) and replacing known values we obtain the values for the resistors R13 and R14:

$$G_{DIF} = \frac{0.5}{\frac{K \cdot T}{q} \cdot \ln(10)} = \frac{0.5}{\frac{1.3806 \times 10^{-23}\left[\frac{J}{°K}\right] \cdot 300[°K]}{1.602 \times 10^{-19}[C]} \cdot 2.3025}$$

$$= \frac{0.5}{0.05952} = 8.4 \tag{11.13}$$

The resistor values chosen for the logarithmic converter preamplifier are 6.2 kΩ for R13 and 2.2 kΩ for R14. So we obtain a $G_{\mathrm{DIF}}$ value of 8.4, as shown in Eq. (11.14).

$$G_{\mathrm{DIF}} = \frac{6.2[\mathrm{K}\Omega] + 2.2[\mathrm{K}\Omega]}{1[\mathrm{K}\Omega]} = 8.4 \tag{11.14}$$

Using Eq. (11.14), the output voltage provided when measuring input currents of 10 pA and 1 mA will be −3.5 and 0.5 V, respectively. Therefore, an offset adjustment of 1.5 V will be required in order to obtain a voltage range between −2 V ($I_{\mathrm{IN}} = 10$ pA) and 2 V ($I_{\mathrm{IN}} = 1$ mA).

Apart from the LCB, other auxiliary circuits to improve the performance of the log current-to-voltage amplifier and to allow its calibration are the following:

1) *Reference current*: a monolithic integrated circuit (REF200, from Burr-Brown Corporation) was used to obtain a precision reference current of 100 μA ($I_{\mathrm{REF}}$). The $I_C$ is configured using two independent 100 μA current sources in series to improve voltage compliance.
2) *Input current*: Drives $I_{\mathrm{IN}}$, the current that flows through the sample under test. The block includes an overcurrent protection based on a single diode and an RC low-pass filter for high-frequency noise attenuation.
3) *Low-pass filter*: This block reduces the high-frequency noise above 150 Hz at the output of the logarithmic converter.
4) *Offset adjustment and (v) adder:* the offset adjustment (which is accessible by the user) allows to adjust the current dynamic range of interest (from 1 pA to 1 mA) to correlate with the input voltage range of the AFM electronics (from −3 to +3 V, in our case). The offset will also allow calibration of the log $I$–$V$ amplifier and compensate for any drift related to temperature changes. The adder module generates the sum of the offset voltage and the voltage provided by the LCB, using a non-inverting summing amplifier, without adding any gain. As shown in Figure 11.4, its implementation is based on a non-inverting summing amplifier. This block is used to sum the LCB output and the offset adjustment output signals. Both signals are driven using 2 V followers in order to provide high impedance. This way, the possible influence between the offset adjustment block and the LCB is avoided. The resistor $R_{24}$ provides the necessary compensation to minimize error due to input bias current in $U_6$ OPAMP. The non-inverting summing amplifier does not add gain, so at the output we obtain directly the sum of the applied input voltages:

$$V_{\mathrm{ADD}} = V_{\mathrm{DIF2}} + V_{\mathrm{ZERO1}} = \frac{K \cdot T}{q} \cdot G_{\mathrm{DIF}} \cdot \ln(10) \cdot \log\left(\frac{I_{\mathrm{IN}}}{I_{\mathrm{REF}}}\right) + V_{\mathrm{ZERO}}$$
$$= 0.5 \cdot \log\left(\frac{I_{\mathrm{IN}}}{10^{-4}}\right) + 1.5 \tag{11.15}$$

5) Fifty hertz notch filter: This module allows the attenuation of the 50 Hz interferences generated by the power mains supply and coupled to everywhere. The filter is based on a twin-T network combined with a follower voltage to improve the quality factor ($Q$) from 0.3 (usual value) to an approximate value of 50. The voltage follower works as a buffer and the $R_{27}$ resistor (50 Ω) provides

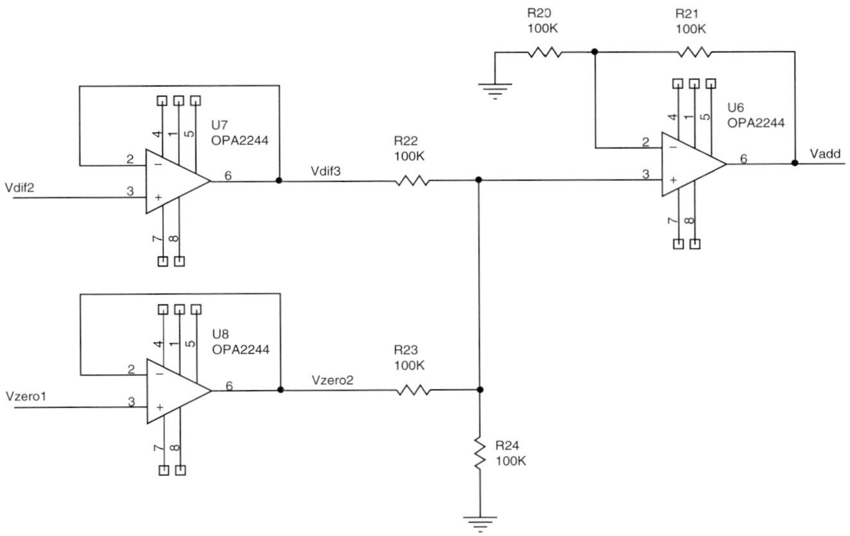


**Figure 11.4** Circuit diagram of the adder block.

the impedance adaptation needed to avoid reflections in the transmission line formed by the coaxial cable that will drive the output voltage signal (the measure) to the CAFM control system.

### 11.2.2 Fabrication and Testing

Other practical considerations for the design implementation of the logarithmic preamplifier are the use of some techniques for the printed circuit board (PCB) layout, as guarding and shielding of the critical parts of the circuit or the air connection between critical parts, which allows low-level measurement devices economically. Noise is the main limitation in the correct operation of analog circuits and the vital issue to consider when measuring low-level current (<1 nA); we could find that the electrical noise generated by the circuit was of the same order of magnitude as the signal intended to be measured. For this reason, electrical noise should be considered during the design and implementation of the electronic system. It is highly recommended that factors that could cause electrical noise be anticipated in order to avoid them or minimize their impact. Some of the possible sources of electrical noise when measuring low-level current could be generated by electrostatic coupling, vibrations, low-impedance sources, and so on. The OPAMPs used in electronic circuits are analog circuits that need to be placed in their own section of the PCB using special layout techniques. The effects of the PCB will be more evident in high-speed analog circuits, but there are common effects affecting all analog circuits. Some of them are described below.

#### 11.2.2.1 Printed Circuit Board

The use of a double-sided layer helped minimize the generation of unwished for leakage currents, which could create electromagnetic fields in the circuit, and

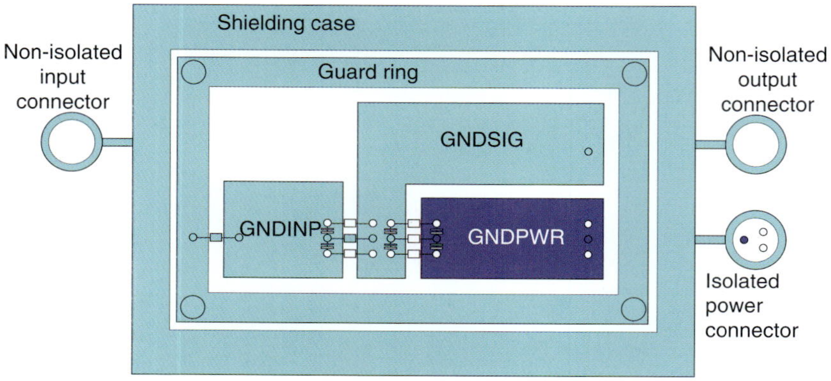


**Figure 11.5** Distribution of grounding planes in the logarithmic preamplifier.

act as a shield for the radiated noise. About grounding, it is important to note that the use of isolated connectors to control internally the connections between the different ground planes and the use of nonisolated connectors to connect directly the ground planes to the Faraday box of the setup has helped minimize the impact of noise sources. As shown in Figure 11.5, the grounding planes used in the logarithmic preamplifier have been designed using three parts, one for the input signal chain, another for the output signal chain, and the third for the power supply circuitry. Low-value resistors have been used to connect the different grounds, avoiding the introduction of leakage currents between the planes and improving conducted noise immunity.

#### 11.2.2.2 Cleaning

Cleaning has been considered on account of the eventual resistance and capacitance originated by flux residues during the PCB assembly. If the circuit gets dirty or dusty, or gets fingerprints on it, the performance could be degraded, especially when the system is intended to measure currents in picoamps region. An effective process to clean the PCB consists of cleaning the surface using isopropyl alcohol and rinsing it with unionized water, and then introducing the PCB into an oven at 60 °C for 10 min. Another option is washing it using your dishwasher, making sure that the circuit does not contain components that cannot get wet [21].

#### 11.2.2.3 Decoupling

Noise can be introduced into the analog circuits through the power lines. Bypass capacitors need to be used in order to avoid or reduce this effect. They provide a low-impedance environment for the analog circuitry. In this work, ceramic capacitors have been used to provide an effective decoupling of high frequencies. For lower frequencies, aluminum or tantalum electrolytic capacitors should be used. To choose the appropriate values, it would be interesting to know the noise frequencies that may affect the circuit and need to be removed. In this work the ceramic capacitors used are 100 nF and the electrolytic capacitors are 10 μF. Another important consideration is tracking as short as possible the connection of these capacitors to the OPAMPs. Therefore, the use of a bottom layer was chosen to place those components under the devices and connect them to the appropriate ground plane.

#### 11.2.2.4 Input and Output Isolation

Some noise sources could affect the circuit through the input and output connections because of the high-frequency limitations of passive components. When frequencies are substantially different from those used in the normal operation of the circuit, the solution can be as simple as using a passive low-pass filter to attenuate undesired frequencies. The radiated noise effects coupled to analog circuitry could be minimized, shielding the system completely by the use of a Faraday case. In this work, different prototypes using different materials have been developed and the use of an aluminum-based Faraday case was enough because the environment was not too noisy.

#### 11.2.2.5 Unexpected Passive Components in the PCB

The copper tracks layout could turn the circuit into a radiated noise source. Therefore, in order to avoid or minimize this effect, the geometric shape of the tracks needs to be considered. On one hand, the use of long tracks should be avoided; and if this is not possible, they should be divided. Furthermore, adding ferrites or low-value resistors is also a common methodology to avoid loop traces [20]. In our design, we avoided the use of 90° angle tracks to avoid possible reflections originated by track width changes, which could distort the signal. Figure 11.6 shows the implemented layout used with the logarithmic preamplifier.

Usually, device fabrication projects demand extremely short deadlines to complete the PCB layout. The digital circuitry has been simulated, but in most cases the final product is still a prototype with the customized characteristics desired by the customer. Digital designers can correct performance requirements by implementing cuts and jumpers, reprogramming gate arrays or flash memories, and go

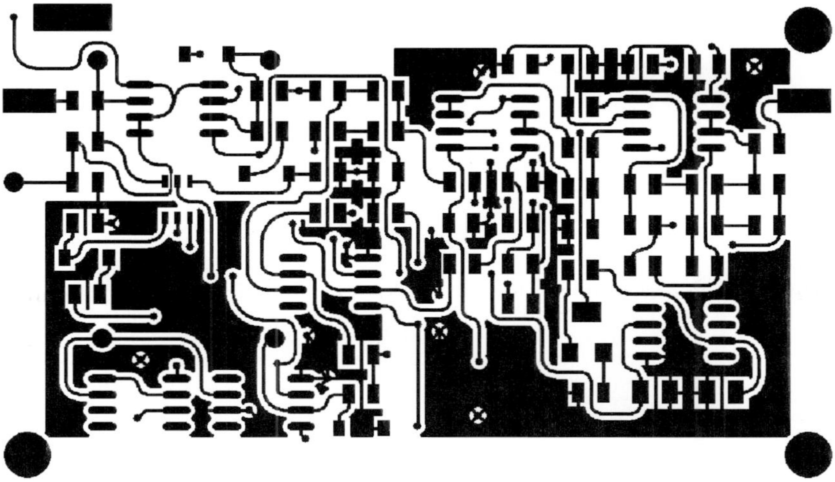

**Figure 11.6** Layout of the PCF of the LogAmpDev07 displaying rounded tracks.

on to the next project. This is not the case with analog circuitry, where common design mistakes cannot be corrected by the cut and jumper method; they may render the entire PCB unusable. Thus, it is really important to understand the problems and the possible preventions or solutions before starting the fabrication, as it can save a lot of development time and money. So it is highly recommended that prototypes of all the analog circuitry are made to study how the problems associated with certain physical phenomenon affect the PCB or the complete system. This was the method used in this work. Using the method, different prototypes of what is generically called LogAmpDev00 have been designed and implemented, with the purpose of studying the different possible options (analog devices) that could form part of the circuit in which this application is based. Particularly, several ultralow bias current OPAMPs and matched paired transistors were used to achieve different logarithmic transfer functions.

The implemented preamplifiers developed in this work offering the best performance were LogAmpDev02, LogAmpDev05, and LogAmpDev06, apart from LogAmpDev07, the name given to the circuit that was implemented in the final setup. LogAmpDev02 and LogAmpDev05 preamplifiers have been implemented using the OPA128 OPAMP of Burr-Brown Co. and the LMC6001 of National Semiconductor Co. As a device that provides the logarithmic transfer function, they have been implemented using the LM394 Dual Match Pair Transistor by National Semiconductor Co. The printed circuits of both devices have been implemented using the guarding technique of the input signal. The major difference between both devices is the size needed for the PCB, as the components used in LogAmpDev02 are larger and consume more space. The LogAmpDev05 is implemented in a smaller area, so it can be used in a smaller nickel-silver Faraday case. The LogAmpDev06 design is based on the LOG112 logarithmic amplifier from Burr-Brown Co., which includes a reference voltage source of 1.5 V. This has been used to implement the offset adjustment block, instead of the MAX6325 of Maxim Semiconductor Co. used in the other devices. Table 11.1 summarizes the different fabrication parameters used for each device.

In the next step the devices are characterized. Table 11.2 shows the output voltages, $V_{OUT}$, provided by the preamplifiers when measuring different input current values, $I_{IN}$. The noise level (in percentage) of the four devices measured is also referenced. As it can be seen, the circuit that offers better SNR is the preamplifier called LogAmpDev07. When the measured input current is 100 pA, the preamplifier could provide us with a measure from 98.65 to 101.4 pA because of the noise. About the linearity, the four devices work appropriately, showing

**Table 11.1** Displaying the differences between the devices fabricated.

| Parameter | LogAmpDev02 | LogAmpDev05 | LogAmpDev06 | LogAmpDev07 |
|---|---|---|---|---|
| Ultralow bias current operational amplifier | AD549 | OPA128 | LOG112 | OPA129 |
| Device providing the logarithmic transfer function | LM394 | LM349 | | HFA3134 |

**Table 11.2** Data to verify the devices.

| Parameter | LogAmpDev02 | LogAmpDev05 | LogAmpDev06 | LogAmpDev07 |
|---|---|---|---|---|
| $V_{OUT}$ [10 pA] | −2.36 V | −1.44 V | −2.32 V | −2.02 V |
| $V_{OUT}$ [100 pA] | −1.69 V | −1.01 V | −1.65 V | −1.52 V |
| $V_{OUT}$ [1 nA] | −1.02 V | −0.58 V | −0.99 V | −1.01 V |
| $V_{OUT}$ [10 nA] | −0.35 V | −0.15 V | −0.32 V | −0.50 V |
| $V_{OUT}$ [100 nA] | 0.32 V | 0.28 V | 0.35 V | 0.00 V |
| $V_{OUT}$ [1 μA] | 0.99 V | 0.71 V | 1.01 V | 0.51 V |
| $V_{OUT}$ [10 μA] | 1.66 V | 1.14 V | 1.68 V | 1.01 V |
| $V_{OUT}$ [100 μA] | 2.33 V* | 1.57 V* | 2.34 V* | 1.52 V* |
| $V_{OUT}$ [1 mA] | 3.00 V | 2.00 V | 3.01 V | 2.03 V |
| $V_{OUT}$ equation | $1.0021 \cdot \log(I_{IN}) + 6.3651$ | $0.8391 \cdot \log(I_{IN}) + 5.6242$ | $0.8600 \cdot \log(I_{IN}) + 5.6100$ | $0.8315 \cdot \log(I_{IN}) + 5.5640$ |
| Noise | 6.6% | 3.4% | 3.6% | 2.7% |

The symbol "*" indicates offset adjustment voltage needed.

small differences between the guarding technique layout and the "air connection" technique. On the other hand, the response of the four devices is repetitive and can be empirically modeled. The resulting $I$–$V$ transfer function for all of them is also shown in Table 11.2.

It can be observed that LogAmpDev05 and LogAmpDev07 have been configured with a gain factor smaller than the others. This is important to avoid the acquisition card of the CAFM system being saturated assuring that the voltage range provided by the preamplifiers does not exceed the limits. However, the gain factor can be modified easily by reducing the differential amplifier gain in the LCB (Figure 11.3). It has to be pointed out that the use of capacitors below 100 pF improves the time response of the system, going from 0.5 to 0.1 s approximately, although it is not a critical factor considering the intended use of the preamplifiers, where the variations in the input current are much slower. In any case, the device that offers a better signal to noise is the circuit LogAmpDev07 and, therefore, it was the preamplifier integrated in the CAFM system.

Before using the logarithmic preamplifier, a validation process was done to test the system performance. No additional calibration process was needed, except for temperature changes, which can be corrected with the offset. A verification procedure is needed to give evidence that the logarithmic preamplifier is performing according to initial specifications. This procedure consists of applying to the input ($I_{IN}$) the entire range of currents intended to be measured and measuring the voltage ($V_{OUT}$) provided by the preamplifier. Therefore, the $I$–$V$ response curve of the preamplifier may be modeled. The application of the entire current range has been done using three smaller current ranges: from 10 pA to 10 nA; from 10 nA to 10 μA; and from 10 μA to 1 mA to obtain good resolution using the instrumentation available in our laboratory. Once the correct operation has been verified, the noise in system needs to be analyzed by injection of a fixed current

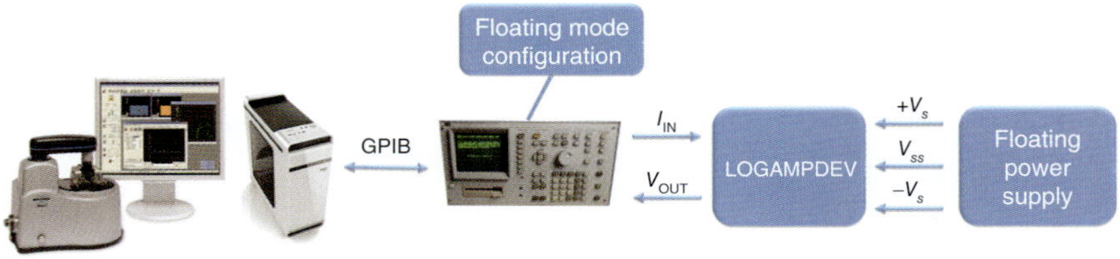


**Figure 11.7** Configuration for the characterization in the answer $I_{IN}$–$V_{OUT}$ of the devices.

and observing the AC output signal using the oscilloscope. Finally, the response time of the device has been studied observing that the time response depends on the current value measured, increasing softly when input current value decreases.

As shown in Figure 11.7, the instrumentation setup for the validation process is based on the Hewlet Packard semiconductor analyzer HP4145B, which has been used to generate the input currents and measure the output voltages. The setup is controlled by a LabView application running in a computer and using GPIB communication. The power supply for the setup is floating configured to allow measurements with an oscilloscope in order to evaluate the total noise of the system.

The HP4145B semiconductor analyzer contains four source monitor units (SMUs) that allow the application of current or voltage while simultaneously measuring voltage or current. In addition, this analyzer has two voltage source units (VSUs) that allow voltage measurements. The $I_{IN}$–$V_{OUT}$ characterization of the devices has to be done configuring the semiconductor analyzer to realize floating measurements, using a floating power supply that maintains the isolation with the earth terminals of the H14145B. The characterization process has been done configuring the logarithmic preamplifier with a differential gain of 12.6 so that the gain factor obtained is about 0.75 V per decade, and also configuring the device with a differential gain of 8.5 obtaining a gain factor of about 0.51 V per decade. The input current versus output voltage characteristics of the preamplifier is shown in Figure 11.7 in a range of currents between 1 mA and 10 pA. Note that the output exhibits a linear behavior, on a semi-logarithmic plot. The response can be modeled using the following equation: $V_{OUT} = 0.5775 \cdot \log (I_{IN}) + 3.8229$. Noise level applying 10 and 100 pA to the input ($I_{IN}$) has a value of, approximately, 9 mV, which represents 2.7% of the signal at the output. Figure 11.8 shows the response time applying low-level currents to the $I_{IN}$ using a 100 pF capacitor in the feedback loop of the logarithmic preamplifier. The time response becomes critical for input current values below 20 pA.

As shown in Figure 11.3, the R15-C2 network is needed to stabilize the OPAMP loop at currents flowing through the transistor. When the current is on the order of microamperes, the transistor adds too much gain to the OPAMP. Generally, the amplifiers work better managing low gains. Figure 11.9 shows how the preamplifier becomes unstable when C2 is not used.

When the input current is low, the transistor adds low gain to the feedback loop and the C2 capacitor value could become too large, causing the device response time to become too large. Figures 11.10 and 11.11 show the response time of

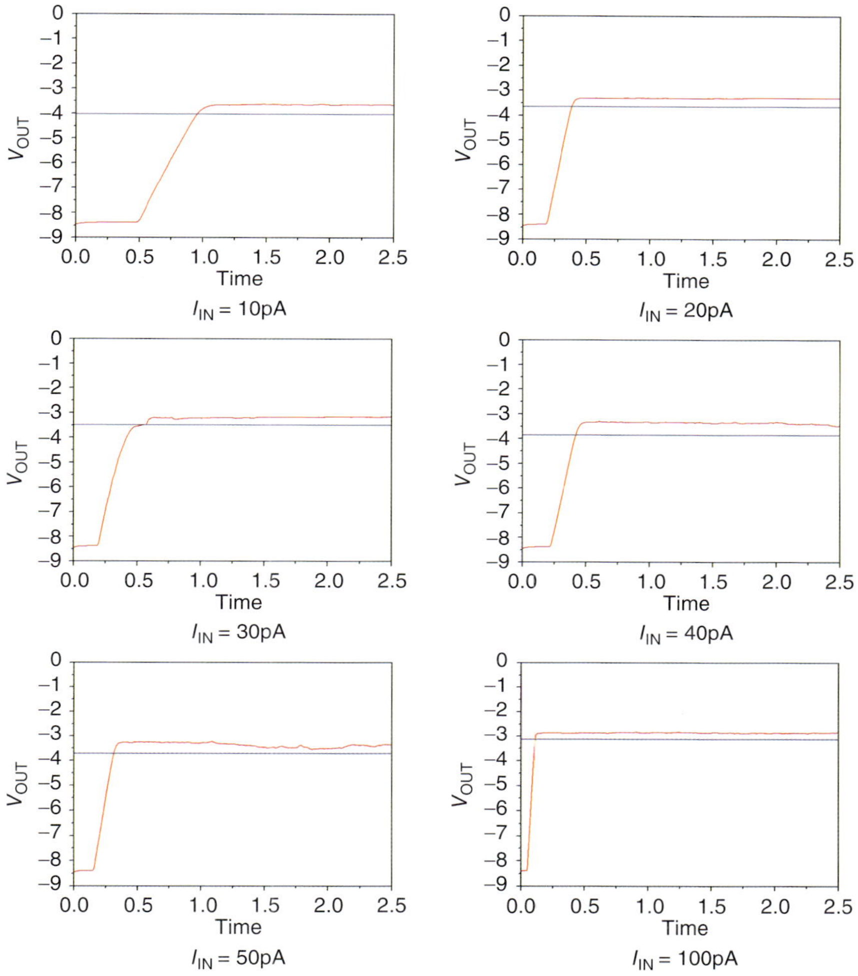


**Figure 11.8** LogAmpDev07 response time with C2 = 100 pF for low-level input currents.

the logarithmic amplifier using C2 values of 220 and 22 pF. For input currents below 50 pA, the higher the capacity used, the slower the system response. Definitely, the preamplifier response becomes faster using a smaller capacitor for C2, although some oscillations appear due to the corresponding increase in the cutoff frequency of the RC feedback loop network. So it could be useful to increase the resistor value of the RC network to avoid this effect.

### 11.2.3 Implementation in a CAFM and Case Study

Once the circuit has been implemented and calibrated, it has been integrated in the atomic force microscope. This new tool has been described in a previous work [22], providing a response time of about 0.1 s for input currents of 10 pA, and a

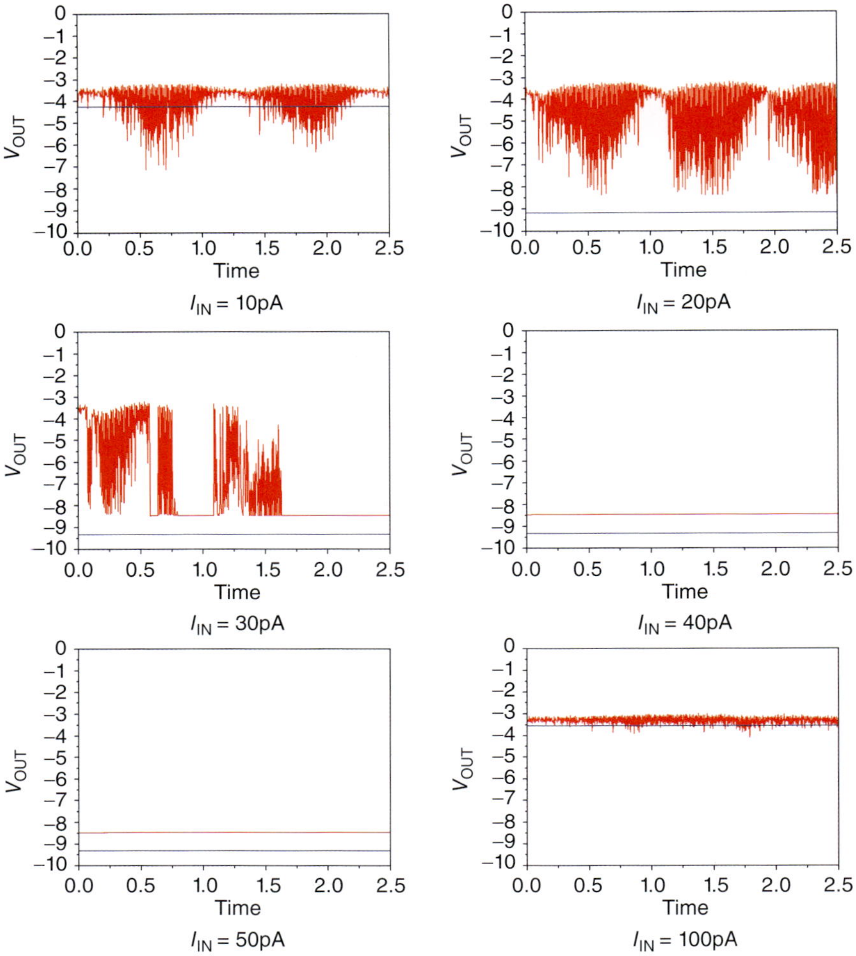


**Figure 11.9** Undesirable operation using LogAmpDev07 without C2.

very low SNR providing high accuracy to the measurements. As mentioned, the lineal *I*–*V* converter of the CAFM can only measure a current range between picoamperes and nanoamperes. This feature limits the study of the degradation process of some dielectric materials. The CAFM control system applies a voltage slope from 0 to 8 V to the substrate. And, the voltage provided by the logarithmic preamplifier has been driven to the acquisition card registering the voltage values. Figure 11.12 shows the setup configuration used for making measurements with the new configuration. The performance of this new setup has been tested by analyzing the electrical properties of 2-nm-thick $SiO_2$ layers (on a Si substrate) of MOS structures subjected to electrical stress. With $SiO_2$ samples, the stress consisted of two consecutive ramped voltage stresses (RVSs), forward and backward ramps, at the same oxide location until BD and the *I*–*V* characteristics

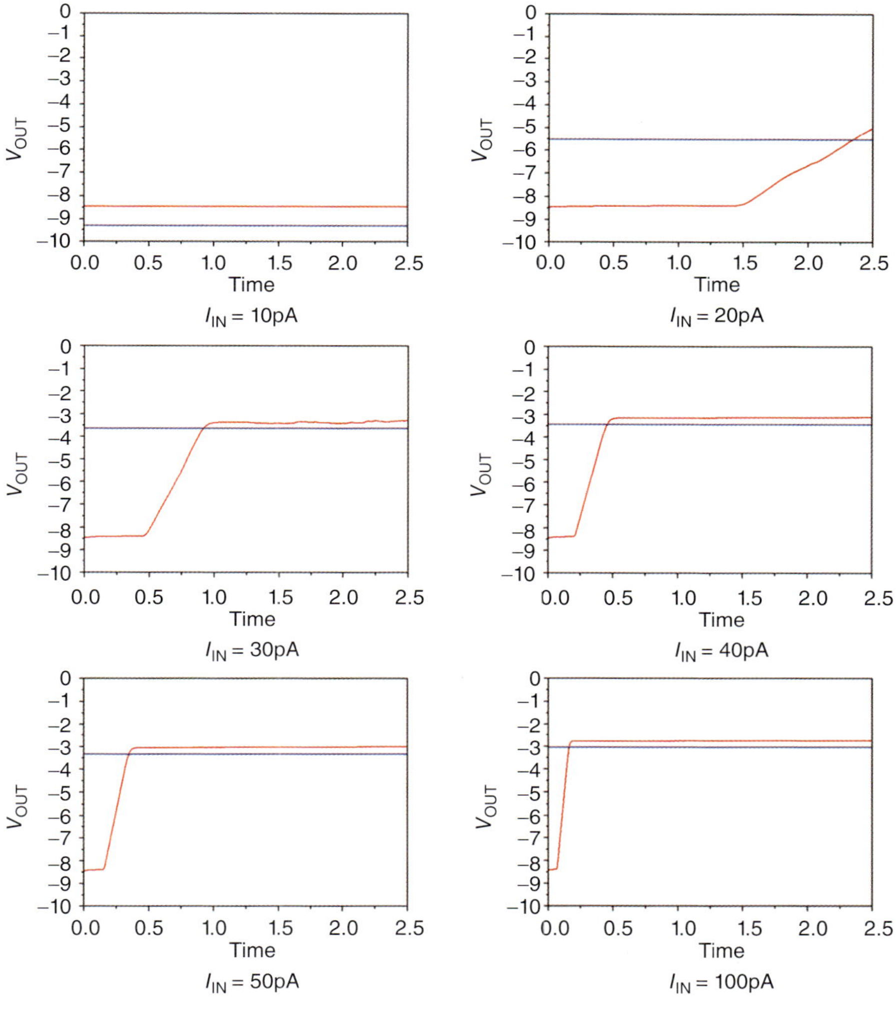


**Figure 11.10** Answer speed depending on the input current with C2 = 220 pF. The response of the amplifier becomes slower than needed in the current range from 10 to 20 pA. On increasing the C2 capacitor, the preamplifier becomes slower when it works in the low current range.

were measured. Figure 11.13 shows the *I–V* characteristic obtained after the forward and backward ramps were applied. The corresponding *I–V* characteristics demonstrate that a BD event was triggered at 8 V. The backward characteristic corresponds to the typical post-BD conduction of this kind of structures, with a current much larger than that measured before BD.

Current images have been obtained (at 5.5 V) of the oxide area where the consecutive RVSs were applied, showing different BD spots. Figure 11.14 shows the current image and the area and maximum current of the three larger spots. The current profile of spot 1 shows a maximum current of $4.1 \times 10^{-7}$ A. The rest of

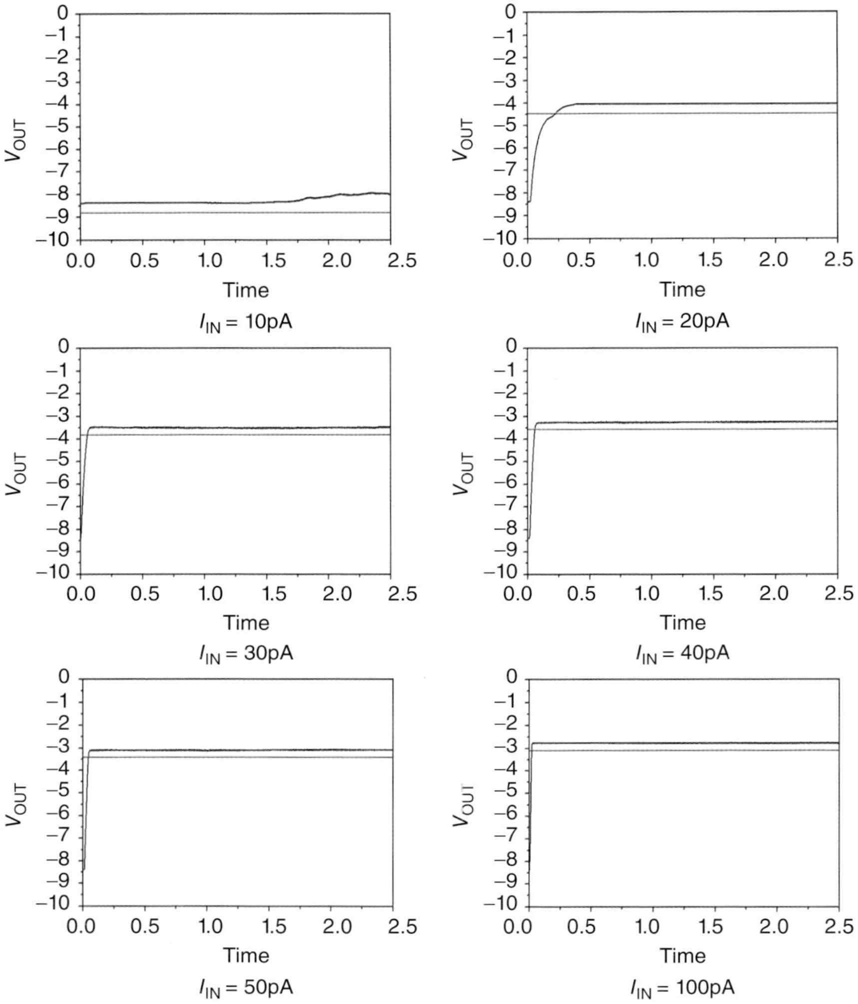


**Figure 11.11** Answer speed depending on the input current with C2 = 33 pF.

the BD spots in the image have been probably induced during the scan since the applied voltage is high.

Note that although the stressed area (i.e., the oxide region under the tip where BD is induced) is on the order of ~300 nm$^2$ [23], the oxide area that has been affected by the BD event can be much larger as observed in the current images, which points out the propagation of the BD event to the surrounding regions. The table shows a clear relation between the BD hardness (maximum current) and the area of the BD spot (i.e., the larger the BD current, the larger the area of the spot). Although this relation was previously pointed out [23], the actual BD currents could not be measured due to the limited setup current range of

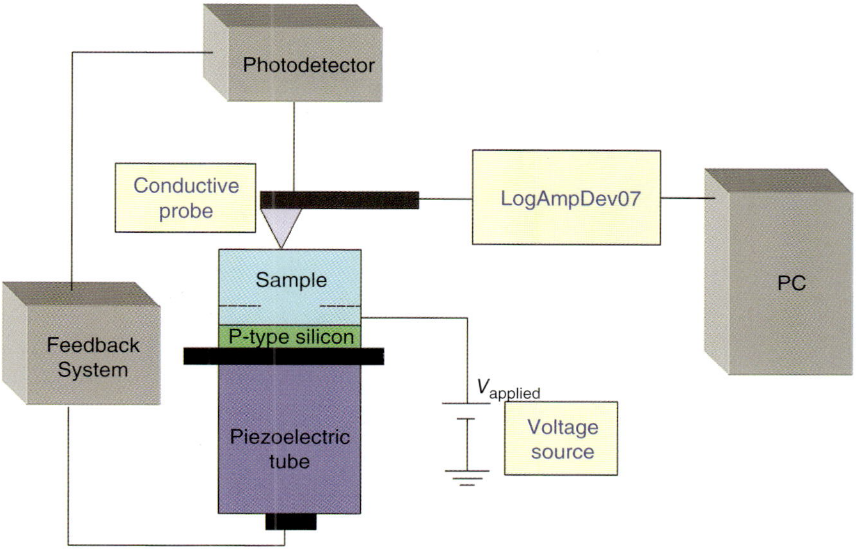


**Figure 11.12** CAFM system configuration with a log *I–V* converter.

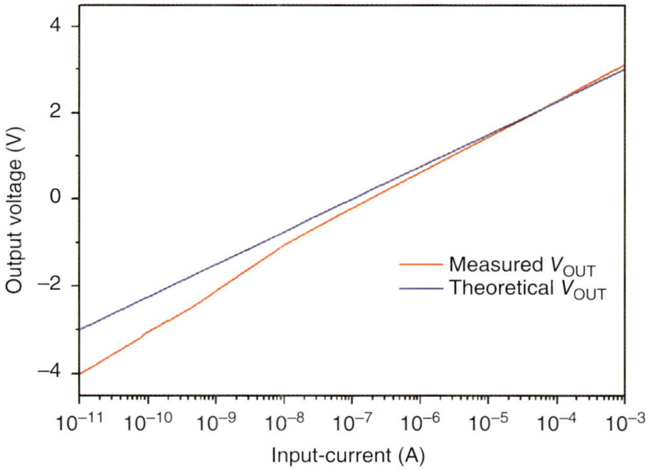


**Figure 11.13** Input current to output voltage characteristic of the logarithmic *I–V* converter.

the standard CAFM used. The wide current dynamic range of the developed setup opens up a new path to investigate in more detail the properties of the BD spot. A similar experiment was performed on a fixed location of a 2.5-nm-thick atomic-layer-deposited $HfO_2$ film (on top of a 1-nm $SiO_2$ and a p-type Si substrate) to induce the BD. Figure 11.15 shows the *I–V* curves measured during the RVSs (forward and backward). As can be observed from the figure, the backward

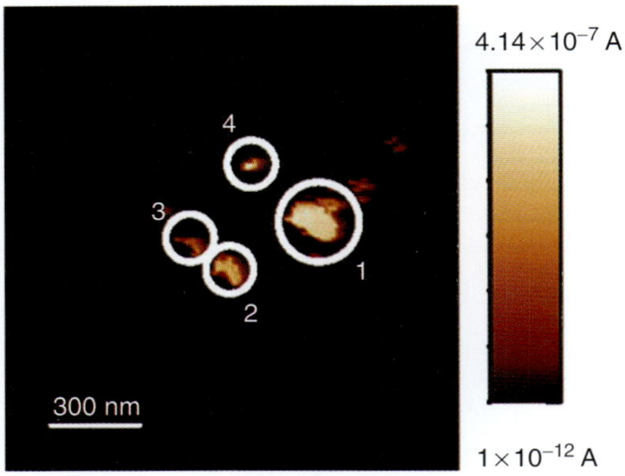


| # Spot | Area ($nm^2$) | Maximum current (A) |
|---|---|---|
| 1 | $3 \times 10^4$ | $4.1 \times 10^{-7}$ |
| 2 | $0.8 \times 10^4$ | $1.4 \times 10^{-7}$ |
| 3 | $0.48 \times 10^4$ | $3 \times 10^{-8}$ |
| 4 | $0.45 \times 10^4$ | $9 \times 10^{-8}$ |

**Figure 11.14** Current image (at 5.5 V) of an oxide area where an RVS was applied until BD. The color bar is in logarithmic scale. Different BD spots are observed. Spot 1 is related to the RVS and the others to the high voltage applied during the scan. The table shows the maximum current and area of the different BD spots registered in the image. Reprinted from [22].

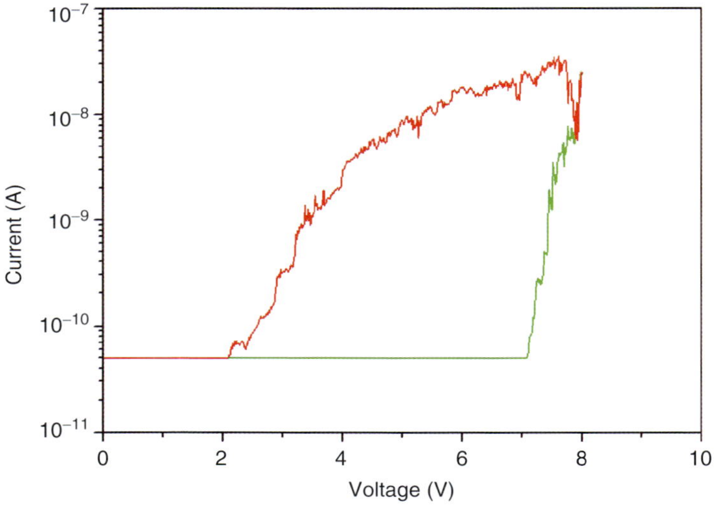


**Figure 11.15** *I–V* curves (forward green, backward red) measured on the $HfO_2$ sample after being stressed with consecutive RVSs. Reprinted from [22].

curve presents larger conductivity at lower voltages, showing, similar to the $SiO_2$ samples, a typical post-BD conduction.

## 11.3 Conclusions

Electric characterization with the atomic force microscope is having an increasing demand among the AFM community to solve the new research topics in nanotechnology like energy harvesting, polymer- or organic-based electronics, or the new advances in the semiconductor industry. In this article are described recent advances in two of the most popular AFM advanced modes such as ResiScope and Soft ResiScope (current/resistance measurements) and high-definition Kelvin force microscopy (HD-KFM) (work-function measurements). ResiScope and Soft ResiScope have been proved to be a new paradigm with clear advantages for conductive measurements due to the high dynamic range in real time, tip and/or sample protection, and reliability. On the other hand, HD-KFM has been demonstrated to provide the highest sensitivity and robustness as it is based on one of the front-end topics in AFM: the multifrequency approach.

A logarithmic preamplifier has been implemented to improve the electrical performance of standard CAFM. The new setup allows one to perform spectroscopic measurements and current maps with a current dynamic range of nine orders of magnitude (from 1 pA to mA). It has been tested on MOS devices with a $SiO_2$ layer and $HfO_2$ samples as the gate dielectric, providing new details on the electrical properties of BD spots. Since the new setup allows electrical tests at the nanoscale equivalent to those traditionally used at wafer level, it can be very useful to the microelectronics community and, especially, to the scientists involved in the reliability of gate dielectrics.

## References


1 Farnesi, M. (2007) Theoretical Investigation of $SiO_2$ Dielectric Breakdown. PhD These. Universita degli Studi di Padova, Italy.

2 Wilk, G.D., Wallace, R.M., and Anthony, J.M. (2001) *J. Appl. Phys.*, **89**, 5243.

3 Toriumi, A. and Satake, H. (1999) The boundary between hard- and soft-breakdown in ultrathin silicon dioxide films. *MRS Proceedings*, **592**, 323. doi: 10.1557/PROC-592-323.

4 Zafar, S., Kumar, A., Gusev, E., and Cartier, E. (2005) Threshold voltage instabilities in high-$\kappa$ gate dielectric stacks. *IEEE Trans. Device Mater. Reliab.*, **5**, 45–64.

5 Lanza, M., Porti, M., Nafría, M., Benstetter, G., Frammelsberger, W., Lodermeier, E., Ranzinger, H., and Jaschke, G. (2007) Influence of the manufacturing process on the electrical properties of thin (<4 nm) Hafnium based high-k stacks observed with CAFM. *Microelectron. Reliab.*, **47**, 1424–1428.

6 Aguilera, L., Porti, M., Nafria, M., and Aymerich, X. (2006) Charge trapping and degradation of $HfO_2/SiO_2$ MOS gate stacks observed with enhanced CAFM. *IEEE Electron Device Lett.*, **27** (3), 157–159.

7 Cimino, S., Pantisano, L., Aoulaiche, M., Degraeve, R., Kwak, H., Groeseneken, G., and Paccagnella, A. (2005) Hot carrier degradation on n-channel HfSiON MOSFETs: effects on the device performance and lifetime. Proceedings. 43rd Annual IEEE International Reliability Physics Symposium, 275–279, 2005.

8 Porti, M. (2003) Caracterització a escala nanomètrica de la degradació i ruptura dielèctrica del $SiO_2$ en dispositius MOS mitjançant C-AFM. PhD These. Universitat Autònoma de Barcelona. Spain.

9 Porti, M., Nafría, M., Aymerich, X., Olbrich, A., and Ebersberger, B. (2002) Electrical characterization of stressed and broken down $SiO_2$ films at a nanometer scale using a Conductive Atomic force microscope. *J. Appl. Phys.*, **91** (4), 2071–2079.

10 Frammelsberger, W., Benstetter, G., Kiely, J., and Stamp, R. (2007) C-AFM-based thickness determination of thin and ultra-thin $SiO_2$ films by use of different conductive-coated probe tips. *Appl. Surf. Sci.*, **253** (7), 3615–3626.

11 Blasco, X., Nafría, M., Aymerich, X., Pétry, J., and Vandervorst, W. (2005) Breakdown spots of ultra-thin (EOT < 1.5 nm) $HfO_2/SiO_2$ stacks observed with enhanced—CAFM. *Microelectron. Reliab.*, **45**, 811–814.

12 Kremmer, S., Wurmbauer, H., Teichert, C., Tallarida, G., Spiga, S., Iemer, C., and Fanciulli, M. (2005) Nanoscale morphological and electrical homogeneity of $HfO_2$ and $ZrO_2$ thin films studied by conducting atomic-force microscopy. *J. Appl. Phys.*, **97** (7), 074315.

13 Fiorenza, P., Polspoel, W., and Vandervorst, W. (2006) Conductive atomic force microscopy studies of thin $SiO_2$ layer degradation. *Appl. Phys. Lett.*, **88**, 222104.

14 Porti, M., Nafría, M., Aymerich, X., Olbrich, A., and Ebersberger, B. (2002) Post-breakdown electrical characterization of ultrathin $SiO_2$ films with Conductive Atomic Force Microscopy. *Nanotechnology*, **13**, 388–391.

15 Dürig, U., Novotny, L., and Michel, B. (1997) Logarithmic current-to-voltage converter for local probe microscopy. *Rev. Sci. Instrum.*, **68** (10), 3814–3816.

16 Mészáros, G., Li, C., Pobelov, I., and Wandlowski, T. (2007) Current measurements in a wide dynamic range-applications in electrochemical nanotechnology. *Nanotechnology*, **18**, 424004.

17 Grifoll-Soriano, J. (2007) *Millora i Aplicacions d'un Microscopi de Forces Atómiques per a la realització de mesures elèctriques*, Dpt. Enginyeria Electrònica Universitat Autónoma de Barcelona, Spain.

18 Pease, R.A. (1999) What's all this logarithmic stuff, anyhow? *Electronic Design*. Article number 6068, June 14.

19 Pease, R.A. (2005) What's all this logarithmic stuff, anyhow? (Part 2). *Electronic Design*. Article number 11301, November 7.

20 Carter, B. (2008) Circuit Board Layout Techniques. Texas Instruments, Literature Number SLOD006A, http://www.ti.com (accessed 21 February 2017)

21 Pease, R.A. (1993) What's all this logarithmic stuff, anyhow? *Electronic Design*. Article number 6136.

22 Aguilera, L., Lanza, M., Porti, M., Grifoll, J., Nafría, M., and Aymerich, X. (2008) Improving the electrical performance of a conductive atomic force microscope with a logarithmic current-to-voltage converter. *Rev. Sci. Instrum.*, **79**, 073701

23 Porti, M., Blum, M.C., Nafria, M., and Aymerich, X. (2002) Imaging breakdown spots in $SiO_2$ films and MOS devices with a conductive atomic force microscope. *IEEE Trans. Device Mater. Reliab.*, **2** (4), 94–101.

# 12

# Enhanced Current Dynamic Range Using ResiScope™ and Soft-ResiScope™ AFM Modes

*Louis Pacheco and Nicolas F. Martinez*

*Concept Scientifique Instruments (CSI), 17 Avenue des Andes, Les Ulis 91940, France*

## 12.1 Introduction

Over the past 30 years, atomic force microscope (AFM) has evolved from a microscope to measure just the surface topography [1] to a wide variety of measurement modes that provide ways to characterize other atomic interactions or physical properties such as magnetic field [2, 3], electrostatic field [4], nanoscale dissipation processes [5, 6], thermal conductivity [7, 8], conductivity [9–11], resistance [12], surface potential [13, 14], piezoresponse [15], and Young's modulus [16–18], among others. Additionally, it can be used in ambient conditions, liquid environment, ultrahigh vacuum, temperature gradients, low temperature, and other external conditions.

Although AFM was designed to characterize topography with nanometric resolution, other measurement modes emerged with the development of electronics, advanced probe functionalization, and probe coatings, for example, electrostatic field microscopy (EFM) and Kelvin force microscopy (KFM). Former coatings of AFM probes for electrical characterization were mainly based on gold or other conductive coatings, but these type of coatings were easily removed after a few scans [19–22]. Nowadays, the most common conductive coating for AFM probes is a thin layer of Pt/Ir, Cr/Pt, or doped diamond probes, as well as new developments based on graphene [23]. The improvement of the conductive coating has led to a wide use of what are usually referred to as *electrical modes*, including conductive atomic force microscope (C-AFM), EFM, KFM, scanning capacitance-AFM [24], and ResiScope™ [25, 26].

Today, these methods have become more important in the characterization of samples such as semiconductors, oxides, and organic layers (for solar cells applications). While the quality of advanced coatings has progressed, the techniques available on the market have stagnated. In this article, we discuss the capabilities and advantages of ResiScope™ in overcoming the intrinsic difficulties of standard C-AFM.

*Conductive Atomic Force Microscopy: Applications in Nanomaterials,* First Edition.
Edited by Mario Lanza.

## 12.2 Conductive AFM

The C-AFM mode allows to characterize simultaneously the conductive behavior of the sample in addition to the topography. The setup is based on what is usually referred to as the contact AFM mode where the probe makes continuous mechanical contact with the sample while it is being scanned over the sample. A conductive probe is used, while a DC bias is applied between the tip and the sample. In order to measure the current that flows through the tip (see Figure 12.1), an assembly consisting of linear or logarithmic amplifiers in series with the probe converts the current signal into volts (see Eqs (12.1) and (12.2)). This signal is displayed as an additional channel while the probe is scanned over the sample. The cantilever deflection is used as feedback signal to map the topography at constant deflection, that is, at constant force (see Figure 12.1). Depending on the order of magnitude of the current generated between the tip and the sample, the user must adapt the value of the amplifier gain. Typical values of linear amplifiers range from 0.1 to 10 nA/V. This setup, although very easy to implement and reliable, has a dynamic range of 3–4 orders of magnitude, therefore is too limited to measure samples with both conductive and isolating domains (i.e., to measure both small and high currents)."

An alternative to extend the range of current detection (electronically) up to 6–7 orders of magnitude is a logarithmic amplifier as shown in Figure 12.2, in which, compared to a linear amplifier, the output resistance $R_2$ is replaced by a semiconductor diode:

$$V_s = \frac{R_2}{R_1} V_e \quad \text{(linear amplifier)} \tag{12.1}$$

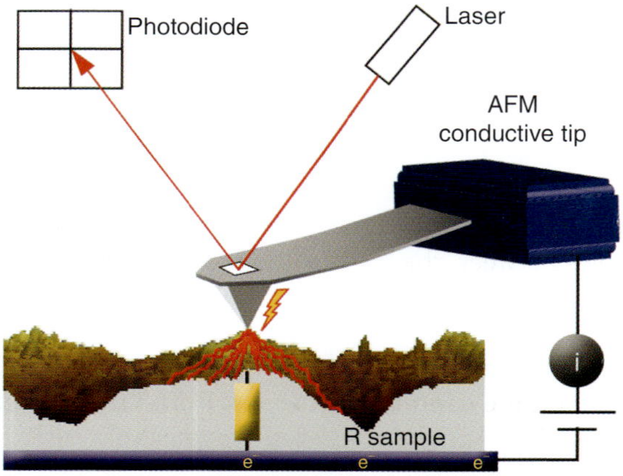


**Figure 12.1** Scheme of the standard conductive AFM (C-AFM) setup. A conductive probe is used to scan the topography in contact mode. A DC bias is applied between the tip and the substrate so that an electric current *i* is induced. This current can be measured with a current-to-voltage converter placed in series with the tip-substrate. The current-to-voltage converter is based on a linear operational amplifier (OPAMP) circuit with a constant gain factor. Typical preset gain factors range from 0.1 to 10 A/V. Current ranges measured in a C-AFM experiment exceed quite often the range provided with a single linear amplifier setup.

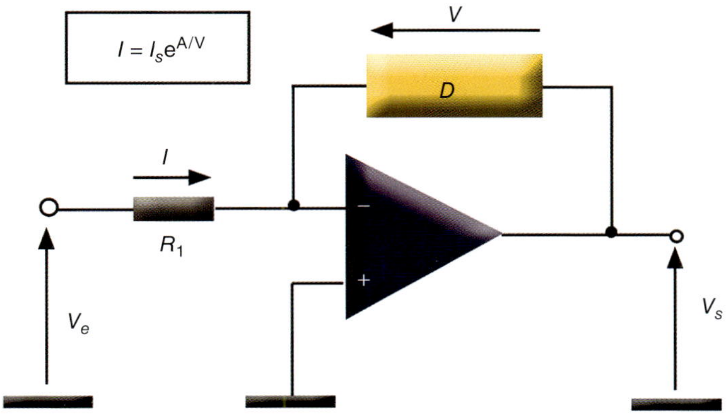


**Figure 12.2** Standard scheme of a logarithmic amplifier. The p–n junction diode is connected to the negative entrance of the OPAMP. As the current characteristic of the diode is exponential, the output voltage $V_s$ has a logarithmic relationship with the input voltage $V_e$. An actual logarithmic amplifier requires more complex electronics to compensate temperature dependence effects of the diode.

$$V_s = A \cdot \ln\left(\frac{V_e}{I_s \cdot R_1}\right) \quad \text{(logarithmic amplifier)} \tag{12.2}$$

where $R_1$ and $R_2$ are the input and output resistances, $A$ and $I_s$ are the current saturation and thermal voltage of the p–n junction diode, respectively. As these parameters are related to the intrinsic features of the semiconductor junction of the diode, they are temperature dependent; and this turns out to be the main drawback of logarithmic amplifiers: they require complex electronic circuitry for temperature compensation and precalibration. Additionally, they require very high integration time for measuring low currents, which would require scanning speeds of about 0.1 lines per second.

The above considerations are related to the electronics required to detect currents that may vary over the sample from a few picoamperes up to microamperes, for example, conducting polymers or organic materials for photovoltaic applications. However, additional physical effects must be taken into account owing to differences on the bias applied between the tip and the sample to detect the current.

One of the effects, is the self-heating of the probe coating due to the Joule effect on samples with high conductivity. As the typical thickness of the thin film conductive coating is 15–30 nm, ant the nano-contact area <1 nm$^2$, the current density becomes too high and the self-induced temperature can reach the melting point of the metal used for the conductive coating. When part of the metal coating evaporates, the conductive measurement is interrupted and the probe has to be replaced by a new one. To avoid this effect, the bias has to be set to very small values (<100 mV), but with highly conductive samples, this might not be enough.

Another phenomenon that may occur in conductive measurement is the bending of the cantilever due to what is the usually referred to as the *bimetallic effect*. The most popular conductive probes consist of silicon of thickness of a few microns with a metallic coating of thickness of a few nanometers. Measuring a sample with significant variations in the current over the substrate (i.e., over conductive and isolant domains) induces temperature variations of the afore mentioned probe coating. The dissimilarities in the thermal expansion coefficient of the silicon and the metallic coating generate stress along the length of the cantilever causing additional bending [27]. The magnitude of this bending depends on the differences between the thermal expansion coefficients of the probe and coating and their respective thickness and Young's modulus [28]. For AFM cantilevers, the relationship between the deflection of the free-end and the temperature may range from 10 to 100 nm/°C. It is important to note that the bimetallic effect is an extended force, that is, it acts along the length of the cantilever, while the interaction force with the sample is a point force localized in the tip at the free-end.

Figure 12.3 shows an example of cantilever bending due to the bimetallic effect. A chip with two conductive cantilevers has been measured with a digital

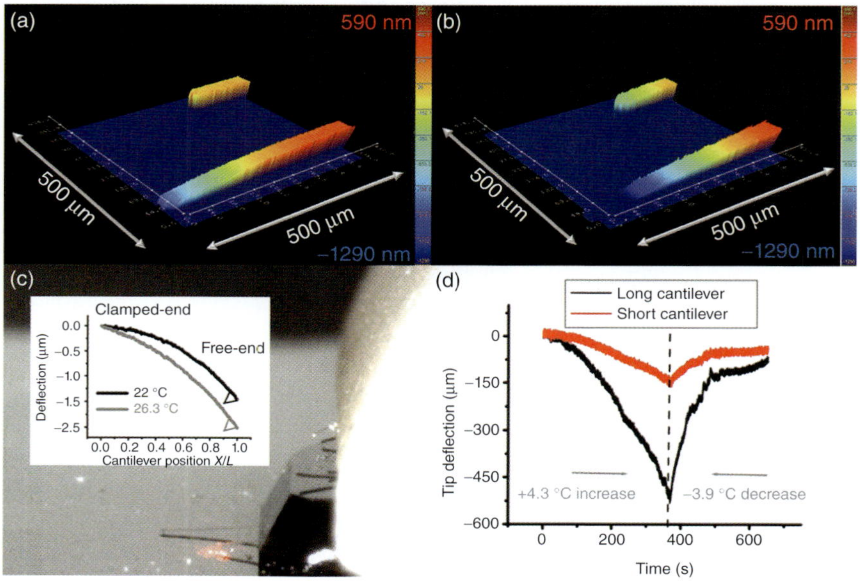


**Figure 12.3** Bimetallic effect of a chip with two Pt/Cr-coated cantilevers with different geometries. (a, b) Three-dimensional measurement of the deflections of the cantilevers at 22 °C (room temperature) and 26.3 °C, respectively. The color scales represent the deflection with respect to the clamped end of the cantilever and have been kept equal in (a) and (b) to show the negative bending with temperature increase of the cantilevers. (c) Photograph with a side view of the same two cantilevers. The inset in (c) shows a profile along the length of the longer cantilever at RT (black line) and 4.3 °C above RT (gray line), respectively. For clarity, the approximate position of the cantilever tip has been added to each profile. (d) Free-end deflection measurement of both cantilevers (red - short cantilever; black - long cantilever) during the temperature cycle. As the temperature increases, the differences in the thermal expansion coefficient of silicon and Pt/Cr metallic coating generates an extended force that bends the cantilevers downward. As the temperature is decreased again, the deflection is restored.

holographic microscope, DHM®, fromLynceeTec (Switzerland). Figure 12.3c shows the lateral view of a chip mounted on a Nano-Observer AFM (CSI, France) before starting a C-AFM measurement. The cantilever geometries are 500 μm × 30 μm × 2.7 μm and 210 μm × 30 μm × 2.7 μm, respectively. The DHM is an interferometric technique which allows to reconstruct a 3D image of a specimen from a single image grab, that is, without scanning, which makes it a live 3D image optical profilometer (CCD camera rate). With the image field provided by the configuration with a 5× objective, 3D measurements with nanometric vertical resolution of both cantilevers were recorded while room temperature was increased and decreased around 4 °C. Figure 12.3a and b, show the deflection of the cantilevers at 22 °C (room temperature) and 26.3 °C (4.3 °C of temperature increment), respectively. The color scale is fixed in both measurements and clearly shows how both of them bend downwards when temperature rises. Figure 12.3d shows a plot with the movement of the free-end for both cantilevers (red corresponds to the short cantilever and black to the long cantilever) during the experiment. The maximum deflection for both cantilevers is −150 and −530 nm, respectively, for a change in temperature of about 4 °C. The negative bending behavior can be explained by the fact that the conductive coating expands more than the silicon probe underneath; this generates a surface stress that forces the cantilever to bend in the negative direction. The inset of Figure 12.3c shows the measured profile along the centered-width of the long cantilever during the experiment. For clarification, a sketch with the approximate position of the tip is added.

## 12.3 ResiScope™ Mode

All the considerations discussed in the previous section reflect the fact that typically the AFM user does not know *a priori* the resistance of the sample; this makes it difficult to control the current that flows through the tip. A different approach to perform AFM current measurements was proposed by a group of the GEEPS Laboratory [12]; instead of carrying out a "direct" local current measurement, they proposed a more complex electronic module that allowed to determine the local resistance as shown in Figure 12.4. This new electronic device allows to control and adjust the appropriate value of the shunt (known resistor) in order to change the measurement range with the advantage at the same time of naturally limiting the current passing through the tip. With this setup, it is possible to manage a current measurement on an actual sample having highly conductive (few tens or hundreds of microamperes) and insulating (few picoamperes) domains by modulating the value of the shunt. This technique is named "ResiScope™."

The schematics of the ResiScope™ module is shown in Figure 12.5. The electrical measurement is made by applying a DC bias between the sample and the C-AFM probe. As an independent measurement, the ResiScope™ measures the sample resistance (and current) through a high-performance amplifier (HPA). During the measurement, the digital signal processor (DSP) chooses the best parameters in real time to optimize the measurement made by the

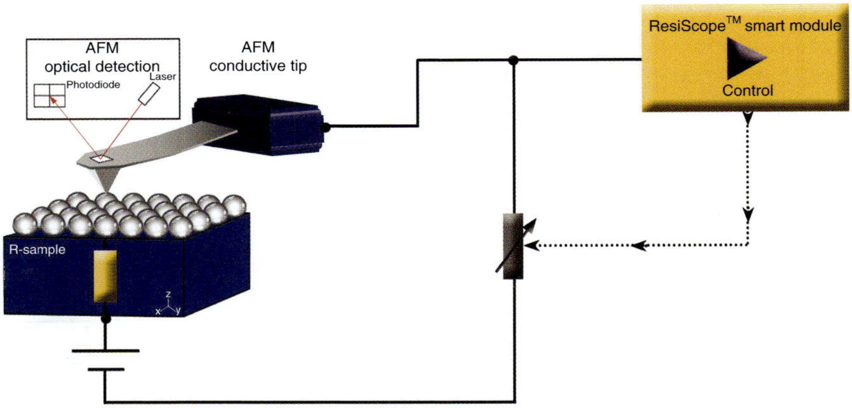


**Figure 12.4** Schematics of the ResiScope™ setup for conductive measurements. A variable resistance is placed in series with the sample. The ResiScope™ module adjusts in real time the amplifier gain and the variable resistance according to the actual value of the resistance of the sample. Conductive measurements performed in this manner allow to minimize the actual current that flows through the sample and have a big dynamic range (100 fA–1 mA).

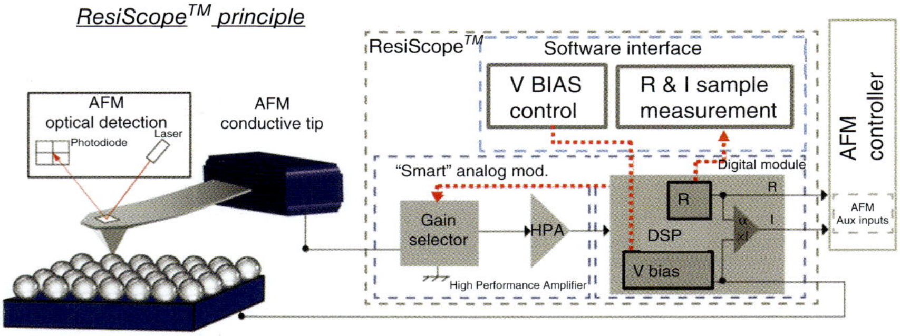


**Figure 12.5** Schematics of the ResiScope™ and Soft-ResiScope™ modules. The ResiScope™ module has three main internal submodules. The "smart" analog module provides the gain selection for the HPA to measure the current and control the variable resistance in series with the sample. The digital module calculates the resistance of the sample and the theoretical current, provided that $V_{bias}$, the total current, and the variable resistance are known. The two magnitudes *R* and *I* (theoretical) can be monitored as auxiliary channels in AFM software. The software interface controls the bias and allows to monitor *R* and *I* (theoretical).

amplifier (HPA). In this manner, as the total current, variable resistance, and bias are known, the digital module calculates the sample resistance *R*. Additionally, it calculates *I* (theoretical) that corresponds to the actual current that would flow through the sample in the absence of a variable resistance in series (i.e., a standard C-AFM setup). The ResiScope™ module provides a wide measuring range (10 orders of magnitude, from 100 fA to 1 mA) and prevents the sample (or the tip) from damage arising from high current flow. This module, with its advanced design and its new concept of measuring, provides several benefits:

1) The actual current that flows through the sample is reduced, thus minimizing associated phenomena such as self-heating of the coating by Joule effect, local oxidation processes [29, 30], controlled environment, and the bimetallic effect.

Figure 12.6 shows a comparison between a standard C-AFM measurement and a ResiScope™ AFM measurement on a sample of micron-size graphene oxide domains over gold substrate. This sample provides a good test for conductive measurements with the ResiScope™ module as it has well-defined conductive and nonconductive domains. In Figure 12.6a and b are shown the topography and current images, respectively, of a standard C-AFM mode (amplifier gain = 1 nA/V) over the oxide graphene/Au sample. The probe used is a DD-FORT-doped diamond probe with micron-size graphene $k = 1.3$ nN/nm (AppNano); the applied force during the image was 14.4 nN. The current image shows typical features of an uncontrolled conductive measurement: on the top part of the image, a stable current image is obtained with well-defined values for the gold (saturated to maximum value 10 nA – red color) and graphene oxide (values <1 nA – blue color). After few lines, the current signal decreases and eventually conductivity is lost. This result is due

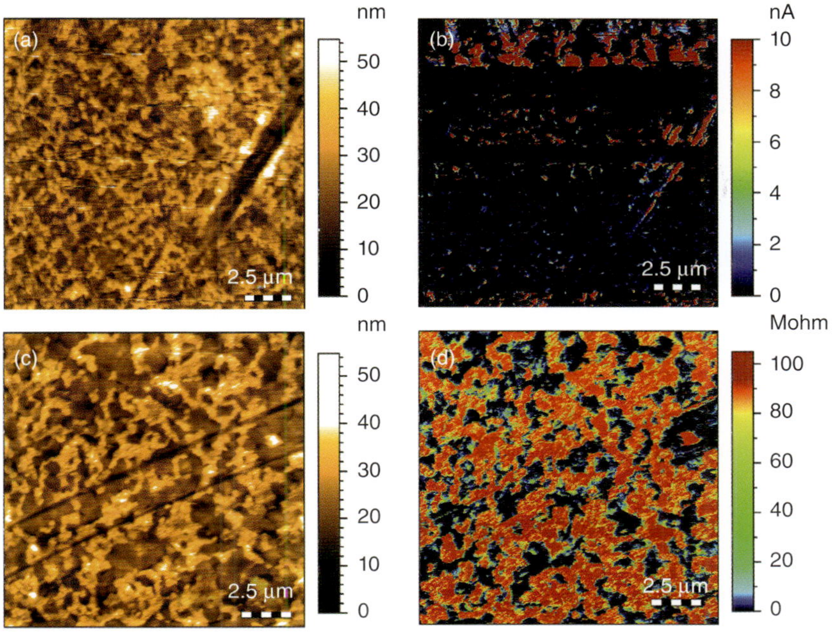


**Figure 12.6** Comparison of conductive measurements on a gold substrate with graphene oxide domains. Topography and current images obtained in standard conductive-AFM mode (a, b). Topography and current images obtained in the ResiScope™ Mode (c, d). The current image shows unstable values of the measured current due to the lack of control in the current that flows through the tip-sample ensemble. On the contrary, the current image obtained in the ResiScope™ mode is stable over the entire scanned area. Additionally, the higher dynamic range allows to measure differences in conductivity on the graphene oxide regions that can be correlated with the thickness of the layer.

to the several aforementioned phenomena (self-heating, bimetallic effect) that affect the sample and probe coating. It can be noticed, for example, that for the unstabilities associated with the bimetallic effect in the topography image, the signal is not completely stable; adjusting the deflection setpoint and the feedback parameters is not enough to compensate the extended force induced by the hot-cold oscillating state.

In Figure 12.6c and d are plotted topography and resistance images of a region similar to that obtained in the ResiScope™ mode. The same cantilever and exerted force as in the images in Figure 12.6a and b were used. A comparison with these reveals the benefits of performing conductive measurements with the ResiScope™ mode: the image is completely stable from the top to the bottom parts of the scanned area. Additionally, the high dynamic range allows to measure the electric signal in the gold areas without saturation (12.6 μA values). This allows to distinguish other domains in the gold areas with less conductivity (around 1.1 μA) which could be probably covered by thin layers of graphene oxide or another less-conductive compound during the growth process.

2) The module has the capability to change fast between different measurement ranges provided in real time. The advantages, as shown in Figure 12.6c and d are a better sensitivity for each current range of the sample together with a high dynamic range.
3) The fast transition of the analog part and high sensitivity of the first amplifier stage (HPA) are added advantages. For these two requirements, it is necessary to design mart electronics that avoid parasitic capacitance problems (gain selector). The electronics of the ResiScope™ module allows to perform electric measurements at a maximum speed of 5 kHz, which is about five times faster than the associated pixel rate of a regular AFM image (1 line/s with 512 points measured on the trace plus 512 points on the retrace).

Figure 12.7 shows two graphs illustrating the high dynamic range and the fast changing rate of the ResiScope™. A set of different resistors with known values was used to simulate a sample. The ResiScope™ was operating in current/

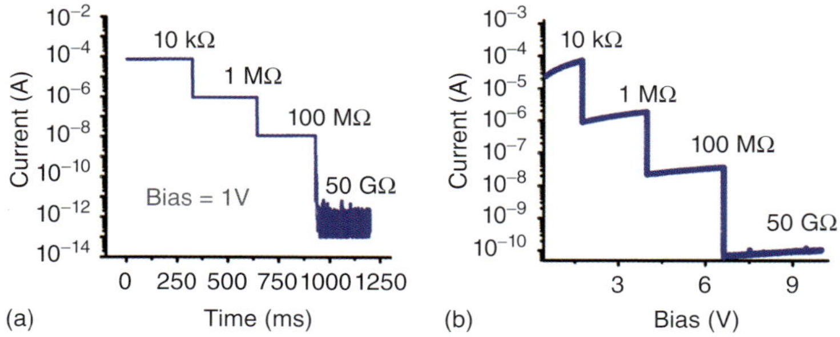


**Figure 12.7** Resistance signal measured with ResiScope™ with a set of several resistors with known values. (a) Logaritmic plot of current versus time for a constant bias of 1 V when the resistor was switched from 10 to 50 GΩ every 250 ms. (b) Logaritmic plot of current versus bias when the resistor was switched from 10 to 50 GΩ while varying the bias from 0 to 10 V. It can be noted that no capacitance effects are seen in the current signal in the transitions between different resistances.

resistance spectroscopy mode; this operating mode is usually referred to as *I–V spectroscopy* in standard C-AFM experiments. The resistor values were 10 kΩ, 1 MΩ, 100 MΩ, and 50 GΩ, respectively. In Figure 12.7a, the measured current values are plotted, while keeping the bias at 1 V. The resistors were switched every 250 ms approximately - the fast velocity can be appreciated as the measured current does not show transients between two resistors. A similar experiment is illustrated in Figure 12.7b, but in this case the bias was varied from 0 to 10 V. In both figures, the current values are plotted in logarithmic scale owing to the high dynamic range of the experiment.

## 12.4 Soft-ResiScope™ Mode

Although the ResiScope™ mode has been proved to improve the standard C-AFM measurements as presented in the previous section, it still has the intrinsic disadvantages of a contact mode, namely, the frictional forces originating from the continuous scan in mechanical contact with the surface. This makes it virtually impossible to extend the use of the ResiScope™ mode to soft samples such as polymers.

Over the years, different approaches have been proposed to overcome this limitation. These approaches are based on the concept of mapping the electrical conductivity of the sample while the tip makes an intermittent contact with its surface. Basically, there are two different concepts of implementing this measurement mode in a similar manner to the already intermittent contact measurement modes for topographic imaging:

1) *Sinusoidal regime.* This approach is based on the dynamic AFM mode, where the cantilever is mechanically excited far below its natural resonance frequency (typically in the frequency range 100–2000 Hz). The vertical displacement of the cantilever at its free end also follows a sinusoidal modulation. When the amplitude of this sinusoidal displacement is high enough, the tip starts to interact mechanically with the substrate in a periodic manner at the bottom part of its sinusoidal movement. This concept is illustrated in Figure 12.8a (top). Figure 12.8a (inset) shows the actual conductive measurements with this concept. The main benefit of this approach is that imaging soft samples is relatively easily done by modulating the amplitude of the movement of the tip. However, the electrical measurement is not quantitative because the force exerted by the tip on the sample is not constant during the contact time.
2) *Linear regime* (*Force spectroscopy volume*). This approach is based on the extensively used force versus distance spectroscopy curves, where the cantilever performs a quasi-static approach-retract cycle toward the sample with constant velocity. With this technique, it is straightforward to choose the force to be exerted on the sample using with the approximation of Hook's law: $F = k{\cdot}z$, where $F$ is the applied force, $k$ the cantilever spring constant, and $z$ the cantilever deflection relative to the rest deflection position. Figure 12.8b illustrates this concept. The main disadvantages of this technique is its intrinsic slowness: with an approach-retract cycle of 1 s, a standard image of $512 \times 512$ measurement points would require 3 days!.

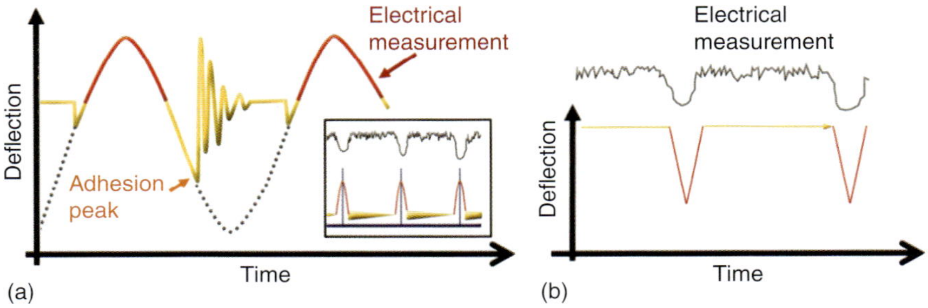


**Figure 12.8** Different approaches to implement conductive measurements with intermittent mechanical contact with the sample. (a) Schematics illustrating the approach based on a sinusoidal regime, typically corresponding with a natural resonance frequency of the cantilever. The inset shows both the cantilever deflection and the electrical signal on a real experiment with this approach. (b) Cantilever deflection and electrical signal on an experiment based on the force spectroscopy curves (force volume mode) where the cantilever performs a quasi-static approach-retract cycle on the surface.

3) *Soft-ResiScope*™. A third approach has recently emerged, combining fast point contact at constant force. This mode is known as *Soft-ResiScope*™ and represents the natural evolution of the ResiScope™ mode. The main difference between the two modes is the lack of frictional forces in the soft-Resiscope™ mode, in which the probe is not scanned on the surface. Additionally, this mode provides the capability of imaging soft samples routinely.

Figure 12.9 illustrates the measurement concept in the Soft-ResiScope™ mode. The AFM probe stays in contact with the sample only for a short period of time with a constant force control. This allows to measure the resistance and the current under the best conditions for quantitative purposes. The tip is then retracted and moved to the next measurement point. It is important to note in Figure 12.8a (bottom) that one critical aspect of conductive measurements in an intermittent contact mode is the pull-off force. The sudden separation of the tip from the surface leads to transient oscillations that decrease exponentially with a decay factor proportional to the ratio $Q/\omega_0$ where $Q$ is the quality factor and $\omega_0$ is the resonance frequency of the cantilever. If this transient phase is too long, the conductive measurement could not have been accurate enough. The Soft-ResiScope™

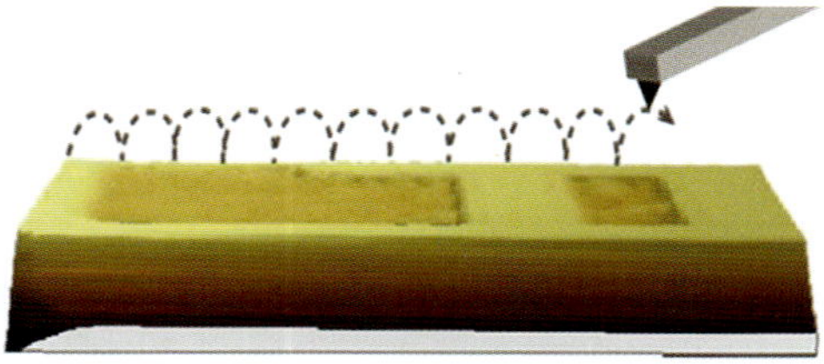

**Figure 12.9** Schematics of the Soft-Resiscope™ mode concept. The cantilever position is modulated at scan rates of a few kilohertz. This modulation accomplishes the requirements of fast contact time and constant applied force during the electrical measurement as illustrated in Figure 12.8.

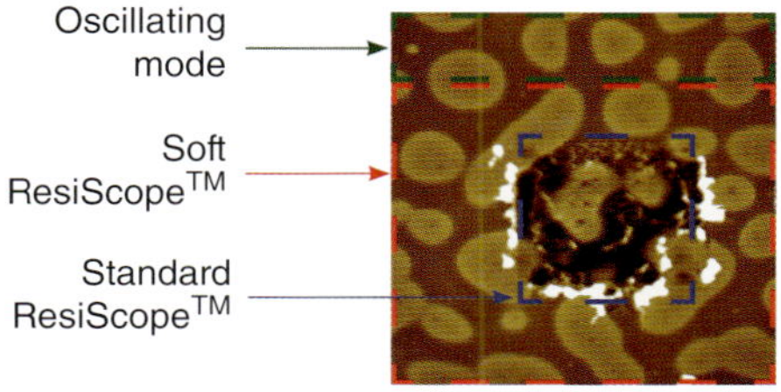


**Figure 12.10** Comparison of the Soft-ResiScope™, ResiScope™, and oscillating modes on a PMMA soft sample. The top part (green dashed rectangle) corresponds to the area imaged in oscillating mode, where the absence of frictional forces does not significantly alter the polymer. The bottom part of the image (red dashed rectangle) has been imaged using the Soft-ResiScope™ mode and provides similar results to those of the oscillating mode. The middle part (blue dashed square) of the image shows a region imaged in contact mode (i.e., the standard ResiScope™ mode) where the polymer has been deformed owing to the frictional forces generated by the cantilever sweep. Image size: 2 μm.

mode can be operated regularly at scan rates of 1 line/s with the same type of cantilevers as those used for standard C-AFM or ResiScope™ mode (1–40 N/m spring constant range, 50–300 kHz frequency range).

In order to illustrate the benefits of imaging fragile samples with the Soft-ResiScope™, a Poly methilmethacrylate (PMMA) polymer blend sample was chosen. This type of polymers are typically imaged in oscillating mode, as this eliminates frictional forces that normally either deform the sample or create permanent damage. Figure 12.10 shows the topography of this sample measured in the oscillating mode (dashed green region), Soft-ResiScope™ mode (dashed red region), and ResiScope™ mode (dashed blue region). A FORT cantilever (AppNano) with $k = 1.7$ nN/nm and $f_0 = 68.3$ kHz was used. The first two modes show similar results in the topographic image with the typical morphology for PMMA estructured in round domains. An applied force of 7.4 nN was used in the Soft-ResiScope™ mode, while an amplitude of 13.6 nm was used in the oscillating mode. However, when imaging in contact mode (using the standard ResiScope™ mode), the same applied force of 7.4 nN clearly damages the sample as some rests of polymer can be seen on the borders of the contact mode scan.

The main difference from the electrical point of view between ResiScope™ and Soft-ResiScope™ is the contact time and friction conditions. While in the ResiScope there is a continuous mechanical contact, in Soft-ResiScope™ the contact is intermittent. This implies, as mentioned earlier, that deflection feedback must be fast and robust to guarantee a quantitative and reliable electrical measurement. To validate quantitatively the electrical measurement of Soft-ResiScope™, a comparison was made between the standard ResiScope™ and Soft-ResiScope™ modes on a well-known sample such as an Static Random Access Memory (SRAM). A DD-FORTA probe (AppNano, USA) with a spring constant of 3 nN/nm and an applied force of 18.3 nN was used to obtain the image in both modes. Figure 12.11a shows a 3D topography plot of the SRAM showing the p-type and n-type Field Effect Transistor (FET) transistors forming the unit bit cell. Figure 12.11b shows the simultaneous resistance image obtained in the

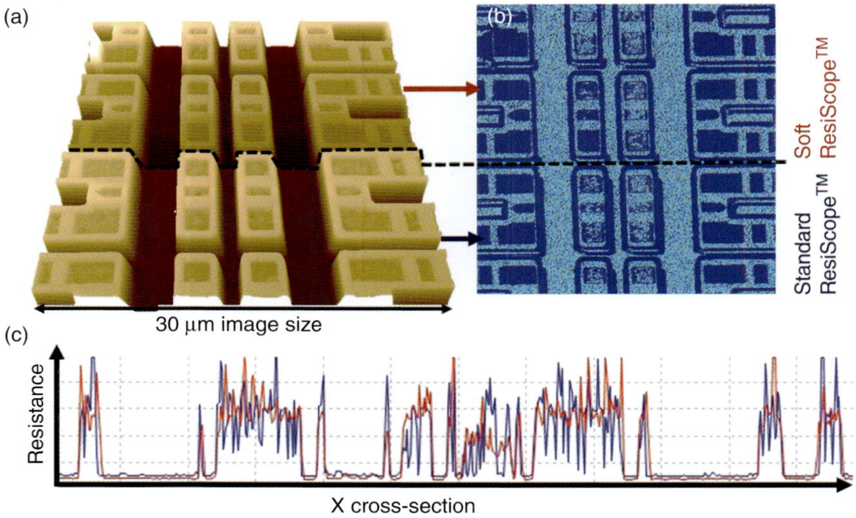


**Figure 12.11** Quantitative similarities of the electrical signal obtained in both ResiScope™ modes. (a, b) Topopgraphy and resistance images on an SRAM memory obtained in Soft-ResiScope™ (top) and standard ResiScope™ (bottom), respectively. The centered black dashed line limits the regions of the image where each one of them was used. (c) Resistance cross section along two equivalent parts of different unit bit cells. Other than the difference in the contact time between the two modes, quantitatively, the resistance values are similar, although the Soft-ResiScope™ mode shows better stability and lateral resolution due to the absence of frictional forces.

Soft-ResiScope™ mode (top part), and in the standard ResiScope™ mode(bottom part). The black dashed line represents the region of transition between both modes and it does not show any significant differences from the electrical point of view. Figure 12.11c depicts the cross-sectional analysis made on the resistance signal, which reveals that the quantitative electrical measurements are comparable in both modes. Additionally, it can be seen in the cross section that the Soft-Resiscope™ mode provides more stable values and better signal on the edges owing to the absence of frictional forces.

Soft electrical materials are becoming more interesting as flexible electronics is based on organic compounds. Electrical characterization of conjugated polymer-based organic solar cells made from poly(3-hexylthiophene) (P3HT) is of interest for this application. Operating in standard C-AFM would damage the organic polymer as already discussed. This is a clear example where the Soft-ResiScope™ mode is the right choice for electrical characterization. Figure 12.12a and b shows topography and resistance images respectively on the P3HT organic solar cell. An ANSCM-PT probe (AppNano, USA) with a spring constant of 3 nN/nm was used for imaging at a constant force of 9.2 nN the topography and resistance images on the active surface of the solar cell. Resistance image shows domains with higher resistance (intense blue regions) in between domains with lower resistance values (light blue domains). This higher resistance areas are correlated to regions where the polymer coating

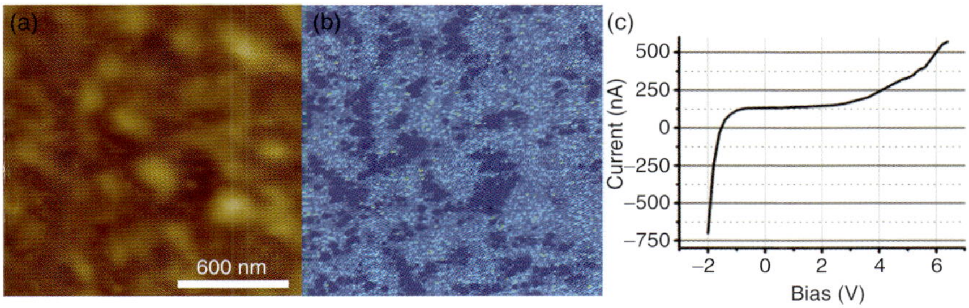


**Figure 12.12** Soft-Resiscope™ mode characterization of an organic solar cell (a) and topography and resistance images on the active surface of the solar cell (b). The resistance image shows domains of higher resistance (intense blue regions) between lower resistance values (light blue domains). The lower resistance areas are correlated to regions where the organic coating partially covers the surface as can be seen in the topography image (the yellow areas correspond to the organic layer, brown areas to partially covered regions).

only partially covers the surface as can be seen in the topographic image (the yellow areas correspond to regions covered with polymer layer, brown areas to those partially covered). In addition to the current/resistance maps, $I$–$V$ spectroscopy curves can be plotted for the regions of interest to determine the electric behavior accurately and model it (see Figure 12.12c).

## 12.5 Conclusions

Electrical characterization with the AFM is finding increasing demand in the AFM community to solve the new research topics in nanotechnology such as energy harvesting, polymer- or organic-based electronics, or new advances in the semiconductor industry. This article shows the advantages of using Resi-Scope and Soft-ResiScope™ modes instead of standard C-AFM to measure current/resistance with nanometric lateral resolution with high dynamic range in real time, as also tip and/or sample protection and reliability. Finally, it has been shown that Soft-ResiScope™ is the best choice to perform conductive measurements on soft samples.

## References

1 Binnig, G., Quate, C.F., and Gerber, C. (1986) *Phys. Rev. Lett.*, **56**, 930.
2 Hartmann, U. (1999) *Annu. Rev. Mater. Sci.*, **29**, 53–87.
3 Kaidatzis, A. and García-Martín, J.M. (2013) *Nanotechnology*, **24**, 165704.
4 Colchero, J., Gil, A., and Baro, A.M. (2001) *Phys. Rev. B*, **64**, 245403.
5 Martínez, N.F., Kaminski, W., Gómez, C.J., Albonetti, C., Biscarinni, F., Pérez, R., and García, R. (2009) *Nanotechnology*, **20**, 434021.
6 García, R., Gómez, C.J., Martínez, N.F., Patil, S., Dietz, C., and Magerle, R. (2006) *Phys. Rev. Lett.*, **97**, 016103–016104.
7 Nonnenmacher, M. and Wickramasinghe, H.K. (1992) *Appl. Phys. Lett.*, **61**, 168.

8 Hinz, M., Marti, O., Gotsmann, B., Lantz, M.A., and Durig, U. (2008) *Appl. Phys. Lett.*, **92**, 3.

9 Lanza, M., Porti, M., Nafría, M., Aymerich, X., Whittaker, E., and Hamilton, B. (2010) *Rev. Sci. Instrum.*, **81** (10), 106110.

10 Nalladega, V., Sathish, S., Jata, K.V., and Blodgett, M.P. (2008) *Rev. Sci. Instrum.*, **79** (7), 073705.

11 Liu, Y., He, J., Kwon, O., and Zhu, D.M. (2012) *Rev. Sci. Instrum.*, **83**, 013701.

12 Houzé, F., Meyer, R., Schneegans, O., and Boyer, L. (1996) *Appl. Phys. Lett.*, **69**, 1975.

13 Sadewasser, S. and Lux-Steiner, M.C. (2003) *Phys. Rev. Lett.*, **91**, 266101.

14 Abraham, D.W., Williams, C., Slinkman, J., and Wickramasinghe, H.K. (1991) *J. Vac. Sci. Technol., B*, **9**, 703.

15 Gruverman, A. (1996) *J. Vac. Sci. Technol., B*, **14** (2), 602.

16 Herruzo, E.T., Perrino, A.P., and Garcia, R. (2014) *Nat. Commun.*, **5**, 3126–3204.

17 Sahin, O., Magonov, S., Chanmin, S., Quate, C.F., and Solgaard, O. (2007) *Nat. Nanotechnol.*, **2**, 507–514.

18 23. Cronin-Golomb, M. and Sahin, O. (2013) *Beilstein J. Nanotechnol.*, **4**, 243–248.

19 Kelley, T., Granstrom, E., and Frisbie, C.D. (1999) *Adv. Mater.*, **11**, 3.

20 O'Shea, S., Atta, R., Murrell, M., and Welland, M. (1995) *J. Vac. Sci. Technol., B*, **13**, 1945.

21 Klein, P. and McEuen, P. (1995) *Appl. Phys. Lett.*, **66**, 2478.

22 Nxumalo, J., Shimizu, D., and Thomson, D. (1996) *J. Vac. Sci. Technol., B*, **14**, 386.

23 Lanza, M., Gao, T., Yin, Z., Zhang, Y., Liu, Z., Tong, Y., Shen, Z., and Duan, H. (2013) *Nanoscale*, **5**, 10816–10823.

24 Martin, Y., Abraham, D., and Wickramasinghe, H. (1988) *Appl. Phys. Lett.*, **52**, 1103.

25 Alvarez, J., Ngo, I., Gueunier-Farret, M., Kleider, J.P., Yu, L., Cabarrocas, P., Perraud, S., Rouvière, E., Celle, C., Mouchet, C., and Simonato, J.P. (2011) *Nanoscale Res. Lett.*, **6**, 110.

26 Alvarez, J., Kleider, J.P., Houze, F., Liao, M.Y., and Koide, Y. (2007) *Diamond Relat. Mater.*, **16**, 1074.

27 Ramos, D., Mertens, J., Calleja, M., and Tamayo, J. (2007) *Sensors*, **7**, 1757–1765.

28 Stoney, G. (1909) *Proc. R. Soc. London, Ser. A*, **82**, 172–175.

29 Avouris, P., Martel, R., Hertel, T., and Sandstrom, R. (1998) *Appl. Phys. A*, **66**, S659.

30 Garcia, R., Calleja, M., and Perez-Murano, F. (1998) *Appl. Phys. Lett.*, **72**, 2295.

# 13

# Multiprobe Electrical Measurements without Optical Interference

*David Lewis, Andrey Ignatov, Sasha Krol, Rimma Dekhter, and Alina Strinkovsky*

*Nanonics Imaging Ltd. Har Hotzvim Hi-Tech Park Beit Bynet, 19 Hartum St. Jerusalem 97775, Israel*

## 13.1 Introduction

This chapter focuses on the nascent technology of multiprobe scanned probe microscopy (SPM) and its applications in advanced SPM studies. These include a wide variety of multiprobe applications including optical, thermal, electrical, nanolithographic, magnetic, and many other important combinations. This chapter specifically highlights electrical applications while providing a glimpse into other multiprobe technologies.

The importance of the development of multiprobe atomic force microscopy (AFM) for nanotechnology has been well recognized. Rocco *et al.* [1] reported that multiprobe SPM is highlighted as an important strategic characterization tool in nanotechnology research in the coming decade. This prediction is evolving at a pace that should make multiprobe SPM one of the most important nanocharacterization tools of the coming era. This is further fueled by the explosion of two-dimensional (2D) materials and devices and their multifunctional characterization needs.

It is important to highlight scanning tunneling microscopy (STM), a subset of SPM, which made several early attempts at multiprobe configurations. Understanding the limitations in these attempts highlights why general applications of SPM multiprobe technology have been limited until now. Such limitations specifically refer to the probe's functionality and geometry as well as the ability to rapidly bring multiple probes into close proximity without damage.

In STM the probe is a simple wire that has a self-contained feedback, the tunneling current, between this wire and a surface that is uniformly conducting. This feedback permits relatively limited multiprobe geometries since if the probes are brought into physical contact the tunneling feedback would be nullified. The importance of physically bringing the probes into contact is essential for one of the main applications of multiprobe technology. This is the ability to measure physical properties as a function of distance on a nanometric scale.

In spite of its limitations, the STM method has been a forerunner of multiprobe applications and gave a glimpse of their importance. Such feedback already provided important information on the electrical properties of the materials being

*Conductive Atomic Force Microscopy: Applications in Nanomaterials*, First Edition.
Edited by Mario Lanza.

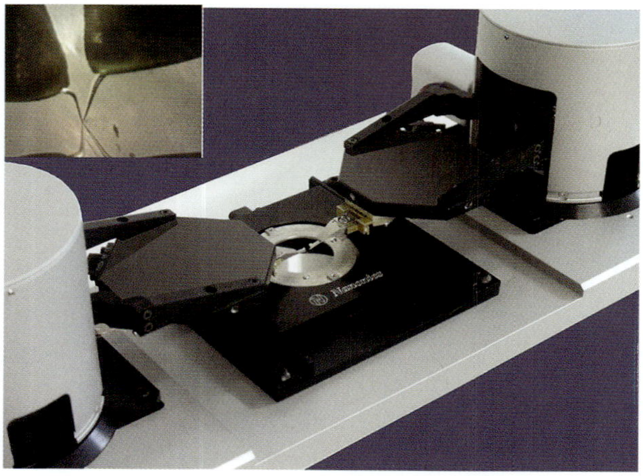

Figure 13.1 The Nanonics Imaging MultiView 4000™ MultiProbe SPM System. Two probes are seen, each on an independent SPM scanner. In the inset image (upper left corner), two probes touching in a multiprobe configuration are shown. Picture provided courtesy of Nanonics Imaging.

studied, but were restricted by the requirement of uniform conductivity of the substrate.

Since both probe geometry and the electrical interference of tunneling probes restrict close physical approach of the probes, there has been the need to use a nanoimaging tool to facilitate probe positioning. Literature has described the use of an online scanning electron microscope (SEM) [2–9] which adds a level of complication, limits work to a vacuum environment, and can lead to damage of the sample's electrical characteristics due to the imposed electron beam imaging.

In 2009 the first paper was published introducing an ambient multiprobe SPM system that addressed all these limitations [10]. This multiprobe platform was based on a patent first filed in 2001 [11]. The patent combines singular technology for both instrument geometry and probes as developed by Nanonics Imaging Ltd. (see Figure 13.1).

The measurements reported in the paper focused on pump probe near-field scanning optical microscopy (NSOM) imaging of plasmonic transport, which could not be addressed with the single-probe technology. The paper demonstrated a multiprobe configuration with NSOM probes and revealed the possibilities that could emanate from such new multiprobe technologies.

An exciting aspect of the measurements in 2009 was the discovery that one probe could come in contact with the second probe without damage. It was even possible to image the second stationary probe with the scanning probe. This addressed the important criteria of allowing multiple probes to come into close contact. This is shown in Figure 13.2 where one probe is seen in the AFM image generated by the second probe. A more recent demonstration by Klein *et al.* [12] showed that the feedback of a single probe was extremely sensitive to a second probe's nanometric presence. This indicated that even correlated feedback between the two probes is possible.

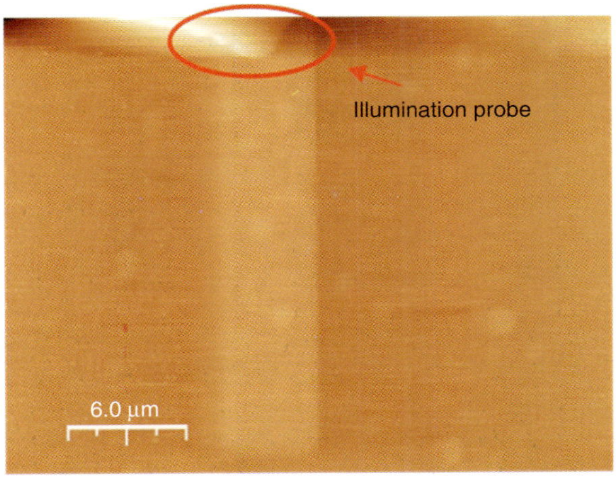


**Figure 13.2** In this image, the scanning probe images the stationary NSOM illumination probe showing that these probes can remain in contact even while one probe images a second one. Image provided courtesy of Prof. Francesco Stellacci (at the time at Massachusetts Institute of Technology (MIT)) who produced this result with the first Nanonics multiprobe system.

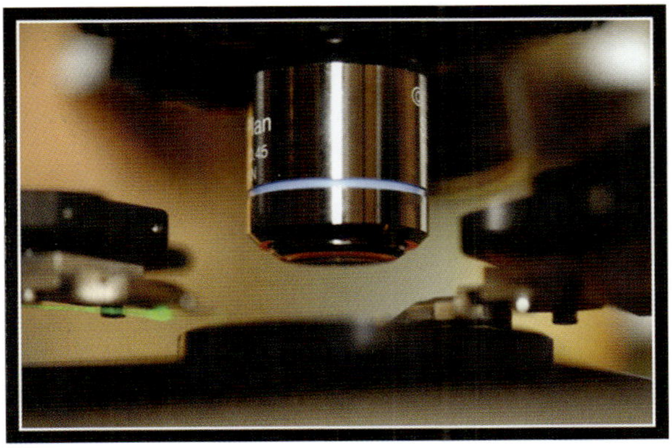

**Figure 13.3** Close-up view of the Nanonics MultiView 4000 system. Two probes in close proximity are shown under an objective highlighting the open optical access, permitting easy viewing, and rapid alignment of multiple probes. A wide variety of high numerical aperture objectives can be used with the system. Picture courtesy of Nanonics Imaging.

An additional important feature was the probe design. In addition to the probes being multiprobe friendly, they also allowed for a clear optical axis from above. This permits rapid placement of the probes in close proximity of one another by optical imaging. Such probe placement is further enhanced by the detecting of one probe's presence by the other probe [12] (Figure 13.3).

Further details on the design of the system and the probes are discussed in the next section.

## 13.2 The Multiprobe Platform: Design and Key Features

The design of the multiprobe platform is based on singular designs for the scanners, the probes, and the feedback mechanism.

### 13.2.1 The Scanner

Cylindrical piezos have been at the heart of modern AFM systems for decades.

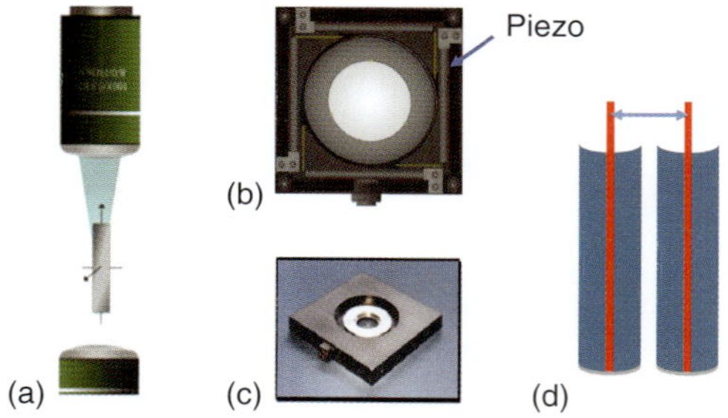


**Figure 13.4** (a) A drawing of a cylindrical piezo, routinely employed in AFM systems, showing why such piezos make optical integration challenging. (b) A drawing of a Nanonics Flat Scanner™ showing the four-piezo design. (c) Picture of Nanonics Flat Scanner™ design which permits both optical access and multiprobe operation. (d) A drawing of two standing cylindrical piezos showing it is impossible to bring the centers (probes) together. Picture and drawings courtesy of Nanonics Imaging.

Such piezos have the ability to effectively bend with the application of voltage in the direction perpendicular to the cylinder axis. This bending can move a sample in the X and Y direction by tens of microns. Movement in Z, which occurs by stretching of the cylinder vertically, is very limited. Therefore, in general, Z movement extent when using such devices is very restricted. In addition, in terms of designing multiprobe systems, the pillar-like design of the piezo scanner in conventional AFM constructions does not allow for easy or close approach of one such scanner to another (see Figure 13.4).

To address these issues, a new design paradigm was developed where cylindrical piezos are placed in a horizontal position. This is an essential feature of the scanner that is used in the multiprobe system and allows for an ultraflat geometry. As a result, the abilities of AFM scanning is broadened significantly.

To understand this aspect of the scanner design, see the diagrammatic representation in Figure 13.4b. The diagram shows four piezos configured in a square geometry (arrow pointing at one piezo). With such a structure, what was previously an X or Y motion now becomes a Z motion. This has the added benefit of solving the Z restriction inherent in upright piezo geometries. X and Y scanning is done with two parallel piezos working in a flexure motion in X, while the two others move the sample in Y. With this new geometry the scanner is very thin, only 7 mm, providing a form factor that permits a variety of unique compact SPM geometries. This compact structure is critical for multiprobe SPM, allowing multiple probes to be mounted on separate scanners as shown in Figure 13.5.

The new design offers several important new functionalities. As can be seen by the opening in the center of the system in Figure 13.5, the configuration offers complete open optical access. This allows easy optical viewing of the sample from both above and below. The probe is exposed, allowing for effective side viewing. Optical access leads to many important combinations such as SPM integration with confocal, fluorescence, and Raman microscopy. For example, the large Z range of the scanner readily allows for effective confocal sectioning. Furthermore, deep-trench imaging, often important in device structures, is easily possible. Finally, a slip–stick motion is made available with

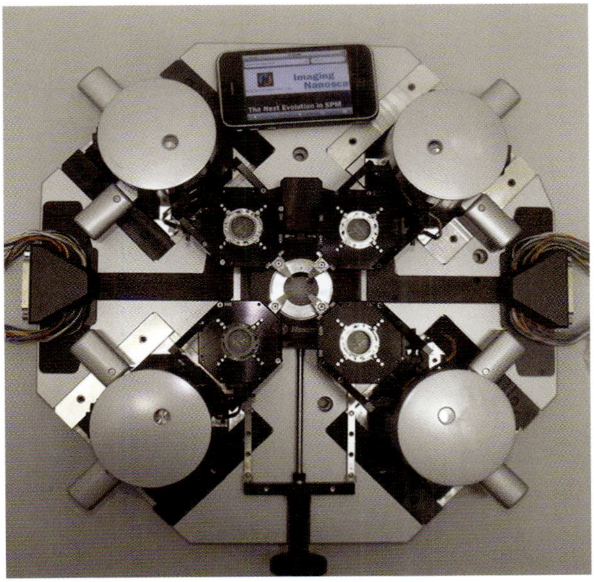


**Figure 13.5** A MultiView 4000™ MultiProbe SPM with four SPM probes. Picture provided courtesy of Nanonics Imaging.

the piezos for large-range sample positioning with a step resolution of 1 μm or larger. Such features allow the multiprobe SPM configuration to be an effective characterization tool for a wide variety of samples and applications.

### 13.2.2 The Probes

An important element of multiprobe design are the probes. Standard AFM probes were never designed with multiprobe operation in mind. The typical AFM probe is made from silicon or silicon nitride with a pyramid design under the cantilever, where the apex is used for imaging. With such probes it is not possible to bring the imaging points of each probe very close to one another. In order to truly realize a multiprobe nature, it is important to have a probe design that can allow bringing the probes easily into close contact.

Glass technology was one of the first to be used to measure forces [13]. This was only a year after glass pulling technology was applied to SPM as an NSOM probe [14]. The Lewis laboratory also pioneered the use of glass probes for AFM imaging [15]. This has now been expanded to a variety of functional glass probes including nanopipette probes for chemical writing [16] and nanoparticle probe tips for producing plasmonic enhancement [17].

Glass probes are also ideal for thermocouple [18], thermal conductivity, and nanothermal light sources [19]. These latter probes are formed from nanowires of platinum enclosed in glass; and the glass can even be coated with metal, which is an ideal structure for the measurement of electrical conductivity. All such functional probes are multiprobe friendly.

This NanoToolKit™ of probes can be purchased today from Nanonics Imaging Ltd, Jerusalem, Israel. Such probes integrate transparently with this multiprobe platform. The extended design of the probe tip permits optical integration

**Figure 13.6** SEM image of a glass AFM probe from Nanonics while controlled pressure is applied, emphasizing the glass probe strength and robustness. Image provided courtesy of Nanonics Imaging.

from above, even with high numerical aperture (NA) objectives, such as a 0.9 NA air objective.

Finally, when glass is pulled to such dimensions, unlike silicon, it is not brittle but very flexible. This is shown in the SEM photograph in Figure 13.6 where controlled pressure is applied to a probe without any damage.

Specifically, for electrical measurements a cantilevered glass probe encasing a metal nanowire was developed. The glass can further be coated by metal to create a true coaxial probe. It is also possible to produce probes of cantilevered metal nanowires without any glass coating. In the next section we explore the most effective method for using these probes in electrical measurements.

### 13.2.3 Feedback of Multiprobe Systems

As noted in the introduction, conventional AFM feedback is based on a laser that bounces off the back of an AFM probe, usually a silicon cantilever, and measures its amplitude as shown in Figure 13.7.

There is an alternate feedback method based on measuring the resonance frequency instead of the amplitude. High Q factor probes are ideal for such frequency feedback, since the resonance frequency of these probes becomes very sharp allowing to accurately know when it is in or even slightly out of resonance. This is especially relevant when one is working in intermittent contact mode. However, in silicon probes the cantilever is very soft, giving the probes a low Q factor. Even silicon probes that have a large force constant such as 300 N/m are not stiff enough to exhibit high Q factors with very sharp resonances. With beam bounce feedback, contact mode is commonly used for electrical measurements. However, contact mode with low Q factor probes can be damaging on both the sample and the probe tip.

For more than a decade, it has been realized that tuning-fork-based probes provide an excellent alternate method of feedback with ultrasensitive characteristics [20] due to their very high Q factor. In this method the resonance frequency of

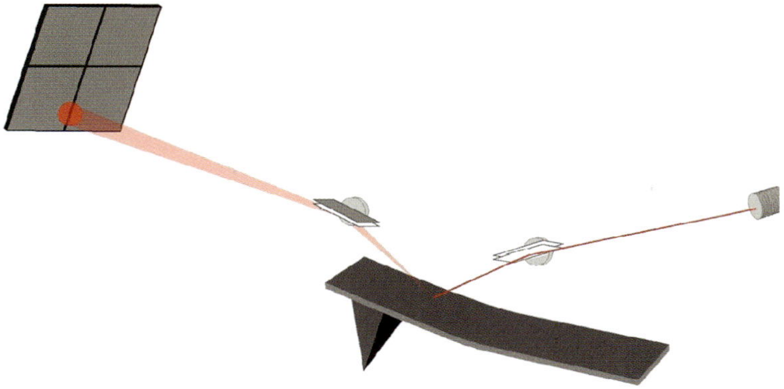

**Figure 13.7** An illustration of a traditional laser-based AFM feedback mechanism where a laser bounces off the probe's back and movement is monitored by a position sensitive detector (PSD).

the tuning fork is the monitored feedback. Today there are accurate equations that correlate the frequency of a tuning fork to the force imposed on a surface [21]. It has been further demonstrated, using the multiprobe platform described in this chapter, that a tuning fork probe can measure the force of a single photon [22]. Thus, tuning fork probes, which operate in intermittent contact mode, with ultrasensitive force feedback and ability to accurately lock-in to the signal, are ideally suited for electrical measurements. Their great sensitivity can alleviate the damage on both the probes and the sample.

Figure 13.8 depicts a tuning fork probe oscillating the cantilever perpendicular to the surface. Even though electrical measurements with one probe have been demonstrated as far back as a decade and a half ago [23], multiprobe measurements have been challenging [24] and have required another step forward in order to ensure repeatability and ease of use.

**Figure 13.8** Depiction of probe mounted on a tuning fork, where the tip is oscillated perpendicular to the sample allowing for "normal force" AFM operation.

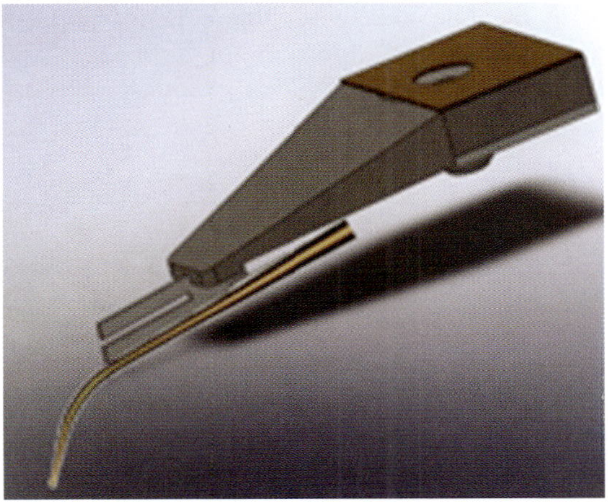

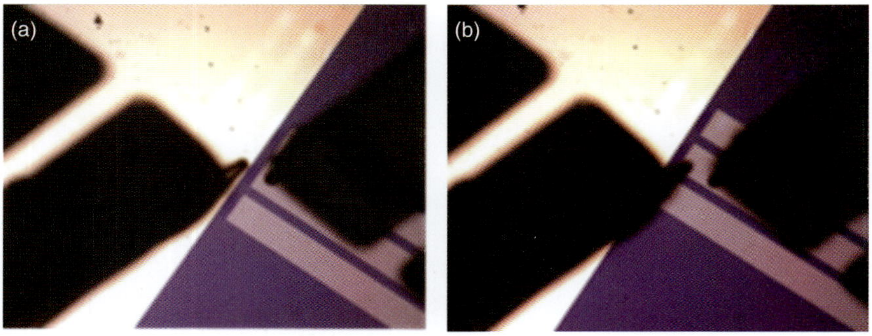


**Figure 13.9** (a) Multiprobe electrical measurement configuration where the tuning fork is rotated 90° and is parallel to the surface. An ideal geometry for multiprobe electrical measurements since the oscillation of the tuning fork is parallel to the surface allowing for a strong contact with the surface without damaging it. (b) Picture showing the two probes positioned on one gold line. Both images were taken through an optical microscope with a 10× objective. Images provided courtesy of Nanonics Imaging.

An ideal configuration with tuning forks for such multiprobe electrical measurements is shown in Figure 13.9. In this configuration, the tuning fork is rotated 90° and is parallel to the surface. This is an ideal geometry for tuning fork electrical measurements since the oscillation of the tuning fork is parallel to the surface on which contact has to be established. The position relative to the surface of the tip does not change in distance. Furthermore, the ultrasmall lateral movement *vis a vis* the surface easily penetrates water layers that are often associated with surfaces under ambient conditions and can prevent effective electrical contact.

This now allows for effective single- and multiprobe electrical measurements with tuning forks. It permits the great sensitivity of tuning forks to be brought effectively into the field of nanoelectronics. Furthermore, the lack of a feedback laser is an advantage for electrical measurements. Specifically, when beam bounce methods are used in a variety of semiconductor and 2D materials, the feedback laser scatters off the cantilever and onto the surface. This can generate carriers in the material producing electrical results with photoconductivity artifacts. The ability to independently impose light perturbations, without artifacts, is very important for future applications of such measurements in terms of both semiconductor systems and 2D materials.

## 13.3 The Present and the Future

Functional multiprobe measurements are only at the start of opening new horizons in SPM in general and more specifically in electrical measurements of nanodevices.

### 13.3.1 AFM Multiprobe Application

Presently, besides a budding group of papers that attempt to achieve multiprobe functional imaging, there are few papers that aim at multiprobe AFM operation.

One such paper using nanonics probes, but with a home-built system, has elegantly shown how multiprobe can be used for metrology studies both in terms of device dimension and in sidewall imaging [25] or without glass probes [26].

### 13.3.2 Optical Multiprobe Operation

NSOM was one of the first applications of multiprobe demonstrated in 2009 by Stellacci *et al.*, as already noted [10]. Apertured NSOM probes have the unique advantage of launching plasmons, as the light coming out of the probe generates all k vectors. Ren *et al.* showed how important it was to be able to launch plasmons with an NSOM probe and detect the propagation of plasmons in a plasmonic structure with a second NSOM probe [27]. Furthermore, Peschel *et al.* [28] have shown effectively the importance of such multiprobe NSOM in the capability to uniquely launch and detect dark plasmons. Dark plasmons are those that propagate via quadrupolar coupling and are transported over a large distance. Finally, Janunts *et al.* have elegantly implemented polarization measurements with such a multiprobe system [29].

It is also of considerable importance in some of these near-field optical measurements that the ability to launch dark and bright plasmons could be done not only at a specific range of wavelengths but also at all energies. Once again, tuning fork multiprobe techniques singularly allow for this to be achieved. It has been shown [30] that with tuning fork feedback, unlike in beam-bounce-based AFM with silicon cantilevers, there is no jump to contact. Jump to contact is a known AFM phenomenon where the cantilever loses control and is drawn into the sample. This usually occurs at the last 20 nm of the tip's approach to the sample. The lack of jump to contact in tuning-fork-based systems lends the ability to switch between tunneling and AFM feedback. Tunneling feedback allows for the use of one electrical probe to excite all plasmons with all energies while a second NSOM probe measures the plasmonic propagation.

Mixing NSOM and AFM probes also has been shown by Tischler *et al.* [31] for fluorescence measurements, with one probe having the maximum resolution of NSOM and the other showing high-resolution AFM. By adding a sharp AFM probe in parallel, the sample can be imaged at the highest AFM resolution. This methodology can be widely applied to any functional probe that is larger than a standard AFM probe.

### 13.3.3 Thermal Measurements

As explained in Section 13.2.2, there are a variety of multiprobe friendly probes available from the NanoToolKit™. Each of the multiprobe measurement protocols described is based on a specific type of these probes.

There are several possible probes that could be used for multiprobe nanothermal measurements. Aimed at heat transport studies, all of these probes are nanowires generally of platinum embedded in glass. A diagrammatic representation is seen in Figure 13.10.

As diagrammatically indicated in this figure, such probes are formed of glass with two platinum nanowires. These wires are either fused at the tip to form a

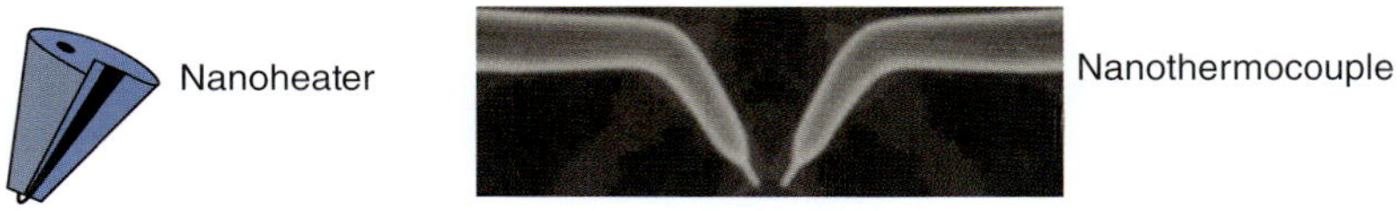


**Figure 13.10** Image of two multiprobe thermal probes. In such a configuration, one nanoheater probe can heat up the sample and the second nanothermocouple probe measures the thermal transport as a function of distance. Image provided courtesy of Nanonics Imaging.

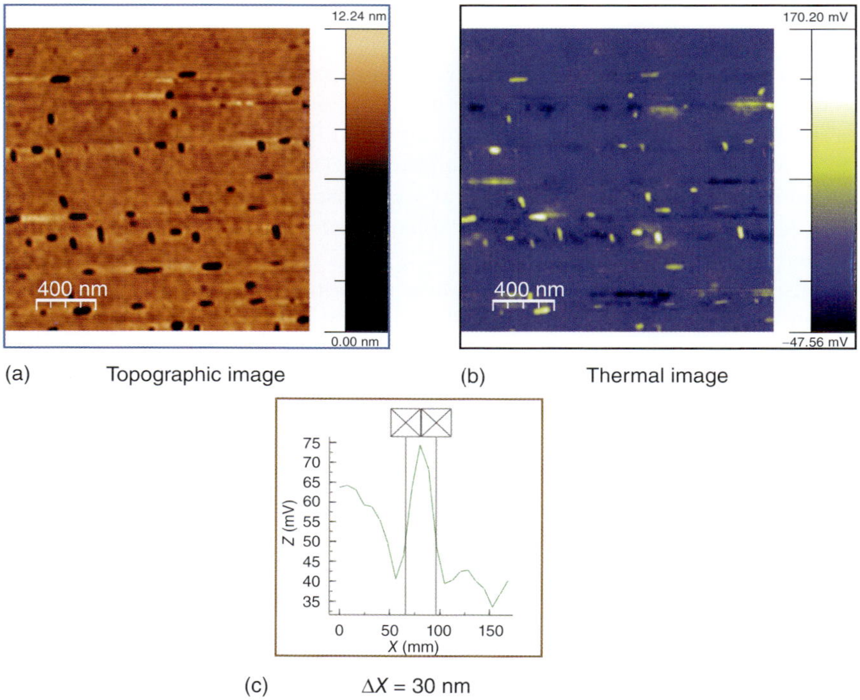


**Figure 13.11** (a) Simultaneously obtained topographic and (b) thermal conductivity images on test sample of voids in a semiconductor chip provided by Sematech. (c) The line-scan across the thermal conductivity image demonstrates a lateral resolution of 30 nm. Images provided courtesy of Nanonics Imaging.

nanoheater or a thermal conductivity probe. A thermal couple where different metals make contact at the tip of the probe is also a possible option.

An example of the resolution of one such probe is shown in Figure 13.11 where semiconductor voids are detected with the thermal conductivity probe and a resolution of 30 nm in the thermal conductivity image is obtained.

The first step in all multiprobe operations is to bring the two probes together. Such an operation can easily be achieved as seen in Figure 13.12. The figure is composed of screenshots that can be seen in a movie at this link: https://www.youtube.com/watch?v=kRewdGQD_l0

This section of the movie shows the process of setting up such a multiprobe measurement by bringing the probes quickly into position.

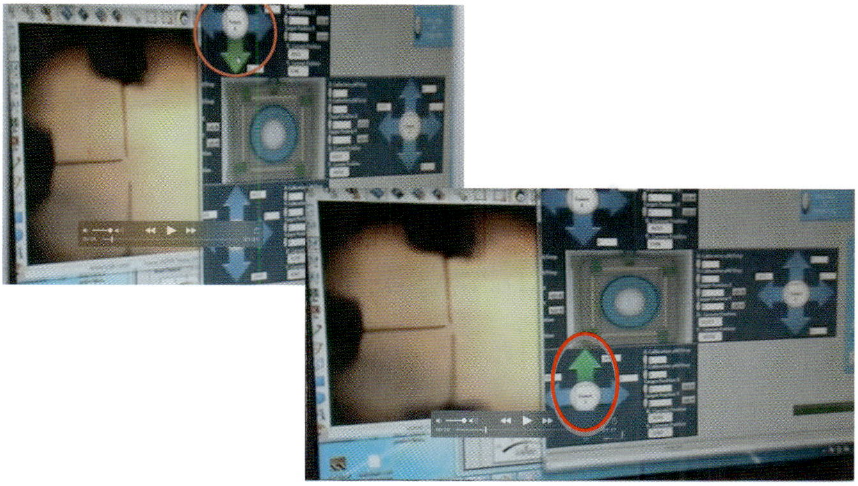

**Figure 13.12** Probes can be easily visualized by a high numerical aperture objective directly from above and brought into close distance to one another via software. Screenshot provided courtesy of Nanonics Imaging.

As shown in Figure 13.12, the controller has hybrid qualities where the feedback is digital, while operations such as large probe motion use online analog controls similar to a joystick. In this figure, the bottom probe is moved by pressing the correct arrow, bringing the probe in the second picture in close enough proximity for multiprobe operation. This movement can be accomplished even when the probes are in feedback. As the probes approach each other on a nanometric scale, the probe feedback indicates this close approach [12]. Using these initial steps, which are general for all such measurements including thermal and optical, such transport measurements are quickly and easily set up.

For the thermal transport measurements, a nanoheater is placed at one end of the sample as a point heater. This probe is maintained at a stationary nonscanning position while a thermocouple probe is scanned into contact with the nanoheater probe. The results of such a measurement are shown in Figure 13.13.

In this measurement, the topography image (Figure 13.13a) shows the ability of one probe to image another probe as noted above in the optical protocols. In this case, the thermocouple probe has scanned the nanoheater (highlighted in the topography image with a red circle). The 2D temperature map shows the distribution of temperature from purple (lower temperature in this color scheme) to red, the hottest regions of the metallic gold strips on silicon, where the nanoheater is sitting. The 3D collage of these images is most interesting since it more distinctly delineates the temperature distribution of the nanoheater.

### 13.3.4 NanoElectrical Transport Measurements

Probably the most exciting possibilities within the context of the subject of this chapter are the nanoelectrical transport capabilities of multiprobe systems. These possibilities can extend from the simple to the more complex.

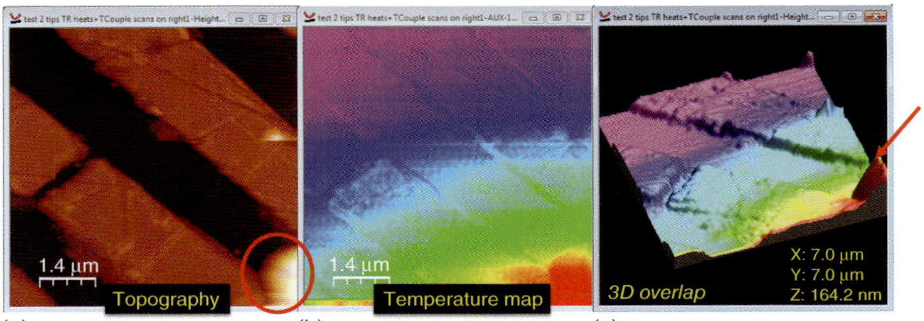


(a) (b) (c)

**Figure 13.13** A scanning thermocouple probe measuring heat transport of gold electrodes on silicon. (a) The topography map where the red circle shows the position of the nanoheater probe imaged by the scanning thermocouple probe. (b) The temperature map. (c) A 3D collage of the temperature map overlayed on the topography. The red arrow highlights the nanoheater probe. Image provided courtesy of Nanonics Imaging.

Multiprobe electrical measurements in the past have been limited and problematic [24]. The issues that were apparent with home-built multiprobe approaches have been widely addressed above. Presently, a variety of platforms are available from room temperature to vacuum operation to variable and ultralow temperature (see Figure 13.14).

These platforms have allowed for a wide spectrum of multiprobe measurements, some of which have been described in this chapter.

In terms of this section, however, it should be mentioned that a multiprobe measurement platform based on beam bounce feedback technology with cantilevered metal wire probes has been available for electrical measurements. This is the system introduced by Multiprobe Inc., Santa Barbara, California. This platform initially could only work in contact mode, potentially causing damage to the samples. Some of these problems were alleviated by solving issues associated with intermittent contact operation. Nonetheless, there is a paucity of general applications as there was little published literature with this system. The focus of this system was primarily semiconductor inspection and the noise floor was about 10 times higher than a standard AFM. Force sensitivity was not addressed since hard contact was required. Today Multiprobe, Inc. has been bought by DCG Systems, Inc. to add to their nanoprobing platform based on non-AFM probe stations for industrial applications.

This platform was distinguished in several ways from the platform that was introduced earlier by Nanonics Imaging Ltd, Jerusalem, Israel. The first difference was the choice of feedback method. In addition, as the Multiprobe system was aimed primarily for semiconductor probing, the noise level was higher than conventional AFM systems. Integration with spectroscopy was not attempted and the lack of a complete set of multiprobe-friendly probes limited other modes of functional imaging. Finally, solutions in vacuum and at variable and ultralow temperature were not designed.

The lack of high feedback sensitivity did not allow for a very close tip–sample contact. On the other hand, conventional tuning fork feedback shows very high

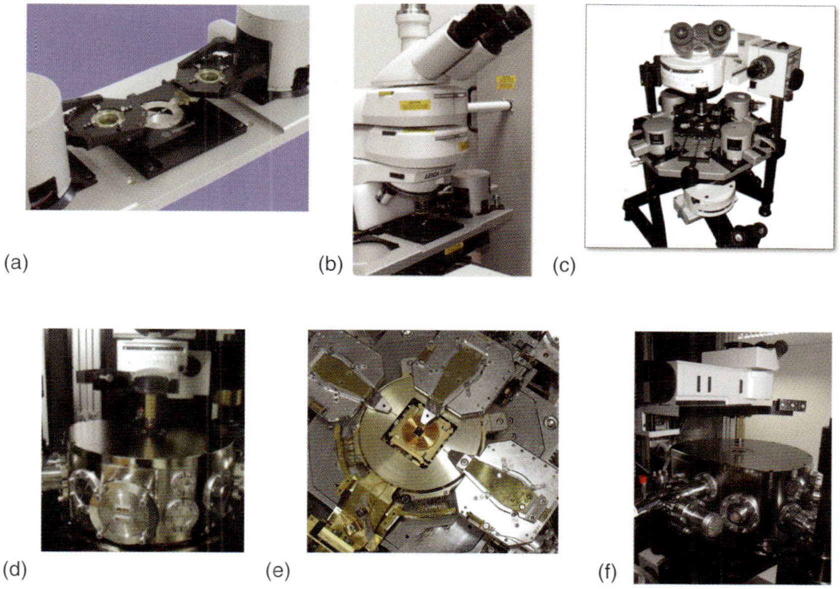

**Figure 13.14** A variety of multiprobe scanning probe microscope platforms from Nanonics Imaging Ltd.: (a) The MultiView 4000™: a two-probe multiprobe room temperature system. (b) Multiprobe system integrated with Raman. (c) The MultiView 4000™ four-probe configuration with full optical integration from above and below. (d) High vacuum multiprobe system with up to four probes and optical integration. (e) The CryoView MP™ multiprobe cryogenic system featuring three SPM probes. (f) The CyroView MP™ multiprobe cryogenic system with full optical integration. Pictures courtesy of Nanonics Imaging.

force sensitivity as demonstrated in the imaging of one probe by another. This, of course, is supported in the literature on tuning forks [20]. It should be noted in this regard that this ultrasensitive force sensitivity, not characteristic of the beam bounce approach, is also not universal for tuning fork feedback. An example is the tuning-fork-based Akiyama probe provided by probe manufacturer Nanonsensors.com. These silicon probes have what is called an ATec probe structure, which means that they have a slightly exposed tip and could possibly be brought together. However, the unique feedback mechanism used causes interference preventing an approach of closer than a couple of microns between the probes. In addition, a very large operational oscillation amplitude (~1 μm), makes such probes suitable primarily for hard samples.

A straightforward application of the multiprobe platform discussed in this chapter is the use of the system for making multiprobe contacts. This is true even for ultrasmall contact pads that could be in the micron range and below.

In the results shown in Figure 13.15, multiple probes make effective contact with graphene pads across graphene channels of different widths and layers. The free optical axis of the system, even in multiprobe operation, allows for online Raman measurements showing alterations in the graphene G′ and 2D band as a function of width and conductivity of the channel. The correlated IV curves are shown in Figure13.15c.

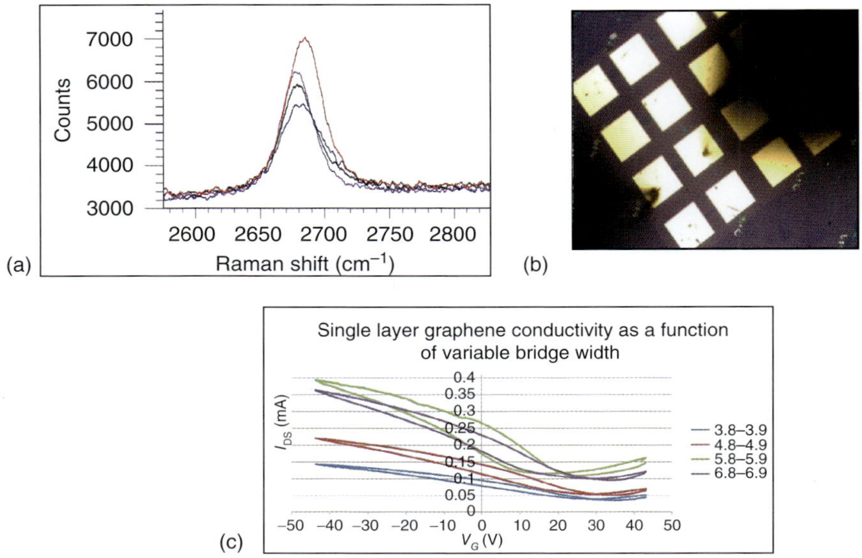


**Figure 13.15** Multiprobe conductivity and Raman measurements of single-layer graphene devices with graphene bridges of variable widths: (a) Raman spectrum shows that the 2D band of the graphene is shifted as a function of the width and conductivity of the graphene channel. (b) Charge-coupled device (CCD) image showing the two electrical probes located on the electrical pads connecting the graphene channels. (c) A voltage of 1 V was applied between the electrical pads using Nanonics electrical probes. The measurements were done between every pair of the electrical pads connected with the graphene channels. Images courtesy of Nanonics Imaging.

Another important example in such multiprobe measurements is the ability to make contact with a carbon nanotube sample, as shown in Figure 13.16.

In Figure 13.16, one sees a SEM of a single-walled carbon nanotube (SWCNT) and the corresponding AFM. The sample is composed of both conducting and semiconducting SWCNT and only the conducting one is seen in the SEM. The AFM shows both SWCNTs that are in close proximity with one another. An optical image of the probes is also seen across the semiconducting SWCNT. An IV curve shows the high repeatability of such measurements.

Still another interesting and challenging example is the measurement of highly conducting graphene nanoribbons grown on a silicon carbide substrate [32]. The conducting nanoribbons occur on the sidewall of a step and are approximately 200 nm in width. Images obtained of such a sample prepared by the group of Prof. Walter deHeer, Georgia Institute of Technology, are seen in Figure 13.17.

These measurements emphasize the importance of multiprobe technology for such new challenging samples. For such diminutive samples, the ability to effectively place two probes one relative to the other is critical. To do this with a single probe and a nanopositioner is considerably more complicated, if even possible. Such an example helps us understand the wider possibilities for electrical measurement and other transport studies as applied to 2D materials with such a multiprobe system.

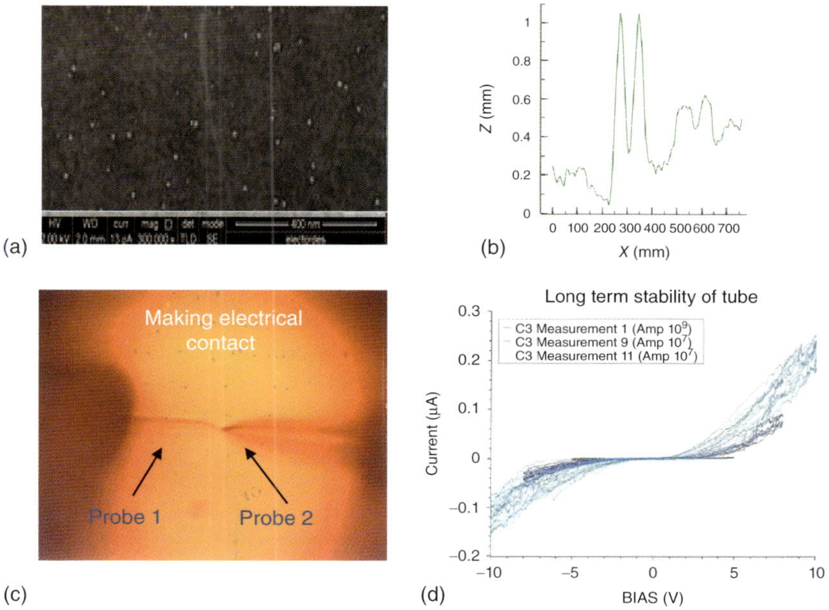


**Figure 13.16** Multiprobe electrical contact with a single-walled carbon nanotube (SWCNT) of 1 nm height. (a) A SEM image of two single-walled carbon nanotubes. (b) AFM cross section showing two SWCNTs one conductive and one nonconductive. (c) CCD image of two electrical probes touching a SWCNT. (d) Semiconductive behavior measured showing repeatability. Images courtesy of Nanonics Imaging.

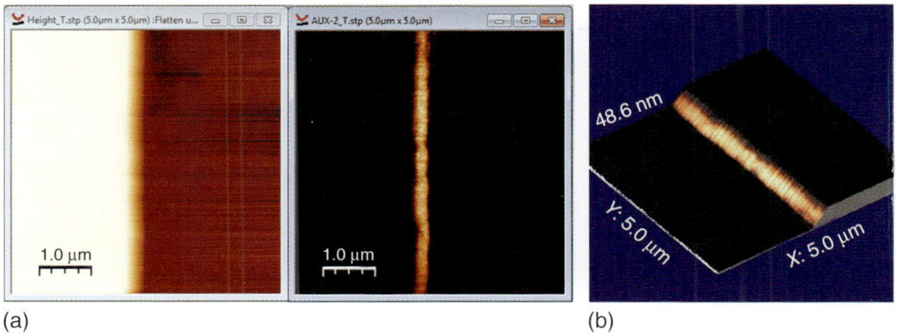


**Figure 13.17** Multiprobe graphene nanoribbon conductivity measurements: The topography is the above left image. (a) Topography of graphene nanoribbon (left) and conductivity measurement. (b) The collage of the two images from (a) is shown above right. Images courtesy of Nanonics Imaging.

### 13.3.5 New Horizons in Multiprobe Measurements

An even more complex and innovative example of multiprobe measurements has recently been reported with the multiprobe vacuum platform [33]. In this study shown in Figure 13.18, a nanopipette probe, from the Nanonics NanoToolKit™, using tuning fork AFM feedback, locally oxidizes a nanowire of Gallium Nitride

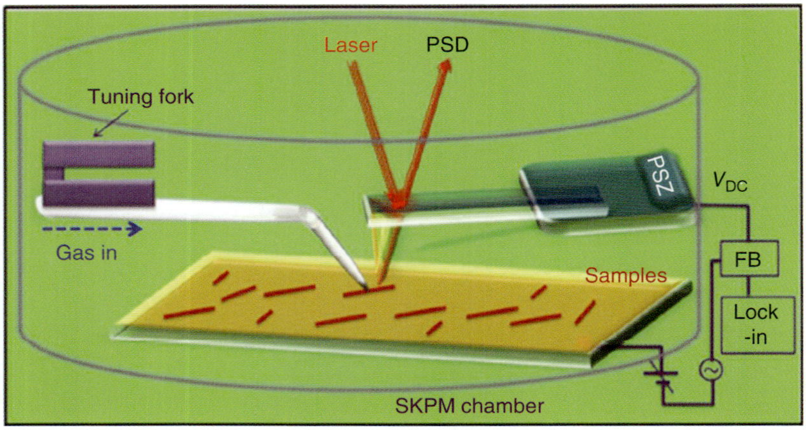


**Figure 13.18** A nanopipette probe from Nanonics NanoToolKit™ using tuning fork AFM feedback, locally oxidizes a nanowire of Gallium Nitride while a second probe using beam bounce feedback is employed online to measure the alterations in the chemical potential with Kelvin probe force microscopy. Image reproduced with permission from [33], Copyright Journal of Physical Chemistry C.

while a second probe using beam bounce feedback is employed online to measure the alterations in the chemical potential with Kelvin probe force microscopy.

The study highlighted the strong influence of the presence of intrinsic or extrinsic nanometric chemical defects on gas sensing applications. This unique application offers a glimpse at the true horizon of multiprobe where a combination of unique and important SPM technologies can be integrated for deep understanding of new materials and samples.

## 13.4 Conclusions

In summary, multiprobe measurements have evolved after a complex development of scanners, probes, and feedback mechanisms. As predicted [1], as applications are proliferating, new horizons of insight into nanometric transport are being reached. Even the initial directions that have been published, as referenced in this chapter, show the important impact multiprobe systems will have on a broad complex of fields from plasmonic transport in optics to thermal transport to chemical sensing and electrical interrogation of nanostructures. In addition, the ability to probe such samples without optical feedback will allow for a deeper understanding of the actual transport process taking place.

Within this latter area, the growing import of 2D materials for future device applications already highlights the importance of multiprobe techniques especially as described in this chapter. These materials demand a variety of characterization techniques including topography, conductivity, and electro-optical measurements. The full optical integration permits critical chemical insights by permitting functional measurements to be coupled to chemical alterations that are occurring in devices during electrical transport.

We expect that this synergy of measurements will bring important understanding to these new and novel materials in the future.

## References


1 Rocco, M.C., Marcin, C.A., and Hersam, M.C. (2010) *Nanotechnology Research Directions for Societal Needs in 2020*, written with the support of the World Technology Evaluation Center, Arlington, VA and supported by the US National Science Foundation., Springer-Verlag, New York, p. 610.

2 Shirak, I., Tanabe, F., Hobart, R., Nagao, T., and Hasegawa, S. (2001) Independently driven four-tip probes for conductivity measurements in ultrahigh vacuum. *Surf. Sci.*, **493**, 633–643.

3 Grube, H., Harrison, B.C., Jian, J., and Boland, J.J. (2001) Stability, resolution, and tip–tip imaging by a dual-probe scanning tunneling microscope. *Rev. Sci. Instrum.*, **72**, 4388–4392.

4 Okamoto, H. and Chen, D. (2001) An ultrahigh vacuum dual-tip scanning tunneling microscope operating at 4.2 K. *Rev. Sci. Instrum.*, **72**, 4398–4403.

5 Guise, O., Mariachi, H., Yates, J.T., Jung, M.-C., Levy, J., and Hanger, J. (2005) Development and performance of the nanoworkbench: a four tip STM for conductivity measurements down to sub micrometer scales. *Rev. Sci. Instrum.*, **76**, 045107.

6 Xiao, L., Xiao-Bo, H., Jun-Ling, L., Li, G., Qing, H., Dong-Xia, S., and Hong-Jun, G. (2005) Four-probe scanning tunneling microscope with atomic resolution for electrical and electro-optical property measurements of nanosystems. *Chin. Phys.*, **14**, 1536.

7 Tamaki, K., Akai-Masaya, M., Saito, A., Agno, M., and Kalahari, Y. (2005) Construction of independently driven double-tip scanning tunneling microscope. *Jpn. J. Appl. Phys., Part 2*, **44**, L120.

8 Yi, W., Kaya, I.I., Altered, I.B., Appel Baum, I., Chen, D.M., and Narayanamurti, V. (2005) Dual-probe scanning tunneling microscope for study of nanoscale metal-semiconductor interfaces. *Rev. Sci. Instrum.*, **76**, 063711.

9 Jaschinsky, P., Coenen, P., Pirug, G., and Voigtländer, B. (2006) Design and performance of a beetle-type double-tip scanning tunneling microscope. *Rev. Sci. Instrum.*, **77**, 093701.

10 Dallapiccola, R., Dubois, C., Gopinath, A., Stellacci, F., and Dal, N.L. (2009) Near-field excitation and near-field detection of propagating surface plasmon polaritons on Au waveguide structures. *Appl. Phys. Lett.*, **94**, 243118.

11 Lewis A, Komissar A, Taha H, Ratner A (2001). Multiple plate tip or sample scanning reconfigurable scanned probe microscope with transparent interfacing of far-field optical microscopes. US Patent 7,047, 796, assigned to Nanonics Imaging Ltd. & Filed, 2001.

12 Klein, A.E., Janunts, N., Tünnermann, A., and Pertsch, T. (2012) Investigation of mechanical interactions between the tips of two scanning near-field optical microscopes. *Appl. Phys, B*, **108**, 737–741.

13 Kishino, T. and Yanagida, Y. (1988) Force measurements by micromanipulation of a single actin filament by a glass needle. *Nature*, **334**, 74–76.
14 Harootunian, A., Betzig, E., Isaacson, M.S., and Lewis, A. (1986) Superresolution fluorescence near-field scanning optical microscopy (NSOM). *Appl. Phys. Lett.*, **49**, 674.
15 Shalom, S., Lieberman, K., Lewis, A., and Cohen, S.R. (1992) A micropipette force probe suitable for near-field scanning optical microscopy. *Rev. Sci. Instrum.*, **63**, 4061.
16 Lewis, A., Kheifetz, Y., Shambrodt, E., and Radko, A. (1999) Edward Khatchatryan and Chaim Sukenik, "Fountain pen nanochemistry: atomic force control of chrome etching,". *Appl. Phys. Lett.*, **75**, 2689–2691.
17 Barsegova, I., Lewis, A., Artium, K., Alexandra, M., Andrei, I., and Chaim, S. (2002) Controlled fabrication of Silver or gold nanoparticle atomic force probes: enhancement of second harmonic generation. *Appl. Phys. Lett.*, **81**, 3461–3463.
18 Fish, G., Bouevitch, O., Kokotov, S., Lieberman, K., Palanker, D., Turovets, I., and Lewis, A. (1995) Ultrafast response micropipette-based submicron thermocouple. *Rev. Sci. Instrum.*, **66**, 3942–3948.
19 Dekhter, R., Khachatryan, E., Kokotov, Y., Lewis, A., Kokotov, S., Fish, G., and Shambrot, Y. (2000) Investigating material and functional properties of static random access memories using cantilevered glass multiple wire force sensing thermal probes. *Appl. Phys. Lett.*, **77**, 4425–4427.
20 Giessibl, F.J. (2003) Advances in atomic force microscopy. *Rev. Mod. Phys.*, **75**, 949.
21 Sader, J.E. and Jarvis, S.P. (2004) Accurate formulas for interaction force and energy in frequency modulation force spectroscopy. *Appl. Phys. Lett.*, **84**, 1801.
22 Kohlgraf-Owens, D.C., Sukhov, S., and Dogariu, A. (2011) Mapping the mechanical action of light. *Phys. Rev. A*, **84**, 011807R.
23 Stadermann, M., Grube, H., Bolandc, J.J., Papadakis, S.J., Falvo, M.R., Superfine, R., and Washburn, S. (2003) Simultaneous atomic force microscopy measurement of topography and contact resistance of metal films and carbon nanotubes. *Rev. Sci. Instrum.*, **74**, 3653–3655.
24 Higuchi, S., Kubo, O., Kuramochi, H., Aono, M., and Nakayama, T. (2011) A quadruple-scanning-probe force microscope for electrical property measurements of microscopic materials. *Nanotechnology*, **22**, 285205–285211.
25 Xie, H., Hussain, D., Yang, F., and Sun, L. (2015) Development of Three-Dimensional Atomic Force Microscope for Sidewall Structures Imaging with Controllable Scanning Density. *IEEE/ASME Trans. Mechatron.*, 1.
26 Xie, H., Hussain, D., Yang, F., and Sun, L. IEEE/ASME (2015) *IEEE/ASME Transactions on Mechatronics*. Manuscript No. TMECH-12-2014-4224.
27 Ren, X., Liu, A., Zou, C., Wang, L., Cai, Y., Sun, L., Guo, G., and Guo, G. (2011) Interference of surface plasmon polaritons from a "point" source. *Appl. Phys. Lett.*, **98**, 201113–201115.
28 Dobmann, S., Kriesch, A., Ploss, D., and Peschel, U. (2014) Near-field analysis of bright and dark modes on plasmonic metasurfaces showing extraordinary suppressed transmission. *Adv. Opt. Mater.*, **2**, 990–999.

**29** Klein, A.E., Janunts, N., Steinert, M., Tünnermann, A., and Pertsch, T. (2014) Polarization-resolved near-field mapping of plasmonic aperture emission by a dual-SNOM system. *Nano Lett.*

**30** Lewis, A., Lev, D., Sebag, D., Hamra, P., Levy, H., Bernstein, Y., Brahami, A., Tal, N., Goldstein, O., and Yeshua, T. (2014) The optical near-field: super-resolution imaging with structural and phase correlation. *Nanophotonics*, **3**, 3–18.

**31** Berezin, S., Kalanoor, B.S., Taha, H., Garini, Y., and Tischler, Y.R. (2014) Multiprobe NSOM fluorescence. *Nanophotonics*, **3**.

**32** Baringhaus, J., Ruan, M., Edler, F., Tejeda, A., Sicot, M., Taleb-Ibrahimi, A., Li, A.-P., Jiang, Z., Conrad, E.H., Berger, C., Tegenkamp, C., and de Heer, W.A. (2014) Exceptional ballistic transport in epitaxial graphene nanoribbons. *Nature*, **506** (7488), 349–354, doi: 10.1038/nature12952.

**33** Patsha, A., Prasana, S., Amirthapandian, S., Prasad, A.K., Das, A., Tyagi, A.K., Cotta, M.A., and Dhara, S. (2015) Localized charge transfer process and surface band bending in methane sensing by GaN nanowires. *J. Phys. Chem. C*, **119**, 21251–21260.

# 14

# KPFM and its Use to Characterize the CPD in Different Materials

*Yijun Xia and Bo Song*

*Soochow University, Suzhou Key Laboratory of Macromolecular Design and Precision Synthesis, Jiangsu Key Laboratory of Advanced Functional Polymer Design and Application, College of Chemistry, Chemical Engineering and Materials Science, Ren'ai Road 199, Suzhou 215123, China*

## 14.1 Introduction

Kelvin probe method [1] takes its name from Lord Kelvin, who firstly introduced it to explain the formation of contact potential differences (CPDs) between metals. Kelvin probe force microscope (KPFM), based on the Kelvin method, measures the CPD between a conducting atomic force microscopy (AFM) tip and a sample, and allows to map the work function ($\phi$) of the sample, which was firstly developed by Nonnenmacher *et al.* [2] Since it was invented in 1991, KPFM has been used extensively as a method to study the nanoscale electrical properties of metal [3], inorganic or organic semiconductors [4], and biological materials [5].

This chapter, focusing on introduction of KPFM, is divided into two parts. The first part introduces principles and theories of KPFM, as well as the effect of measurement mediums, mainly ambient and ultrahigh vacuum (UHV) conditions, on the results of KPFM. The second part emphasizes the application of KPFM in different fields.

## 14.2 Kelvin Probe Force Microscopy

### 14.2.1 Basic Principle of Kelvin Probe Force Microscopy

The CPD for metal surfaces can be ascribed to the difference of the $\phi$ of two materials [2, 6]. The $\phi$ of the metallic conductor, defined as the energy difference between the Fermi level and the vacuum level [7], can be influenced by adsorption of molecules with different dipole moments [6]. CPD can be measured by using a vibrating capacitor with the Kelvin method, in which two conductors are arranged as a parallel plate capacitor with a small spacing. A periodic vibration at a defined frequency ($\omega$) between the two conductors generates a current $i_{(t)}$, given by [2]

$$i_{(t)} = V_{\mathrm{CPD}}\omega\Delta C \cos(\omega t) \tag{14.1}$$

*Conductive Atomic Force Microscopy: Applications in Nanomaterials*, First Edition.
Edited by Mario Lanza.

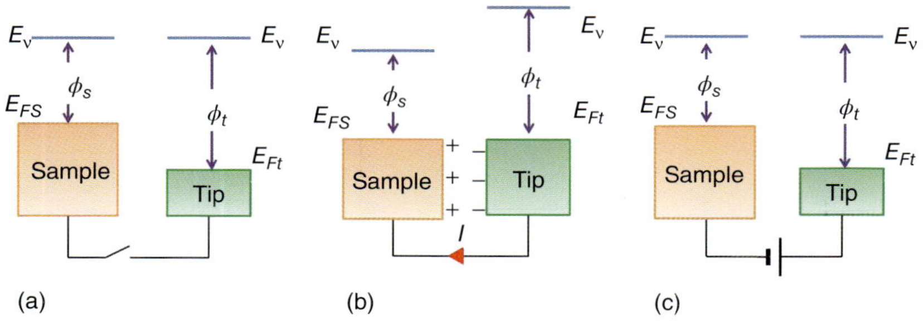


**Figure 14.1** Electronic energy levels of the sample and AFM tip for three cases: (a) tip and sample are separated, (b) tip and sample are in electrical contact, and (c) external bias ($V_{DC}$) is applied between tip and sample to nullify the CPD. Herein, $E_V$ is the vacuum energy level, $E_{FS}$ and $E_{Ft}$ Fermi energy levels of the sample and tip, respectively.

where $\Delta C$ refers to the change of capacitance. Then an additional voltage ($V_{DC}$) applied to the plates is controlled so that no current is induced by the vibration. In this way, the CPD can be measured by adjusting the additional bucking voltage. Since the size of the plates is too small to generate a sufficient current in the microscopic scale, KPFM does not measure the current flow, but rather it records the electrostatic–force interaction between the tip and the sample in order to improve the poor sensitivity [8]. In the realistic measurement of KPFM, a cantilever of AFM probe with a sharp tip at one end is served as a force sensor.

The KPFM measures the CPD between an AFM tip and a sample. So the $V_{CPD}$ between the tip and sample is [9]

$$V_{CPD} = -(\phi_{tip} - \phi_{sample})/e, \tag{14.2}$$

where $\phi_{tip}$ and $\phi_{sample}$ are the work functions of the tip and the sample, respectively, and $e$ is the elementary charge. According to Eq. (14.2), the $\phi_{sample}$ can be calculated using a calibrated tip with a known $\phi$. Figure 14.1 shows the energy level diagram of the sample surface and tip. The energy levels of the tip and sample surface when separated by a distance $z$ and not electrically connected are depicted (Figure 14.1a). As soon as the tip and the sample surface are electrically wired together, electrons will flow from the material with lower $\phi$ (weak electron binding) to the one with higher $\phi$ (strong electron binding), as shown in Figure 14.1b. As a consequence, the tip and the sample surface will be charged, and a $V_{CPD}$ will form. As shown in Figure 14.1c, an electric field between tip and sample due to these charges can be then easily detected, which is nullified through applying an external potential $V_{DC}$.

In addition to the compensation $V_{DC}$ between tip and sample, an AC-voltage, $V_{AC}\sin(\omega_{AC}t)$, at the frequency of $\omega_{AC}$ is applied. Considering the tip–sample system as a capacitor, the electrostatic force ($F_{es}$) between the tip and sample can now be expressed as [9]

$$F_{es(z)} = -\frac{1}{2}\Delta V^2 \frac{\delta C(z)}{\delta z}, \tag{14.3}$$

where $z$ is the distance between the tip and the sample surface and $\partial C(z)/\partial z$ is the capacitance gradient of the tip–sample system, $\Delta V$ is the potential difference between $V_{CPD}$ and the voltage applied to the tip. $\Delta V = V_{tip} \pm V_{CPD} = (V_{DC} \pm V_{CPD}) + V_{ac}\sin(\omega t)$, where the "+" and "−" refer to the external bias applied to the sample and the tip, respectively. Thus, the $F_{es}$ can be divided into three parts: [9]

$$F_{DC} = -\frac{\partial C(z)}{\partial z}\left[\frac{1}{2}(V_{DC} \pm V_{CPD})^2\right] \quad (14.4)$$

$$F_{\omega} = -\frac{\partial C(z)}{\partial z}(V_{DC} \pm V_{CPD})V_{AC}\sin(\omega t) \quad (14.5)$$

$$F_{2\omega} = \frac{\partial C(z)}{\partial z}\frac{1}{4}V_{AC}^2[\cos(2\omega t) - 1] \quad (14.6)$$

The DC part $F_{DC}$ (Eq. 14.4) contributes to the topographical signal, $F_{\omega}$ with frequency $\omega$ (Eq. 14.5) is used to measure the $V_{CPD}$, and $F_{2\omega}$ can be used for capacitance microscopy [8, 10].

### 14.2.2 KPFM Operational Modes: AM- and FM-Mode

KPFM is primarily based on the AFM system and operated under the noncontact mode (nc mode). In nc mode, the cantilever is oscillated at, or close to, its resonance frequency, and the change of the oscillation is monitored in dependence of the tip–sample interaction [11]. In order to detect the change, two different modes can be applied, namely, amplitude modulation technique (AM mode) [12] and frequency modulation technique (FM mode) [12a, 13]. The schematic setup of AFM with the main electronic components for KPFM is shown in Figure 14.2.

For the AM mode, $F_{\omega}$ is directly measured from the amplitude of the cantilever oscillation at the AC frequency $\omega$. Since KPFM measures topography

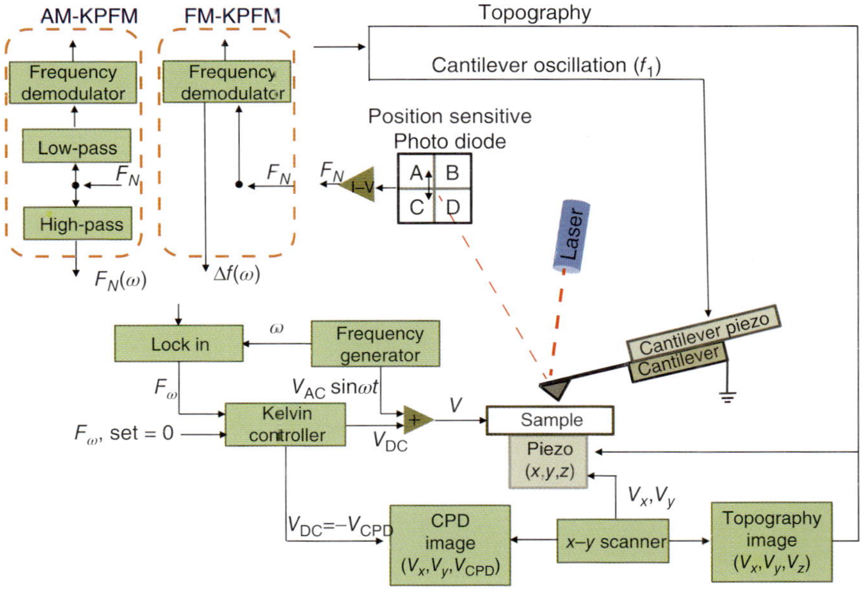


**Figure 14.2** Schematic setup of AFM with the main electronic components for the AM-mode and FM-mode detection in KPFM.

concurrently with $V_{CPD}$ using an AFM tip, it is required to separate the topographical signal from the $V_{CPD}$ measurement. Frequently, this is realized in the two-pass mode, where in the first scan line the topography is obtained and then retraced with the tip lifted up by several tens of nanometers; while a $V_{AC}$ at the resonance frequency is applied for detection of $V_{CPD}$. A mechanically vibrated cantilever generally has several resonance peaks in the oscillation amplitude-frequency spectrum and the frequency of the second resonance peak is about six times that of the first resonance frequency [12a]. Thus, a more elegant way is to tune the AC frequency to facilitate separation of the two signals. When topography is measured by the oscillation at the first resonance frequency of the AFM tip, $V_{CPD}$ is measured by the amplitude of the oscillation at the second resonance frequency of the AFM tip at the same time. Then the oscillation at $\omega$ is amplified by the quality factor ($Q$). The response time ($\tau$) of the KPFM system to measure CPD depends on the Q and the second resonance ($f$): $\tau = \frac{Q}{\pi f}$. However, in AM mode the lateral energy resolution is reduced by an averaging effect of the tip and the cantilever due to the detection of long-range electrostatic force.

For the FM mode, the applied $V_{AC}$ induces a modulation of the electrostatic force, which results in an oscillation of the frequency shift $\Delta f_0$ at the frequency $\omega$ of the $V_{AC}$ [14]. Compared to the detection of topography, which is measured by the amplitude of the oscillation, $V_{CPD}$ is determined by the oscillation at the resonance frequency of the AFM tip. The measured signal of FM mode is approximately proportional to the force gradient [8]:

$$\Delta f_0(\omega) \propto \frac{\partial F_\omega}{\partial z} = \frac{\partial^2 C}{\partial z^2}(V_{DC} \pm V_{CPD})V_{AC}\sin(\omega t) \tag{14.7}$$

The determination of electrostatic force is considered long-range detection, whereas the determination of electrostatic force gradient is short-range detection. The electrostatic interaction takes place mainly between the tip apex and sample surface in the FM-mode KPFM [9]. However, the electrostatic interaction between sample and AFM probe includes the influence of both the tip and the cantilever in the AM mode. Consequently, the spatial resolution of AM-mode KPFM is reduced by an averaging effect between the tip and the cantilever, and FM-mode KPFM has the better spatial resolution than the AM mode [12a, 15].

### 14.2.3 KPFM Measurement, at Ambient or UHV Conditions

Under ambient conditions, water molecules, particularly hydroxyl groups in it, may interact with the surface of the sample [16]. This may increase $\phi$ by the dipoles generated by the aligned water molecules. In air, several layers of water generally exist at surfaces, and thus the $\phi$ under ambient conditions will always be higher than that with respect to the $\phi$ under UHV conditions. As long as the same samples are exposed to UHV, most of the depolarizing water molecules will desorb, leaving behind the other molecules firmly attached to the sample and thus considerably diminishing the depolarization effect caused by water. What is more, imaging in UHV also prevents electrical discharges on the surfaces of insulators [14a]. As a consequence, compared to ambient conditions, KPFM measurement under UHV conditions may be a closer representation of the real surface $\phi$ [17].

## 14.3 Applications of KPFM

### 14.3.1 KPFM on Conventional Inorganic Materials

#### 14.3.1.1 Metallic Nanostructures

Theoretical calculations show that the electronic structure of Au depends closely on its particle size and shape [18]. The highest occupied molecular orbital (HOMO) and lowest unoccupied molecular orbital (LUMO) energy levels are also dependent on the features of the Au cluster's geometry [18a]. The size difference usually leads to fluctuations in the mean value of the electrochemical potentials. Salmeron *et al.* [19] find that the $\phi$ of Au nanoparticles (NPs) is size dependent and deviates strongly from that of the bulk Au. The $\phi$ of Au NPs can also be influenced by chemical changes of the underlying substrate. Therefore, Au NPs could be used for chemical and biological sensors, whose environmentally sensitive charge state can be read out by $\phi$ measurements.

Figure 14.3a and b shows the topography and corresponding CPD mapping of Au nanostructures grown on a heavily $n$-doped Si substrate pre-covered with an insulating monolayer film of alkyl chain molecules. In the CPD image, the NPs appear as depressions (negative CPD). A plot of the CPD as a function of NP diameter (Figure 14.3e) shows that the $\phi$ decreases when the NP diameter decreases and deviates substantially from that of the bulk Au ($\phi_{Au} \approx 5.1$ eV). Since Au NPs are much different from their bulk counterparts in terms of optical and electrical properties due to their small size, Au NPs were used for improving the photocatalytic activity of ZnO photocatalysts [20]. KPFM is applied to explore the role of Au NPs in the enhancement of the photoactivity of ZnO under ultraviolet (UV) irradiation.

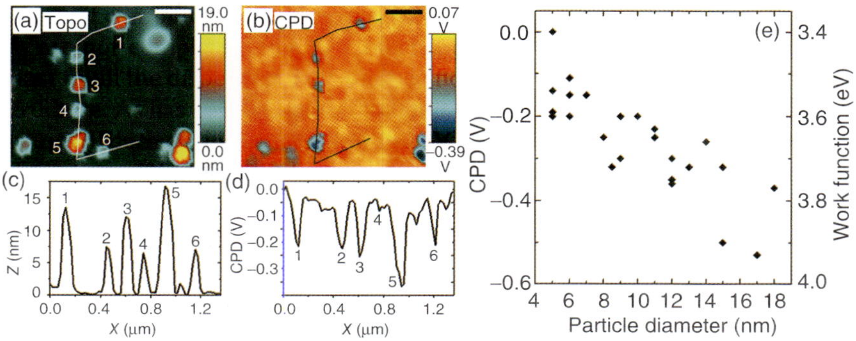


**Figure 14.3** Work function versus size of NPs. Simultaneously obtained (a) topography and (b) CPD images of NPs. The cross-section profiles of (c) topography and (d) CPD. (e) CPD (left scale) and $\phi$ (right scale) as a function of NP diameter. The $\phi$ was calibrated with the value determined by UV photoemission spectroscopy. Reprinted with permission from Ref. [19]. Copyright (2015) American Chemical Society.

Metallic nanostructures supported on thin oxide films such as $TiO_2$ are known to exhibit extraordinary properties in catalysis and sensor applications [21]. In catalytic reactions, it is important to investigate the charge transfer between metallic nanostructures and supported oxide films [22]. Morita *et al.* [21b] supplied new evidence supporting the charge-transfer formation from the $TiO_2$ surface to the Au cluster. The higher CPD value relative to the $TiO_2$ surface infers charge transfer from the $TiO_2$ surface to the Au cluster, thus providing a new experimental signature and indicating the presence of negative charges locally distributed over Au clusters.

Compared to the topography image, CPD mapping provides more details concerning the developed surface topography [3a, 23]. Some small features that are difficult to be seen in the topography image, due to a large variation in the image Z direction, are easily recognized with the help of the CPD signal. Krok *et al.* [23] performed KPFM imaging of nanostructures formed during a deposition of Au on semiconductor surfaces. They found that Au islands of lateral edge sizes ranging from 8 to 15 nm resulted in almost the same CPD.

In comparison to Au, Cu is cheaper and it possesses an increased electrical conductivity. However, the surface of Cu is subject to oxidation, even at room temperature. KPFM measurements enabled to distinguish between the different types of Cu oxide with nanometer resolution and to correlate the oxidation states to local topography features [3b].

#### 14.3.1.2 Semiconductor Surfaces

Researches on atomic-resolution KPFM on semiconductor surfaces are dominated by Si surfaces [24], but also include other surfaces such as GaAs [25], InSb [26], and $MoS_2$ [27]. KPFM can be used to investigate surface defects [24a], step edge [4b], and grain boundaries (GBs) [28].

Topographic and CPD images of InSb nanodots grown on GaAs substrates were measured in Rosenwaks's research [26]. Although a clear correlation between the two images is observed in Figure 14.4, the dark rings around the dots observed in the CPD image and the CPD peak heights do not match the work-function difference between InSb and GaAs, which is expected due to the composition changes. An additional CPD peak on each side outside of the dot area is also analyzed by sectional profiles (Figure 14.4c). The value of the peak is consistent with the estimated CPD signal variations by considering the strain and the composition changes within and around individual nanodots.

KPFM also is a powerful technique in characterizing two-dimensional materials such as $MoS_2$. Li *et al.* [29] studied the dependence of surface potential of $MoS_2$ nanoflakes on the number of layers before and after annealing by KPFM. The surface potential of pristine $MoS_2$ nanoflakes decreased with increasing the thickness; while after annealing, the trend was opposite and the screening length became smaller. These results were qualitatively explained by a charge-transfer model with the built-in electric field induced by trapped charges.

One of the most interesting and potentially useful applications of atomic-resolution KPFM is the visualization of single-defect formation on the surface. A vacancy at the Si (111) $7\times7$ surfaces was observed at an atomic scale by KPFM by Tadashi *et al.* [24a] As shown in Figure 14.5a, AFM can image a Si atom

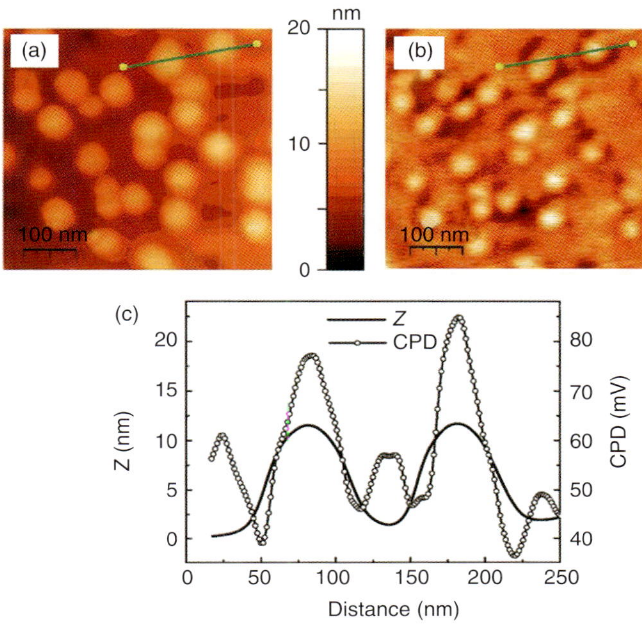


**Figure 14.4** (a) Topography and (b) KPFM measurements of InSb nanodots grown on GaAs substrate. (c) Sectional profiles of topography (solid line) and CPD (or surface potential, gray line marked with empty circles), along the green lines on images (a) and (b). The typical CPD profile of a single nanodot consists of a sharp deep, surrounded by a higher region. Reprinted with permission from Ref. [26]. Copyright (2007) American Chemical Society.

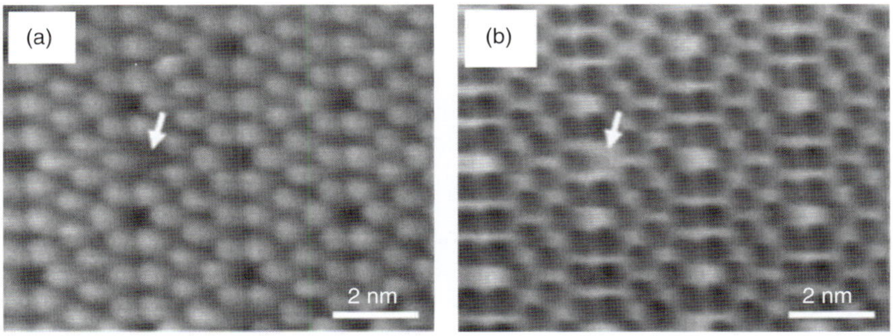


**Figure 14.5** (a) The AFM and (b) KPFM images of the Si (1 1 1) 7 × 7 surface including a point defect. Reprinted with permission from Ref. [24a]. Copyright (2002) The Japan Society of Applied Physics.

vacancy, which is indicated by the arrow. Also, in the KPFM image, the vacancy is clearly resolved, as seen in Figure 14.5b. The imaging of a point defect proves that a true atomic-scale lateral resolution is achieved.

In addition, KPFM can be adopted to study defects larger than vacancies, such as step edges or GBs. Glatzel *et al.* [4b] investigated the surfaces of UHV-cleaved *n*- and *p*-type doped GaAs and GaP, and focused on the $\phi$ variations at step edges. For *p*-type materials, they found an increase and for *n*-type doped semiconductors a decrease of the $\phi$ at step edges. Lanza *et al.* [28b] used KPFM to study the dependence of the electrical properties of $Al_2O_3$ stacks for flash memories on the annealing temperature. The CPD change – a sudden increase of CPD around the crystallization temperature, observed by KPFM could be related to some kind of defects or trapping sites generated after the crystallization – may be related to the GBs. The conductive characteristics of the GBs in polycrystalline $HfO_2$ gate dielectrics were also investigated by Lanza *et al.* [28a] KPFM measurements were used to assess the charge state of defects in these structures. The regions of GBs show higher CPD signals than the nanocrystal grains' regions. A higher CPD signal can be interpreted as downward band bending, which indicates a larger concentration of positive charges at a given location.

### 14.3.2 KPFM on Organic Adsorbates on Surfaces

KPFM can also be applied in observation of the interaction properties of an adsorbate on a surface [30]. The potential change provides information about the tip–adsorbate interactions corresponding to the tip–surface interactions, indicating the electron exchange or dangling bonds [9]. An average CPD value can also be measured, representing the true surface potential.

Organic and printed electronics technologies require conductors with a $\phi$ which is sufficiently low to facilitate the transport of electrons in and out of various optoelectronic devices. The surface of the conductive electrode is coated with an ultrathin layer of a material that is chemically or physically adsorbed onto the conductor surface, in order to create strong interface and/or molecular dipoles that induce a vacuum-level shift and modify the $\phi$ of the conductor. Based on this approach, Zhou *et al.* [31] modified indium tin oxide (ITO) surface with polymers containing simple aliphatic amine groups such as polyethylenimine ethoxylated (PEIE) and polyethylenimine (PEI) to reduce the $\phi$ of conductors. Instead of using pure aliphatic organic materials, Cao *et al.* [32] reported poly [(9,9-bis(3′-(*N*,*N*-dimethylamino)propyl)-2,7-fluorene)-alt-2,7-(9,9-dioctylfluorene)] (PFN), which was used as the surface modifier of ITO to affect $\phi$.

To get an insight investigation into the effect of topology of the molecules on the surface $\phi$, Song *et al.* [7] investigated the thin films formed by *linear*- or *branched*-PEI (*l*- or *b*-PEI) on ITO electrode, and used the surface dipole ($\Delta\phi$) to interpret the change of $\phi$ upon the modification of these thin layers. The adsorption of PEI, due to the transfer of protons from hydroxyl groups to amino groups, generates an opposite dipole to $\Delta\phi$ as illustrated in Figure 14.6, thus resulting in a decrease of the $\phi$. In order to enhance the solvent resistance of the modified layer on the surface, the ITO substrates were covalently modified with surface-grafting hyperbranched polymer formed by aziridine as well as 3-aminopropyltriethoxysilane [33]. The $\phi$ of ITO before and after modification with different materials and methods was assessed by KPFM. They found that, after the modification, in spite of the materials and methods, the $\phi$ decreases dramatically.

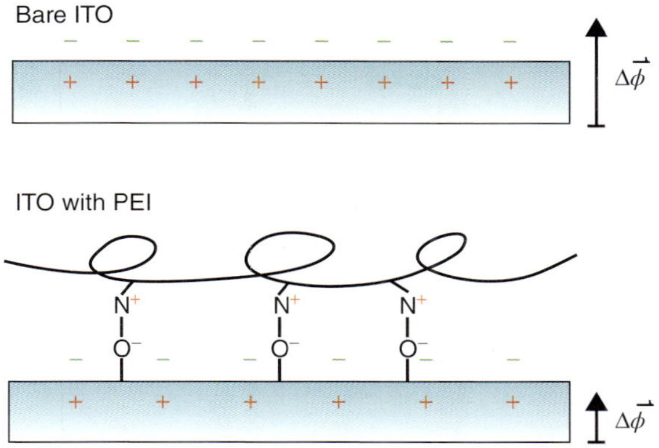


**Figure 14.6** Charge density distribution and surface dipoles with/without modification of PEIs at the adjacent of the surface. Reprinted from with permission from Ref. [7]. Copyright (2015) Elsevier.

KPFM has also been exploited to study complex biological architectures such as DNA. CPD images revealed the oriented DNA strands on Si and glass substrates, recorded using a hexadecanethiol-modified gold tip. Fujihira *et al.* [5a] found that stretched DNA strands have higher CPD compared to the bare substrate surface. The cylindrical symmetry of DNA suggests that the CPD would not be affected significantly by DNA adsorption on the substrate. The result suggests that the dipole moments formed between the surface water layer and the negatively charged DNA strands have a direction from the substrate toward air.

Surface properties and electrical charges are critical factors elucidating cell interactions on biomaterial surfaces [34]. The surface potential distribution of organic polypyrrole–hyaluronic acid (PPy–HA) substrates was assessed using KPFM. AFM topography images and their corresponding surface potential maps of uncharged (+0 mV, Figure 14.7a and b) and charged (+200 mV, Figure 14.7c and d) PPy-HA obtained from KPFM are presented in Figure 14.7. Compared with the uncharged films, the surface potential of the charged films showed a more uniform distribution of the surface potential with no clear correlation with the topography (Figure14.7d).

### 14.3.3 Characterization of the Electrical Properties of Nanoscaled Devices

#### 14.3.3.1 Junctions and Heterostructrues

Recently, KPFM has been recognized as a tool to characterize *p–n* junctions [35] and heterostructures [36], due to its high spatial resolution in measuring the surface potential.

The *p–n* junction is a fundamental building block in modern electronic circuits. Xu *et al.* [35d] reported graphene *p–n* junctions formed by a one-step thickness-dependent surface treatment of mono-/bi-layer graphene steps. The junction electronic properties are systemically studied by means of KPFM, which reveals the spatial potential distribution and field-effect response along the junction interface.

An extreme example was demonstrated by Kang *et al.* [37] through connecting two junctions in one nanorod. Specifically, Mg-doped GaN nanorods were grown on undoped *n*-type GaN nanorods uniaxial on monolayer graphene. The formed *p–n* nanorods show a Schottky behavior with a turn-on voltage of 3 V. The *p*

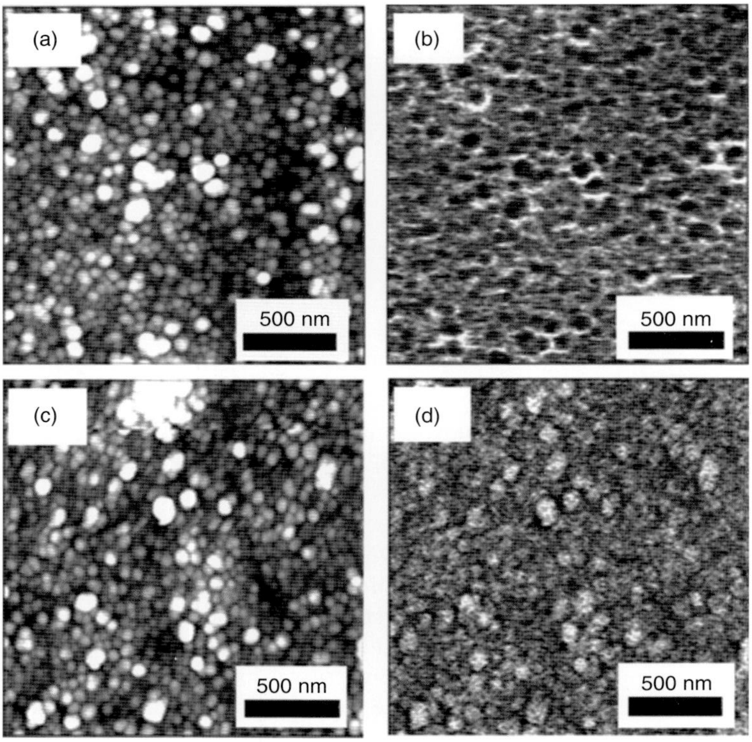


**Figure 14.7** (a, c) Height and (b, d) surface potential images for the PPy-HA film before and after charging, respectively. Reprinted with permission from Ref. [34a]. Copyright (2013) American Chemical Society.

and *n* regions of the *p*–*n* junction in a single GaN nanorod can be visualized with distinctive contrast by KPFM. Furthermore, the CPD measured across the brighter *n*-region and darker *p*-region was around 175 and 72 mV, respectively. The potential difference in the $V_{CPD}$ values ascertains the formation of the *p*–*n* junction in the entire nanorod.

ZnO nanostructures are promising elements for future electronic and optoelectronic devices. In Yang's work [38], the undoped/As-doped homojunction ZnO nanorods were grown on Si (1 1 1) substrates by self-assembly. They assessed the electronic properties of As-doped homojunction ZnO nanorods by KPFM. The topography shows no specific difference in the vicinity of the doping transition region. On the contrary, the brightness contrast appears obviously near the middle region of the nanorod in the KPFM image. The brighter region corresponds to the undoped ZnO part, whereas the darker region corresponds to the As-doped ZnO part in the homojunction structure.

The charge recombination rate in poly(3-hexylthiophene) (P3HT)/$TiO_2$ nanorod solar cells is demonstrated to correlate to the interfacial properties of bulk heterojunction between these two materials [39]. The surface charge of the bulk heterojunction is investigated by KPFM under dark conditions and under

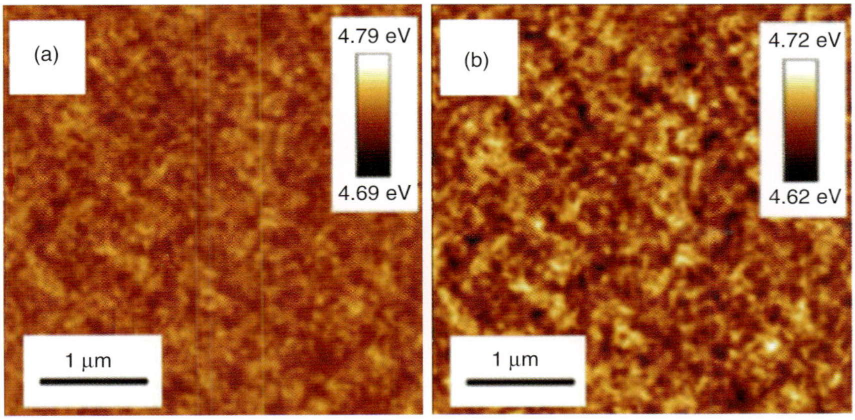


**Figure 14.8** KPFM surface potential images (a) in the dark and (b) under illumination for the active layer of the device. Reprinted with permission from Ref. [39]. Copyright (2011) American Chemical Society.

illumination. In the dark, $TiO_2$ nanorod-rich regions possess smaller potential than the surrounding P3HT-rich regions (Figure 14.8a). Upon illumination, the excitons are separated at the interfaces, which enhance the contrast of the two phases. The $TiO_2$- nanorod-rich regions become more negatively charged, and the surrounding P3HT-rich regions become more positively charged. This allows easier identification of $TiO_2$ nanorod-rich regions from the P3HT matrix (Figure 14.8b).

#### 14.3.3.2 Transistors

Experimental techniques providing valuable electrical, mechanical, and processing properties are critical to the development of field-effect transistors (FETs). [4e, 40] In this regard, KPFM is deemed an ideal tool for transistor structures [41], since it offers two-dimensional profiles of surface potentials.

Using KPFM, Salmeron *et al.* [42] studied the local charge trapping states at the $SiO_2$-oligothiophene (with five repeat units, denoted by D5TBA) interface in a FET, where $SiO_2$ acts as the gate dielectric. KPFM reveals surface potential inhomogeneities within the D5TBA monolayer. The inhomogeneities should correlate with the structure of the D5TBA monolayer. The results indicate that D5TBA forms both crystalline and amorphous domains on the substrate. The negatively charged carboxylic group and the positively charged alkyl chain ends of the D5TBA pointed toward and against the surface, respectively. The surface potential image clearly resolves the different structures of the D5TBA monolayer film on the oxide surface, as the amorphous domains could be less densely packed than the crystalline domains, and results in a smaller electric dipole moment. In the FET structure, surface potential distribution of D5TBA also depends on the gate bias. When negative gate bias is applied, hole carriers can be injected into the D5TBA channel, and hence the surface potential of the channel remains close to that of the source/drain electrodes. On the contrary, with positive gate bias, the hole carriers in the channel are depleted while no electrons are injected, and hence the channel becomes insulating and follows the gate bias.

Si nanowires (NWs) have played a key role in the development of nanoelectronic devices [43] and have been integrated in FET device structures [44]. Introducing defect states into the energy bandgaps has been demonstrated to affect the mobility of Si NWs. Ragan *et al.* [45] applied KPFM to examine how the surface electronic structure in the vicinity of defects and impurities is influenced under the conditions of current flow across Si NWs. The research shows that different defect types behave differently for local decreases in surface potential.

#### 14.3.3.3 Solar Cells

Thin-film solar cells based on polycrystalline inorganic materials contain an abundance of GBs. The role of these GBs is still under discussion [46]. KPFM has been employed to study the electrostatic properties of GBs, as well as the possible presence of charges [47]. In addition to the solar cells based on inorganic semiconductors, organic solar cells based on polymers or small molecules make up another big portion of the solar cell family [48]. The photoactive layers are often composed of donor and acceptor materials [49]. For a good charge separation and transport, interdigitated phase separation of the donor and acceptor materials is required [50]. KPFM becomes useful in terms of determining the intermixing of the two materials [4c, d, 51].

The past 5 years have witnessed the uniquely rapid emergence of a new class of solar cells based on organic–inorganic hybrid perovskite [52]. GBs are present in polycrystalline thin-film solar cells. They could be benign or detrimental to the performance of solar cells. Yun *et al.* [46a] present efficient charge separation and collection at the GBs measured by KPFM in $CH_3NH_3PbI_3$ film in a $CH_3NH_3PbI_3/TiO_2$/fluorinated tin oxide (FTO) heterojunction structure. They observe the presence of a potential barrier along the GBs under dark conditions and higher photovoltage along the GBs compared to the grain interior under illumination.

Figure 14.9a shows a spatial topography mapping of the surface of the heterojunction structure. Most of the grains, varying in size between 50 and 500 nm, are usually larger than the film thickness. In Figure 14.9b, individual grains are clearly distinguishable and GBs have lower CPD values compared with the CPD within the grains. GBs in perovskite oxides are often found to be depleted and form space-charge regions due to the existence of oxygen vacancies at the GBs. Figure 14.9c–e shows CPD mappings under various illumination conditions of the same area. Figure 14.9f plots the average CPDs as a function of light intensities. The CPD increases with light intensity, indicating that photogenerated carriers are efficiently separated at the $CH_3NH_3PbI_3/TiO_2$ heterojunction.

KPFM measurements on organic solar cells are performed both in the dark and under illumination to observe the surface $\phi$ changes in Watanabe's research [51a]. The surface potential distribution in the active layer of solar cells indicates the formation of spatial distribution of donor and acceptor.

The experiment was performed on delicately designed strips formed by donor and acceptor materials. Figure 14.10 shows the AFM and KPFM images of the strips both in the dark and under illumination. The KPFM image shows

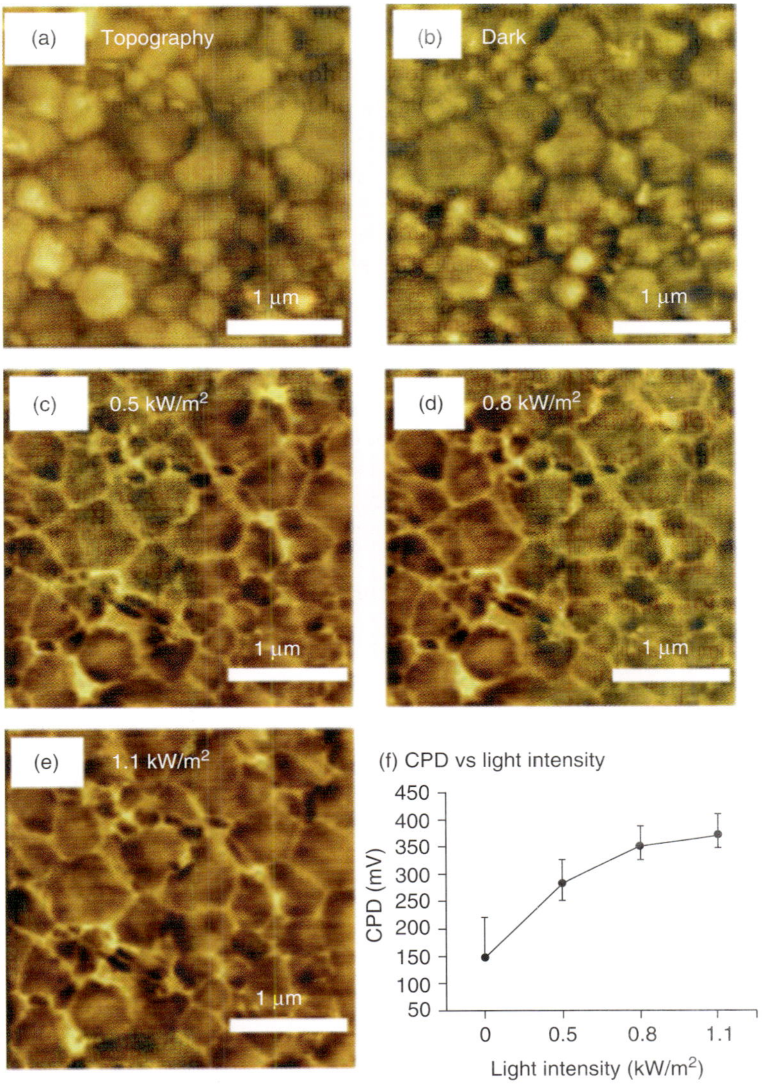


**Figure 14.9** KPFM measurements performed on a $CH_3NH_3PbI_3/TiO_2/FTO$ structure. (a) Topography and (b) CPD image taken in the dark. (c–e) CPD mappings under various laser illumination intensities at a wavelength of 500 nm. (f) Intensity dependence of CPD of the sample. Reprinted with permission from Ref. [46a]. Copyright (2015) American Chemical Society.

the periodic surface potential modulation based on the donor/acceptor spatial distribution in the films. After being irradiated by a white light, the morphology of the film was not affected, whereas the distribution of surface potential made a significant difference. After being irradiated for 40 min, the film showed a smaller depth of the surface potential modulation. The difference

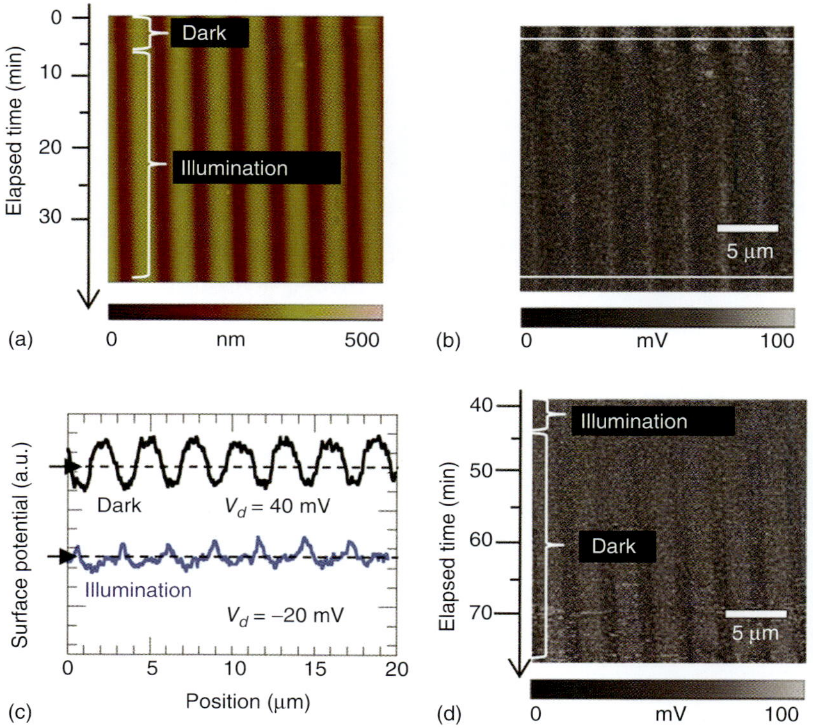


**Figure 14.10** (a) Topography mapping, (b) KPFM images, and (c) their profiles in the dark and under illumination from a white-light source. (d) KPFM image of the grating after the white light was turned off. Adapted with permission from Ref. [51a]. Copyright (2014) American Chemical Society.

can be expressed numerically by $V_d = V_u - V_l$, where $V_u$ and $V_l$ are the surface potentials of the upper and lower regions of the film, respectively. The decrease of $V_d$ after illumination indicates that charge generation occurs over the entire area of the film and the electrons migrate from donor to acceptor regions. As displayed in Figure 14.10d, the surface potential distribution gradually returned to the pre-illuminated state after the white light was turned off. This reversibility further confirms the charge generation under illumination and charge collection.

In the research of organic solar cells, evaluating the stability of photoactive materials against degradation under illumination is one of the major issues. Applications of KPFM in characterizing materials and interfaces at a length scale of nanometers could lead to new insights into degradation processes. Berger *et al.* [51b] investigated the photo-induced changes in the surface potential of P3HT: phenyl-C61-butyric acid methyl ester (PCBM) active layer by KPFM. The results reveal that the degraded part has a lower surface potential as compared to the nondegraded areas in darkness. This difference is in accordance with

$p$-doping of the P3HT with oxygen that shifts the Fermi level up. Furthermore, under illumination, the surface potential of the degraded area becomes slightly higher than the nondegraded area, as oxygen doping increases the charge carrier concentration in P3HT:PCBM for degraded samples.

## 14.4 Conclusion and Outlook

In the past decades, KPFM has been developed to impact surface/material science and a variety of semiconductor industries. Besides present basic principles, this chapter highlighted the potential of KPFM to gain insight into the structural and electronic properties of functional surfaces and interfaces. Throughout this chapter, details of KPFM instrumentation and operational principles are described, including comparison of the AM and FM modes, simultaneous mapping of topography, and potential (or $\phi$) positions. Regarding the introduction of the applications, KPFM is adopted to characterize the electrical properties of surfaces of metals and semiconductors. In addition, KPFM also shows an interesting advantage in the study of devices including junctions, heterostructrues, FETs, and solar cells.

Although KPFM is a powerful technique, challenges still exist for several applications. (i) The values of $\phi$ measured by KPFM differ from the ones measured by UV photoemission spectroscopy [33]. In most of the cases, the measurement in atmosphere was used to explain the difference, but we still believe the measurement mechanisms are significant. The comparison needs extensive investigation. (ii) For similar type of samples with the same adsorbate, $\phi$ shows considerable differences [7]. For example, we have found that when varying the topology and the thickness of PEI, $\phi$ values of ITO also change a lot, as shown in Table 14.1. It is not convincible using only dipole moments of molecules to make an interpretation.

By extracting the electrical properties and supplying highly spatial resolution, KPFM plays a more important role in the development of electronic devices. With accumulation of the experiences in collecting and analyzing the data, KPFM has developed into a mature characterization method. We anticipate that its applications will potentially open pathways for the study of design and performance of new devices based on semiconductor nanostructures.

**Table 14.1** The $\phi$ values of ITO substrates determined by KPFM as a function of the thickness of PEI film.

| | | 0 nm | 2 nm | 3.5 nm | 5.5 nm | 10 nm | 20 nm |
|---|---|---|---|---|---|---|---|
| $\phi$ (eV) | $l$-PEI | 4.75 | 4.37 | 4.39 | 3.37 | 4.33 | 4.32 |
| | $b$-$PEI_{25K}$ | 4.75 | 4.41 | 4.30 | 4.30 | 4.37 | 4.42 |
| | $b$-$PEI_{750K}$ | 4.75 | 4.47 | 4.38 | 4.33 | 4.31 | 4.46 |

## Acknowledgment

The work was supported by the National Natural Science Foundation of China (21204054), a project funded by the Priority Academic Program Development of Jiangsu Higher Education Institutions and State and Local Joint Engineering Laboratory for Novel Functional Polymeric Materials.

## References

1 Kelvin, L. (1898) On electric equilibrium uranium and an isolated metal in its neighourhood. *Philos. Mag.*, **46**, 82–120.

2 Nonnenmacher, M., O'Boyle, M.P., and Wickramasinghe, H.K. (1991) Kelvin probe force microscopy. *Appl. Phys. Lett.*, **58** (25), 2921–2923.

3 (a) Goryl, M., Kolodziej, J.J., Krok, F., Piatkowski, P., Such, B., and Szymonski, M. (2005) Epitaxial nanostructures assembled on InSb (0 0 1) by submonolayer deposition of gold. *Microelectron. Eng.*, **81** (2), 394–399; (b) Berthold, T., Benstetter, G., Frammelsberger, W., Rodríguez, R., and Nafría, M. (2015) Nanoscale characterization of copper oxide films by Kelvin probe force microscopy. *Thin Solid Films*, **584**, 310–315; (c) Barth, C. and Henry, C.R. (2006) Gold nanoclusters on alkali halide surfaces: charging and tunneling. *Appl. Phys. Lett.*, **89** (25), 252119; (d) Min, B., Wallace, W., Santra, A., and Goodman, D. (2004) Role of defects in the nucleation and growth of Au nanoclusters on SiO2 thin films. *J. Phys. Chem. B*, **108** (42), 16339–16343; (e) Valden, M., Lai, X., and Goodman, D.W. (1998) Onset of catalytic activity of gold clusters on titania with the appearance of nonmetallic properties. *Science*, **281** (5383), 1647–1650.

4 (a) Xu, M., Liang, T., Shi, M., and Chen, H. (2013) Graphene-like two-dimensional materials. *Chem. Rev.*, **113** (5), 3766–3798; (b) Glatzel, T., Sadewasser, S., Shikler, R., Rosenwaks, Y., and Lux-Steiner, M.C. (2003) Kelvin probe force microscopy on III–V semiconductors: the effect of surface defects on the local work function. *Mater. Sci. Eng., B*, **102** (1), 138–142; (c) Spadafora, E.J., Demadrille, R., Ratier, B., and Grévin, B. (2010) Imaging the carrier photogeneration in nanoscale phase segregated organic heterojunctions by Kelvin probe force microscopy. *Nano Lett.*, **10** (9), 3337–3342; (d) Dabirian, R., Feng, X., Ortolani, L., Liscio, A., Morandi, V., Muellen, K., Samori, P., and Palermo, V. (2010) Micron-sized 6,6 -phenyl C61 butyric acid methyl ester crystals grown by dip coating in solvent vapour atmosphere: interfaces for organic photovoltaics. *Phys. Chem. Chem. Phys.*, **12** (17), 4473–4480; (e) Kehrer, L.A., Feldmeier, E.J., Siol, C., Walker, D., Melzer, C., and von Seggern, H. (2013) A new method to invert top-gate organic field-effect transistors for Kelvin probe investigations. *Appl. Phys. A*, **112** (2), 431–436.

5 (a) Kwak, K.J., Yoda, S., and Fujihira, M. (2003) Observation of stretched single DNA molecules by Kelvin probe force microscopy. *Appl. Surf. Sci.*, **210** (1), 73–78; (b) Birkenhauer, E. and Neethirajan, S. (2014) Surface

potential measurement of bacteria using Kelvin probe force microscopy. *J. Vis. Exp.*, **93**, e52327; (c) Lee, I., Chung, E., Kweon, H., Yiacoumi, S., and Tsouris, C. (2012) Scanning surface potential microscopy of spore adhesion on surfaces. *Colloids Surf., B*, **92**, 271–276; (d) Moores, B., Hane, F., Eng, L., and Leonenko, Z. (2010) Kelvin probe force microscopy in application to biomolecular films: frequency modulation, amplitude modulation, and lift mode. *Ultramicroscopy*, **110** (6), 708–711; (e) Palermo, V., Palma, M., and Samori, P. (2006) Electronic characterization of organic thin films by Kelvin probe force microscopy. *Adv. Mater.*, **18** (2), 145–164.

6 Fujihira, M. (1999) Kelvin probe force microscopy of molecular surfaces. *Annu. Rev. Mater. Sci.*, **29** (1), 353–380.

7 Yan, L., Song, Y., Zhou, Y., Song, B., and Li, Y. (2015) Effect of PEI cathode interlayer on work function and interface resistance of ITO electrode in the inverted polymer solar cells. *Org. Electron.*, **17**, 94–101.

8 Sadewasser, S. and Glatzel, T. (2011) *Kelvin probe force microscopy: measuring and compensating electrostatic forces*, vol. 48, Springer Science & Business Media.

9 Melitz, W., Shen, J., Kummel, A.C., and Lee, S. (2011) Kelvin probe force microscopy and its application. *Surf. Sci. Rep.*, **66** (1), 1–27.

10 Hochwitz, T., Henning, A.K., Levey, C., Daghlian, C., Slinkman, J., Never, J., Kaszuba, P., Gluck, R., Wells, R., Pekarik, J., and Finch, R. (1996) Imaging integrated circuit dopant profiles with the force-based scanning Kelvin probe microscope. *J. Vac. Sci. Technol., B*, **14** (1), 440–446.

11 Martin, Y., Williams, C.C., and Wickramasinghe, H.K. (1987) Atomic force microscope–force mapping and profiling on a sub 100-Å scale. *J. Appl. Phys.*, **61** (10), 4723–4729.

12 (a) Glatzel, T., Sadewasser, S., and Lux-Steiner, M.C. (2003) Amplitude or frequency modulation-detection in Kelvin probe force microscopy. *Appl. Surf. Sci.*, **210** (1), 84–89; (b) Kikukawa, A., Hosaka, S., and Imura, R. (1996) Vacuum compatible high-sensitive Kelvin probe force microscopy. *Rev. Sci. Instrum.*, **67** (4), 1463–1467.

13 (a) Barth, C. (2015) in *Defects at Oxide Surfaces*, vol. 58 (eds J. Jupille and G. Thornton), Springer International Publishing, pp. 215–239; (b) Kitamura, S.I. and Iwatsuki, M. (1998) High-resolution imaging of contact potential difference with ultrahigh vacuum noncontact atomic force microscope. *Appl. Phys. Lett.*, **72** (24), 3154–3156.

14 (a) Kitamura, S., Suzuki, K., and Iwatsuki, M. (1999) High resolution imaging of contact potential difference using a novel ultrahigh vacuum non-contact atomic force microscope technique. *Appl. Surf. Sci.*, **140** (3), 265–270; (b) Kitamura, S.i., Suzuki, K., and Iwatsuki, M. (1998) Observation of silicon surfaces using ultrahigh vacuum noncontact atomic force microscope. *Jpn. J. Appl. Phys.*, **37** (6S), 3765; (c) Kalinin, S.V. and Gruverman, A. (2007) *Scanning probe microscopy: electrical and electromechanical phenomena at the nanoscale*, Vol, Springer Science & Business Media, p. 1.

15 Zerweck, U., Loppacher, C., Otto, T., Grafström, S., and Eng, L.M. (2005) Accuracy and resolution limits of Kelvin probe force microscopy. *Phys. Rev. B: Condens. Matter*, **71** (12), 125424.

16 Castellani, M., Winkler, S., Bröker, B., Baumgarten, M., Müllen, K., and Koch, N. (2013) Work function increase of transparent conductive electrodes by solution processed electron acceptor molecular monolayers. *Appl. Phys. A*, **114** (2), 291–295.
17 (a) Kikukawa, A., Hosaka, S., and Imura, R. (1995) Silicon pn junction imaging and characterizations using sensitivity enhanced Kelvin probe force microscopy. *Appl. Phys. Lett.*, **66** (25), 3510–3512; (b) Sadewasser, S., Glatzel, T., Shikler, R., Rosenwaks, Y., and Lux-Steiner, M.C. (2003) Resolution of Kelvin probe force microscopy in ultrahigh vacuum: comparison of experiment and simulation. *Appl. Surf. Sci.*, **210** (1), 32–36; (c) Lee, S. (2015) Surface potential measurements of 2× 1 reconstructed Si(001) using UHV Kelvin probe force microscopy. *Surf. Sci.*, **641**, 16–22.
18 (a) Qian, H., Zhu, M., Wu, Z., and Jin, R. (2012) Quantum sized gold nanoclusters with atomic precision. *Acc. Chem. Res.*, **45** (9), 1470–1479; (b) Batista, R.J.C., Mazzoni, M.S.C., and Chacham, H. (2010) First-principles investigation of electrochemical properties of gold nanoparticles. *Nanotechnology*, **21** (6), 065705.
19 Zhang, Y., Pluchery, O., Caillard, L., Lamic-Humblot, A.-F., Casale, S., Chaba, Y.J., and Salmeron, M. (2015) Sensing the charge state of single gold nanoparticles via work function measurements. *Nano Lett.*, **15** (1), 51–55.
20 Lin, W.-H., Wu, J.-J., Chou, M.M.C., Chang, Y.-M., and Yoshimura, M. (2014) Charge transfer in Au nanoparticle-nonpolar ZnO photocatalysts illustrated by surface-potential-derived three-dimensional band diagram. *J. Phys. Chem. C*, **118** (34), 19814–19821.
21 (a) Haruta, M., Yamada, N., Kobayashi, T., and Iijima, S. (1989) Gold catalysts prepared by coprecipitation for low-temperature oxidation of hydrogen and of carbon monoxide. *J. Catal.*, **115** (2), 301–309; (b) Chung, H.J., Yurtsever, A., Sugimoto, Y., Abe, M., and Morita, S. (2011) Kelvin probe force microscopy characterization of TiO2 (110)-supported Au clusters. *Appl. Phys. Lett.*, **99** (12), 123102; (c) Meier, D.C. and Goodman, D.W. (2004) The influence of metal cluster size on adsorption energies: CO adsorbed on Au clusters supported on TiO2. *J. Am. Chem. Soc.*, **126** (6), 1892–1899.
22 Jiang, D.-e., Overbury, S.H., and Dai, S. (2011) Interaction of gold clusters with a hydroxylated surface. *J. Phys. Chem. Lett.*, **2** (10), 1211–1215.
23 Krok, F., Sajewicz, K., Konior, J., Goryl, M., Piatkowski, P., and Szymonski, M. (2008) Lateral resolution and potential sensitivity in Kelvin probe force microscopy: towards understanding of the sub-nanometer resolution. *Phys. Rev. B: Condens. Matter*, **77** (23), 235427.
24 (a) Tadashi, S. and Keiji, N. (2002) Atom-resolved imaging of the potential distribution at Si (111) 7×7 surfaces. *Jpn. J. Appl. Phys.*, **41** (10B), L1178; (b) Moraru, D., Samanta, A., Tyszka, K., Le The, A., Muruganathan, M., Mizuno, T., Jablonski, R., Mizuta, H., and Tabe, M. (2015) Tunneling in systems of coupled dopant-atoms in silicon nano-devices. *Nanoscale Res. Lett.*, **10** (1), 372.
25 (a) Melitz, W., Shen, J., Lee, S., Lee, J.S., Kummel, A.C., Droopad, R., and Yu, E.T. (2010) Scanning tunneling spectroscopy and Kelvin probe force microscopy investigation of Fermi energy level pinning mechanism on InAs

and InGaAs clean surfaces. *J. Appl. Phys.*, **108** (2), 023711; (b) Ladutenko, K.S., Ankudinov, A.V., and Evtikhiev, V.P. (2010) On the accuracy of quantitative measurements of the local surface potential. *Tech. Phys. Lett.*, **36** (3), 228–231; (c) Szymonski, M., Goryl, M., Krok, F., Kolodziej, J.J., and de Mongeot, F.B. (2007) Metal nanostructures assembled at semiconductor surfaces studied with high resolution scanning probes. *Nanotechnology*, **18** (4), 395701.

26 Shusterman, S., Raizman, A., Sher, A., Paltiel, Y., Schwarzman, A., Lepkifker, E., and Rosenwaks, Y. (2007) Nanoscale mapping of strain and composition in quantum dots using Kelvin probe force microscopy. *Nano Lett.*, **7** (7), 2089–2093.

27 (a) Wang, Q.H., Kalantar-Zadeh, K., Kis, A., Coleman, J.N., and Strano, M.S. (2012) Electronics and optoelectronics of two-dimensional transition metal dichalcogenides. *Nat. Nanotechnol.*, **7** (11), 699–712; (b) Ochedowski, O., Marinov, K., Scheuschner, N., Poloczek, A., Bussmann, B.K., Maultzsch, J., and Schleberger, M. (2014) Effect of contaminations and surface preparation on the work function of single layer MoS2. *Beilstein J. Nanotechnol.*, **5** (1), 291–297.

28 (a) Iglesias, V., Lanza, M., Zhang, K., Bayerl, A., Porti, M., Nafría, M., Aymerich, X., Benstetter, G., Shen, Z.Y., and Bersuker, G. (2011) Degradation of polycrystalline HfO2-based gate dielectrics under nanoscale electrical stress. *Appl. Phys. Lett.*, **99** (10), 103510; (b) Lanza, M., Porti, M., Nafria, M., Aymerich, X., Benstetter, G., Lodermeier, E., Ranzinger, H., Jaschke, G., Teichert, S., and Wilde, L. (2011) Conductivity and charge trapping after electrical stress in amorphous and polycrystalline devices studied with AFM-related techniques. *IEEE Trans. Nanotechnol.*, **10** (2), 344–351.

29 Li, Y., Xu, C.-Y., and Zhen, L. (2013) Surface potential and interlayer screening effects of few-layer MoS2 nanoflakes. *Appl. Phys. Lett.*, **102** (14), 143110.

30 (a) Sharifi, F., Bauld, R., and Fanchini, G. (2013) Acridine orange as a biosensitive photovoltaic material. *J. Appl. Phys.*, **114** (14), 144504; (b) Saito, N., Hayashi, K., Sugimura, H., Takai, O., and Nakagiri, N. (2002) Surface potential images of self-assembled monolayers patterned by organosilanes: ab initio molecular orbital calculations. *Surf. Interface Anal.*, **34** (1), 601–605.

31 Zhou, Y., Fuentes-Hernandez, C., Shim, J., Meyer, J., Giordano, A.J., Li, H., Winget, P., Papadopoulos, T., Cheun, H., Kim, J., Fenoll, M., Dindar, A., Haske, W., Najafabadi, E., Khan, T. M., Sojoudi, H., Barlow, S., Graham, S., Bredas, J-L., Marder, S. R., Kahn, A., and Kippelen, B. (2012) A universal method to produce low-work function electrodes for organic electronics. *Science*, **336** (6079), 327–332.

32 He, Z., Zhong, C., Su, S., Xu, M., Wu, H., and Cao, Y. (2012) Enhanced power-conversion efficiency in polymer solar cells using an inverted device structure. *Nat. Photonics*, **6** (9), 591–595.

33 Song, Y., Yan, L., Zhou, Y., Song, B., and Li, Y. (2015) Lowering the work function of ITO by covalent surface grafting of aziridine: application in inverted polymer solar cells. *Adv. Mater. Interfaces*, **2** (1), 1400397.

34 (a) Pelto, J.M., Haimi, S.P., Siljander, A.S., Miettinen, S.S., Tappura, K.M., Higgins, M.J., and Wallace, G.G. (2013) Surface properties and interaction

forces of biopolymer-doped conductive polypyrrole surfaces by atomic force microscopy. *Langmuir*, **29** (20), 6099–6108; (b) Martin, Y., Abraham, D.W., and Wickramasinghe, H.K. (1988) High-resolution capacitance measurement and potentiometry by force microscopy. *Appl. Phys. Lett.*, **52** (13), 1103–1105.

35 (a) Schmidt, H., Habicht, S., Feste, S., Mueller, A.-D., and Schmidt, O.G. (2013) Kelvin probe force microscopy for characterizing doped semiconductors for future sensor applications in nano- and biotechnology. *Appl. Surf. Sci.*, **281**, 24–29; (b) Saraf, S. and Rosenwaks, Y. (2005) Local measurement of semiconductor band bending and surface charge using Kelvin probe force microscopy. *Surf. Sci.*, **574** (2), L35–L39; (c) Doukkali, A., Ledain, S., Guasch, C., and Bonnet, J. (2004) Surface potential mapping of biased pn junction with kelvin probe force microscopy: application to cross-section devices. *Appl. Surf. Sci.*, **235** (4), 507–512; (d) Wang, X., Xie, W., Chen, J., and Xu, J.-B. (2014) Homo- and hetero- p-n junctions formed on graphene steps. *ACS Appl. Mater. Interfaces*, **6** (1), 3–8.

36 (a) Liu, C., Hwang, Y.J., Jeong, H.E., and Yang, P. (2011) Light-induced charge transport within a single asymmetric nanowire. *Nano Lett.*, **11** (9), 3755–3758; (b) Lochthofen, A., Mertin, W., Bacher, G., Furitsch, M., Bruederl, G., Strauss, U., and Haerle, V. (2008) Microscopic investigation of InGaN/GaN heterostructure laser diode degradation using Kelvin probe force microscopy. *J. Phys. D: Appl. Phys.*, **41** (13), 135115.

37 Ryu, S.R., Ram, S.D.G., Lee, S.J., Cho, H.-d., Lee, S., Kang, T.W., Kwon, S., Yang, W., Shin, S., and Woo, Y. (2015) Vertical current-flow enhancement via fabrication of GaN nanorod p-n junction diode on graphene. *Appl. Surf. Sci.*, **347**, 793–798.

38 Ben, C.V., Cho, H.D., Kang, T.W., and Yang, W. (2012) Surface potential measurement of As-doped homojunction ZnO nanorods by Kelvin probe force microscopy. *Surf. Interface Anal.*, **44** (6), 755–758.

39 Zeng, T.-W., Ho, C.-C., Tu, Y.-C., Tu, G.-Y., Wang, L.-Y., and Su, W.-F. (2011) Correlating interface heterostructure, charge recombination, and device efficiency of poly(3-hexyl thiophene)/TiO2 nanorod solar cell. *Langmuir*, **27** (24), 15255–15260.

40 (a) Tsai, C.-C., Chiang, P.-L., Sun, C.-J., Lin, T.-W., Tsai, M.-H., Chang, Y.-C., and Chen, Y.-T. (2011) Surface potential variations on a silicon nanowire transistor in biomolecular modification and detection. *Nanotechnology*, **22** (13), 395701; (b) Afsharimani, N. and Nysten, B. (2013) Scanning probe microscopy study of electronic properties in alkyl-substituted oligothiophene-based field-effect transistors. *Vacuum*, **90**, 17–24.

41 (a) Henning, A., Klein, B., Bertness, K.A., Blanchard, P.T., Sanford, N.A., and Rosenwaks, Y. (2014) Measurement of the electrostatic edge effect in wurtzite GaN nanowires. *Appl. Phys. Lett.*, **105** (21), 213107; (b) Afsharimani, N. and Nysten, B. (2013) Electronic properties of dioctylterthiophene-based organic thin-film transistors: a Kelvin probe force microscopy study. *Thin Solid Films*, **536**, 295–301.

42 Zhang, Y., Ziegler, D., and Salmeron, M. (2013) Charge trapping states at the SiO2-oligothiophene monolayer interface in field effect transistors studied by Kelvin probe force microscopy. *ACS Nano*, **7** (9), 8258–8265.

43 Cui, Y. and Lieber, C.M. (2001) Functional nanoscale electronic devices assembled using silicon nanowire building blocks. *Science*, **291** (5505), 851–853.

44 Quitoriano, N.J. and Kamins, T.I. (2008) Integratable nanowire transistors. *Nano Lett.*, **8** (12), 4410–4414.

45 Bae, S.S., Prokopuk, N., Quitoriano, N.J., Adams, S.M., and Ragan, R. (2012) Characterizing defects and transport in Si nanowire devices using Kelvin probe force microscopy. *Nanotechnology*, **23** (40), 405706.

46 (a) Yun, J.S., Ho-Baillie, A., Huang, S., Woo, S.H., Heo, Y., Seidel, J., Huang, F., Cheng, Y.-B., and Green, M.A. (2015) Benefit of grain boundaries in organic-inorganic halide planar perovskite solar cells. *J. Phys. Chem. Lett.*, **6** (5), 875–880; (b) Jiang, C.-S., Yang, M., Zhou, Y., To, B., Nanayakkara, S.U., Luther, J.M., Zhou, W., Berry, J.J., van de Lagemaat, J., Padture, N.P., Zhu, K., and Al-Jassim, M.M. (2015) Carrier separation and transport in perovskite solar cells studied by nanometre-scale profiling of electrical potential. *Nat. Commun.*, **6**, 8397.

47 Jiang, M., Wu, J., Lan, F., Tao, Q., Gao, D., and Li, G. (2015) Enhancing the performance of planar organo-lead halide perovskite solar cells by using a mixed halide source. *J. Mater. Chem. A*, **3** (3), 963–967.

48 (a) Günes, S., Neugebauer, H., and Sariciftci, N.S. (2007) Conjugated polymer-based organic solar cells. *Chem. Rev.*, **107** (4), 1324–1338; (b) Thompson, B.C. and Fréchet, J.M.J. (2008) Polymer–fullerene composite solar cells. *Angew. Chem. Int. Ed.*, **47** (1), 58–77.

49 Jiao, W., Ma, D., Lv, M., Chen, W., Wang, H., Zhu, J., Lei, M., and Chen, X. (2014) Self n-doped [6,6]-phenyl-C61-butyric acid 2-((2-(trimethylammonium)ethyl)-(dimethyl)ammonium) ethyl ester diiodides as a cathode interlayer for inverted polymer solar cells. *J. Mater. Chem. A*, **2** (35), 14720–14728.

50 Hoppe, H. and Sariciftci, N.S. (2004) Organic solar cells: an overview. *J. Mater. Res.*, **19** (07), 1924–1945.

51 (a) Watanabe, S., Fukuchi, Y., Fukasawa, M., Sassa, T., Kimoto, A., Tajima, Y., Uchiyama, M., Yamashita, T., Matsumoto, M., and Aoyama, T. (2014) In situ KPFM imaging of local photovoltaic characteristics of structured organic photovoltaic devices. *ACS Appl. Mater. Interfaces*, **6** (3), 1481–1487; (b) Sengupta, E., Domanski, A.L., Weber, S.A.L., Untch, M.B., Butt, H.-J., Sauermann, T., Egelhaaf, H.J., and Berger, R. (2011) Photoinduced degradation studies of organic solar cell materials using Kelvin probe force and conductive scanning force microscopy. *J. Phys. Chem. C*, **115** (40), 19994–20001; (c) Venkatesan, S., Chen, J., Ngo, E.C., Dubey, A., Khatiwada, D., Zhang, C., and Qiao, Q. (2015) Critical role of domain crystallinity, domain purity and domain interface sharpness for reduced bimolecular recombination in polymer solar cells. *Nano Energy*, **12**, 457–467.

52 Chen, L.-C., Chen, J.-C., Chen, C.-C., and Wu, C.-G. (2015) Fabrication and properties of high-efficiency perovskite/PCBM organic solar cells. *Nanoscale Res. Lett.*, **10** (1), 1020–1025.

# 15

# Hot Electron Nanoscopy and Spectroscopy (HENs)

*Andrea Giugni[1], Bruno Torre[1], Marco Allione[1], Gerardo Perozziello[2], Patrizio Candeloro[2], and Enzo Di Fabrizio[1,2]*

[1] *SMILEs Lab, Physical Science and Engineering Division (PSE), King Abdullah University of Science and Technology (KAUST), Ibn Al-Haytham Bldg., Thuwal 23955-6900, Saudi Arabia*
[2] *Departement of Experimental Clinics, Bionem Lab, University Magna Graecia, Campus "Salvatore Venuta" Viale Europa, 88100 Germaneto-Catanzaro, Italy*

## 15.1 Introduction

The fields of plasmonics [1] and nanoelectronics [2] have attracted the interest of a wide audience over the past two decades triggered by the disclosure of many fundamental scientific breakthroughs. New ideas have led to fascinating experimental results in many fields in physics and biology, pushing to a new level the knowledge and control on the processes that rule optoelectronic devices and life science at the nanoscale [3, 4].

Scanning probe microscopy (SPM) has greatly improved since its first implementation [5], now offering several specialized techniques that can reveal important insights into the topographical, mechanical, electronic, and magnetic properties of materials at the micro- and nanoscale [6].

At the same time, Raman spectroscopy can be combined with SPM setup [7] reaching the sub-diffraction limit resolution and providing detailed molecular information, in some cases supplementing those obtained by other techniques, and in others adding original insights into the study of materials [8].

In the field of surface plasmon photonics, a plasmonic device is a tool that, taking advantage of enhanced interactions of light with metallic nanostructures, shows some new special abilities. In particular, due to their intrinsic electromechanical nature [9] the so-called surface plasmon polaritons (SPPs) can overcome both limits in the fundamental optical diffraction (Rayleigh criterion) and in the energy transport mean free path of the electrons in the metal. This provides an effective way to guide and concentrate energy at the nanoscale, where specific energy transfer processes take place. Thanks to field intensification [10] and nonlocal effects, efficient plasmonic substrates have been proposed for enhanced spectroscopies [11] and selective photodetection [12]. Beside the photonic generation channel, an SPP can decay in hot carriers (*hc*), which can be used for transport measurements in parallel with optical characterization. Energy confinement and an effective hot charges generation rate

*Conductive Atomic Force Microscopy: Applications in Nanomaterials*, First Edition.
Edited by Mario Lanza.

have driven plasmonic-assisted chemical catalysis [13] and offered new tools in material characterization [14]. The strong coupling, polarization sensitivity, and nonlinearities [15] in plasmonic devices made possible applications in lasing media [16], beam shapers [11, 17], and ultrafast devices [18] as well as metamaterials [19].

In the past years, quantum mechanical nature of SPPs and medium was included in the theory, getting new insights into the description of the propagation and scattering processes of electrons and phonons. This allows, in a quantum mechanical framework, the description of the decay process into charge pairs, known as hot electrons ($he^-$) and hot holes, in a confined space [20] and in bulk media [21], where energy distribution, decay channels, relaxation times, and mean free paths can be retrieved by *ab initio* calculations [21, 22]. First principle quantum description allows accounting for material-specific features, like interband transitions in noble metals, which are precluded in simplified jellium model, thus providing insight into the dynamics of plasmonic *hc* generation.

From an optical point of view, the property of localized plasmons and SPPs generates the near-field light, with wavelength well below the diffraction limit, with a local field enhancement [23, 24] that has opened the way to spectroscopic analysis of tiny amounts of matter, down to the single molecule detection limit [25].

In this context, we have recently proposed a multi-spectroscopy scanning probe apparatus [14] for the simultaneous investigation of chemical, structural, and charge transport properties at nanoscale, exploiting synergic phenomena. These are the low-loss propagation through adiabatic plasmonic compression of the SPPs that induces a field enhancement (energy focusing) [23, 26] with maximal momentum transfer at the interface, and the effect of quantum confinement on the $he^-$ generation rate (improved responsivity). All these terms allow the control of the extraction rate of hot charges in a nanometric confined space.

The functional element is a plasmonic device fabricated on an atomic force microscopy (AFM) probe, composed of a photon to SPP coupler (a grating or a photonic cavity) and a final conical tip where energy compression and charge generation occur. The injection direction, along the normal to the interface between the tip and the sample, provides the maximal photonic contribution to the *hc* momentum exchange and increases the SSP conversion rate at the nanostructure apex. The energy confinement in a small volume (<10 nm in depth) at the tip–sample interface maximizes the hot charges production rate and their injection probability before their thermalization in the metal. At the interface between the nanometric tip apex and the semiconductor, the momentum conservation rule relaxes, allowing the hot electron to access the required phase space for crossing the barrier. Further, the observed radiative scattering at the tip apex generates a sub-diffraction optical probe, which can be used with the hot electron current source for combined transport and optical spectroscopy. Such a configuration does not require any voltage bias for the injection of the charge through the Schottky barrier (SB) and benefits from the reduced depletion region (a few nanometers linear length and depth) that is jointly responsible for the high spatial resolution and current sensitivity.

The surface of the tip is made of gold, also ensuring good stability and biocompatibility. The customized tip is fully compatible with all AFM cantilevers and can be used in many of the operational modes of the instrument. In particular, the transport of *hcs* with high localization and collection efficiency through an SB was first accomplished in contact mode. An ultimate efficiency exceeding 30% and a lateral resolution of 3 nm for a Schottky Au–GaAs junction is reported. In the present contribution, we include our updated results on hot electron AFM nanoscopy.

We have organized the chapter to include a brief description of different laser coupling methods with guided SPP modes at the surface of a cone. We show some devices, their electromagnetic simulations, and their optical characterization. A theoretical section illustrates the optical and quantum description of the hot charge generation rate as obtained for the SPP propagation along the nanocone in adiabatic compression. We also show some experimental results concerning the application of the hot electron nanoscopy and spectroscopy (HENs) in the so-called Schottky configuration, highlighting the sensitivity and the nanoscale resolution of the technique. The comparison with Kelvin probe and other electric AFM techniques points out the intrinsic advantages of the HENs. In the end, some further insights are given about the possibility of exploiting HENs with a pulsed laser at the femtosecond time scale without significant pulse broadening and dispersion.

## 15.2 Coupling Schemes

According to their dispersion relations, there is a momentum mismatch between the photon and the SPP having the same energy, as shown in Figure 15.1a. Several coupling configurations have been proposed to couple the freely propagating photon with the SPP, fulfilling the energy and momentum conservation rules for the interaction process. The first constraint fixes the frequency, while the second determines the amount of momentum that must be provided, once the frequency-dependent SPP wave-vector dispersion is known [27]. An effective coupling is also responsible for an efficient hot electron generation.

As SPP launchers [28], we consider two methods: a diffraction grating consisting of a periodic modulation of the surface of the metallic device, based on the phenomenon named after W. Wood, and a photonic cavity of proper symmetry. Other configurations are possible, such as the use of a high-index glass prism, commonly used in the Kretschmann–Raether [29, 30] or the Otto [31] configurations.

The complete device consists of an AFM cantilever where a grating, acting as a coupler, is engraved by means of a focused ion milling process. The coupler is placed in the proximity of a micrometric metallic cone whose apex can be fabricated with a radius of curvature down to 5 nm, as shown in Figure 15.3b. According to our scattering geometry, which fixes the incident angle $\theta_{\text{in}} \sim 36°$ with respect to the normal of the grating surface, we calculated the grating period $P$ allowing to compensate the momentum mismatch (Figure 15.1a) from

$$\frac{2\pi}{P} = \Delta K_z = K_z^{\text{SPP}} - K_z^{\text{Ph}} = \frac{\omega}{c}\sqrt{\frac{\varepsilon_d \varepsilon_m}{\varepsilon_d + \varepsilon_m}} - \frac{2\pi}{\lambda} \sin \theta_{\text{in}} \tag{15.1}$$

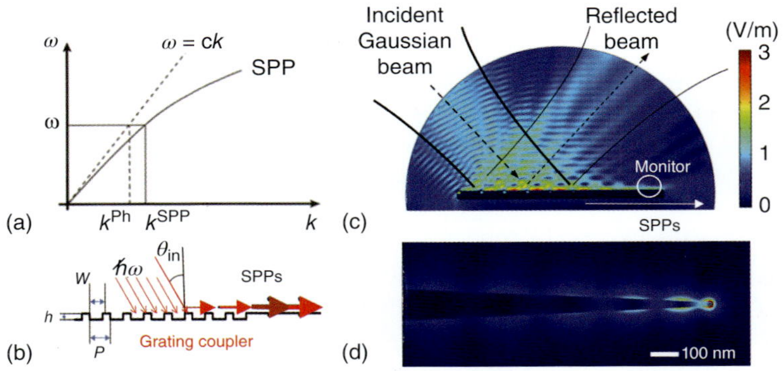


**Figure 15.1** (a) The graph shows the momentum mismatch $K^{SPP} - K^{Ph}$ between the dispersion curve of an SPP mode and photon at a fixed angular frequency $\omega$. (b, c) Represent, respectively, the grating coupling geometry and the corresponding EM simulation. There are indicated the four meaningful parameters for the optimal coupling and the monitor position used for the energy flow analysis described in the text. The free parameters are the groove-pitch *P*, the height *h*, the width *w*, and the incidence angle $\theta_{in}$. (d) The E field amplitude in the propagation-induced adiabatic compression of a TM0 mode along a plasmonic taper.

being $\lambda$ the free space wavelength, $K_z^{SPP}$ and $K_Z^{Ph}$ the wavevector of the plasmon and of the photon along the *Z* propagation direction, respectively, and $\varepsilon_m$ and $\varepsilon_d$ the frequency-dependent permittivity of gold and the dielectric material, respectively.

A 2D numerical simulation for six equally spaced grooves, performed using the COMSOL Multiphysics finite element method software, allowed us to optimize all the three significant parameters of the grating, that are, the period *P*, the height *h*, and the width *w* of the grooves for the geometrical configuration [34, 35]. The coupler, sketched in Figure 15.1b, was then realized on the flat sides of the tip, as shown in Figure 15.3b and c. Our simulations in Figure 15.1b show that about 10% of the incident radiation transfers to the propagating SPPs. We evaluated by means of Poynting vector analysis the efficiency of the coupling as the ratio between the energy flowing along the metal surface and the incident one, as indicated in Figure 15.1c. To estimate the propagation of SPPs to the tip apex, we performed a complete 3D electromagnetic simulation of the whole device, with the optimal parameters calculated for a wavelength of 670 nm ($P = 1550$ nm, $h = 150$ nm, $w = 220$ nm). We considered a Gaussian beam with $E_0 = 1$ V/m on the axis, and the whole device was immersed in a dielectric medium of refractive index $n = 1$. We inserted five layers of perfect matching layers (PMLs) around the simulation volume, far away from the metal structures to avoid spurious effects between the evanescent waves and the PML. The electromagnetic simulation allows estimating the amplitude of the electric field at the apex (tip curvature ~12 nm, as the experimental value of the tip used for the measurement reported in Section 15.5) of the conical structure extending about 2.5 μm over the AFM pyramidal tip. The cone promotes the axial symmetric TM0 propagating mode, a transverse magnetic mode with a radially polarized electric field with respect to the cone axis [36]. The result, reported in Figure 15.1c as a zoomed image of the tip, reproduces the so-called adiabatic compression phenomenon, which describes the slowing down to zero of both group and phase velocity of the SPP proceeding along the cone toward the tip apex, with a temporal logarithmic dependence [23].

Another possible coupling configuration, favorable from an experimental point of view, relies on the use of a photonic crystal (PhC) slab cavity coaxial with the cone and in contact with its base [37]. In this case, the cone can be fabricated directly on a tipless AFM cantilever. Optimal coupling for the TM0 mode along the cone surface can be obtained with the use of a so-called PhC-H1 structure, a PhC cavity with a hexagonal periodicity of holes in the dielectric and with a single missing hole that defines the optical cavity center. This PhC-H1 selects a radial mode when illuminated with linear polarized light, for example, through a microscope objective, as shown in the drawing in Figure 15.3a.

In Figure 15.2a–e we report the electromagnetic simulation of the cavity proposed in Ref. [37] when illuminated with 633 nm, while in Figure 15.3a is the

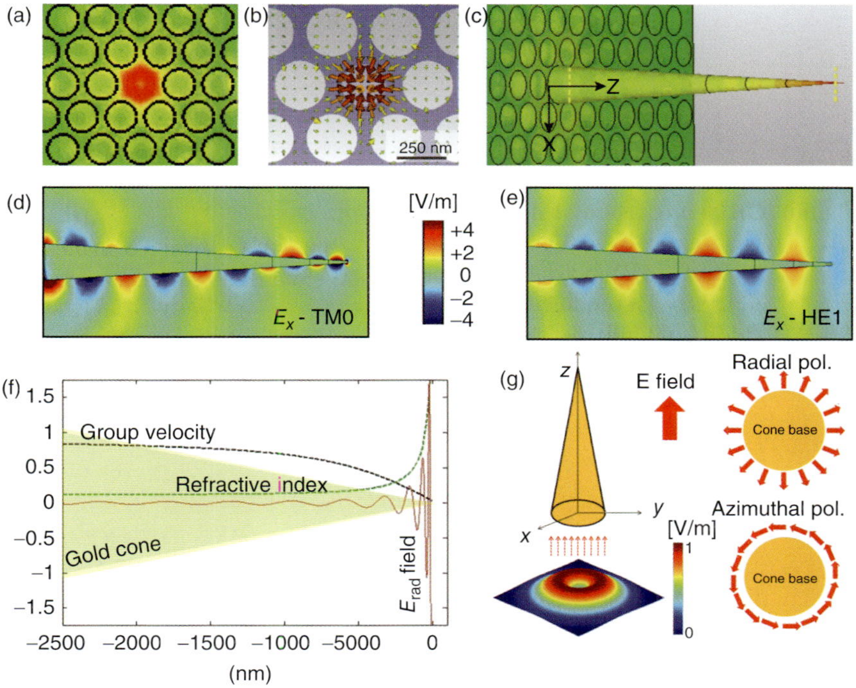


**Figure 15.2** (a) Total electric field at $z = z_0$ (slab thickness) for the photonic crystal slab of Figure 15.3a supporting radial modes at $\lambda = 633$ nm. (b) Vectorial plot of the electric field at $z_0$ evidencing the radial TM0 structure of the mode. (c) Full SPP propagation and field calculation for a silver nanocone on H1 radial cavity (d, e) radial component of the electric fields of TM0 and HE1 modes propagating along the conical tip. Only for the TM0 case is observable the focusing and the progressive reduction of the effective wavelength. (f) Simulated $E_r$ field for a TM0 mode in adiabatic compression, there are graphed also the group velocity, with respect to light speed in vacuum, and the effective refractive index (10×). (g) Coupling scheme of a radial polarized mode at the cone base. (a–c) Reproduced with permission from [37], copyright OSA 2011. (d–g) Reproduced with permission from [33], copyright IOP 2014.

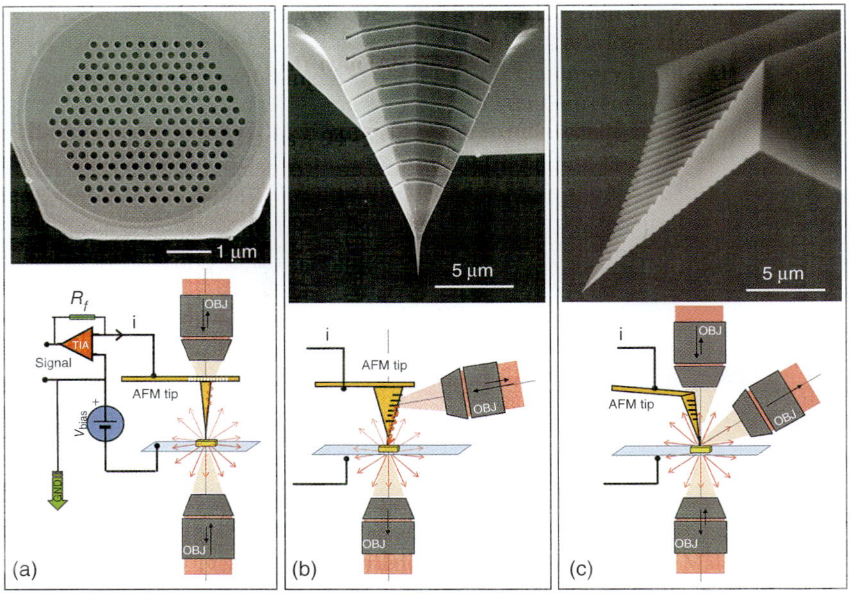


**Figure 15.3** Devices promoting adiabatic compression and corresponding possible coupling geometries. (a) SEM micrograph of PhC-H1 realized on a silicon AFM cantilever used to generate an axially symmetric mode. Using the illumination geometry indicated, the H1 photonic cavity allows to couple a nanocone with laser beam, squeezing the spot dimension below the diffraction limit. (b, c) SEM micrographs of devices for non-local excitation of the tip apex mediated by a grating coupler. Alternative configurations for maximal sample optical accessibility are displayed. In the pictures, the black arrow in correspondence of the objectives specifies if they are for excitation detection or both. From polarization considerations the configuration in (b) is to be preferred. (a) Reproduced with permission from [32], copyright NPG 2010. (c) Reproduced with permission from [33], copyright IOP 2014.

scanning electron microscopy (SEM) image of a PhC example. In this case, the material was silicon with a refractive index $n_{633\text{nm}} = 3.917 + i\,0.0122$. The periodicity was $P = 290$ nm with hole radius $R = 0.4^*P$ and slab thickness $z_0 = 162$ nm. The finite difference time domain (FDTD) simulation (CST package) of the devices clarifies the symmetric structure of the sustained pure radial mode. In particular, the electric field distribution of a TM0 mode is shown in Figure 15.2a and its vectorial structure in Figure 15.2b. The simulated cone is made up of silver, $\varepsilon = -14.469 + i\ 1.094$, and has a base diameter of about 300 nm and a height of 2.5 μm. For a PhC-H1, the field has a pure TM0 structure, both at the base of the cone, where the SPP is launched, and at the apex. Figure 15.2d and e shows the value $E_z$ of the electric field along the cone when either a radially polarized source or a plane wave impinges on the base of the cone. For both cases, the direction of propagation of the illumination is parallel to the axis of the metallic cone, and the laser spot size matches the cone base. As expected, the final field of the SPP and its mode strongly depend on the source symmetry. The plane wave generates an HE1 mode that does not show any compression.

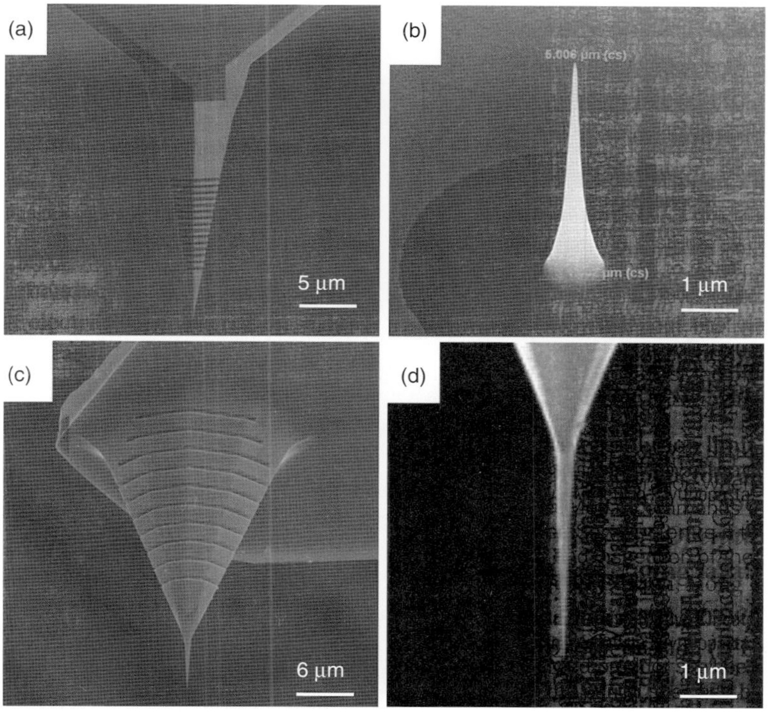


**Figure 15.4** Scanning electron micrographs of three different devices used for non-local excitation of the tip apex. (a) SEM image illustrating the fabrication of the grating for back propagation of SPPs at the wavelength 670 nm, as in Figure 15.3c. (b) SEM image of the sharp cone grown on a suspended thin $Si_3N_4$ membrane. (c, d) One of the devices and the detailed image of its apex used for the hot electron imaging reported in this paper. The devices were prepared by focused $Ga^+$ ions beam milling and electron-induced deposition, preceded and followed by gold sputtering. (c, d) Reprinted with permission from [14], copyright NPG 2013.

Figure 15.2f shows the characteristic effective wavelength, the group velocity, the refractive index, and the radial electric field of an SPP that undergoes adiabatic compression in the material.

For both structures, after successful coupling, the SPP propagates with a negligible back reflection and dissipation with a progressive enhancement of the electric field and an increase of the wave vector while approaching the apex of the cone.

In the optical configuration shown in Figure 15.3a, it is also possible to directly illuminate the edge–base of the cone with a radial polarized beam. In this case, the cone can be fabricated on a tipless AFM cantilever, as shown in Figure 15.4b. The same edge relaxes the wavevector matching conditions, while the shape of the optical mode and its polarization symmetry match directly the cone geometry and the field symmetry constrains, as shown in its optical characterization in Figure 15.7.

## 15.3 Plasmonic Device and Optical Characterization

When the size of our plasmonic devices approaches the bulk skin depth, the size that defines the geometry of the device becomes the only important length scale parameter. At this point, the SPP localization depends only on our ability to manufacture small objects. The theoretical picture at the base of this concept is detailed in Section 15.4.

The fabrication of the devices, reported in Figure 15.7, required a focused $Ga^+$ ion milling (focused ion beam (FIB)) at 30 keV to fabricate the coupling grating and the photonic cavity, followed by an intermediate gold/silver coating process and a subsequent electron-beam-induced deposition at 5 keV, to grow the cone using a Pt-C precursor gas. The final step consisted in a noble metal deposition of ~50 nm on the device by sputtering, with a custom procedure that maintains the tip apex radius in the range of a few nanometers. These steps provide an effective coupling between tip and cone. The control of the groove width is ±5 nm, while the depth was realized with a reproducibility of ±15 nm. SEM images of some examples of the final devices obtained for different coupling geometries are reposted in Figure 15.3 and 15.4.

Important aspects to control are the symmetry of the apex, the homogeneity of the coating, the final radius of curvature, and the lack of asperities, which act as unwanted scattering centers in the proximity of the tip apex. Moreover, a smooth connection between the grating launcher, usually fabricated on a side of the commercial tip, and the base of the cone has to be obtained to minimize the impedance mismatch along the SPP propagation direction. Finally, the tip shape has to be preserved during scanning; therefore, the cone hardness and the metal coverage at the apex, which are the elements that define the Schottky contact [38], have to be realized with high uniformity and control over material growth. For the same reason, the use of very low spring constant cantilevers, about 0.1 N/m, is suggested, since lower loads are more easily achievable while working in contact mode.

Once these parameters have been optimized, it is possible to reach a final efficiency very close to the theoretical limit, around 30%, and $he^-$ responsivity in the order of nA/μW, with a very good scanning stability, comparable to the one obtained with commercial probes used in other electrical mode configurations.

An optical characterization follows the fabrication phase. To routinely check the optical performances, we have built up an *ad hoc* setup constituted by three optical microscopes along $X$, $Y$, $Z$ orthogonal axes, all imaging the same spatial volume where a tip can be positioned (Figure 15.5f). The tip stage has six degrees of freedom ($X$, $Y$, $Z$, $\theta$, $\phi$, $\alpha$) and the laser illumination, with tunable polarization, is aligned coaxially with one of the optical paths. Each microscope has a calcite displacer to image the two polarizations of the scattered light at the same time. In general, a calcite displacer should be avoided in white-light imaging, since it acts as a dispersive prism for the displaced beam. Nevertheless, the use of monochromatic excitation maintains the same focal plane, resulting in a parallel acquisition of the two different polarizations for the image, and thus providing a good solution for the polarization analysis of the light scattered from

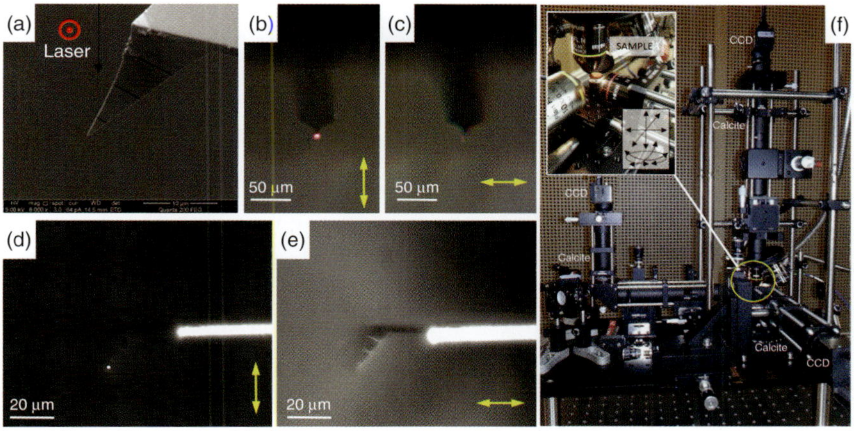


**Figure 15.5** (a) SEM image of a grating coupler realized on a nose-type tip for top illumination. Direction and polarization of exciting laser is also shown. (b–e) Refers to the test of the grating coupling on nose-type tip (Nanosensors ATEC-contAu cantilever) for $\lambda = 633$ nm top side illumination. (b, c) Optical images of the tip as seen from the front and analyzed for two hortogonal polarizations as indicated by yellow arrows. (d, e) Side views of the tip. A background far field white light illuminates the overall tip and cantilever profiles. (f) Photograph of the optical setup; inset shows the sample area.

the device. In Figures 15.5 and 15.6, we report the SEM micrographs and optical characterization of two devices similar to the ones used for the experiment reported below.

Figure 15.7 illustrates the excitation of the axially symmetric SPP mode with 647-nm laser on a silver cone realized on suspended $Si_3N_4$ membranes. In this case, a specific optical setup for the control of the polarization of the incident beam and the nanopositioning of the cone has been realized; it is illustrated schematically in Figure 15.7d. The large apertures on both sides of the metal act as diaphragms for the focused impinging beam, allowing to locally optimize the radial mode profile matching the cone base edge (Figure 15.7c). By changing the polarization of the incident beam, it is possible to move continuously from pure radial to pure azimuthal polarization, while monitoring at the same time the SPP propagation along the cone by imaging the tip plane, as shown in Figure 15.7f–h.

## 15.4 Theoretical Section

The absorption of a photon by a metal surface, as described, can generate an SPP [39]. An SPP is defined as a nonequilibrium collective electron gas excitation in the metal, oscillating at the electromagnetic frequency, that propagates confined along the metal–dielectric interface for tens of micrometers before some scattering event damps it, allowing for its decay into energized charges or a new photon. Due to the nature of the coupling, the optically generated SPP has intrinsically low momentum resulting in a nonfavorable condition for the SPP $he^-$ scattering. This gives to SPPs an exceptionally long mean free propagation length, compared to the one of a single electron.

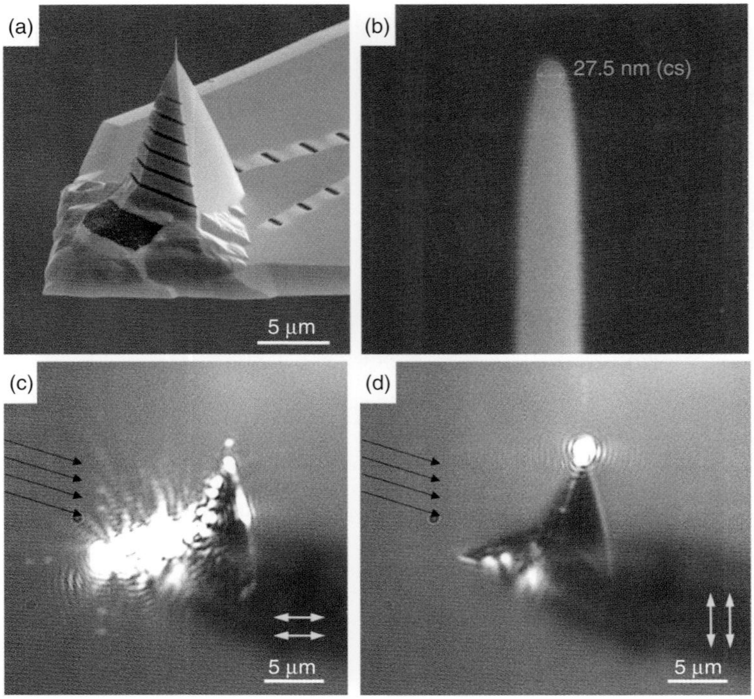


**Figure 15.6** (a) Full device structure fabricated considering optimal value of the simulated parameters for a wavelength of 633 nm and illumination direction as depicted in (c, d). (b) Detail of the apex of the cone of (a). (c, d) Optical characterization showing, respectively, the horizontal or the vertical polarization in excitation and in analysis at the wavelength of 633 nm, as indicated by the arrows. Different scattering intensities reflect the prevailing mechanism for the two polarizations: scattering in (c) and coupling in (d). A background far-field white-light illumination allowed capturing the overall tip profile. The vertical excitation, correctly, induces a strong scattering from the apex, confirming the SPP propagation described in the text. The device shown here was realized on Micromash CSC38 cantilever. The cone apex has an estimated radius of curvature <15 nm.

Coupling, propagation, and damping of an SPP are ruled by complex dynamics, strongly depending on the intimate nature of the medium. The intrinsic nature of SPPs physically bridges different length scales, from the micrometric to nanometric: by nanostructuring the specimen at those length scales, it is therefore possible to show different aspects of this complexity to underline the huge richness of phenomena and valuable experimental potential for different applications. The dynamic processes that regulate *hc* generation in noble metals, their propagation, and possibly their extraction are attracting an increasing interest in both the theoretical research and applied fields. Furthermore, another important emerging research technique is the realization of a multi-spectroscopies probe, with improved efficiency for the investigation of the matter at the nanoscale.

Despite very significant experimental and theoretical efforts, a comprehensive description of the processes that determine the measure of a net photocurrent is still not clearly given at the tip of the adiabatic nanocone. Here we recall classical

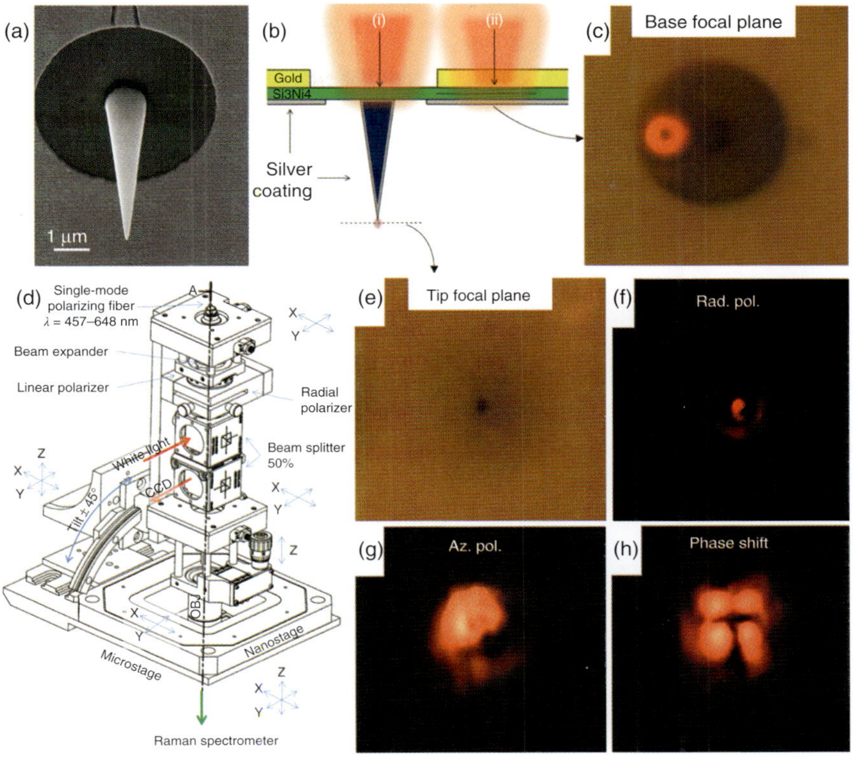


**Figure 15.7** (a) SEM micrograph of a silver cone, fabricated by FIB and sputtering techniques, on $Si_3N_4$ membrane (500-nm thick) and supported by Si wafer promoting adiabatic compression. (b) Corresponding coupling geometry. Direct coupling at the base requires the dimension matching between spot and cone base diameters. In (d), we report the mechanical design of the optical setup used to excite SPPs along the cone by means of direct illumination with a radial polarized optical mode. The main optical elements are indicated in the picture. The system allows converting continuously a linear polarization into a radial or azimuthal one. A fine nano-position of the sample allows checking the robustness of this type of coupling. Two nonpolarizing beam splitter cubes accommodate the optical path for a charge-coupled device (CCD) imaging system and a 50× objective allows focusing light on the back of the cone. A second objective collects the images from the bottom, facing the tip of the cone, as shown in (c, e–h). (c, e) The focal planes, collected in white light, at the base and top of the cone. (c) The characteristic donut intensity profile of the circularly polarized spot at 647 nm in its focal plane. (f–h) The scattered light pattern as can be collected from the tip focal plane. When radial polarization is used, the coupled SPPs propagate adiabatically to the apex originating a point source useful for spectroscopic applications.

and quantum mechanical descriptions that are relevant to the explanation of the phenomena [20, 21, 23, 40–50].

### 15.4.1 Semiclassical Considerations

A classical electromagnetic model (EM) assumes the continuity of the medium and describes a noble metal in a jellium framework as a free-electron-like system, with a parabolic dispersion curve for the conduction electrons. The adoption of experimental complex frequency-dependent dielectric function allows for the description of the interaction between the electromagnetic field and the metallic guide, introducing geometrical factors.

The EM description of a tapered plasmonic guide, terminating with apex dimension of a few nanometers, has led to the so-called adiabatic nanocompression idea, introduced for the first time in a pioneering work by Stockman [23] in 2004. Such a transport mechanism determines the concentration of the SPP energy at the tip, with a corresponding local field enhancement by about three orders of magnitude. The spatial confinement breaks the diffraction limit and opens up the experimental capability to harvest energy along the nanostructure, in the form of SPPs, and to bring it up to the very tip apex with negligible losses. All these phenomena cooperate to realize a sensor, which is capable of probing materials at the apex by means of both electromagnetic field and electronic transport.

We can assume, in general, that the creation of $N$ electron–hole pairs along the taper, resulting from the damping of SPPs can be described by means of its rate [23]

$$\dot{N} = \mathrm{d}N/\mathrm{d}t = 2k'' \frac{P}{\hbar\omega} \Delta z \tag{15.2}$$

where $\omega$ is the angular frequency, $k''$ is the imaginary part of the complex wave vector $k$, and $P$ is the optical power of the SPP. $\Delta z$ is the distance of the electron mean free path, ~20 nm in our case (gold in the near-infrared range (NIR)). For a taper with a nanoscale radius $k_0 R \ll 1$, $k_0$ being the plasmon wavevector on a flat surface, the SPP dispersion relations are well described by the analytically approximated formula

$$k = k_0 n(R) \approx \frac{1}{R}\left[-\frac{\varepsilon_m}{2\varepsilon_d}\left(\ln\sqrt{-\frac{4\varepsilon_m}{\varepsilon_d}} - \gamma\right)\right]^{-1/2} \tag{15.3}$$

where $R$ is the radius of curvature of the tip, $\varepsilon_m$ and $\varepsilon_d$ are the metal and air dielectric functions, respectively, and $\gamma$ is the Euler constant. It is then possible to predict that the generation rate increases as $R^{-1}$ in the last few tens of nanometers close to the tip apex.

Equivalently, we can evaluate the SPP attenuation length in a metallic conical structure at the tip. The skin depth, $l_{\text{skin}}$, that reflects the evanescent nature of the SPP at the interface in the perpendicular direction to the metal surface, is commonly expressed for a Drude metal by the formula

$$l^i{}_{\text{skin}} = \frac{\lambda_0}{2\pi}\left(\frac{-\varepsilon_i^2}{\varepsilon_m + \varepsilon_d}\right)^{-1/2} \tag{15.4}$$

where the index $i = m$ and $d$ accounts for the metal or the dielectric material.

Considering that $|\varepsilon_m| \gg 1$, this expression leads to a great concentration of the SPP near the interface. Recalling that $\varepsilon = n^2$ for $\mu = 1$, neglecting absorption and using the expression of $k$, the penetration depth at the tip apex can be expressed as $l^{m}{}_{\text{skin}} \sim \frac{\lambda_0}{2\pi}\left(\frac{1}{-\varepsilon_m}\right)^{1/2} = \frac{\lambda_0}{2\pi n(R)} \cong R$, which states that an additional strong confinement takes place, reducing the penetration of the SPP to only a few nanometers below the surface.

Within a bulk picture for the electron scattering, we assume that each electron of the metal can be excited from any level of the conduction band with the same probability and that the $he^-$ are generated isotropically. Under this hypothesis, the probability for an $he^-$ to be transferred from the metal into the semiconductor is

$$\eta = \frac{1}{2n}\int_0^{\pi/2} d\theta \sin\theta \int_0^{\varepsilon_F} d\varepsilon\, \rho(\varepsilon)\Theta((\varepsilon + \hbar\omega)\cos^2\theta - (\varepsilon_F + e\phi_b)), \quad (15.5)$$

where $\theta$ is the angle to the normal to the junction through which the electron is emitted (Figure 15.8d reports a schematic represetation of how such angle is determined), $\varepsilon$ is the initial energy of the electron, $\rho(\varepsilon)$, and $\varepsilon_F$ are the density of states and the Fermi energy of the metallic coating of the tip, respectively (i.e., SPP taper), $e\phi_b$ the height of the SB (see Figure 15.8b), and $\Theta$ is the Heaviside unit-step function. The factor one-half takes into account that only half of all the hot electrons propagate through the junction, while the second half does not contribute to the Schottky current as they propagate in the wrong direction. This equation states that only a part $(\varepsilon + \hbar\omega)\cos^2\theta$ of the energy of electrons (the motion in the direction of the junction) is to be compared with the height of the SB.

Taking into account that in the NIR range $e\phi_b$, $\hbar\omega \ll \varepsilon_F$, some useful approximations can be done, the density of states can be taken at the Fermi surface, that is, we can set $\rho(\varepsilon) \approx \rho(\varepsilon_F)$, and we can write down for the electron energy $\varepsilon$ and the incidence angle $\theta$ the inequality $\cos\theta \geq 1 - \frac{\hbar\omega - e\phi_b}{2\varepsilon_F}$. In this limit, the probability distribution can be integrated to obtain the efficiency $\eta$ for the electron transport through the Schottky junction as

$$\eta = \frac{(\hbar\omega - e\phi_b)^2}{4\varepsilon_F \hbar\omega} \cdot \Theta(\hbar\omega - e\phi_b), \quad (15.6)$$

which has the same quadratic dependence on electron energy as the well-known Fowler formula [51]. The previous result is obtained under the assumption of the conservation of the electron momentum in the plane of the junction. If such conservation is relaxed at the junction by roughness and curvature at the very tip of the nanostructure, then the probability of the electron transfer is significantly increased. In this case, we have

$$\eta = \frac{1}{2n}\int_0^{\pi/2} d\theta \sin\theta \int_0^{\varepsilon_F} d\varepsilon\, \rho(\varepsilon)\Theta(\varepsilon + \hbar\omega - (\varepsilon_F + e\phi_b)) \quad (15.7)$$

The integration now is trivial, and we obtain

$$\eta = \frac{\hbar\omega - e\phi_b}{2\hbar\omega} \cdot \Theta(\hbar\omega - e\phi_b), \quad (15.8)$$

which is much greater with respect to the one of Eq. (15.6), since $2\varepsilon_F \gg \hbar\omega - e\phi_b$.

Finally, responsivity $\rho$ and the total current $J$ across the Schottky junction can be related as

$$J = e\eta\dot{N} = \rho P \quad (15.9)$$

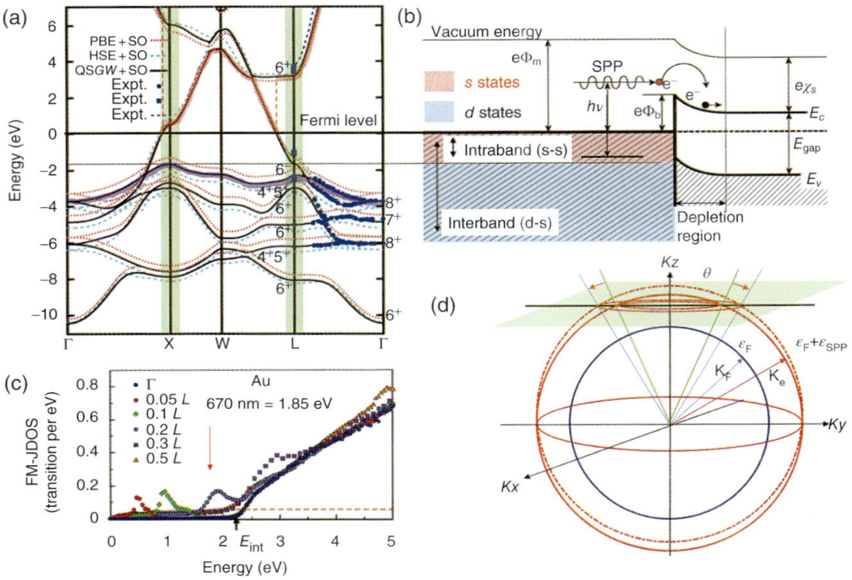

**Figure 15.8** (a) Band structure of gold calculated within different many-body perturbation theory approximation. The zero of energy has been set at the Fermi level. The QSGW-SO, solid black lines, theoretical result best fit various experimental data from literature. Purple line specifies 5d topmost occupied band, while red lines indicate 6sp-like lowermost empty bands. Green vertical bands, about ±one-tenth of BZ, representing the possible wavevector contribution provided by adiabatic compression. (b) Band scheme at metal/n-type semiconductor interface. We show the distribution of the conduction and valence bands in the semiconductor and energies of s and d states of gold. We also indicate the $he^-$ generation process with respect to Fermi level. (c) Finite momentum-joint density of states (FM-JDOS evaluated for different accessible electron wavevectors as a function of the energy. At about 2.1 eV, the interband energy indicates the onset of a characteristic linear dependence with the energy, typical of free electrons. In particular, a momentum of about 1/10 BZ allows for the efficient generation of energetic electron with energy around 1 eV (green data). (d) Fermi sphere for the free electron model (blue). Schottky barrier (green plane) and the maximal electron energy after the SPP absorption (solid red circle). A rigid 1/10 BZ shift in the direction of the interface is also displayed (dotted red line). Panel a adapted with permission from Ref. [50], copyrighted by the American Physical Society. Panel c is reproduced with permission from Ref. [21], copyright American Institute of Physics 2015.

From these, we calculated the electron injection efficiency $\eta$ and responsivity $\rho$ at the tip by considering the two distinct experimental regimes described in the momentum exchange relations (15.6) and (15.7). As in the experiments, we considered a sample of n-GaAs ($N_d = 3.6*10^{16}$ cm$^{-3}$), the Fermi energy level $\varepsilon_F = 5.53$ eV of gold, and the dielectric function from Ref. [52], resulting in an SB $e\Phi_b = 0.865$ eV. The radius of curvature of the tip and the electron mean free path were set to a conservative value of $R = 20$ nm and $\Delta z = 25$ nm, the SPP optical power to 1 μW, deriving from a 10% of the laser power. The calculated transfer efficiency and responsivity are shown in Figure 15.12 as functions of wavelength $\lambda_0$, proving the quantitative agreement with the experimental values. Dotted and continuous blue lines in Figure 15.12a and b are pure $he^-$ contributions to the current when either the relation (15.6) or (15.7) is considered in the calculations, while the continuous red line in Figure 15.12b accounts also for the photovoltaic contribution. We estimated the last term in the so-called point contact geometry limit that, departing from the planar model, describes the nanojunction in spherical coordinates and retrieves a depletion region extension as a spheroid with a remarkably shorter radius than the extension of the classical depletion region commonly used for an extended contact.

### 15.4.2 Quantum Mechanical Considerations

Experimental [53] and theoretical [21, 22] results show that the decay of SPPs with energy lower than the interband threshold in gold leads to the generation of long-living *hcs* with long isotropic mean free paths of up to 40 nm and energy within the 1–2 eV range above $\varepsilon_F$. Moreover, promoting an SPP (<2 eV) of wave vector of about 1/10 of the Brillouin zone (BZ), as could be obtained from the expression of $k$ reported above, we could expect a maximum in the efficiency for *hc* generation with an energy above 1 eV, that are well suited for $he^-$ scanning probe instrumentation, as can be deduced from Figure 15.8c.

Commonly, optically generated SPPs have wave vectors of 1/1000 of BZ. For them a decay process in *hcs* is highly unfavorable. Adiabatic propagation of a TM0 mode along a conical structure can easily increase plasmon momentum by a large factor while approaching the apex. This compression mechanism strongly depends on the last few tens of nanometers from the apex of the cone. For this reason, a proper fabrication procedure is crucial to obtain a sharp tip shape, as suggested by the theoretical model.

An important aspect in the adiabatic compression process is that the SPP skin depth reduces proportionally with the cone radius, meaning that at the apex a strong confinement is expected at the surface, with a behavior quite different from that observed for plasmonic thin films. The bulk SPP skin depth (13 nm at $\lambda = 670$ nm for gold) can rescale several times down to a few nanometers reaching the apex.

Even if commonly called free electrons, highly energetic carriers in noble metals are not completely free, but Bloch waves and their energy and momentum must fulfill the dispersion relation specific to the particular material and its band structure. As has been proposed for gold, phonon scattering contributes primarily to SPP decay in low-energy $he^-$ (<2 eV) [21], differently from the common use of the Auger model. A similar behavior has been proposed for other noble metals.

The study of the band structure of gold, calculated in Ref. [50] using *ab initio* many-body perturbation theory (MBPT) compares several possible schemes in the approximation of the GW framework (G stands for the single particle Green's function and W for the dynamically screened Coulomb potential) including all relativistic effects, at least for the single particle. The quasiparticle self-consistent GW (QSGW) scheme, when taking into account all single-particle relativistic effects, particularly the spin-orbit (SO) coupling introduced on top of the scalar-relativistic approach, provides a theoretical band structure that better fits the experimental findings. This MBPT model results in optimal agreement with experimental data from different techniques like the angle-resolved ultraviolet photoelectron spectroscopy shown in Figure 15.8a [50]. This QSGW-SO model lowers the 5d bands, and globally achieves a considerable opening of the 5d–6sp interband gap up to 0.8 eV, minimizing the discrepancy with the experiment. The QSGW-SO results are reported in Figure 15.8a along with experimental values. For our purpose, this work quantifies the band structure of gold and supports the interpretation of *ab initio* MBPT calculation on SPP decay and energy spectrum of the generated *hcs*, as previously discussed.

### 15.4.3 Quantum Confinement

Here we consider the effects of quantum confinement on the energy distribution of *hcs* [20] and on the injection across an SB. This contribution is considered fundamental in the description of experimental data for isolated nanoparticles and small clusters. For example, it can explain the enhanced photovoltaic effects, the Plasmon-induced $H_2$ dissociation by catalytic effect of *hcs* [13], the size dependence of fluorescence and energy spectrum of *hcs* in quantum dots [54], and nanorods [55, 56]. For this reason, we do not exclude its effect on the metal at the nanosized apex of the device. It is plausible that SPP spatial confinement also relaxes the wave vector constraint under a quantum mechanical point of view, that is, the wave vector of the hot electron particle can access a larger phase space. This fact contributes to match with the momentum requirement of 1/10 BZ, that is possible for about a spatial confinement of $\Delta x \sim 3$ nm. In this case, the occupation probability will change in favor of high-energy electrons.

The work of Govorov *et al.* [20] shows how an efficient generation of *hcs* can be obtained in small nanocrystals (≤20 nm) due to quantum confinement. In

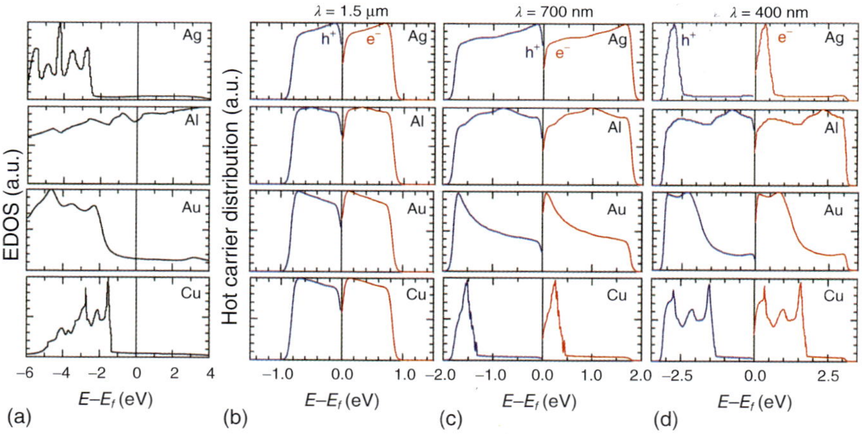


**Figure 15.9** Electron density of states (EDOS) and hot carrier distributions for common plasmonic materials: Ag, Al, Au, and Cu. (a) EDOS. Except Al (free electron-like), all exhibit a much higher density of states below the Fermi level. Under monochromatic illumination, HC distributions are created from incident photons with wavelengths: (b) 1.5 μm (0.83 eV), (c) 700 nm (1.78 eV), and (d) 400 nm (3.11 eV). Low photon energies yield relatively uniform HC distributions. Upon higher energy, illumination peaks begin to appear due to high densities of occupied states below the Fermi level for Ag, Au, and Cu. Under 700 nm illumination, Au and Cu are more efficient for hole extraction than electron, being the hot hole distribution peaked farther from the Fermi level. Reproduced with permission from [57]. Copyright OSA 2015.

this case, energy and momentum distributions for *hcs* shift toward higher values, largely improving the extraction yield. This work shows that photoexcited large nanocrystals produce a large number of low-energy hot electrons (few meV) and only a relatively small number of highly energized charges (>1 eV), resulting in a not favorable condition for current extraction. Differently, an efficient generation and the subsequent injection of high-energy *hcs* can be realized exciting small nanocrystals, in particular when the optically induced electric current is along the smallest dimensions and it is normal to the collecting barrier interface.

These results are in agreement with those obtained for the bulk from *ab initio* calculations [21, 22], that is, also in the case of localized SPPs in nanocrystals, intraband transitions appear to be preferable for the generation of $he^-$ dominating the absorption for $\lambda > 600$ nm, while interband transitions preferably generate energetic holes. This implies significant differences in the carrier generation and absorption spectra, as shown in Figure 15.9.

All these contributions together increase the final quantum efficiency of the device with respect to a bulk or to a classical nanowire.

## 15.5 HENs Measurements: Plasmon-Assisted Current Maps and Ultimate Spatial Resolution

Details about the experimental setup and the current to voltage I–V characterization of an Au/n-type GaAs nanojunction can be found in Refs [14, 33]. Here, we briefly recall the experimental setup architecture: we modulate the amplitude of the laser light while scanning in contact mode, as shown in Figures 15.10 and 15.13, measuring simultaneously the tip–sample current. Therefore, it is possible to discriminate and acquire at the same time the classical conduction and the $he^-$ map, for any arbitrary polarization of the junction, being the *hc* signal the only one synchronous with the laser. Here, the *hcs* generated by plasmon damping enhance the internal photoemission (IPE) [58], a process that improves largely the signal-to-noise ratio with respect to the case of conventional microjunctions. The sharp apex of the cone determines a scanning nanocontact that, electrically, is correctly described by the so-called point contact geometry [59, 60]. While reducing the contact size, this three-dimensional description of charge transport increasingly deviates from the one-dimensional SB model, already at a modest doping level in the semiconductor. The corresponding I–V curve reflects the specific dependence on the size and shape of the junction, while the thickness of the barrier no longer depends solely on the doping level. For these reasons, the current follows the local electronic structure on a spatial length of the same order. A characteristic length, to be compared with the diode size $d$, is $L_c = (2\varepsilon_r\varepsilon_0 (V_{\text{built-in}} + V_{\text{bias}})/eN_{e-})^{1/2} = 106$ nm for our case, that marks the transition from the classical (microscale) to the point contact geometry regime (nanoscale) [60].

In particular, there are two peculiar phenomena of general character that control the charge transport in a nano-sized junction that have to be taken into account. First, the full depletion approximation is no longer valid [61], since in this case the depleted volume, responsible for charge screening, tends to a spheroidal shape centered at the contact point. Consequently, this space charge

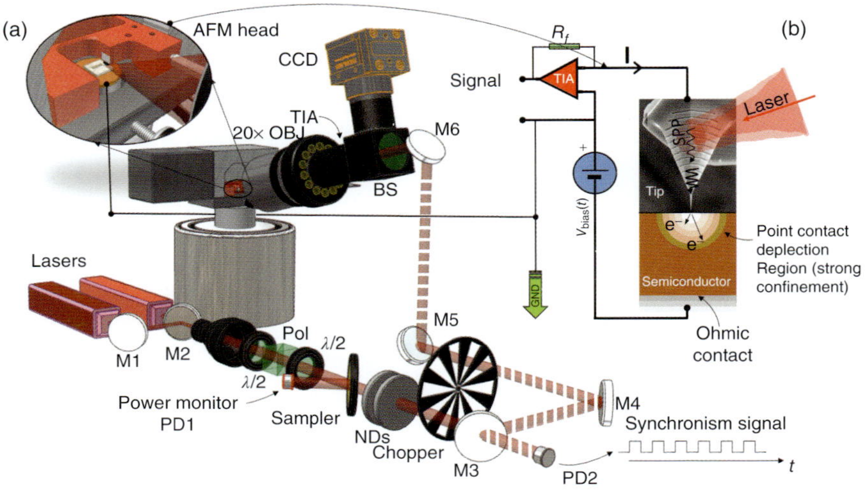


**Figure 15.10** Experimental setup for photocurrent imaging. (a) Complete optical layout. Vertically polarized CW laser excitation at 670, 980, and 1060 nm (about 10 μW of optical power on the coupling grating) is amplitude modulated with a mechanical chopper at ~250 Hz and focused on the AFM tip. Additional elements are used to control the intensity and the polarization of the incident radiation. A transimpedance amplifier (TIA) amplifies the junction current. An external circuit is used to bias the junction to obtain the I–V curves, while a lock-in amplifier and a data acquisition (DAQ) board are used to acquire and record the signals. A confocal CCD allows a real-time monitoring of the scanning and illumination conditions. Inset: close-up image of the sample area. (b) Basic circuit scheme.

limited regime modifies deeply the field potential and controls the dynamic current performances. The second critical point concerns the assumptions of charge distribution continuity and homogeneity in the depleted region. Here, the meaningful length scale is the average distance among dopants, $N_d^{-1/3}$. In the cases reported below, $N_d$ is ~1.15 $10^{17}$ $cm^{-3}$, (referring to measures in Figures 15.11 and 15.13) we calculate $N_{e-}^{-1/3} = 23$ nm, resulting to be in the same scale of the point contact.

From an instrumental point of view, this localized charge transport enhances the sensitivity of the measure down to a single dopant component.

### 15.5.1 Hot Electron Mapping

In Figure 15.11 we report the results of Ref. [14] that show the topography and corresponding photocurrent maps on two different surfaces obtained by local anodic oxidation and ion implantation in an n-type GaAs sample, excited at 647 and 980 nm. The oxide pattern is written by AFM local anodic oxidation in air using high field discharge at room temperature at controlled humidity (~40% RH, +4 V sample bias and at 4 μm/s writing speed), in contact mode (set point <10 nN). Topography and $he^-$ intensity maps are simultaneously recorded with the same plasmonic tip and are shown in the overlaid 3D measure in Figure 15.11c. The photocurrent image is obtained in contact mode and under $N_2$ atmosphere (~5% RH). This measurement allows extracting the resolution

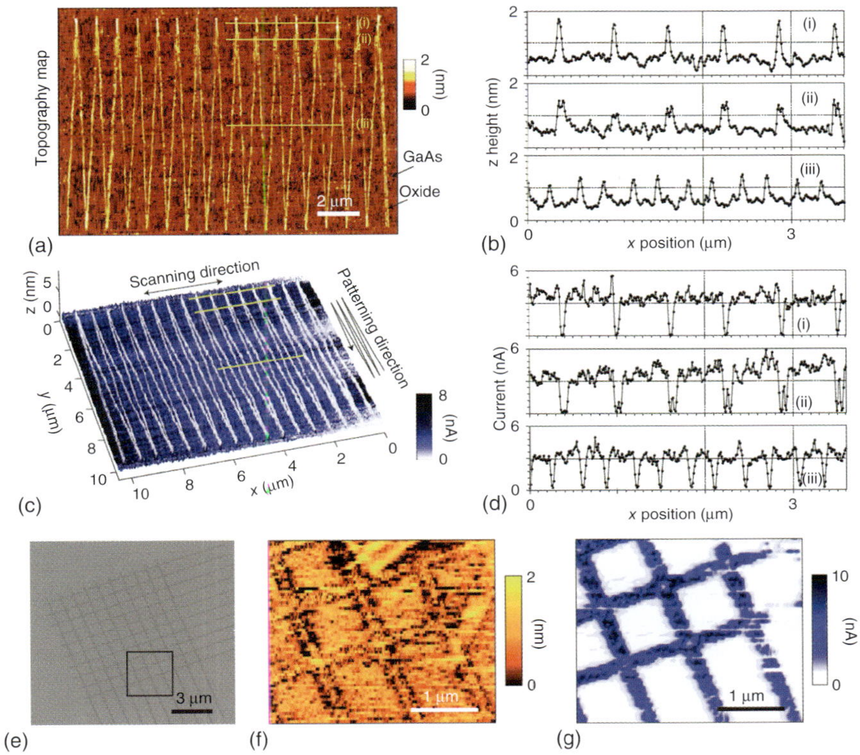


**Figure 15.11** HENs maps of locally oxidized and ion-implanted patterned samples. (a, b) High-resolution AFM topography and height profiles of a continuous oxide pattern deposited on GaAs. The profiles (indicated by yellow lines in the topography map) are not deconvolved for the tip shape. (c) Photocurrent imaging overlaid on the 3D surface reconstruction shows simultaneously the current and topographic resolution. The zig–zag profile allows a direct check of experimental resolutions from line profiles (ii) of (b, d). (e) SEM image of the GaAs sample after implantation of $Ga^+$ ions. The black rectangle indicates the area relative to the topography and plasmonic $he^-$ maps, generated at 980 nm laser excitation ($E_{GAP} >$ 1.26 eV $> e\Phi_b$), shown in (f, g). Reprinted with permission from [14], copyright NPG 2013.

and sensitivity of the technique that can be routinely achieved, as can be inferred from the line profiles in the map. The reproducibility of the profiles in trace and retrace demonstrates the stability of this scanning probe technique. Figure 15.11f and g shows the topography and photocurrent maps of the GaAs substrate reported in the SEM micrograph of Figure 15.11e. In this case, we implanted $Ga^+$ ions in the surface by FIB, creating a grid pattern (40 pA, 100 ns/pt, 30 keV, single pass). Using these fabrication parameters, the minimal dose of ions is used to locally modulate the electrical conductivity, with a negligible modification of surface topography, which remains essentially flat. The measure in Figure 15.11g reveals a conductance cross section of the implanted line, which reflects the typical spreading of implanted dose. We remark that this measure was performed at 980 nm (1.26 eV), for which a photovoltaic contribution for the GaAs is not expected, being the excitation energy smaller than the bandgap ($E_{GAP} \approx 1.45$ eV,

Si doped n-type GaAs, $N_d \sim 10^{17}$ cm$^{-3}$). Finally, we stress that for both samples the standard conductive electric map does not reveal any special feature within the sensitivity limits of the technique.

### 15.5.2 Hot Electron Resolution Limit

To quantify the ultimate spatial resolution and sensitivity of the technique we have prepared an *ad hoc* sample. An n-type GaAs substrate was spin coated with a thin poly(methyl methacrylate) (PMMA) layer (<40 nm) and subsequently patterned by means of electron beam lithography in aligned lines of variable width (from 70 to 100 nm), and interdistances of 30 nm. An overexposure and a controlled developing process allowed obtaining a polymer thinning over the whole area. The process resulted in a resist coverage with randomly dispersed nanometer-sized holes generally less than 1 nm in depth, where the substrate is accessible. This sample allows testing the experimental resolution limit by measuring the point spread function for hot electron injection through a SB.

The sample was scanned in small areas at high spatial resolution. The $he^-$ map of Figure 15.13d clearly reveals repeated lines and holes features. To test the reproducibility, the $he^-$ current was acquired both during trace and retrace pass while scanning the surface.

To correctly estimate the ultimate spatial resolution achievable at usual speeds, we switched from a lock-in amplifier (LIA) detection scheme, requiring an average over several optical cycles (introducing an additional broadening $\tau_{\mathrm{LIA}} = 30$ ms ~9 optical cycles), to a post-processing analysis of the current data recorded at a high sampling rate. After a spectral analysis, we applied a digital infinite impulse response (IIR) low-pass filter with about 1 kHz bandpass to remove spurious high-frequency noise from the signal. Then, as shown in Figure 15.13b and c,

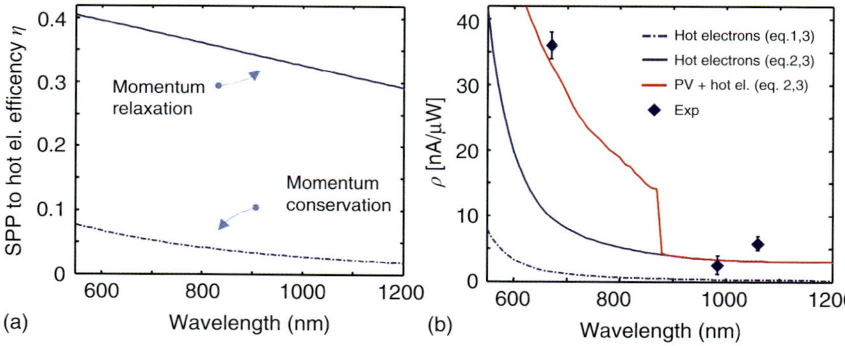


**Figure 15.12** (a) Calculated transmission efficiency $\eta$ of the $he^-$ across the n-type GaAs/Au Schottky junction as a function of wavelength. The two curves refer to the different limit of the momentum conservation and relaxation. (b) Experimental device responsivity $\rho$ (blue diamond) at 670, 980, and 1060 nm and its theoretical values, as calculated from Eq. (15.8). The two momentum exchange regimes generate the blue dotted line, Eq. (15.6), and blue continuous line, Eq. (15.7). The red line adds the photovoltaic contribution, as obtained from literature data, to the responsivity calculation of Eq. (15.8) combined to Eq. (15.7). Adapted with permission from [14], copyright NPG 2013.

the current was reconstructed point by point as $\langle I_{on}\rangle - \langle I_{off}\rangle$, for each optical period. The use of this matched filter allows for a further increase in $S/N$ ratio, maintaining the time resolution and giving the true spatial resolution while scanning at a reasonable speed. We selected the values to be averaged (colored point

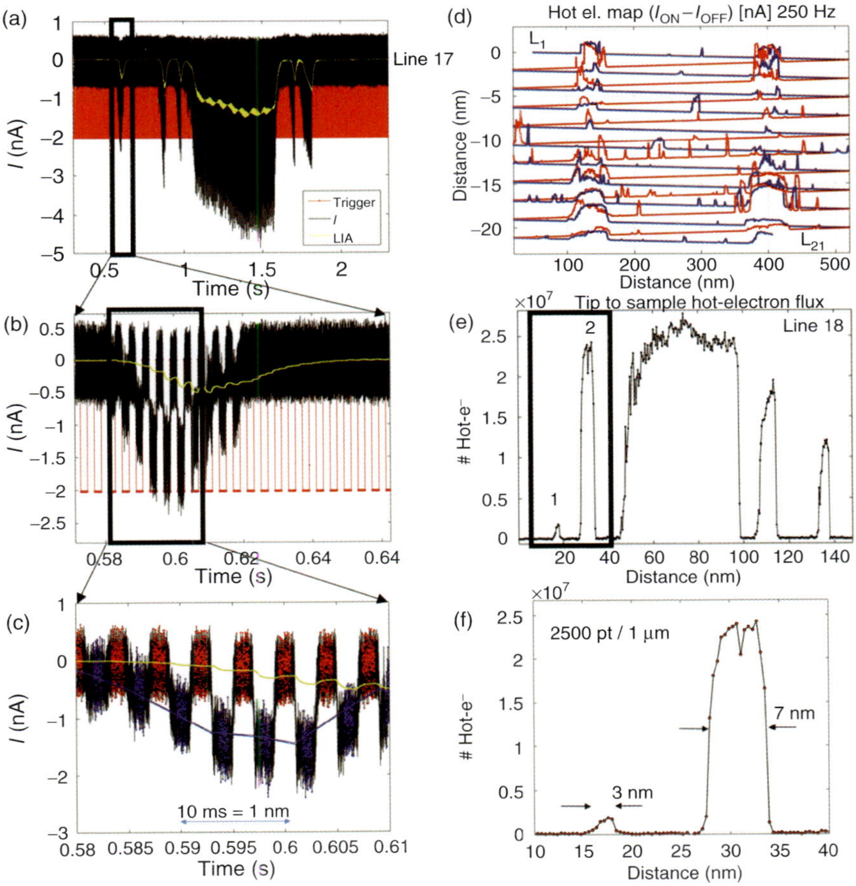


**Figure 15.13** Experimental resolution. (a) A 2 s record of the current stream while scanning the patterned n-type GaAs/Au junction, at 670 nm. (b) The graph refers to an acquisition time of 0.1 s. Here we show on and off states on the current flow (black line) and the laser trigger events (red lines). The yellow line, reported for the sake of comparison, is the output signal of the lock-in. (c) Few optical cycles of acquisition time. Here we have indicate the data points selected to calculate the $I_{on}$ (blue) and $I_{off}$ (red) local values as described in the text. The blue broken line connects consecutive points from averaged cycles, demonstrating the optimal reconstruction of the temporal profile even for high-dynamic signal, differently from the slower LIA detection procedure. (d) $\langle I_{on}\rangle - \langle I_{off}\rangle$ surface reconstruction. Blue and red lines display AFM trace and retrace, demonstrating a correct agreement. The first ($L_1$) and the last ($L_{21}$) lines of this reconstruction are indicated. (e) Detail of the spatial reconstructed signal of line $L_{18}$, in terms of number of $he^-$ per pixel. The detail of the first and second peaks, reported in (f), allows the direct estimation of the ultimate experimental resolution, as FWHM of peak 1 or as 10–90% of the rising front of peak 2, which is 3 nm.

in Figure 15.13c) by excluding the transient regimes in the signal, synchronizing it with the modulated laser intensity. In this way, we were able to pin down a point every 0.4 nm, cutting out almost all the noise contribution while preserving the maximal part of the signal. Figure 15.13e and f demonstrates an ultimate full-width-at-half-maximum (FWHM) resolution of 3 nm, preserved over repeated lines during both trace and retrace.

## 15.6 Kelvin Probe, HENs, and Electrical Techniques

In this section, we compare the technical advances proposed by HENs in terms of chemical sensitivity and spatial resolution, also in comparison with other electrical applications of AFM microscopy, such as scanning Kelvin probe microscopy (SKPM), sensitive to the same parameters, and conductive AFM. Further references to these last techniques can be found in more detail in a dedicated chapter on the topic.

### 15.6.1 SKPM Theoretical Frame: a Short Introduction

The most diffused technique providing a direct measurement of the contact potential between two surfaces is referred to as Kelvin probe, after its inventor Lord Kelvin, who first postulated this method in 1861. The first implementation on an AFM setup, by employing a metallic tip at the end of a cantilever, dates back to 1991 [62], when a nanometric spatial resolution (50 nm) with a remarkable sensitivity (0.1 V) was first demonstrated. Since then, continuous efforts to improve both parameters to their ultimate extent have been made.

The built-in electric field between a sample surface and a reference electrode depends on many physical properties of the materials. A major contribution is given by the chemical potential mismatch resulting from the difference between the Fermi and the vacuum level in the two materials when they are electrically connected to create a junction.

The energy required by electrons to be extracted from a material is given by

$$\phi = E_{\text{Vac}} - \varepsilon_F \tag{15.10}$$

where $E_{\text{Vac}}$ is the vacuum. At thermal equilibrium, the Fermi levels of two electrically connected materials pair, giving origin to a potential difference

$$-e\Delta V_{\text{CPD}} = \phi_1 - \phi_2 = E_{\text{Vac1}} - E_{\text{Vac2}} = -e(V_1 - V_2) \tag{15.11}$$

which represents the work done to bring an electron from one side of the junction to infinity and back to the second side.

By adding an external voltage $V_{\text{ext}} = V_{\text{DC}} + V_{\text{AC}}(\omega_{\text{el}})$, it is possible to measure the capacitive coupling between the two electrodes, synchronously measuring macroscopic properties such as the transport current or the electrostatic force. Minimal coupling will occur when $V_{\text{DC}} = V_{\text{CPD}}$, therefore, by externally tuning a DC bias, it is possible to measure the contact potential between the two electrodes. In AFM, the cantilever already provides an extremely sensitive force sensor [63], which can be used to measure the capacitive force.

The measurement technique relies on the following strategy: line by line the surface is scanned two times. During the first pass, a topography map is acquired in AC mode to get the morphology of the surface. In the second one, the same line is repeated at a different height $z'(x, y) = z_{topo}(x, y) + \Delta z$ while applying $V_{ext}$ to the tip.

The resulting tip–sample force is

$$F_{tot}(z', t) = k(z(t) - z')\sin(\omega t) - F_{Pauli} - F_{vdW} - \frac{1}{2}\frac{dC}{dz}(V_{ext} - V_{CPD})^2 \tag{15.12}$$

where $F_{Pauli}$ and $F_{vdW}$ are the short-range repulsive (topography) and long-range attractive van der Waals forces, respectively, and $k(z(t) - z')$ is the withdrawal force due to the cantilever elastic constant $k$ for an oscillation amplitude $A = z(t) - z'$. The last term is the capacitive coupling given by the total tip–sample voltage, which brings in the contact potential. During the second pass, the tip voltage is modulated by $V_{AC}$ to induce a tip oscillation at a frequency $\omega_{el}$ (usually slightly different from the mechanical resonance frequency of first pass) with oscillation amplitude $A_{el}$ that can be remarkably different from the first pass one. A proper second pass elevation $\Delta z$ allows to cancel the Pauli contribution and to maximize the sensitivity to the capacitance coupling with respect to the van der Waals forces at the same time.

The capacitive force can be rewritten as

$$-\frac{1}{2}\frac{dC}{dz}(V_{ext} - V_{CPD})^2 = -\frac{1}{2}\frac{dC}{dz}\left[(V_{DC} - V_{CPD})^2 + 2(V_{DC} - V_{CPD}) \times V_{AC}\sin(\omega_{el}t) + \frac{V_{AC}{}^2}{2}[1 - \cos(2\omega_{el}t)]\right] \tag{15.13}$$

This has three main components in the force frequency spectrum, one DC, one at frequency $\omega_{el}$, and one at $2\omega_{el}$. An additional feedback loop adjusts $V_{DC} = V_{CPD}$ to cancel the force component at $\omega_{el}$ point by point, providing a local measurement of the contact potential. At the same time, the capacitive coupling $-\frac{1}{2}\frac{dC}{dz}V_{AC}^2$ can be also measured from the second harmonic, obtaining the scanning capacitance signal, which is a widely used technique in microelectronics to estimate surface oxidation states.

For a direct comparison with HENs, we have performed a measurement that provides the resolution achievable with SKPM on a sample like the one used for the data in Figure 15.11.

A new GaAs (n-type $N_d = 10^{16}$) substrate was freshly cleaned in HCl 37% before the nano-oxidation procedure. We used commercial PtIr-coated Asylum electrilever ($K = 2$ N/m, nominal tip radius = 30 nm, 75 kHz) on an Asylum MFP-3D AFM microscope. Writing parameter and experimental condition were 9 V sample bias, oxidative current 3 nA, and writing speed 500 nm/s in environment humidity of 45%. The oxidized pattern was then read by scanning at 90° the surface. The results of topography and SKPM potential map 5 × 5 μm$^2$ area are reported in Figure 15.14c and d, with details of a smaller scan in

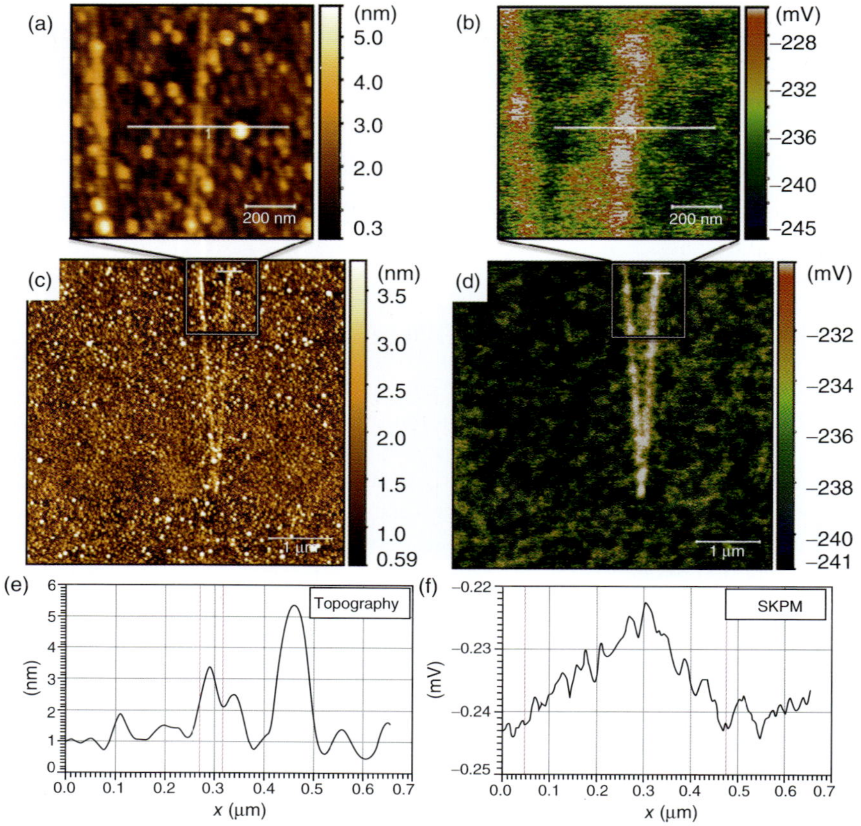


**Figure 15.14** SKPM characterization of an oxide-patterned GaAs sample (n-type $N_d = 10^{16}$) with nanoxidtive regime at 9 V sample bias and oxidative current 3 nA in relative environment humidity of 45%. (a, b) Detail images of the topography and SKPM maps shown in (c, d). Here topography reveals features not accessible with SKPM. (e, f) The graphs report the line profiles indicated with white lines in the zoomed area.

Figure 15.14a and b. From the line profiles of Figure 15.14e and f, corresponding to the high-resolution zoomed areas Figure 15.14a and b, it is possible to evaluate the spatial resolution of the technique with respect to the topography one. The not deconvoluted topography reveals features as large as 46 nm, measurable between markers in Figure 15.14e, while SKPM results are limited to a line resolution not better than 200 nm.

This technique has many drawbacks and implications:

1) The sensitivity of the measurement strongly depends on the distance $\Delta z$ set during the second pass. For best resolution and sensitivity, since the closest tip–sample distance should be $\Delta z - A_{el} > 0$, $A_{el}$ is usually by far lower than $A_{topo}$, and both tip radius $R$ and $z$ depend through the relation $A_{el} = \frac{V_{AC}}{2k}\frac{dC(R,z)}{d}$, the optimal distance has to be tuned during the measurements, also with $\Delta z < 0$.

2) A remarkably high resolution and surface sensitivity can be achieved in frequency modulation mode. This is due to the reduced oscillation amplitude allowing a second pass a few nanometers off the surface. This is commonly performed in an ultrahigh vacuum (UHV) environment where such a technique is more stable [64]. However, the UHV environment is usually less compatible with biological samples, reducing the field of application for this technique.
3) A tip modification, usually occurring during first pass due to the contact with the sample, changes its surface potential. This affects the final results in three different ways:
   a) The tip can pick up tiny amounts of materials from the sample (mostly occurring for soft/organic specimens), changing sensitively the reference work function during a scan.
   b) Due to environmental humidity and the applied voltage, the tip can oxidize.
   c) The tip can modify its shape by plastic deformation or material loss. This, again, affects the results by changing tip radius, that is, capacitance, and, to a minor extent, by redefining local work function due to different geometry. This effect can be important in case of heavy damage to coated tips, when the metal is peeled off, leaving the silicon substrate exposed.
4) As discussed, $C$ depends on $z$ and $R$ in the first approximation. To be more precise, one should consider the true shape of the electric field between tip and sample giving $\boldsymbol{F}(x,y,z) = -e\boldsymbol{E}(x,y,z) = e\nabla V$, which more properly accounts for the tip and sample geometries and compositions. The shape of the electric field $\boldsymbol{E}$ gives the effective resolution and sensitivity, worsened by the spreading of the field at the sample surface.
5) The material of the tip also technically plays an important role: aside from high conductivity, geometrical and chemical stability imposes main requirements on the tip. For these reasons Pt-, PtIr-, PtSi-, or Ir-coated silicon tips are widely used, ending up with a tip radius usually around 30 nm. For sharper doped Si tips, the additional oxidation effects have to be taken into account. For biological materials, a good choice would rely on hydrophobic tips, such as conductive PtC-coated ones, that are less prone to material pick up due to hydrophobic/hydrophilic repulsion.
6) Moreover, SKPM is an inherently surface-sensitive technique. The dependence on surface contaminations and charge distribution cannot be neglected. Due to the presence of fixed surface charges in trap states or embedded in native oxide on top of the sample surface, an additional potential appears as an offset on the $V_{\mathrm{CPD}}$ as $V_q = qN_F\ t_{\mathrm{ins}}/\varepsilon_{\mathrm{ins}}$. Here, $q$ is the elementary charge, $N_F$ is the charge density, $t_{\mathrm{ins}}$ is the thickness of the insulating layer separating the tip from the substrate, and $\varepsilon_{\mathrm{ins}}$ is the average dielectric constant. Assuming a surface charge in the order of $N_F = 10^{10}$–$10^{11}$ cm$^{-2}$, values which are commonly achieved for high-quality microfabrication oxide, one gets $V_q = 5$–$50$ mV which is a relatively high value for the offset in the measurements.
7) Tip oxidation and contamination have to be considered. Pre- and after-measurement calibrations on a reference sample, such as freshly cleaved highly oriented pyrolytic graphite, have to be routinely employed for accurate measurements [65].

### 15.6.2 HENs

As introduced before, HENs is a charge transport AFM technique performed in contact mode. The net charge transfer is strongly dependent on the initial photon energy, on the metal of the tip, and on the local electronic properties of the sample at the nanometer scale. Here we compare HENs to different AFM electrical modes in terms of spatial resolution, sensitivity, and dynamics.

#### 15.6.2.1 Spatial Resolution

HENs provides a contrast on the local Fermi level of the tip and the conduction band of the sample and its local modification due to adsorbed materials, surface charge, and doping, but with a remarkably higher spatial resolution. The reason is that while in SKPM the final resolution is affected by the second pass elevation $\Delta z$ giving the relatively high broadening of the electric field at the surface, in HENs the final resolution is practically limited just by the area of the junction. In contact mode, the effective tip–sample contact can be limited to a few nanometers in diameter, similar to the case of conductive AFM and scanning tunneling microscope (STM) in air. The main difference is that no external polarization bias is needed to drive the current since the plasmon compression gives the energy and the momentum to overcome the SB at the interface. In contact AFM or STM, an external bias of a few Volts can easily generate an electric field exceeding the dielectric breakdown of nanomaterials at the interface [66], and induce oxidation and/or metal electromigration from the tip. This results in both tip and sample modification, as well as junction instability, phenomena frequently occurring while operating under environmental conditions. Apart from these practical aspects, adiabatic compression of evanescent plasmonic modes provides a strong localization of the electric field at the tip, stronger than the quasi-static field other electrical AFM modalities rely on. The electromagnetic simulations in Figure 15.15 show how in a plasmonic tip the electric field is distributed around the apex. Here the field is naturally concentrated in a spatial region limited to a few nanometers around the tip, different from the quasi-static case, where a relatively high broadening is given by $E = -\nabla V$. Then in HENs, the point contact area is further localized in the junction region once the contact is established.

#### 15.6.2.2 Sensitivity and Specificity

In HENs, the generated hot electrons bring the SPP momentum associated to the adiabatic compression. This allows them to be selectively injected into the semiconductor with efficiency depending on crystallographic directions. Another important difference is in the energy distribution of the carriers involved in the process. In conductive AFM and STM/STS spectroscopy, for instance, by applying an external voltage the Fermi level of the tip is aligned with different energy states of the substrate. The current comes from the facing of occupied states on the tip and empty states on the sample or vice versa, allowing tuning the current and its direction by means of an external bias. By increasing the bias, the number of states entering in the conduction process increases, involving states even many electronvolts far from the Fermi level, whose number

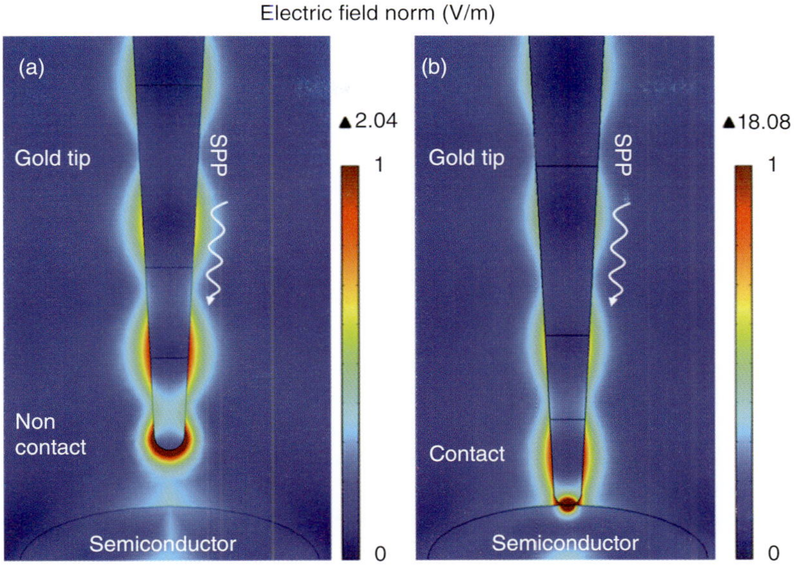


**Figure 15.15** Electric field norm of a TM0 mode along the conical tip ($R_{tip}$ = 25 nm), excited at 670 nm. (a) Simulation results for the case of no contact between the metal tip and the semiconductor. (b) Tip in contact with the semiconductor. A further enhancement of the induced adiabatic SPP focusing at the tip apex is observed. In both cases, fields are normalized to the input amplitude, $E_0 = 1$ V/m. The numerical simulation clarifies the effect of the contact on the plasmonic field at the tip apex. Higher field enhancement at the Au tip is observed when the contact is established, due to the high value of the refractive index of the semiconductor ($n \sim 4.4$ in our case). Reprinted with permission from [14], copyright NPG 2013.

is given by the integral of the occupancy probability over the allowed energies $P(E) \propto D(E - eV) f(E - eV)$ ($D$ and $f$ are the density of states and the Fermi distribution). Since the density of states in metals is described by a quite broad distribution, the number of electrons contributing to conduction increases with V as the integral over the states. The two current spectroscopies, I–V and the more energy-resolved $\frac{dI}{dV}$, allow to evaluate the bandgap, the energy levels distribution, and the gaps between the highest occupied and lowest unoccupied molecular orbitals in materials and molecules by probing them with a continuum of energy states of the tip up to the Fermi level. Similarly, the applied voltage in SKPM mode is optimized to adjust the Fermi levels in the two materials (tip and sample), providing the measure of the contact potential. In both cases, the electrons at the Fermi level, which by definition is an equilibrium state, play a key role in the measurements. On the other hand, HENs relies on *hcs*, which by definition are nonequilibrium (excited) states in electronic distribution. In this process the decay of a plasmon produces *hcs* in excited states, whose occupancy probability and energy distribution $P(E)$ can be calculated following the work of Gong and Munday [57] as

$$P(E) \propto D(E - E_{Ph}) f(E - E_{Ph}) D(E) f(E) \tag{15.14}$$

According to this formula, an electron in the half-filled band can be excited up to an energy $E$ which is set experimentally by the light source. This produces a hot electron-hole pair, whose energy distribution is reported in Figure 15.9 for the most commonly used plasmonic metals and energies.

These results allow for some considerations:

1) Selecting the external energy, that is, laser wavelength, it is possible to control the $hc$ energy density in the metal to have a sharp peak overcoming the junction barrier. In this case, only a narrow distribution of excited states contributes to the electronic transport and probes the energy states of the sample. This goes toward a highest energy resolution for spectroscopy studies and energy-resolved imaging.
2) The basic process of charge transfer can occur at zero bias. This allows testing energy levels without an external DC electric field that, for standard conductive sharp tips very close to the surface, can easily exceed an electrical field of $10^9$–$10^{10}$ V/m. This is believed to be a great advantage for many samples, such as thin oxidation layers, heterojunctions, and biological molecules since no perturbation or shrinking is induced. This is a very promising aspect when Raman measurements are combined with current transport measurements in a multiprobe configuration.
3) In HENs, the addition of a small modulation voltage can be used to refine the local density of states measurement of the sample by tuning the energy of the hot charges. This technique therefore could be used in parallel with other electrical modes such as SKPM or STS.
4) As shown in Figure 15.9 and further explained in the literature [21, 57], the transport is due not only to hot electrons but also to hot holes, since the plasmon decays into a carrier pair. According to the material and the incident photons, either hot electron or hot hole current can prevail. This second case, for instance, occurs for Cu and Au for both excitation wavelengths of $\lambda = 700$ and 400 nm and for Ag at $\lambda = 400$ nm [21]. This opens up additional possibilities for spectroscopies.
5) A shaping of electron density can be achieved in HENs also with the combination of different excitation wavelengths at the same time.
6) In the classical electrical mode, the detectable dynamics is usually quite limited (well below megahertz range). HENs process is triggered by laser excitation and the mechanism underneath is intrinsically a nonequilibrium one, hence the dynamics can be pushed to the very limit of the state-of-the-art detection systems for current acquisition (trans-impedance amplifiers).
7) According to recent literature [57, 67], new material can be explored to realize plasmonic tips, offering the possibility of optimizing and tuning HENs for different applications. Remarkably, most of them offer mechanical properties similar to the ones commonly used for electrical tip coatings (Young Modulus $E_Y \approx 60$–$130$ GPa), appearing therefore well suited for contact probes fabrication.

## 15.7 Fast Pulses in Adiabatic Compression for Hot Electron Generation

One of the trends that has emerged in recent years among the applications of SPP adiabatic compression in plasmonic nanocones is the use of these tools in combination with pulsed lasers to access a faster dynamic. Pulsed laser sources are already fundamental in many fields of research as well as in many industrial, medical, and biological applications. Nevertheless, their use in microscopy- and microscopy-related applications remains limited by diffraction phenomena when combined with far-field focusing tools as microscope objectives. Coupling with modes that experience adiabatic compression in a plasmonic tapered waveguide is certainly a promising strategy to overcome this limitation. This type of application has already been explored in literature and the possibility of coupling ultrashort laser pulses in metallic cones squeezing the optical excitation to a nanometric-sized spot has been previously demonstrated [68, 69].

The interest in this particular application, due to the fact that it is possible to achieve nanometric-size excitation, derives from the possibility of having both high peak intensity and time-resolved excitation, hence leading to the long-searched-for goal of a coherent control of electromagnetic field at nanometer-large and femtosecond-limited scale [70].

A major concern which arises while dealing with ultrashort laser pulses in optical media, is the time broadening and pulse reshaping due to group delay dispersion that different frequencies accumulate over propagation in any dispersive medium. This does not seem to be a major concern in plasmonic propagation in the adiabatic compression regime, since an electromagnetic pulse propagating along a metallic nanocone experiences modest group velocity dispersion and hence travels with an almost unperturbed temporal profile [68, 69].

Particular attention, in this case, has to be paid to the coupling mechanism. In previous paragraphs, we described the main excitation techniques that we use to convert photons into surface-bounded plasmons. The different techniques would provide different results in the case of ultrafast excitation for both what concerns coupling efficiency and for what concerns the time dispersion. Direct coupling of radially polarized light beams at the base of a metallic cone seems to be a promising way to control pulse distortion. This method is weakly sensitive to wavelength spreading, resulting in being weakly affected by the broad spectrum of the light source. Simulations on this kind of structure indicate a substantial preservation of the laser pulse after propagation along a 10-μm gold cone [33] and that this coupling efficiency can be extremely effective, even approaching 90% [71].

Coupling through a grating, on the other side, has a natural limit, as it is intrinsically wavelength dependent. Many authors, anyway, explored this kind of configuration, since it happens to be the most practical one in many experimental arrangements. They managed to find interesting solutions to compensate for the large spectral range covered by these excitation sources, mostly based on the use of linearly chirped gratings [68, 72]. Nevertheless, it has been demonstrated that coupling of 10-fs laser pulses onto a gold nanocone via a normal nonchirped grating preserves substantially unchanged the pulse time-frequency profile. Also, in this case, numerical analysis demonstrates the poor distortion introduced by the grating itself, and the authors were able to measure directly the frequency-resolved autocorrelation of the plasmonic pulse at the apex of the nanocone [69].

Applications of nanofocusing of pulsed lasers are still at a very early stage. Efficient nano-localized second-harmonic generation [68], nonlinear optical measurements [73], and high-harmonic generation in tapered metallic hollow conical waveguides [18] have already been demonstrated. The possibility of extending time-resolved techniques down to nanoscale opens up countless potential applications.

Another outcome is the inherent enhancement of the electromagnetic field at the cone apex, which triggers promising applications of nonlinear optics in fields such as nanoimaging and nanocharacterization. Tip-enhanced electromagnetic field with high spatial and temporal resolution has already demonstrated to provide insight into infrared molecular spectroscopy [74]. The application to nonlinear Raman [75–78] of adiabatically compressed pulsed excitation will certainly represent a significant improvement to imaging techniques at the nanoscale, particularly in the field of biology and bio-imaging.

With respect to *hc* generation, application of pulsed laser to metallic cones represents an interesting perspective from different points of view. From one side the possibility of performing time-resolved nonlinear characterization in parallel to *hc* measurements and optical spectroscopies, such as enhanced Raman, paves the way for parallel multiprobe nanosized characterization of the sample. Time-resolved *hc* injection could have interesting applications in understanding important phenomena like zitterbewegung in two-dimensional material [79, 80] and nonlinear plasmon-induced photocatalisys [81]. Moreover, the possibility of obtaining free-space electron emission from excited cones in adiabatic compression, exploiting multiphoton absorption of ultrashort pulses to overcome the metal work function has been recently demonstrated [82].

## 15.8 Conclusion

In this paper, we revised our recent results on hot electron SPM showing also our ultimate spatial resolution limit of 3 nm and compared them with the literature to give perspectives and possible additional implications. In our research, we are aware of the unicity of our approach and having in our hands a novel multipurpose method for investigating materials at nanoscale and to femto-attosecond dynamics. In the past years we demonstrated the full compatibility of adiabatic compression with AFM techniques. Now our effort is devoted to merge AFM force spectroscopy, light-enhanced spectroscopy, and hot electron spectroscopy. Applied to molecules, we foresee the possibility of obtaining novel structural information in a wide range of environmental conditions never explored before.

## Acknowledgments

We thank our colleagues at BIONEM Lab of Magna Graecia University, Catanzaro, and former colleagues at Nanostructures Division of IIT, Genoa, Italy, for the contribution they have given to this research activity in the past years.

We acknowledge financial support from (i) King Abdullah University of Science and Technology for OCRF-2014-CRG grant, (ii) the Italian Ministry of Health under project nos. GR-2010-2320665 and GR-2010-2311677, and (iii) the European project EUROMBR (grant no. 608104).

## References

1 Shahbazyan, T.V., Stockman, M.I. (2014) *Plasmonics: Theory and Applications*, Springer Netherlands.
2 Dragica Vasileska, S.M.G. (ed.) (2011) *Nano-Electronic Devices*, Nano-Electronic Devices. Semiclassical and Quantum Transport Modeling, Springer, New York.
3 Xie, Z., Yu, W., Wang, T., Zhang, H., Fu, Y., Liu, H., Li, F., Lu, Z., and Sun, Q. (2011) Plasmonic nanolithography: a review. *Plasmonics*, **6**, 565–580.
4 Kumar, R. and Kumar, N. (2014) *Nanotechnology and Nanomaterials in the Treatment of Life-threatening Diseases*, Elsevier Inc., 225 Wyman Street, Waltham, MA.
5 Binnig, G., Quate, C.F., and Gerber, C. (1986) Atomic Force Microscope. *Phys. Rev. Lett.*, **56**, 930–933.
6 Kalinin, S.V., Gruverman, A. (2011) *Scanning Probe Microscopy of Functional Materials*. Springer.
7 Anderson, M.S. (2000) Locally enhanced Raman spectroscopy with an atomic force microscope. *Appl. Phys. Lett.*, **76**, 3130–3132.
8 Schlücker, S. (2014) Surface-enhanced Raman spectroscopy: concepts and chemical applications. *Angew. Chem. Int. Ed.*, **53**, 4756–4795.
9 Barnes, W.L. (2006) Surface plasmon-polariton length scales: a route to sub-wavelength optics. *J. Opt. A: Pure Appl. Opt.*, **8**, S87–S93.
10 Coluccio, M.L., Gentile, F., Das, G., Nicastri, A., Perri, A.M., Candeloro, P., Perozziello, G., Proietti Zaccaria, R., Gongora, J.S.T., Alrasheed, S., Fratalocchi, A., Limongi, T., Cuda, G., and Di Fabrizio, E. (2015) Detection of single amino acid mutation in human breast cancer by disordered plasmonic self-similar chain. *Sci. Adv.*, **1**, e1500487.
11 Yu, N. and Capasso, F. (2014) Flat optics with designer metasurfaces. *Nat. Mater.*, **13**, 139–150.
12 Atwater, H.A. and Polman, A. (2010) Plasmonics for improved photovoltaic devices. *Nat. Mater.*, **9**, 205–213.
13 Mukherjee, S., Libisch, F., Large, N., Neumann, O., Brown, L.V., Cheng, J., Lassiter, J.B., Carter, E.A., Nordlander, P., and Halas, N.J. (2012) Hot electrons

do the impossible: plasmon-induced dissociation of H2 on Au. *Nano Lett.*, **13**, 240–247.

14 Giugni, A., Torre, B., Toma, A., Francardi, M., Malerba, M., Alabastri, A., Proietti Zaccaria, R., Stockman, M.I., and Di Fabrizio, E. (2013) Hot-electron nanoscopy using adiabatic compression of surface plasmons. *Nat. Nanotechnol.*, **8**, 845–852.

15 Kauranen, M. and Zayats, A.V. (2012) Nonlinear plasmonics. *Nat. Photonics*, **6**, 737–748.

16 Huang, J., Liu, C., Zhu, Y., Masala, S., Alarousu, E., Han, Y., and Fratalocchi, A. (2016) Harnessing structural darkness in the visible and infrared wavelengths for a new source of light. *Nat. Nanotechnol.*, **11**, 60–66.

17 Lin, J., Genevet, P., Kats, M.A., Antoniou, N., and Capasso, F. (2013) Nanostructured holograms for broadband manipulation of vector beams. *Nano Lett.*, **13**, 4269–4274.

18 Park, I.-Y., Kim, S., Choi, J., Lee, D.-H., Kim, Y.-J., Kling, M.F., Stockman, M.I., and Kim, S.-W. (2011) Plasmonic generation of ultrashort extreme-ultraviolet light pulses. *Nat. Photonics*, **5**, 677–681.

19 Rizza, C., Di Falco, A., Scalora, M., and Ciattoni, A. (2015) One-dimensional chirality: strong optical activity in epsilon-near-zero metamaterials. *Phys. Rev. Lett.*, **115**, 057401.

20 Govorov, A.O., Zhang, H., Demir, H.V., and Gun'ko, Y.K. (2014) Photogeneration of hot plasmonic electrons with metal nanocrystals: quantum description and potential applications. *Nano Today*, **9**, 85–101.

21 Bernardi, M., Mustafa, J., Neaton, J.B., and Louie, S.G. (2015) Theory and computation of hot carriers generated by surface plasmon polaritons in noble metals. *Nat. Commun.*, **6**, 7044.

22 Sundararaman, R., Narang, P., Jermyn, A.S., Goddard Iii, W.A., and Atwater, H.A. (2014) Theoretical predictions for hot-carrier generation from surface plasmon decay. *Nat. Commun.*, **5**, 5788.

23 Stockman, M.I. (2004) Nanofocusing of optical energy in tapered plasmonic waveguides. *Phys. Rev. Lett.*, **93**, 137404.

24 Stockman, M.I., Kling, M.F., Kleineberg, U., and Krausz, F. (2007) Attosecond nanoplasmonic-field microscope. *Nat. Photonics*, **1**, 539–544.

25 Katrin, K., Harald, K., Irving, I., Ramachandra, R.D., and Michael, S.F. (2002) Surface-enhanced Raman scattering and biophysics. *J. Phys. Condens. Matter*, **14**, R597.

26 Proietti Zaccaria, R., Alabastri, A., De Angelis, F., Das, G., Liberale, C., Toma, A., Giugni, A., Razzari, L., Malerba, M., Sun, H.B., and Di Fabrizio, E. (2012) Fully analytical description of adiabatic compression in dissipative polaritonic structures. *Phys. Rev. B*, **86**, 035410.

27 Ritchie, R.H., Arakawa, E.T., Cowan, J.J., and Hamm, R.N. (1968) Surface-plasmon resonance effect in grating diffraction. *Phys. Rev. Lett.*, **21**, 1530–1533.

28 Baron, A., Devaux, E., Rodier, J.-C., Hugonin, J.-P., Rousseau, E., Genet, C., Ebbesen, T.W., and Lalanne, P. (2011) Compact antenna for efficient and unidirectional launching and decoupling of surface plasmons. *Nano Lett.*, **11**, 4207–4212.

29 Kretschmann, E. (1972) The angular dependence and the polarisation of light emitted by surface plasmons on metals due to roughness. *Opt. Commun.*, **5**, 331–336.

30 Foley Iv, J.J., Harutyunyan, H., Rosenmann, D., Divan, R., Wiederrecht, G.P., and Gray, S.K. (2015) When are surface plasmon polaritons excited in the Kretschmann-Raether configuration? *Sci. Rep.*, **5**, 9929.

31 Otto, A. (1968) Excitation of nonradiative surface plasma waves in silver by the method of frustrated total reflection. *Z. Physik*, **216**, 398–410.

32 De Angelis, F., Das, G., Candeloro, P., Patrini, M., Galli, M., Bek, A., Lazzarino, M., Maksymov, I., Liberale, C., Andreani, L.C., and Di Fabrizio, E. (2010) Nanoscale chemical mapping using three-dimensional adiabatic compression of surface plasmon polaritons. *Nat. Nanotechnol.*, **5**, 67–72.

33 Giugni, A., Allione, M., Torre, B., Das, G., Francardi, M., Moretti, M., Malerba, M., Perozziello, G., Candeloro, P., and Fabrizio, E.D. (2014) Adiabatic nanofocusing: spectroscopy, transport and imaging investigation of the nano world. *J. Opt.*, **16**, 114003.

34 Lalanne, P., Hugonin, J.P., and Rodier, J.C. (2005) Theory of surface plasmon generation at nanoslit apertures. *Phys. Rev. Lett.*, **95**, 263902.

35 Lalanne, P., Hugonin, J.P., and Rodier, J.C. (2006) Approximate model for surface-plasmon generation at slit apertures. *J. Opt. Soc. Am. A*, **23**, 1608–1615.

36 Proietti Zaccaria, R., De Angelis, F., Toma, A., Razzari, L., Alabastri, A., Das, G., Liberale, C., and Di Fabrizio, E. (2012) Surface plasmon polariton compression through radially and linearly polarized source. *Opt. Lett.*, **37**, 545–547.

37 De Angelis, F., Zaccaria, R.P., Francardi, M., Liberale, C., and Di Fabrizio, E. (2011) Multi-scheme approach for efficient surface plasmon polariton generation in metallic conical tips on AFM-based cantilevers. *Opt. Express*, **19**, 22268–22279.

38 Canale, C., Torre, B., Ricci, D., and Braga, P. (2011) in *Atomic Force Microscopy in Biomedical Research* (eds P.C. Braga and D. Ricci), Humana Press, pp. 31–43.

39 Raether, H. (1988) *Surface Plasmons on Smooth and Rough Surfaces and on Gratings*, Springer Berlin Heidelberg, Berlin.

40 Burtscher, H., Muller, U., and Schmidt-Qtt, A. (1988) Photoemission from small metal spheres: a model calculation using an enhanced three-step model. *Phys. Rev. B*, **38**, 7814.

41 Scales, C. and Berini, P. (2010) Thin-film Schottky barrier photodetector models. *IEEE J..Quantum Electron.*, **46**, 633–643.

42 Petrin, A.B. (2012) Two regimes of nanofocusing of the surface plasmon-polariton wave at the vertex of the metal micropoint. *High Temp.*, **50**, 15–20.

43 Knight, M.W., Wang, Y., Urban, A.S., Sobhani, A., Zheng, B.Y., Nordlander, P., and Halas, N.J. (2013) *Nano Lett.*, **13**, 1687.

44 Wu, J.C.K., McBride, J.R., and Lian, T. (2015) Efficient hot-electron transfer by a plasmon-induced interfacial charge-transfer transition. *Science*, **349**, 632–635.

45 Hubenthal, F. (2013) Increased damping of plasmon resonances in gold nanoparticles due to broadening of the band structure. *Plasmonics*, **8**, 1341–1349.

46 Govorov, A.O. and Zhang, H. (2015) Kinetic density functional theory for plasmonic nanostructures: breaking of the plasmon peak in the quantum regime and generation of hot electrons. *J. Phys. Chem. C*, **119**, 6181–6194.

47 Esteban, R., Borisov, A.G., Nordlander, P., and Aizpurua, J. (2012) Bridging quantum and classical plasmonics with a quantum-corrected model. *Nat. Commun.*, **3**, 825.

48 Akbari, A., Tait, R.N., and Berini, P. (2010) Surface plasmon waveguide Schottky detector. *Opt. Express*, **18**, 8505–8514.

49 Gramotnev, D.K., Vogel, M.W., and Stockman, M.I. (2008) Optimized nonadiabatic nanofocusing of plasmons by tapered metal rods. *J. Appl. Phys.*, **104**, 034311.

50 Rangel, T., Kecik, D., Trevisanutto, P.E., Rignanese, G.M., Van Swygenhoven, H., and Olevano, V. (2012) Band structure of gold from many-body perturbation theory. *Phys. Rev. B*, **86**, 125125.

51 Fowler, R.H. (1931) The analysis of photoelectric sensitivity curves for clean metals at various temperatures. *Phys. Rev.*, **38**, 45–56.

52 Palik, E.D. (1985) *Handbook of Optical Constants of Solids. Part II*, Academic Press, New York.

53 Manjavacas, A., Liu, J.G., Kulkarni, V., and Nordlander, P. (2014) Plasmon-induced hot carriers in metallic nanoparticles. *ACS Nano*, **8**, 7630–7638.

54 Zheng, J., Zhang, C., and Dickson, R.M. (2004) Highly fluorescent, water-soluble, size-tunable gold quantum dots. *Phys. Rev. Lett.*, **93**, 077402.

55 Rowland, C.E., Fedin, I., Zhang, H., Gray, S.K., Govorov, A.O., Talapin, D.V., and Schaller, R.D. (2015) Picosecond energy transfer and multiexciton transfer outpaces Auger recombination in binary CdSe nanoplatelet solids. *Nat. Mater.*, **14**, 484–489.

56 Rohani, S., Quintanilla, M., Tuccio, S., De Angelis, F., Cantelar, E., Govorov, A.O., Razzari, L., and Vetrone, F. (2015) Enhanced luminescence, collective heating, and nanothermometry in an ensemble system composed of lanthanide-doped upconverting nanoparticles and gold nanorods. *Adv. Opt. Mater.*, **3**, 1606–1613.

57 Gong, T. and Munday, J.N. (2015) Materials for hot carrier plasmonics [Invited]. *Opt. Mater. Express*, **5**, 2501–2512.

58 Afanas'ev, V.V. (2008) *Internal Photoemission Spectroscopy*, Elsevier.

59 Donolato, C. (2004) Approximate analytical solution to the space charge problem in nanosized Schottky diodes. *J. Appl. Phys.*, **95**, 2184–2186.

60 Smit, G.D.J., Rogge, S., and Klapwijk, T.M. (2002) Scaling of nano-Schottky-diodes. *Appl. Phys. Lett.*, **81**, 3852–3854.

61 Sze, S.M. and Ng, K.K. (2007) *Physics of Semiconductor Devices*, Wiley.

62 Nonnenmacher, M., O'Boyle, M.P., and Wickramasinghe, H.K. (1991) Kelvin probe force microscopy. *Appl. Phys. Lett.*, **58**, 2921–2923.

63 Henning, A.K., Hochwitz, T., Slinkman, J., Never, J., Hoffmann, S., Kaszuba, P., and Daghlian, C. (1995) Two-dimensional surface dopant profiling in silicon using scanning Kelvin probe microscopy. *J. Appl. Phys.*, **77**, 1888–1896.

64 Zerweck, U., Loppacher, C., Otto, T., Grafström, S., and Eng, L.M. (2005) Accuracy and resolution limits of Kelvin probe force microscopy. *Phys. Rev. B*, **71**, 125424.

65 Lorenzoni, M., Brandi, F., Dante, S., Giugni, A., and Torre, B. (2013) Simple and effective graphene laser processing for neuron patterning application. *Sci. Rep.*, **3**, 1954.

66 Petrin, A.B. (2010) Thermionic-field emission of electrons from conical metal points. *High Temp.*, **48**, 305–314.

67 West, P.R., Ishii, S., Naik, G.V., Emani, N.K., Shalaev, V.M., and Boltasseva, A. (2010) Searching for better plasmonic materials. *Laser Photonics Rev.*, **4**, 795–808.

68 Berweger, S., Atkin, J.M., Xu, X.G., Olmon, R.L., and Raschke, M.B. (2011) Femtosecond nanofocusing with full optical waveform control. *Nano Lett.*, **11**, 4309–4313.

69 Schmidt, S., Piglosiewicz, B., Sadiq, D., Shirdel, J., Lee, J.S., Vasa, P., Park, N., Kim, D.-S., and Lienau, C. (2012) Adiabatic nanofocusing on ultra-smooth single-crystalline gold tapers creates a 10-nm-sized light source with few-cycle time resolution. *ACS Nano*, **6**, 6040–6048.

70 Stockman, M.I., Faleev, S.V., and Bergman, D.J. (2002) Coherent control of femtosecond energy localization in nanosystems. *Phys. Rev. Lett.*, **88**, 067402.

71 Mohammadi, A. and Agio, M. (2012) Light scattering under nanofocusing: towards coherent nanoscopies. *Opt. Commun.*, **285**, 3383–3389.

72 Bouillard, J.S., Vilain, S., Dickson, W., Wurtz, G.A., and Zayats, A.V. (2012) Broadband and broad angle SPP antennas based on plasmonic crystals with linear chirp. *Sci. Rep.*, **2**, 829.

73 Kravtsov, V., Ulbricht, R., Atkin, J.M., and Raschke, M.B. (2015) Plasmonic Nano-Focused Four-Wave Mixing. CLEO: QELS_Fundamental Science, Optical Society of America, 2015, pp. JTh5B. 5.

74 Xu, X.G. and Raschke, M.B. (2013) Near-field infrared vibrational dynamics and tip-enhanced decoherence. *Nano Lett.*, **13**, 1588–1595.

75 Min, W., Freudiger, C.W., Lu, S., and Xie, X.S. (2011) Coherent nonlinear optical imaging: beyond fluorescence microscopy. *Annu. Rev. Phys. Chem.*, **62**, 507–530.

76 Saar, B.G., Freudiger, C.W., Reichman, J., Stanley, C.M., Holtom, G.R., and Xie, X.S. (2010) Video-rate molecular imaging in vivo with stimulated Raman scattering. *Science*, **330**, 1368–1370.

77 Evans, C.L. and Xie, X.S. (2008) Coherent anti-Stokes Raman scattering microscopy: chemical imaging for biology and medicine. *Annu. Rev. Anal. Chem.*, **1**, 883–909.

78 Freudiger, C.W., Min, W., Saar, B.G., Lu, S., Holtom, G.R., He, C., Tsai, J.C., Kang, J.X., and Xie, X.S. (2008) Label-free biomedical imaging with high sensitivity by stimulated Raman scattering microscopy. *Science*, **322**, 1857–1861.

**79** Romera, E. and de los Santos, F. (2009) Revivals, classical periodicity, and zitterbewegung of electron currents in monolayer graphene. *Phys. Rev. B*, **80**, 165416.

**80** Rusin, T.M. and Zawadzki, W. (2007) Transient Zitterbewegung of charge carriers in mono- and bilayer graphene, and carbon nanotubes. *Phys. Rev. B*, **76**, 195439.

**81** Yakovlev, V., Kruchinin, S., Paasch-Colberg, T., Stockman, M., and Krausz, F. (2016) in *Ultrafast Dynamics Driven by Intense Light Pulses* (eds M. Kitzler and S. Gräfe), Springer International Publishing, pp. 295–315.

**82** Müller, M., Kravtsov, V., Paarmann, A., Raschke, M.B., and Ernstorfer, R. (2016) A nanofocused plasmon-driven sub-10 femtosecond electron point source, *ACS Photonics*, **3**, 611.

# Index



*Conductive Atomic Force Microscopy: Applications in Nanomaterials*, First Edition.
Edited by Mario Lanza.

## l



## m



## n